91

D0674197

Methods of
Enzymatic Analysis

Methods of
Enzymatic Analysis

Third Edition

Editor-in-Chief: Hans Ulrich Bergmeyer
Editors: Jürgen Bergmeyer and Marianne Graßl

© VCH Verlagsgesellschaft mbH, D-6940 Weinheim (Federal Republic of Germany), 1986

Distribution:
VCH Verlagsgesellschaft, P.O. Box 1260/1280, D-6940 Weinheim (Federal Republic of Germany)
USA and Canada: VCH Publishers, 303 N.W. 12th Avenue, Deerfield Beach FL 33442-1705 (USA)

ISBN 3-527-26052-8 (VCH Verlagsgesellschaft)   ISBN 0-89573-242-4 (VCH Publishers)

Methods of Enzymatic Analysis

Third Edition

Editor-in-Chief: Hans Ulrich Bergmeyer
Editors: Jürgen Bergmeyer and Marianne Graßl

Volume XII
Drugs and Pesticides

Editorial Consultant: Michael Oellerich

VCH

Editor-in-Chief:
    Prof. Dr. rer. nat. Hans Ulrich Bergmeyer
    Hauptstraße 88
    D-8132 Tutzing
    Federal Republic of Germany

Language Editor:
    Prof. Donald W. Moss, Ph. D., D. Sc.
    42 Greenways
    Esher, Surrey
    KT10 OQD
    United Kingdom

Editorial Consultant:
    Prof. Dr. Michael Oellerich
    Kahlendamm 21 b
    D-3000 Hannover 51
    Federal Republic of Germany

Editors:
    Dr. rer. nat. Jürgen Bergmeyer
    In der Neckarhelle 168
    D-6900 Heidelberg
    Federal Republic of Germany

    Dr. rer. nat. Marianne Graßl
    Frauenchiemsee-Straße 20
    D-8000 München 80
    Federal Republic of Germany

Note

The methods published in this book have not been checked experimentally by the editors. Sole responsibility for the accuracy of the contents of the contributions and the literature cited rests with the authors. Readers are therefore requested to direct all enquiries to the authors (addresses are listed on pp. XIX – XXIII).

Previous editions of "Methods of Enzymatic Analysis":
1st Edition 1963, one volume
    2nd printing, revised, 1965
    3rd printing, 1968
    4th printing, 1971

2nd Edition 1974, four volumes
    2nd printing, 1977
    3rd printing, 1981

Previous editions of "Methoden der enzymatischen Analyse":
1. Auflage 1962, one volume
2. neubearbeitete und erweiterte Auflage 1970, two volumes
3. neubearbeitete und erweiterte Auflage 1974, two volumes

Library of Congress Card No. 84-105641

Deutsche Bibliothek, Cataloguing-in-Publication Data
Methods of enzymatic analysis / Ed.-in-chief: Hans Ulrich Bergmeyer. Eds.: Jürgen Bergmeyer and Marianne Graßl. – Weinheim; Deerfield Beach, FL: VCH
    Dt. Ausg. u.d.T.: Methoden der enzymatischen Analyse
NE: Bergmeyer, Hans Ulrich [Hrsg.]
Vol. XII. Drugs and Pesticides / Ed. consultant: Michael Oellerich. – 3rd edit. – 1986
    ISBN 3-527-26052-8 (Weinheim, Basel)
    ISBN 0-89573-242-4 (Deerfield Beach)
NE: Oellerich, Michael [Bearb.]

© VCH Verlagsgesellschaft mbH, D-6940 Weinheim (Federal Republic of Germany), 1986.
All rights reserved (including those of translation into other languages). No part of this book may be reproduced in any form – by photoprint, microfilm, or any other means – nor transmitted or translated into a machine language without written permission from the publishers. Registered names, trademarks, etc. used in this book, even when not specifically marked as such, are not to be considered unprotected by law.
Production Manager: Heidi Lenz
Composition: Krebs-Gehlen Druckerei, D-6944 Hemsbach
Printing: Hans Rappold Offsetdruck GmbH, D-6720 Speyer
Bookbinding: Josef Spinner, D-7583 Ottersweier
Printed in the Federal Republic of Germany

NIHSS
QUBML
2 6 JUN 1990
BELFAST

Preface to the Series

"Methods of Enzymatic Analysis" appeared for the first time in 1962 as a one-volume treatise in German. Several updated and improved editions in English and German have been published since then. The latest English edition appeared in 1974.

In the meantime, enzymatic analysis has continued to find new applications, refinements and extensions at a pace that justifies – indeed, demands – the preparation of a new and completely revised edition. However, the field has grown so enormously that it can no longer be surveyed adequately by one person. Fortunately, therefore, I am supported in this new enterprise by Dr. M. Graßl, who is highly experienced in biochemical analysis, and Dr. J. Bergmeyer, who represents the younger generation of biochemists.

With the 1974 edition of "Methods of Enzymatic Analysis" as a starting point for our work towards the new edition, it soon became obvious that many chapters had to be eliminated, re-written or added. Moreover, the increased number of analytes that can now be determined enzymatically and of enzymes regularly requiring analysis, especially in the clinical laboratory, together with the emergence of an entirely new field of application through the technique of the enzyme-immunoassay, demanded a new arrangement and subdivision of the contents, if the vast range of material was to be dealt with properly and lucidly.

The result is the plan of the work printed on the page opposite the title page of this volume. Of course, it would be impossible to publish a whole series such as this at one moment and still maintain an equal degree of topicality for all contributions. Therefore, we decided to produce the series at a pace of several volumes per year. The volumes will not necessarily appear in their numerical order, but will be made available as they can be planned and completed.

As before, the purpose of the work is to provide reliable descriptions of well-developed procedures of enzymatic analysis in the broadest sense of the term. Special efforts are being made to arrange every chapter, and to co-ordinate the contents of all chapters, in such a way that the volumes are useful as laboratory manuals for daily work.

Internationally-agreed enzyme nomenclature as well as quantities and units correlating with the "Système International d'Unités" are used wherever possible in order to make statements and data unambiguous and comparable over time and space.

All contributions are and will be written in English: however, contributors come from all over the world and their manuscripts naturally show various versions of English. These have to be harmonized in style and spelling in order to achieve uniformity throughout the series without, we hope, entirely eliminating each author's personal approach. Professor Donald W. Moss has kindly agreed to undertake this task. We agreed with him to use modern English spelling, but to try to minimize differences between British and American practice. We hope that this will be considered as a fair solution and one which will make the series accessible to as wide a readership as possible.

Thanks are due to the authors in the first place for responding so readily to our invitations, for writing their chapters so diligently within a short time and for communicating their experience and expertise. We are also indebted to all colleagues who gave their advice and to Professor Moss for accepting the task of language editor. Finally I wish to record my gratitude to Verlag Chemie for the fruitful and excellent co-operation during all stages from planning to production.

Tutzing, February 1983   Hans Ulrich Bergmeyer

Preface to Volume XII

This volume is dedicated to the analysis of drugs and pesticides. Although several physicochemical methods are well established for the determination of these compounds, the introduction of enzyme-immunoassays has opened new possibilities.

In the case of drugs, we distinguish between "Drugs Monitored during Therapy" and "Drugs of Abuse and of Toxicological Relevance".

Therapeutic drug monitoring has become increasingly important during the last ten years. The quantitative determination of serum concentrations is clinically useful, in particular for drugs that are dangerously toxic, that have a narrow therapeutic range, or show considerable intra- and inter-individual variations in their pharmacokinetic properties. In far more cases than is generally realized, the serum concentrations of such drugs move outside the therapeutic range unless dosage is matched to the individual needs of the patient. Enzyme-immunoassays are rapid and easy to perform. Therefore, they are most suitable for measuring drug concentrations in emergencies (e.g. in patients with suspected drug overdosage) or when rapid dose-adjustments are required (e.g. in critically ill patients). The availability of enzyme-immunoassays for therapeutic drug monitoring has thus improved the efficacy and safety of pharmacotherapy.

For most methods described under "Drugs of Abuse and of Toxicological Relevance", a semi-quantitative result or even a "yes-or-no" answer is sufficient. The collection of procedures presented here can only give an indication of the current status of enzymatic and enzyme-immunological drug analysis. This part also includes a chapter on the measurement of drug residues in animals. More extended applications of enzyme-immunoassays in food analysis will certainly be demanded in future.

The third part of this volume contains methods for the determination of pesticides. In the age of environmental awareness and with the availability of methods that allow the detection and measurement of these compounds in the femtomole range, it is surprising that more publications have not appeared describing the application of enzyme-immunoassays in this field. By including this topic in the series, the editors hope to stimulate analysts involved in environmental investigations to design specific and simple enzyme-immunoassays.

This volume completes the series of twelve on Methods of Enzymatic Analysis. The editors are indebted to the authors of this volume for their willingness to adapt their manuscripts to the special requirements of this series. Thanks are also due to the editorial consultant, Professor M. Oellerich, for his continuous assistance.

It is now five years since we started to plan and collect topics for this new edition of Methods of Enzymatic Analysis. Originally, the work was expected to comprise six volumes. When Volume I appeared in April 1983, it was obvious that enzymatic analysis had grown so much since the last edition (1974), that ten volumes with 400 to 600 pages each would be required. Finally, in 1985 we had again to expand the series, to 12 volumes, because of the exciting progress in the use of enzyme-immunoassays. We are most grateful to all subscribers for accepting these changes to the original programme.

Now, Volume XII completes a collection of 562 working instructions, averaging 56 each in Volumes III to XII. In addition, there are 68 chapters discussing fundamental aspects in Vol. I, describing statistics and the handling of specimens, samples, and biochemical reagents in Vol. II, and dealing with supplementary or introductory topics in Volumes III, V, IX, X and XII. A total of 702 authors have contributed to the series. Approximately one half are co-authors, which means that many colleagues contributed more than one chapter. Because of that great readiness to participate, we were able to accumulate the knowledge of specialists from 29 countries: a comprehensive list will appear in the Cumulative Index and List of Contributors which will complete the series.

Again, we would like to express our gratitude to all authors, to the reviewers for their kind appreciation and constructive criticism, to the editorial consultants for their excellent advice, and to the staff of the publisher and printer. Last but not least my thanks are due to my wife, Ingrid Bergmeyer, for the patience and understanding with which she has enabled me to devote my time and efforts to this undertaking.

Tutzing, April 1986   Hans Ulrich Bergmeyer

Contents

Contents of Volumes I – XII

(Chapter Headings only)

Volume V
Enzymes 3: Peptidases, Proteinases and Their Inhibitors

Volume VI
Metabolites 1: Carbohydrates

Volume VII
Metabolites 2: Tri- and Dicarboxylic Acids, Purines, Pyrimidines and Derivatives, Coenzymes, Inorganic Compounds

Volume VIII
Metabolites 3: Lipids, Amino Acids and Related Compounds

Volume IX
Proteins and Peptides

Volume X
Antigens and Antibodies 1

Volume XI
Antigens and Antibodies 2

Volume XII
Drugs and Pesticides

Contributors

Atkinson, Tony
Microbial Technology Laboratory
Centre for Applied
Microbiology and Research
Porton Down
Salisbury, Wiltshire SP4 0JG
United Kingdom

p. 266, 362, 374

Benner, Jill P.
Imperial Chemical Industries plc
Plant Protection Division
Jealott's Hill
Research Station
Bracknell, Berkshire RG12 6EY
United Kingdom

p. 451

Bergmeyer, Hans Ulrich
Boehringer Mannheim GmbH
Biochemical Research Center
Bahnhofstr. 9 – 15
D-8132 Tutzing
Federal Republic of Germany

p. 404

Brimfield, Alan A.
Department of Pharmacology
Uniformed Services University
F. Edward Hébert School of Medicine
4301 Jones Bridge Road
Bethesda, MD 20814-4799
U.S.A.

p. 426

Campbell, R. Stewart
Microbial Technology Laboratory
Centre for Applied
Microbiology and Research
Porton Down
Salisbury, Wiltshire SP4 0JG
United Kingdom

*p. 266, 269, 278, 291, 303, 310, 317,
325, 332, 339, 347, 355, 362, 374*

Castro, Albert
Department of Pathology (R-40)
University of Miami
Medical School
P.O. Box 016960
1600 N.W. 10th Avenue
Miami, FL 33101
U.S.A.

p. 381

Christensen, Michael L.
Pharmacokinetics and
Pharmacodynamics Section
St. Jude Children's
Research Hospital
332 North Lauderdale
Memphis, TN 38105
U.S.A.

p. 134

Chubb, S. A. Paul
Department of Clinical Biochemistry
Addenbrooke's Hospital
Hills Road
Cambridge, CB2 2QR
United Kingdom

p. 374

Evans, William E.
Pharmacokinetics and
Pharmacodynamics Section
St. Jude Children's
Research Hospital
332 North Lauderdale
Memphis, TN 38105
U.S.A.

p. 2, 134

Fleeker, James
North Dakota State University
of Agriculture and Applied Science
Fargo, ND 58105
U.S.A.

p. 391

Fujiwara, Kunio
Faculty of Pharmaceutical Sciences
Nagasaki University
Bunkyo-machi 1 – 14
Nagasaki 852
Japan

p. 187

Hammond, Peter M.
Microbial Technology Laboratory
Centre for Applied
Microbiology and Research
Porton Down
Salisbury, Wiltshire SP4 0JG
United Kingdom

p. 266, 362, 374

Haven, Mary C.
Department of Pathology and
Laboratory Medicine
College of Medicine
University of Nebraska
Medical Center
42nd and Dewey Avenue
Omaha, NB 68105
U.S.A.

p. 216

Hock, Bertold
Fakultät für Landwirtschaft
und Gartenbau
Lehrstuhl für Botanik der
Technischen Universität München
D-8050 Freising-Weihenstephan
Federal Republic of Germany

p. 438

Horváth, László
Institute of Isotopes of the
Hungarian Academy of Sciences
P.O. Box 77
H-1525 Budapest
Hungary

p. 406

Huber, Sigmund J.
Fakultät für Landwirtschaft
und Gartenbau
Lehrstuhl für Botanik der
Technischen Universität München
D-8050 Freising-Weihenstephan
Federal Republic of Germany

p. 438

Hunter, Jr., Kenneth W.
Departments of Pediatrics and
Preventive Medicine Biometrics
Uniformed Services University
F. Edward Hébert School of Medicine
Bethesda, MD 20814-4799
U.S.A.

p. 426

Jochum, Marianne
Abteilung für Klinische Chemie
und Klinische Biochemie in der
Chirurgischen Klinik Innenstadt
der Universität München
Nußbaumstr. 20
D-8000 München 2
Federal Republic of Germany

p. 257

Kitagawa, Tsunehiro
Faculty of Pharmaceutical Sciences
Nagasaki University
Bunkyo-machi 1 – 14
Nagasaki 852
Japan

p. 187, 200

Kleinhammer, Gerd
Boehringer Mannheim GmbH
Biochemical Research Center
Bahnhofstr. 9 – 15
D-8132 Tutzing
Federal Republic of Germany

p. 112

Kovács Huber, Gyöngyi
Research Institute for
Heavy Chemical Industries
P.O. Box 160
H-8201 Veszprem
Hungary

p. 412

Lenz, David E.
Basic Pharmacology Branch
Pharmacology Division
U.S. Army Medical Research
Institute of Chemical Defense
Aberdeen Proving Ground, MD 21010
U.S.A.

p. 426

Leone, Leonello
Ospedale Infantile
Regina Margherita
Servizio di Recerche Cliniche
e Microbiologiche
Piazza Palonia, 94
I-10126 Torino
Italy

p. 33, 56, 69, 82

Li, Thomas M.
Syntex Medical Diagnostics
3221 Porter Drive
Palo Alto, CA 94304
U.S.A.

p. 95

Marks, Vincent
Department of Biochemistry
St. Luke's Hospital
Guildford, Surrey GU1 3NT
United Kingdom

p. 232

Mattersberger, Hans
Boehringer Mannheim GmbH
Biochemical Research Center
Bahnhofstr. 9 – 15
D-8132 Tutzing
Federal Republic of Germany

p. 112

Monji, Nobuo
Genetics Systems Corporation
3005 1st Avenue
Seattle, WA 98121
U.S.A.

p. 381

Mould, Graham P.
Department of Biochemistry
St. Luke's Hospital
Guildford, Surrey GU1 3NT
United Kingdom

p. 232

Müller-Esterl, Werner
Abteilung für Klinische Chemie
und Klinische Biochemie in der
Chirurgischen Klinik Innenstadt
der Universität München
Nußbaumstr. 20
D-8000 München 2
Federal Republic of Germany

p. 246

Niewola, Zbigniew
Imperial Chemical Industries plc
Central Toxicology Laboratory
Alderley Park
Macclesfield, Cheshire SK10 4TJ
United Kingdom

p. 451

Oellerich, Michael
Institut für Klinische Chemie
Medizinische Hochschule Hannover
Konstanty-Gutschow-Str. 8
D-3000 Hannover 61
Federal Republic of Germany

p. 2, 5

Perucca, Emilio
Dipartimento di Medicina Interna
e Terapia Medica
Università di Pavia
Piazza Botta, 10
I-27100 Pavia
Italy

p. 33, 56, 69, 82

Price, Christopher P.
Department of Clinical Biochemistry
Addenbrooke's Hospital
Hills Road
Cambridge CB2 2QR
United Kingdom

*p. 266, 269, 278, 291, 303, 310, 317,
325, 332, 339, 347, 355, 362, 374*

Reinauer, Hans
Diabetes-Forschungsinstitut
Auf'm Hennekamp 65
D-4000 Düsseldorf 1
Federal Republic of Germany

p. 103

Singh, Prithipal
Didetek, Inc.
1057 Sneath Lane
San Bruno, CA 94066
U.S.A.

p. 23

Standefer, Jim
Department of Pathology
University of New Mexico
Albuquerque, NM 87131
U.S.A.

p. 172

Ueda, Clarence T.
Department of Pharmaceutics
College of Pharmacy
University of Nebraska
Medical Center
42nd and Dewey Avenue
Omaha, NB 68105
U.S.A.

p. 216

Wędzisz, Anna
Medical Academy in Lodz
Institute of Environmental
Research and Bioanalysis
Narutowicza 120A
90-145 Lodz
Poland

p. 418

Wenk, Markus
Kantonsspital Basel
Universitätskliniken
Department Innere Medizin
Abteilung für Klinische Pharmakologie
CH-4031 Basel
Switzerland

p. 159

Zysset, Thomas
Abteilung für Klinische Pharmakologie
Universität Bern
Inselspital
Murtenstr. 35
CH-3010 Bern
Switzerland

p. 23

1 Drugs Monitored during Therapy

1.1   Introduction

William E. Evans and Michael Oellerich

Before drug assays became routinely available, optimal drug dosages were usually determined empirically by trial and error. Within the past three decades, analytical methods have been developed that allow for rapid and accurate quantitation of drug concentrations in biological fluids. This, coupled with the apparent relationships between pharmacological effect and serum concentration for many drugs, has led to the widespread monitoring of serum drug concentrations (therapeutic drug monitoring) to match drug dosage to individual patient needs.

Use of serum concentrations for optimizing therapy is of greatest value for drugs which may produce toxic effects at dosages close to those required for therapeutic effects. Characteristics of drugs for which monitoring is recommended include drugs with a narrow therapeutic range (e. g., aminoglycosides, theophylline, digoxin, anti-arrhythmics, etc.), unpredictable dose-concentration relationship (e. g., phenytoin), predictable serum concentration-response relationship (e. g., anti-arrhythmics, theophylline), drugs administered on a chronic basis for diseases with only occasional clinical symptoms (e. g., anticonvulsants, theophylline) and special situations such as high-dose methotrexate where adjustment of leucovorin rescue is based on serum methotrexate concentrations.

Therapeutic drug monitoring may also be useful in evaluating the appropriateness of drug therapy when clinical response to a specific drug is complicated by the co-administration of other drugs with similar pharmacological effects. Significant inter-patient variability in pharmacokinetic parameters has been demonstrated for many drugs and, for some drugs, significant intra-patient variability may also occur. Such unpredictable variability indicates that a standard dosage, even adjusted for body weight or body surface area, will result in serum drug concentrations outside the usual therapeutic range for many patients. This has led to the individualization of therapy based on serum concentrations of many drugs, including those listed in Table 1.

Table 1. Drugs currently monitored

Acetaminophen	Kanamycin	Propranolol
Amikacin	Lidocaine	Quinidine
Ampicillin	Lithium	Salicylate
Caffeine	Methotrexate	Sisomycin
Carbamazepine	Netilmicin	Streptomycin
Chloramphenicol	Nortriptyline	Theophylline
Ciclosporin	Phenobarbital	Tobramycin
Digitoxin	Phenytoin	Tocainide
Digoxin	Primidone	Valproic Acid
Disopyramide	Procainamide	Vancomycin
Ethosuximide	N-Acetylprocainamide	Viomycin
Gentamicin		

Although a relationship between dose and pharmacological effect can often be demonstrated, generally the factors that affect dosage-serum concentration relationships can be more readily overcome with therapeutic drug monitoring, than those factors influencing serum concentration-pharmacological effect relationships (Table 2).

Table 2. Selected factors affecting dose-concentration relationship

Environmental
   diet
   concomitant drugs

Patient-specific
   age
   pregnancy
   sex
   genetic disease processes

Bioavailability (oral, intramuscular or subcutaneous injection)

Patient compliance

Serum concentration depends on drug dose, and on patient medication compliance, drug absorption, distribution, biotransformation and excretion. Drug concentration at the site of action also depends on regional blood flow, protein binding, and transport mechanisms. In addition, the intensity of pharmacological effect may be altered by other drugs, disease states and the age of the patient. Routine methods of drug assay quantitate total drug concentration in serum or plasma and do not quantitate drug concentration at the receptor site. Drug not bound to proteins in serum is thought to interact with specific receptors and evoke the pharmacological effects. For highly protein-bound drugs (i. e., > 80%), some investigators advocate measurement of unbound (free) drug concentrations in serum when the extent of binding is unpredictable or highly variable.

A "therapeutic range" for a drug can be defined as that concentration range within which the probability of the desired pharmacological effect is relatively high and the probability of drug-related toxicity is relatively low. Thus, even though the likelihood of the desired effect is increased if the drug concentration is within this range, some patients may not respond and may require either lower or higher drug concentrations to achieve the desired response without the toxicity.

Interpretation of serum concentration data is a vital part of therapeutic drug monitoring. The drug concentration is an important part of the clinical evaluation, but must be interpreted in conjunction with the patient's current clinical condition and the specific therapeutic goals which have been established for each individual. Other important factors include the duration of therapy, concurrent drug therapy, assay accuracy, and time that blood specimens were obtained in relation to when the dose was administered. Also, within specific drug classes additional factors should be considered when interpreting drug concentrations (i. e., for antibiotics: site of infections, immune status of the patient and the minimum inhibitory concentration of the infecting organism).

There are currently many assay methods available for quantitating drug concentrations in biological fluids. Many of the techniques are comparable, each with its own advantages and limitations. Chromatographic methods include gas-liquid chromatography (GLC) and high-pressure liquid chromatography (HPLC). These techniques can achieve a high specificity and therefore may be used for the development of reference methods. A further advantage of these procedures is the ability to quantitate multiple drugs and metabolites. However, these methods are relatively complex and require considerable technical expertise to achieve acceptable accuracy and precision.

Microbiological assay methods are available for certain drugs, such as antibiotics. They are inexpensive and easy to perform. However, there are inherent technical problems which can alter drug assay results, such as interference from concomitant antibiotic therapy.

Today, immunoassays, e.g., enzyme-multiplied immunoassay technique (EMIT), fluorescence polarization immunoassay (FPIA), radioimmunoassay (RIA), are used in many laboratories for serum drug determinations. Among these assays, homogeneous non-isotopic immunoassays in particular have gained increasing importance for therapeutic drug monitoring. Some of these techniques will be discussed in detail in the following chapters. These immunoassays have made drug analysis feasible for most laboratories, and offer advantages such as speed, sensitivity and ease of operation, and require minimal technical expertise. The main problem inherent in immunoassays is potential cross-reactivity with structurally similar compounds, especially drug metabolites, as will be discussed.

In summary, substantial advances in the technology for analyzing drug concentrations in biological fluids, coupled with more clearly defined pharmacodynamic relationships and improved pharmacokinetic approaches to individualizing drug dosages, have led to the widespread use of therapeutic drug monitoring.

1.2   Theophylline

3,7-Dihydro-1,3-dimethyl-1*H*-purine-2,6-dione

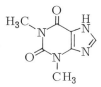

Michael Oellerich

General

Theophylline was first described in 1888 by *Kossel,* who extracted the drug from tea leaves [1]. The drug was mainly used as a diuretic and cardiac stimulant until, in 1922, *Hirsch* reported on the first successful clinical use of theophylline in bronchial asthma [2]. Today, theophylline is chiefly used as a bronchodilator in the treatment of obstructive respiratory disorders and as an analeptic drug in premature infants with apnoea. The relaxant effects of theophylline in bronchial and other smooth muscle preparations [3] presumably result from an inhibition of phosphodiesterase activity, with a subsequent increase in cyclic adenosine 3′,5′-monophosphate (cAMP).

About 90% of theophylline is metabolized by the liver to relatively inactive metabolites, mainly through oxidation to 1,3-dimethyluric acid, and by *N*-demethylation of both this metabolite and the parent drug to 1-methyluric acid and 3-methylxanthine, respectively [4]. Approximately 10% of an administered theophylline dose is excreted unchanged in the urine. In neonates, about 10% of the drug is metabolized to caffeine (1,3,7-trimethylxanthine) and about 50% is excreted as unchanged theophylline. Small amounts of theophylline are produced in man by demethylation from caffeine [5]. Dose-dependent kinetics of theophylline disposition have been observed by various authors [4].

The half-life of elimination of theophylline from serum shows large inter-individual variations [4]. It ranges from about 20 to 30 hours in premature neonates, to 3 to 4 hours in children (1 – 17 years) and 3 to 12 hours in adults (non-smoking). Various factors can influence the clearance of theophylline, such as diseases, aberrant diets, smoking habits and concurrent drugs [4]. The apparent volume of distribution at steady state averages 0.45 l/kg (range: 0.30 to 0.70 l/kg) [4]. Neonates show a slightly larger volume of distribution. Only about 50% of theophylline is bound to serum proteins [6]. Reduced protein-binding of the drug is found in premature neonates and patients with liver cirrhosis. The concentration of theophylline in saliva is about 40% lower than in plasma [7].

Theophylline is rapidly and completely absorbed from liquids and plain tablets. Sustained-release formulations with adequate bio-availability are available [4, 8, 9].

The main reasons for monitoring serum theophylline concentrations are the large inter-individual differences in pharmacokinetics, the narrow therapeutic range and the potential for life-threatening adverse effects. It is advisable to monitor serum theophylline levels in the following clinical situations: suspected overdose, absence of expected therapeutic effect, unknown pre-medication with theophylline, continuous i.v. infusion, continued oral medication, especially in patients with certain concurrent illnesses (e.g. cardiac decompensation, hepatic cirrhosis, viral respiratory-tract infections), changing smoking habits or persistent side-effects. In addition, dose prediction methods may be helpful in individualizing theophylline dosage [9, 10].

Application of method: in clinical chemistry, pharmacology and toxicology.

Substance properties relevant in analysis: theophylline (M_r = 180.17) is an amphoteric drug (pK_a = 8.8 at 25 °C; pK_b = 13.7). The solubility of theophylline in water is relatively low (8.3 g/l). Soluble theophylline salts used in commercial preparations exist at physiological pH only as mixtures with the various bases (e.g. ethylenediamine, choline and sodium glycinate).

Methods of determination: numerous methods of measuring serum theophylline concentrations have been developed: chromatographic procedures (high-pressure liquid chromatography, gas chromatography, gas chromatography-mass spectrometry, thin-layer chromatography), homogeneous immunoassays (enzyme-multiplied immunoassay technique, substrate-labelled fluorescent immunoassay, prosthetic-group-label immunoassay, fluorescence polarization immunoassay, nephelometric inhibition immunoassay, particle-enhanced turbidimetric inhibition immunoassay), heterogeneous immunoassays (enzyme-linked immunosorbent assay, fluorescence immunoassay, radioimmunoassay) and other tests (fluorimetry, isotachophoresis, UV-spectrophotometry after extraction). Immunoassays and high-pressure liquid chromatography are the methods which are currently most widely used.

Recently, a highly specific high-pressure liquid chromatographic procedure has been described [11] which appears to be very useful for establishing the accuracy of other procedures in comparative studies, or as a confirmatory analysis in cases of suspected interference.

For routine monitoring of serum theophylline concentrations, homogeneous enzyme- and fluorescence polarization immunoassays have so far proved to be reliable and most convenient [12].

A competitive heterogeneous enzyme-immunoassay in which alkaline phosphatase (EC 3.1.3.1) is used as marker enzyme (Endab Theophylline Enzyme Immunoassay Kit, *Immunotechn. Corp.*), and a reagent-strip method (Seralyzer® Aris Theophylline, *Ames*) based on the prosthetic-group label immunoassay [13, 14] have recently become commercially available. Moreover a non-instrumental test-strip method has been developed, which is based on the principles of capillary immunochromatography and enzyme channelling [14a]. Furthermore, a multilayer enzyme-immunoassay element and a slide assay have been described, the latter is based on the inhibition of alkaline phosphatase from beef liver by theophyllin [14b, 14c]. In a further competitive heterogeneous enzyme-immunoassay, the activity of the marker enzyme, catalase (EC 1.11.1.6), bound to an antibody-coated membrane is determined by an oxygen-sensitive electrode [15]. Recently, a ligand displacement immunoassay (LIDIA®) has been described, which is based on the reaction of theophylline with a pre-formed solid-phase: β-galactosidase-theophylline conjugate [16]. So far, the performance of most of these enzyme-immunoassays has not been fully validated and clinical experience with these tests is lacking.

International reference method and standards: neither standardization at the international level, nor the existence of reference standard materials is known so far.

1.2.1   Determination with Enzyme-multiplied Immunoassay Technique

Assay

Method Design

The method was developed by *Gushaw et al.* [17] based on the enzyme-multiplied immunoassay technique (EMIT®) first described by *Rubenstein et al.* [18].

Principle

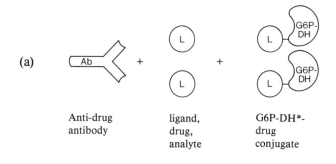

(a)

Anti-drug   ligand,   G6P-DH*-
antibody   drug,   drug
                   analyte   conjugate

---

* D-Glucose-6-phosphate: NADP$^+$ 1-oxidoreductase, EC 1.1.1.49.
   Note: no special EC number exists for the NAD and NADP-dependent enzyme from *Leuconostoc mesenteroides*.

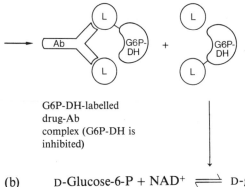

G6P-DH-labelled
drug-Ab
complex (G6P-DH is
inhibited)

(b)   D-Glucose-6-P + NAD$^+$ $\rightleftharpoons$ D-glucono-δ-lactone-6-P + NADH + H$^+$.

The activity of the marker enzyme, glucose-6-phosphate dehydrogenase G6P-DH, is reduced if the enzyme-theophylline conjugate (E-L) is coupled to the antibody (Ab). Theophylline (L) in a sample competes with G6P-DH-labelled theophylline (E-L) for the binding sites of an antibody (Ab) present in limited concentration. The activity of G6P-DH measured as increase in absorbance per unit time, $\Delta A/\Delta t$, is related to the theophylline concentration of the sample (cf. Vol. I, chapter 2.7, p. 244). The method has been mechanized in various analytical systems [19].

Selection of assay conditions and adaptation to the individual characteristics of the reagents: the assay is designed for rapid analysis and requires no separation step (results of samples from patients available within about 15 min).

Highly specific anti-theophylline antibodies have been produced by immunizing sheep with a bovine γ-globulin conjugate of 1-methyl-3-(3'-carboxypropyl)xanthine [20, 21]. Antibodies are raised in sheep, which are first inoculated with an emulsion of 1 mg of this antigen in complete *Freund's* adjuvant and subsequently receive monthly inoculations of the same amount of antigen in incomplete *Freund's* adjuvant [21].

The addition of 1-methylxanthine to these antibodies reduces cross-reactivity to this compound without adversely affecting cross-reactivities of other compounds [20]. The amount of 1-methylxanthine added to the reagents depends on the degree of cross-reactivity to this compound observed with the antibodies used.

Enzyme-drug conjugate is obtained by coupling 1-methyl-3-(3'-carboxypropyl)-xanthine to G6P-DH from *Leuconostoc mesenteroides* (cf. p. 27). Interference from endogenous G6P-DH is avoided by use of the coenzyme NAD, which is converted only by the bacterial enzyme.

The optimal quantities and concentrations of sample and reagents are dependent on the characteristics of the antibodies present in the antiserum.

The desired assay range is from 0.5 to 40.0 mg/l. The serum sample is 50 µl. These parameters being fixed, the optimum ratio of antibody to enzyme-conjugate must be determined in the presence of different theophylline concentrations within the desired assay range.

First, a prospective enzyme-conjugate concentration is chosen, to which gradually increasing amounts of the corresponding antibody are added. Enzymic activity is increasingly inhibited as the amount of antibody is raised. A maximum inhibition of 90 to 100% of the enzymic activity can be achieved by addition of a sufficient amount of antibody [21]. In order to assess the ability of the system to discriminate among serum theophylline concentrations likely to be encountered during theophylline therapy, the following optimization procedure is used: calibrators containing 0.0, 2.5, 10.0 and 40.0 mg/l are assayed according to the regular assay protocol (cf. p. 12) at a fixed enzyme-conjugate concentration and with different amounts of antibody. The zero calibrator serves as a reagent blank and all measurements are corrected for this value. The corrected readings obtained for each calibrator are plotted against the amount of antibody added. The antibody quantity is then selected according to the following considerations:

– an optimum response should be obtained at a point within the assay range at which assay sensitivity is most desirable (e.g. at 10 mg theophylline per litre),

– the assay response between theophylline calibration values reflecting the discriminatory power of the assay configuration should attain a maximum possible value.

As the cross-reactivity to related methylxanthines may vary with the amount of antibody present in the assay, the specificity also plays an important role in selecting the final antibody titre. Furthermore the ratios of antibody to enzyme-conjugate must be compared at various conjugate concentrations to select an optimum quantity of both reagents [21].

The first step of the assay procedure (addition of the antibody/substrate solution to the diluted sample) is carried out at room temperature. Immediately after the subsequent addition of the conjugate, the reaction mixture is incubated for 45 s in the spectrophotometer flow-cell at 30 °C and pH 8.0. During this period the immuno- and enzymatic reactions take place simultaneously. 15 s after the start of the incubation, absorbance readings are made over a period of 30 s. There is a slight decrease in the rate of the reaction with time, which is presumably due to a changing interaction between antibody and the enzyme-drug complex [22]. The results are calculated from the change in absorbance over a 30 s measurement period.

Equipment

Spectrophotometer or spectral-line photometer capable of exact measurement at 339 or Hg 334 nm. It should be equipped with a thermostatted flow-cell which consistently maintains 30.0 ± 0.1 °C throughout the working day. Work rack and disposable 2 ml beakers with conical bottom. Pipetter-diluter capable of delivering 50 ± 1 µl sample and 250 ± 5 µl buffer with a relative standard deviation less than 0.25%. Delivery must be of sufficient force to ensure adequate mixing of the components. The spectrophotometer should be connected to a suitable data handling device. Laboratory centrifuge.

Reagents and Solutions

Purity of reagents: G6P-DH from *Leuconostoc mesenteroides* should be of the best available purity (cf. chapter 1.3, p. 27). Chemicals are of analytical grade.

Preparation of solutions* (for 100 determinations): all solutions in re-purified water (cf. Vol. II, chapter 2.1.3.2).

1. Tris buffer (Tris, 55 mmol/l; Triton X-100, 0.1 ml/l; pH 8.0):

   dissolve 1.996 g Tris base in 140 to 160 ml water. Add 0.03 ml surfactant Triton X-100. Adjust to pH 8.0 with HCl, 1.0 mol/l. Make up with water to 300 ml.

   Alternatively, dilute concentrated buffer provided with each kit to 200 ml with water.

2. Antibody/substrate solution (γ-globulin**, ca. 50 mg/l; G-6-P, 66 mmol/l; NAD, 40 mmol/l; 1-methylxanthine***, ca. 1 − 2 mg/l; Tris, 55 mmol/l; pH 5.0):

   reconstitute available lyophilized preparation with 6.0 ml water.

3. Conjugate solution (G6P-DH†, ca. 1 kU/l; Tris, 55 mmol/l; pH 8.0):

   reconstitute available lyophilized preparation with 6.0 ml water. This reagent must be standardized to match the antibody/substrate solution (2) (cf. p. 8).

4. Theophylline standard solutions (2.5 to 40 mg/l):

   dissolve 100 mg theophylline in 100 ml Tris buffer (1). Dilute 0.4 ml with theophylline-free human serum containing sodium azide (5 g/l) and Thimerosal (0.5 g/l) to 10 ml (40 mg/l). Dilute this solution 1 + 1, 1 + 3, 1 + 7 and 1 + 15 with theophylline-free human serum, yielding the following concentrations of theophylline: 20.0, 10.0, 5.0, 2.5 mg/l. Take pure serum as zero calibrator.

   Alternatively, reconstitute each of the available lyophilized standard preparations with 1 ml water.

** Reagents, calibrators and buffer are commercially available from *Syva Co.,* Palo Alto, U.S.A. The solutions contain sodium azide.*
** Standardized preparation from immunized sheep.
*** Amount depending on antibody quality.
† Coupled to 4-(3'-carboxypropyl)-1-methylxanthine.

Stability of solutions: after reconstitution, the antibody/substrate solution (2), conjugate solution (3) and standards (4) must be stored at room temperature (20°C to 25°C) for at least one hour before use. The reagents are then stable for 12 weeks, if stored at 2°C to 8°C. The buffer solution (1) can be used for 12 weeks when stored at room temperature.

Procedure

Collection and treatment of specimen: it is important that the specimen is collected at the appropriate time after administration of the dose. During long-term oral therapy, blood specimens should be taken in the "steady-state", i.e. after treatment with a constant dose over at least 4 half-lives. The specimen is then drawn at the time of the peak concentration (e.g. about 4 – 6 h after administration of products with sustained release properties) and/or immediately before administration of the next dose, depending on the clinical indication. The intake of beverages containing caffeine need not be restricted before sampling, if the antibody employed shows a sufficient specificity (cf. p. 14).

Serum, plasma or saliva can be used in this assay. Blood cannot be used. Acceptable anticoagulants are heparin, EDTA, and oxalate. Only 50 µl sample is required per assay. No evidence of interference was obtained when sera from slightly haemolyzed, lipaemic or icteric blood were assayed [23]. Results obtained with sera from severely haemolyzed or lipaemic blood should be confirmed with new clear samples. Saliva specimens should be collected after stimulation of saliva flow by chewing Parafilm®. The usual precautions have to be taken to ensure good dental hygiene to avoid contamination of saliva samples with the drug. Centrifugation of saliva specimens before analysis is recommended.

Stability of the substance in the sample: samples should be stored at 2°C to 8°C upon collection. Sera can be stored at room temperature, 4°C or – 20°C for 12 weeks without appreciable loss.

Assay conditions: measurements of samples and zero calibrator in duplicate. Single determinations are made of the remaining standard solutions.

A pre-dilution of samples and standard solutions is required: mix one part (0.05 ml) sample or standard solution (4) with 5 parts (0.25 ml) Tris buffer (1).

Incubation time 45 s (in the spectrophotometer flow-cell); 0.90 ml; 30°C; measurement during the last 30 s against water; wavelength 339 or Hg 334 nm; light path 10 mm.

Establish a calibration curve under exactly the same conditions (within the series). The zero calibrator assay serves as a reagent blank; all measurements are corrected for this value.

Measurement

Pipette into a disposable beaker:			concentration in assay mixture
pre-diluted sample or standard solution (4)		0.05 ml	theophylline up to 370 µg/l
Tris buffer	(1)	0.25 ml	Tris 52 mmol/l
antibody/substrate solution	(2)	0.05 ml	sheep serum variable
			γ-globulin ca. 2.8 mg/l
			G-6-P 3.7 mmol/l
			NAD 2.2 mmol/l
			1-methylxanthine variable
			ca. 60 – 110 µg/l
Tris buffer	(1)	0.25 ml	
conjugate solution	(3)	0.05 ml	G6P-DH* variable
			ca. 56 U/l
Tris buffer	(1)	0.25 ml	
mix well; immediately aspirate into the clean spectrophotometer flow-cell; after a 15 s delay read ΔA for a further incubation time of 30 s.			

* Coupled to 3-(3'-carboxypropyl)-1-methylxanthine.

If ΔA per 30 s ($\Delta A/\Delta t$) of the sample is greater than that of the highest standard, dilute the sample further with Tris buffer (1) and re-assay.

Calibration curve: plot the corrected $\Delta A/\Delta t$ readings of the standards *versus* the corresponding theophylline concentrations (mg/l). By use of special graph paper* matched with the reagents, a linear calibration curve is obtained (Fig. 1). The construction of this paper is based on the logit-log function. So far the logit-log model (cf. Vol. I, p. 243) has proved to be useful to fit calibration curves of this EMIT assay [22, 24].

Calculation: read the theophylline mass concentration, ρ (mg/l), corresponding to the mean corrected $\Delta A/\Delta t$ reading of the sample from the calibration curve. To convert from mass concentration to substance concentration, c (µmol/l), multiply by 5.5503 (molecular weight is 180.17).

---

* Supplied by *Syva*; for using this graph paper multiply each value of $\Delta A/\Delta t$ by 3.667.

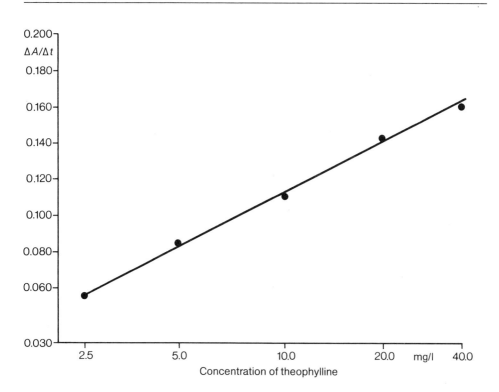

Fig. 1. Typical calibration curve for the assay of theophylline in serum.

Validation of Method

Precision, accuracy, detection limit and sensitivity: in order to attain an adequate precision with this assay the use of partially or fully mechanized analytical systems is recommended [19]. With appropriate mechanization, between-days RSDs range from 1.8 to 5.9% at theophylline concentrations between 2.5 and 40.0 mg/l [12, 23].

The recovery of theophylline added to drug-free pooled human serum ranges from 97 to 102% at concentrations of 5.0 – 35.0 mg/l [12, 23]. The accuracy has been proved by comparison with HPLC methods [11, 12]. The results obtained by EMIT with serum samples from patients treated with theophylline agreed very well with those determined by a highly specific HPLC method (EMIT *vs.* HPLC: y = 1.00 x − 0.09 mg/l, r = 0.99, *n* = 59) [11].

Data on the detection limit defined according to *Kaiser* [25] apparently have not so far been established. The lower limit of the working range is at about 0.5 mg/l as indicated by precision values observed at low theophylline concentrations [26]. Sensitivity depends on the shape of the calibration curve and is highest on the steepest part of the curve, plotted on linear graph paper.

Source of error: sera from highly lipaemic blood give falsely decreased theophylline concentrations.

Specificity: the specificity depends on the quality of the antiserum. A common method of assessing specificity is to add to drug-free serum potentially cross-reacting substances (e.g. for theophylline, other methylxanthines such as caffeine). The concentration of these compounds is determined which is necessary to produce a $\Delta A/\Delta t$ equivalent to that observed with 2.5 mg theophylline per litre. If a concentration of $\geqslant 500$ mg/l is found, significant interference by the substance is unlikely. When interpreting cross-reactivity data the concentrations of the assayed substances expected in serum specimens from patients should be considered. With commercially available EMIT reagents, no significant cross-reactions of caffeine, paraxanthine, major theophylline metabolites and various other methylxanthines were observed [12, 23]. Only 1-methylxanthine showed a marked cross-reaction at a concentration of 25 mg/l. From our experience samples obtained from uraemic patients treated with theophylline can be accurately analyzed by use of this assay, despite high serum concentrations of theophylline metabolites.

Therapeutic ranges: usually, serum (plasma) theophylline concentrations between 8 and 20 mg/l (44 to 111 µmol/l) provide substantial improvements in pulmonary function with minimal toxicity [27]. In premature infants, serum (plasma) concentrations of 6 to 11 mg/l (33 to 61 µmol/l) appear to be effective in decreasing the frequency of apnoeic episodes [28].

Appendix

Preparation of G6P-DH-3-(3'-carboxypropyl)-1-methylxanthine conjugate: the *N*-hydroxysuccinimide ester method results in a satisfactory conjugate [20, 21].

Prepare an enzyme solution containing 2.8 mg (27 nmol, ca. 1.82 kU) G6P-DH, 10 mg G-6-P-Na, 20 mg NADH, disodium salt, and 300 µl carbitol [2-(2-ethoxyethoxy) ethanol] in 1 ml sodium carbonate buffer (100 mmol/l, pH 9.0) at 0°C,

- add to this solution with stirring, in seven 10 µl portions, the *N*-hydroxysuccinimide ester of 3-(3'-carboxypropyl)-1-methylxanthine (0.36 mol/l), previously prepared in *N,N*-dimethylformamide [20]; this process takes about 2 h,

- then dialyze this solution five times at 4°C against Tris buffer (55 mmol/l, pH 7.9).

The drug-enzyme conjugate obtained reportedly retains 29% of the original enzyme activity. A maximum inhibition of 90 – 100% of the remaining activity was observed, when antibodies to 3-(3'-carboxypropyl)-1-methylxanthine were added [21]. The amount of enzyme-conjugate used in the assay is based upon optimization procedures which match this reagent with the antibody/substrate reagent (cf. p. 8).

1.2.2   Determination with
Substrate-labelled Fluorescent Immunoassay

Assay

Method Design

The method was developed by *Li et al.* [29] based on the principle of substrate-labelled fluorescent immunoassay [30].

Principle

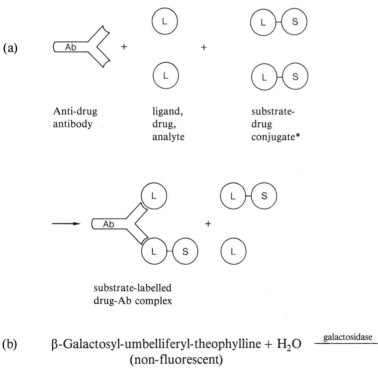

(a)

Anti-drug   ligand,   substrate-
antibody   drug,   drug
                   analyte   conjugate*

substrate-labelled
drug-Ab complex

(b)   β-Galactosyl-umbelliferyl-theophylline + H_2O $\xrightarrow{\text{galactosidase}}$
                    (non-fluorescent)

β-galactose + umbelliferyl-theophylline .
                    (fluorescent)

---

* e.g., β-galactosyl-umbelliferyl-theophylline, non-fluorescent.

8-(3-Aminopropyl)-theophylline is labelled with a fluorigenic enzyme substrate, 7-β-galactosylcoumarin-3-carboxylic acid. This conjugate* (S-L) is non-fluorescent and competes with theophylline (L) in the sample for a limited amount of antibody (Ab). When the substrate-theophylline conjugate (S-L) is bound by the antibody, it cannot be hydrolyzed by β-galactosidase**. The amount of conjugate available for reaction with galactosidase and the resulting increases in fluorescence intensity per unit time, $\Delta F/\Delta t$, are related to the concentration of theophylline in the sample (cf. Vol. I, chapter 2.7, p. 244). The method has been mechanized in various analytical systems [19].

Selection of assay conditions and adaptation to the individual characteristics of the reagents: the assay is designed for rapid analysis and requires no separation step (results of samples from patients are available within about 35 min).

Anti-theophylline antibodies have been obtained by immunizing New Zealand white rabbits with a bovine serum albumin conjugate of 8-(3-carboxypropyl)-theophylline [29]. Recently, a specific monoclonal antibody to theophylline has been produced in mice using hybridoma techniques [31].

The substrate-theophylline conjugate (8-[3-(7-β-galactosylcoumarin-3-carboxyamido) propyl] theophylline) is prepared by coupling 8-(3-aminopropyl) theophylline to 7-β-galactosylcoumarin-3-carboxylic acid (cf. p. 21). After enzymatic hydrolysis, the umbelliferyl moiety at pH 8.3 exhibits an excitation maximum at 398 nm and a fluorescence maximum at 452 nm.

The reaction mixture should contain sufficient antibody to decrease the production of fluorescence in the absence of theophylline to 10% of that observed without addition of antibody [29, 32]. The assay range is from 1.0 to 40.0 mg/l.

The assay is carried out at room temperature, pH 8.3; fluctuations of temperature should not exceed ±3°C during the assay. The reaction is initiated by addition of substrate-theophylline conjugate. During the subsequent incubation for exactly 20 min, the immuno- and enzymatic reactions take place simultaneously. Fluorescence intensity is then measured in each cuvette.

Equipment

Fluorimeter capable of measuring fluorescence intensity with excitation and emission wavelengths of 400 and 450 nm, respectively. Polystyrene disposable cuvettes suitable for the fluorimeter. Glass test tubes 12 mm × 75 mm for sample and calibrator dilutions; stopwatch; pipetter-diluter capable of delivering 50 μl sample and 500 μl or 2.5 ml buffer. Laboratory centrifuge.

---

* β-Galactosyl-umbelliferyl-theophylline, or 8-[3-(7-β-galactosylcoumarin-3-carboxyamido) propyl] theophylline.
** β-D-Galactoside galactohydrolase, EC 3.2.1.23.

The fluorimeter should be checked periodically for linearity of the relationship of readings with concentrations. Use a solution of 7-hydroxycoumarin-3-[N-(2-hydroxy-ethyl)] carboxamide* in Bicine buffer (50 mmol/l, pH 8.3) at concentrations of 0.0, 1.0, 5.0, 10.0 and 15.0 nmol/l.

Reagents and Solutions

Purity of reagents: β-galactosidase from *Escherichia coli* should be of the best available purity. Chemicals are of analytical grade.

Preparation of solutions** (for 100 determinations): all solutions in re-purified water (cf. Vol. II, chapter 2.1.3.2).

1. *N,N*-Bis(2-hydroxyethyl)glycine (Bicine) buffer (Bicine, 1.0 mol/l; NaN$_3$, 308 mmol/l; pH 8.3):

   dissolve 4.08 g *N,N*-bis(2-hydroxyethyl) glycine and 0.5 g sodium azide in 20 ml water. Adjust to pH 8.3 with NaOH, 12.5 mol/l. Make up with water to 25 ml. Filter to remove particles. 25 ml of this buffer is provided with each kit.

2. Diluted Bicine buffer (Bicine, 50 mmol/l; NaN$_3$, 15 mmol/l):

   mix one part (1) with 19 parts water.

3. Antibody/enzyme solution (antibody***; β-galactosidase, 3 kU†/l; NaN$_3$, 15.4 mmol/l; Bicine, 50 mmol/l; pH 8.3):

   5 ml of this solution are provided with each kit.

4. Substrate-theophylline conjugate solution (β-galactosyl-umbelliferyl-theophylline 0.55 μmol/l; formate, 30 mmol/l; pH 3.5):

   5 ml of this solution is provided with each kit.

---

   * Provided by *Ames*.
  ** Reagents, calibrators and buffer are commercially available as a kit from *Ames Division, Miles Laboratories,* Elkhart, U.S.A. The solutions contain sodium azide.
 *** Standardized preparation (cf. 16), concentration is titre-dependent.
   † Enzyme activity is defined as in *Wong et al.* [32]: one unit of enzyme activity hydrolyzes 1.0 μmol 2-nitrophenyl-β-D-galactoside per min, when galactosidase is assayed at 25 °C in Bicine buffer, 50 mmol/l, pH 8.2, containing 2-nitrophenyl-β-D-galactoside (3.0 mmol/l) and sodium azide (15.4 mmol/l). The molar absorption coefficient at λ = 415 nm of the product of this reaction, 2-nitrophenol, is 4.27 × 10^2 l × mol^{-1} × mm^{-1}.

5. Theophylline standard solutions (10 to 40 mg/l):

prepare theophylline standard solution (40 mg/l) as described in chapter 1.2.1, p. 10. Dilute this solution 1 + 1/3, 1 + 1 and 1 + 3 with theophylline-free human serum, yielding the following concentrations of theophylline: 30.0, 20.0, 10.0 mg/l. Take pure serum as zero calibrator.

Standard solutions containing the indicated theophylline concentrations are provided with each kit.

Stability of solutions: reagents and calibrators are stable for 12 months, if stored at 2°C to 8°C. Diluted Bicine buffer (2) is stable for 6 months at 2°C to 23°C, and diluted calibrators are stable for 2 weeks at 2°C to 8°C. The substrate-theophylline conjugate solution (4) should be protected from light.

Procedure

Collection and treatment of specimen: serum or plasma can be used in this assay. Acceptable anticoagulants are heparin and EDTA. Only 50 µl sample is required per assay. No evidence of interference was obtained when sera from slightly haemolyzed, lipaemic or icteric blood were assayed. Results obtained with sera from severely haemolyzed or lipaemic blood should be confirmed with new clear samples. The intake of beverages containing caffeine need not be restricted before sampling, if the antibody employed shows a sufficient specificity (cf. p. 14). For further information about specimen collection cf. chapter 1.2.1, p. 11.

Stability of the substance in the sample: cf. chapter 1.2.1, p. 11.

Assay conditions: measure standards and samples in duplicate under identical conditions (within series). When analyzing series of samples, initiate reaction at 30-s intervals. All solutions must be allowed to reach room temperature before performing the assay.

A pre-dilution of samples and standard solutions is required: mix one part (0.05 ml) sample or standard solution (5) with 50 parts (2.5 ml) diluted Bicine buffer (2).

Incubation time 20 min; ambient temperature; 1.65 ml; excitation wavelength 400 nm; emission wavelength 450 nm.

Establish a calibration curve under exactly the same conditions (within series).

If $\Delta F/\Delta t$ of the sample is greater than that of the highest standard, dilute the pre-diluted sample further with Bicine buffer (2) and re-assay.

Calibration curve: plot mean increase in fluorescence per 20 min, $\Delta F/\Delta t$, of duplicates *versus* the corresponding theophylline concentrations (mg/l) on linear graph paper. Fig. 2 shows a typical calibration curve.

Measurement

Pipette into a reaction cuvette:			concentration in assay mixture
antibody/enzyme solution	(3)	0.05 ml	β-galactosidase 91 U/l rabbit sera or monoclonal antibody variable*
diluted Bicine buffer	(2)	0.50 ml	Bicine 48.5 mmol/l NaN_3 14.9 mmol/l
pre-diluted sample or standard solution (5)		0.05 ml	theophylline up to 24 µg/l
diluted Bicine buffer	(2)	0.50 ml	

substrate-theophylline
   conjugate solution
	(4)	0.05 ml	8-[3-(7-β-galactosyl-
			coumarin-3-carboxyamido)
			propyl] theophylline
			     16.7 nmol/l
			 formate 909 µmol/l

diluted Bicine buffer	(2)	0.50 ml	
mix well, incubate for exactly 20 min; measure change in fluorescence** ΔF.			

 * Amount of antibody sufficient to decrease the fluorescence to 10% of that observed in the absence of antibody.
** Adjust fluorimeter reading to zero by use of a cuvette containing diluted Bicine buffer, and to 90% of full scale with the highest calibrator cuvette incubated for about 19 min after addition of the substrate-theophylline conjugate (4).

Various curve-fitting methods are available for the evaluation of the results from enzyme-immunoassays (cf. Vol. I, p. 243). The use of a simple calculator programme based on linear interpolation between multiple points has been recommended for this assay [29].

Calculation: read or calculate theophylline mass concentrations, ρ (mg/l), from the calibration curve using the mean $\Delta F/\Delta t$ value of each sample. To convert from mass concentration to substance concentration, c (µmol/l), multiply by 5.5503 (cf. p. 12).

Validation of Method

Precision, accuracy, detection limit and sensitivity: for concentrations within the major part of the calibration curve, between-days relative standard deviations less than 8.7% were observed [12].

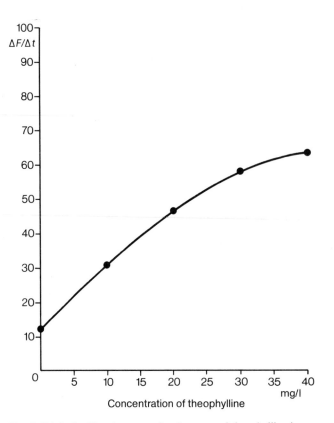

Fig. 2. Typical calibration curve for the assay of theophylline in serum.

The recovery of theophylline added to drug-free pooled human serum ranged from 91 to 108% at concentrations of 5.0 to 30.0 mg/l [12]. The accuracy has been proved by comparison with HPLC and GLC methods [12, 29]. The results obtained with samples from patients by the substrate-labelled fluorescent immunoassay (SLFIA) using a monoclonal antibody agreed very well with those determined by a highly specific HPLC method [11] (SLFIA vs. HPLC: y = 0.99 x + 0.15 mg/l, r = 0.98, $n = 59$).

The assay has a detection limit* of 0.5 mg/l. Sensitivity is highest on the steepest part of the calibration curve.

Sources of error: sera from highly lipaemic blood give falsely elevated theophylline concentrations. Certain fluorescent compounds such as triamterene may cause false elevations of theophylline concentrations. Correction for the presence of fluorescent compounds, cf. chapter Phenytoin, 1.4.1.2. An interference of β-galactosidase from human serum was not observed [30].

---

* Detection limit is defined as mean value plus 3 standard deviations obtained with the zero calibrator in a series of 20 determinations.

Specificity: the specificity depends on the quality of the antibody, which should be carefully checked (cf. chapter 1.2.1, p. 14). With a monoclonal antibody currently used in the *Ames* TDA® Theophylline kit, no significant cross-reactions of caffeine, major theophylline metabolites and various other methylxanthines are observed. Samples from uraemic patients treated with theophylline can be accurately analyzed with this assay despite high concentrations of theophylline metabolites. However, antisera previously employed showed marked cross-reactions with caffeine and other methylxanthines [12, 29].

Therapeutic ranges: cf. chapter 1.2.1, p. 14.

Appendix

Preparation of β-galactosyl-umbelliferyl-theophylline conjugate: the procedure described by *Li et al.* [29] results in a satisfactory conjugate.

- Cool a mixture of 7-β-galactosylcoumarin-3-carboxylic acid (1.47 g, 4 mmol), triethylamine (404 mg, 4 mmol), and dry dimethyl formamide (40 ml) to $-10\,°C$ while stirring under argon,
- add 546 mg (4 mmol) isobutyl chloroformate to this solution,
- after 10 min add an additional 404 mg triethylamine and 949 mg (4 mmol) 8-(3-aminopropyl) theophylline to the flask containing the mixed anhydride,
- stir this mixture for 30 min at $-10\,°C$, then allow the mixture to warm to room temperature,
- combine the reaction mixture with 10 g silica gel and remove dimethyl formamide under greatly reduced pressure,
- place impregnated silica gel on top of a column of silica gel (170 g) and elute with anhydrous ethanol,
- combine fractions 41 to 475 and evaporate to give 545 mg of a yellow solid,
- dissolve this solid in water, filter, and concentrate to 20 ml,
- discard precipitate, and elute the filtrate with water on a 25 mm × 570 mm column of Sephadex LH-20, collecting 15 ml fractions,
- combine fractions 18 to 23, evaporate, and recrystallize the residue from water to give 55 mg (2% yield) of the conjugate as a light yellow solid (m.p. $190-192\,°C$).

References

[1] *A. Kossel,* Über eine neue Base aus dem Pflanzenreich, Ber. dtsch. chem. Ges. *21,* 2164–2167 (1888).
[2] *S. Hirsch,* Klinischer und experimenteller Beitrag zur krampflösenden Wirkung der Purinderivate, Klin. Wochenschr. *1,* 615–618 (1922).
[3] *W. R. Kukovetz, G. Pöch, S. Holzmann,* Cyclic Nucleotides and Relaxation of Vascular Smooth Muscle, in: *P. M. Vanhoutte, I. Leusen* (eds.), Vasodilatation, Raven Press, New York 1981, pp. 339 to 353.

[4] *L. Hendeles, M. Weinberger, G. Johnson,* Theophylline, in: *W. E. Evans, J. J. Schentag, W. J. Jusko* (eds.), Applied Pharmacokinetics, Applied Therapeutics, San Francisco 1980, pp. 95 – 138.

[5] *M. J. Arnaud, C. Welsch,* Theophylline and Caffeine Metabolism in Man, in: *N. Rietbrock, B. G. Woodcock, A. H. Staib* (eds.), Theophylline and Other Methylxanthines, Vieweg, Braunschweig 1982, pp. 135 – 148.

[6] *A. Niemann, M. Oellerich, G. Schumann, G. W. Sybrecht,* Determination of Theophylline in Saliva, Using Fluorescence Polarization Immunoassay (FPIA), J. Clin. Chem. Clin. Biochem. *23*, 725 – 732 (1985).

[7] *Ch. Knott, M. Bateman, F. Reynolds,* Do Saliva Concentrations Predict Plasma Unbound Theophylline Concentrations? A Problem Re-examined. Br. J. Clin. Pharmacol. *17*, 9 – 14 (1984).

[8] *S. Wemhöner, M. Oellerich, G. Sybrecht,* Optimierung der Therapie mit Theophyllin-Präparaten bei obstruktiven Ventilationsstörungen. 1. Bioverfügbarkeit und Pharmakokinetik verschiedener Theophyllin-Präparate. Prax. Pneumol. *35*, 36 – 41 (1981).

[9] *C. Ranke, G. Schmidt, M. Oellerich, G. W. Sybrecht,* Optimierung der Therapie mit Theophyllin-Präparaten bei obstruktiven Ventilationsstörungen. 3. Vorhersage der Theophyllin-Clearance mit einer einfachen pharmakokinetischen Methode und Vergleich der Bioverfügbarkeit verschiedener Theophyllinretard-Präparate. Prax. Klin. Pneumol. *37*, 99 – 108 (1983).

[10] *J. H. Wilkens, H. Neuenkirchen, G. W. Sybrecht, M. Oellerich,* Individualizing Theophylline Dosage: Evaluation of a Single-Point Maintenance Dose Prediction Method, Eur. J. Clin. Pharmacol. *26*, 491 – 498 (1984).

[11] *G. Schumann, I. Isberner, M. Oellerich,* Highly Specific HPLC Method for the Determination of Theophylline in Serum, Fresenius Z. Anal. Chem. *317*, 677 (1984).

[12] *M. Oellerich, P. Hannemann, W. R. Külpmann, M. Beneking, G. W. Sybrecht,* Die Bestimmung von Theophyllin im Serum mit radioaktivitätsfreien Immunotests und Hochdruck-Flüssigkeits-Chromatographie, Internist *23*, 641 – 646 (1982).

[13] *R. J. Tyhach, P. A. Rupchock, J. H. Pendergrass, A. C. Skjold, P. J. Smith, R. D. Johnson, J. P. Albarella, J. A. Profitt,* Adaptation of Prosthetic-group-label Homogeneous Immunoassay to Reagent-strip Format, Clin. Chem. *27*, 1499 – 1504 (1981).

[14] *R. Lindberg, K. Ivaska, K. Irjala, T. Vanto,* Determination of Theophylline in Serum with the Seralyzer®Aris Reagent Strip Test Evaluated, Clin. Chem. *31*, 613 – 614 (1985).

[14a] *R. F. Zuk, V. K. Ginsberg, T. Houts, J. Rabbie, H. Merrick, E. F. Ullman, M. M. Fischer, C. C. Sizto, S. N. Stiso, D. J. Litman,* Enzyme-immunochromatography – A Quantitative Immunoassay Requiring no Instrumentation, Clin. Chem. *31*, 1144 – 1150 (1985).

[14b] *A. L. Thunberg, G. M. Dappen, B. A. Burdick, J. L. Daiss, S. J. Danielson, J. B. Findlay, P. H. Frickey, P. G. Hosimer, G. J. McClune, M. W. Sundberg, K. J. Sanford,* Multilayer Enzyme Immunoassay Element for Theophylline, Clin. Chem. *31*, 911 (1985).

[14c] *G. E. Norton, J. B. Mauck,* Development of a Kodak Ektachem Clinical Chemistry Slide Assay for Theophylline, Clin. Chem. *31*, 911 (1985).

[15] *H. Itagaki, Y. Hakoda, Y. Suzuki, M. Haga,* Drug Sensor: An Enzyme Immunoelectrode for Theophylline, Chem. Pharm. Bull. *31*, 1283 – 1288 (1983).

[16] *J. A. Hinds, C. F. Pincombe, R. K. Kanowski, S. A. Day, J. C. Sanderson, P. Duffy,* Ligand Displacement Immunoassay: A Novel Enzyme Immunoassay Demonstrated for Measuring Theophylline in Serum, Clin. Chem. *30*, 1174 – 1178 (1984).

[17] *J. B. Gushaw, M. W. Hu, P. Singh, J. G. Miller, R. S. Schneider,* Homogeneous Enzyme Immunoassay for Theophylline in Serum, Clin. Chem. *23*, 1144 (1977).

[18] *K. E. Rubenstein, R. S. Schneider, E. F. Ullman,* "Homogeneous" Enzyme Immunoassay. A New Immunochemical Technique, Biochem. Biophys. Res. Commun. *47*, 846 – 851 (1972).

[19] *M. Oellerich,* Enzyme Immunoassay: a Review. J. Clin. Chem. Clin. Biochem. *22*, 895 – 904 (1984).

[20] *P. Singh, M. W. Hu, J. B. Gushaw, E. F. Ullman,* Specific Antibodies to Theophylline for Use in a Homogeneous Enzyme Immunoassay, J. Immunoassay *1*, 309 – 322 (1980).

[21] *J. Chang, S. Gotcher, J. B. Gushaw, R. H. Gadsden, Ch. A. Bradley, Th. C. Stewart,* Homogeneous Enzyme Immunoassay for Theophylline in Serum and Plasma, Clin. Chem. *28*, 361 – 367 (1982).

[22] *D. N. Dietzler, N. Weidner, V. L. Tieber, J. M. McDonald, C. H. Smith, J. H. Ladenson, M. P. Leckie,* Adaptation of the EMIT Theophylline Assay to Kinetic Analyzers: the Relationship of Reaction Kinetic to Calculation Procedures, Clin. Chim. Acta *101*, 163 – 181 (1980).

[23] *M. Oellerich, G. W. Sybrecht, R. Haeckel,* Monitoring of Serum Theophylline Concentrations by a Fully Mechanized Enzyme Immunoassay (EMIT), J. Clin. Chem. Clin. Biochem. *17*, 299 – 302 (1979).

[24] *R. Cook, D. Wellington,* Data Handling for EMIT® Assays, Syva, Palo Alto, U.S.A. (1980).

[25] *H. Kaiser,* Zum Problem der Nachweisgrenze, Fresenius Z. anal. Chem. *209*, 1 – 18 (1965).

[26] *R. Wettengel, M. Oellerich, J. Schnitker,* Vergleichende Untersuchung der Theophyllin-Serum-Konzentrationen und der bronchospasmolytischen Wirkung von Cholintheophyllinat (Euspirax®) und Theophyllin-Äthylendiamin (Euphyllin retard®) bei Patienten mit obstruktiver Atemwegs-erkrankung. Prax. Pneumol. *33*, 1125 – 1131 (1979).

[27] *J. W. Jenne, E. Wyze, F. S. Rood, F. M. McDonald,* Pharmacokinetics of Theophylline: Application to Adjustment of the Clinical Dose of Aminophylline, Clin. Pharmacol. Ther. *13*, 349 – 360 (1972).

[28] *G. Giacoia, W. J. Jusko, J. Menke, J. R. Koup,* Theophylline Pharmacokinetics in Premature Infants with Apnea, J. Pediatr. *89*, 829 – 832 (1976).

[29] *T. M. Li, J. L. Benovic, R. T. Buckler, J. F. Burd,* Homogeneous Substrate-Labeled Fluorescent Immunoassay for Theophylline in Serum, Clin. Chem. *27*, 22 – 26 (1981).

[30] *J. F. Burd, R. C. Wong, J. E. Feeney, R. J. Carrico, R. C. Boguslaski,* Homogeneous Reactant-Labeled Fluorescent Immunoassay for Therapeutic Drugs Exemplified by Gentamicin Determination in Human Serum, Clin. Chem. *23*, 1402 – 1408 (1977).

[31] *H. M. Clements, S. G. Thompson, R. A. Ott,* Theophylline Substrate-Labeled Fluorescent Immunoassay Using Monoclonal Antibody, Clin. Chem. *29*, 1176 (1983).

[32] *R. C. Wong, J. F. Burd, R. J. Carrico, R. T. Buckler, J. Thoma, R. C. Boguslaski,* Substrate-Labeled Fluorescent Immunoassay for Phenytoin in Human Serum, Clin. Chem. *25*, 686 – 691 (1979).

1.3 Caffeine

3,7-Dihydro-1,3,7-trimethyl-1*H*-purine-2,6-dione

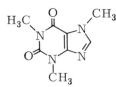

Thomas Zysset and Prithipal Singh

General

Caffeine is a naturally occurring compound found in a number of plants such as the coffee-bean, the tea-leaf, the kolanut and the ilex plant [1]. In our society, it is consumed as an ingredient of food, beverages or drugs, mostly for its mildly stimulating effect. The average *per-capita* caffeine intake in the U.S.A. is about 200 mg/day [1 – 3]. Like theophylline caffeine has been found to be effective in the treatment of apnoea in premature infants [4]. A recent report has confirmed the bronchodilating effect of caffeine in asthmatics [5]. This effect may be due at least in part to an inhibition of phosphodiesterase activity, with a subsequent increase in cyclic adenosine 3′,5′-monophosphate (cAMP) [6]. The main caffeine metabolites in human plasma are paraxanthine (1,7-dimethylxanthine; about 74% of a caffeine dose), theobromine (3,7-dimethylxanthine; about 6%) and theophylline (1,3-dimethylxanthine; about 9%) [7]. About 1.2 to 3.7% of a caffeine dose is excreted unchanged in urine [7, 8]. The pharmacokinetics appear to be dose-independent at the levels achieved during "normal" caffeine consumption [8, 9].

The pharmacokinetic parameters of caffeine in healthy adult volunteers are as follows: complete absorption after oral administration takes about 45 minutes [8]; elimination half-life from serum is 3.8 h (range 2.1 to 4.7 h); systemic clearance is 2.0 ml/min per kg (range 1.3 to 3.6 ml/min per kg), and the apparent volume of distribution corresponds to the volume of body water (0.64 l/kg, range 0.48 to 0.97) [10, 11]. Smoking, liver disease and drugs may profoundly influence the clearance of caffeine [10 – 14]. In cirrhotic liver disease, half-life was 17.5 h (4.5 to 50.2 h) and clearance was 0.67 ml/min per kg (0.14 to 1.29 ml/min per kg) [11]. In contrast to adults, caffeine elimination in the newborn is mainly determined by urinary excretion [4, 15]. Consequently, the plasma half-life of caffeine in both premature and full-term newborns is approximately 4 days. Binding to serum proteins is about 40%. The concentration in saliva is about 70% of the total serum concentration. The pharmacokinetics of caffeine evaluated in plasma and saliva showed a close correlation [9, 16, 17].

The main reason for monitoring plasma caffeine levels is its slow elimination in newborns, in whom caffeine is used for treatment of apnoea [4]. Despite an apparently wider therapeutic index for caffeine, compared with theophylline, cases of caffeine intoxication in newborns have been reported [18, 19]. In addition, these patients metabolize theophylline to caffeine in varying amounts [20]. Monitoring only theophylline levels during theophylline therapy might, therefore, lead to underestimation of pharmacologically active compounds.

A new and promising indication for the measurement of plasma or saliva caffeine levels is their use for the assessment of liver function [10, 11, 14]. Both plasma caffeine clearance and half-life showed highly significant changes in patients with liver disease compared with normal controls. Since almost all persons in our society consume caffeine in some form, this quantitative liver function test is simple, noninvasive and also suitable for use in children [21].

Application of method: in clinical chemistry, in pharmacology and in toxicology.

Substance properties relevant in analysis: caffeine (M_r = 194.19) is a very weak base with a pK_a of 0.61 at 40 °C and a pK_a of 14.0 at 25 °C [22]. The solubility in water is 1 g in 46 ml. Since caffeine is non-ionized over a large pH range, the fraction of caffeine not bound to protein diffuses into saliva, where also it can be assayed [9, 16].

Methods of determination: the methods of choice for the quantitative determination of caffeine are either gas chromatography (GLC) [23], high-pressure liquid chromatography (HPLC), radioimmunoassay [24], or enzyme-immunoassay (Emit®) procedures. The chromatographic methods are very sensitive and highly specific and, with a normal-phase HPLC method, it is possible to detect simultaneously the three main caffeine metabolites theophylline, theobromine and paraxanthine in plasma besides the parent compound [25]. For measurement of caffeine alone (without its metabolites), a reversed-phase HPLC method may be an easier approach [26, 27].

For determination of large series of small volumes of patients' sera, an enzyme-immunoassay, as described here, may be preferable. The method is simple, rapid and accurate, and since it is a homogeneous immunoassay (requiring no separation step) it can easily be automated [28].

International reference method and standards: neither standardization at the international level nor the existence of reference standard materials is known at this time.

Assay

Method Design

The method is based on the enzyme-multiplied immunoassay technique (EMIT®) first described by *Rubenstein et al.* [29].

Principle: as described in detail on pp. 7, 8.

The activity of the marker enzyme, glucose-6-phosphate dehydrogenase, G6P-DH, is reduced if the enzyme-caffeine conjugate (E-L) is coupled to the antibody (Ab). Caffeine (L) in a sample competes with G6P-DH-labelled caffeine (E-L) for a limited concentration of antibody (Ab). The activity of G6P-DH measured as increase in absorbance per unit time, $\Delta A/\Delta t$, is related to the caffeine concentration of the sample (cf. Vol. I, chapter 2.7, p. 244). The method has been mechanized in various analytical systems.

Selection of assay conditions and adaptation to the individual characteristics of the reagents: the anti-caffeine antibodies are produced in sheep from an immunogen prepared by conjugating 1,7-dimethyl-3-(3'-carboxypropyl)xanthine to bovine γ-globulin [30]. The animals are inoculated with an emulsion of 1 mg of the immunogen with complete *Freund's* adjuvant. Subsequent booster injections are carried out with the same amount of the immunogen in incomplete *Freund's* adjuvant every month. The animals are bled every month and the IgG fraction of the antibodies is used for the assay. Highly specific polyclonal antibodies have been produced by this method.

Enzyme-drug conjugate: the enzyme conjugate is obtained by coupling 1,7-dimethyl-3-(3′-carboxypropyl)xanthine to glucose-6-phosphate dehydrogenase (G6P-DH) from *Leuconostoc mesenteroides* by conventional methods. Interference from the endogenous G6P-DH is avoided since the coenzyme, NAD, is converted only by the bacterial enzyme used in the assay.

Optimum ratio of antibody to enzyme conjugates: the optimum quantities and concentration of sample and enzyme-drug conjugates in an assay medium is dependent on the characteristic of the antibodies used. Since the desired assay range of 0.5 to 30.0 mg/l for caffeine and the sample size of 50 µl per assay are fixed, the optimum ratio of the concentrations of the antibodies and the enzyme conjugates used in assays must be determined. This is done as follows:

An appropriate enzyme conjugate concentration showing a maximum $\Delta A/\Delta t \cong$ 0.600 under the assay conditions, is chosen and is gradually titrated with increasing amounts of the antibody. Its enzymic activity is gradually inhibited with increasing concentration of the antiserum. A maximum inhibition of about 90 to 100% of the activity is often achieved by adding excess antibodies. This provides the approximate amount of antibodies for use in the assay.

In order to optimize the assay system a fixed amount of the enzyme conjugate is used and calibrators containing caffeine in concentrations of 0.0, 1.0, 2.0, 3.0, 7.0, 15.0, and 30.0 mg/l, in the drug-free human serum, are assayed according to the standard EMIT protocol using different amounts of antibodies. The zero calibrator serves as the reagent blank and all readings are corrected for this value. The corrected $\Delta A/\Delta t$ values for each calibrator are plotted against the amount of antibodies used. From these titration curves the antibody quantity is then chosen which provides (a) the maximum separation between different calibrators, (b) optimum response in the assay at the most desirable concentration of 7 mg/l and (c) minimum cross-reactivity with different cross-reactants.

The ratio of the concentrations of the antibodies and the enzyme conjugates is compared at different conjugate concentrations using the above criteria, before finalizing the exact concentrations and quantities of these reagents for use in the assay. As the cross-reactivity to related methylxanthines may vary with the amount of antibody present in the assay, the specificity also plays an important role in selecting the final antibody titre.

Assay conditions and procedure are as described in detail on p. 16.

Equipment

Spectrophotometer or spectral-line photometer capable of exact measurement at 339 or Hg 334 nm. It should be equipped with a thermostatted flow-cell which consistently maintains 30.0 ± 0.1 °C throughout the working day. Work rack and disposable 2 ml beakers with conical bottoms. Pipetter-diluter capable of delivering 50 ± 1 µl sample and 250 ± 5 µl buffer with a relative standard deviation less than 0.25%.

Delivery must be of sufficient force to ensure adequate mixing of the components. The spectrophotometer should be connected to a suitable data-handling device. Laboratory centrifuge.

Reagents and Solutions

Purity of reagents: G6P-DH from *Leuconostoc mesenteroides* should be of the best available purity: specific activity at least 650 U/mg at 30°C, G-6-P and NAD as substrates (supplier e.g. *Sigma, Boehringer Mannheim*). Chemicals are of analytical grade.

Preparation of solutions* (for 50 determinations): all solutions in re-purified water (cf. Vol. II, chapter 2.1.3.2).

1. Tris buffer (Tris, 55 mmol/l; Triton X-100, 0.1 ml/l; pH 8.0):

   dissolve 1.996 g Tris base in 140 to 160 ml water; add 0.03 ml Triton X-100 and adjust to pH 8.0 with HCl, 1 mol/l; make up with water to 300 ml.

   Alternatively, dilute the buffer concentrate provided with each kit according to the manufacturer's specifications.

2. Antibody/substrate solution (γ-globulin**, ca. 50 mg/l; G-6-P, 66 mmol/l; NAD, 40 mmol/l; Tris, 55 mmol/l; pH 5.2):

   reconstitute available lyphilized preparation with 3.0 ml water.

3. Conjugate solution (G6P-DH***, ca. 1 kU/l; Tris, 55 mmol/l; pH 6.2):

   reconstitute available lyophilized preparation with 3.0 ml water according to specifications. This reagent must be standardized to match antibody/substrate solution (2) (cf. p. 26).

4. Caffeine standard solutions (1 to 30 mg/l):

   dissolve 100 mg caffeine in 100 ml Tris buffer (1); dilute 0.3 ml and 0.7 ml, respectively, with caffeine-free human serum (containing 5 g sodium azide per l and 0.5 g Thimerosal per l) to 10 ml (30 and 70 mg/l), respectively. Dilute the 30 mg/l concentration 1 + 1, 1 + 9 and 1 + 29, the 70 mg/l solution 1 + 9 with caffeine-free human serum (cf. above), yielding the following concentrations of caffeine: 15.0, 3.0, 1.0, 7.0 mg/l. Take pure serum as zero calibrator.

   Alternatively, reconstitute each of the available lyophilized standard preparations with 1 ml water.

---

* Reagents, calibrators and buffer are commercially available from *Syva*, Palo Alto, CA. U.S.A. The solutions contain sodium azide.

** Standardized preparation from sera from immunized sheep.

*** Coupled to 1,7-dimethyl-3-(3'-carboxypropyl)xanthine.

Stability of solutions: after reconstitution the antibody/substrate solution (2), conjugate solution (3) and standards (4) must be stored at room temperature (20 °C to 25 °C) for at least one hour before use. The reagents are then stable for 12 weeks, if stored at 2 °C to 8 °C. The buffer solution (1) can be used for 12 weeks when stored at room temperature.

Procedure

Collection and treatment of specimen: it is important to collect the specimen at an appropriate time following dosing. During long-term therapy, blood should be taken after achieving a "steady-state" concentration. In pre-term infants, it is recommended to draw a blood specimen 60 minutes after administration of an intravenous loading dose and then prior to the administration of the maintenance doses. Serum, plasma and saliva can be used in this assay. Blood cannot be used. Acceptable anticoagulants are heparin, EDTA and oxalate. Only 50 µl sample is required per assay. No evidence of interference was obtained when sera from haemolyzed, lipaemic or icteric blood were assayed at concentrations up to 8 g haemoglobin per litre, 11.3 mmol triglycerides per litre, and 513 µmol bilirubin per litre [31].

Saliva specimens should be collected according to the following procedure: the mouth is carefully rinsed with water. Three minutes later, collection of saliva is started. To stimulate salivary flow, Parafilm® may be chewed. If the saliva samples are not analyzed immediately, refrigeration or freezing may be necessary to prevent bacterial growth. Centrifugation of saliva specimens prior to analysis is recommended.

Stability of caffeine in the sample: at 4 °C, the samples are stable for several weeks, frozen samples are stable for at least 6 months. Serum samples may be stored at room temperature for several days. However, refrigeration or freezing of saliva samples as soon as possible is recommended to prevent bacterial growth.

Assay conditions: measurements of samples and zero calibrator (caffeine-free serum) in duplicate. Single determinations are made of the standards.

A pre-dilution of samples and standard solutions is required: mix one part (0.05 ml) sample or standard solution (4) with 5 parts (0.25 ml) Tris buffer (1).

Incubation time 45 s (in the spectrophotometer flow-cell); 0.90 ml; 30 °C; measurement during the last 30 s against water; wavelength 339 or Hg 334 nm; light path 10 mm.

Establish a calibration curve with solution (4) instead of sample under exactly the same conditions (within the series). The zero calibrator assay serves as a reagent blank; all measurements are corrected for this value.

Measurement

Pipette into a disposable beaker:			concentration in assay mixture	
pre-diluted sample or standard solution (4)		0.05 ml	caffeine	up to 278 µg/l
Tris buffer	(1)	0.25 ml	Tris	52 mmol/l
antibody/substrate solution	(2)	0.05 ml	sheep serum variable	
			γ-globulin	ca. 2.8 mg/l
			G-6-P	3.7 mmol/l
			NAD	2.2 mmol/l
Tris buffer	(1)	0.25 ml		
conjugate solution	(3)	0.05 ml	G6P-DH* variable	
				ca. 56 U/l
Tris buffer	(1)	0.25 ml		
mix well; immediately aspirate into the clean spectrophotometer flow-cell; after a 15 s delay, read absorbance over a period of 30 s.				

* Coupled to 1,7-dimethyl-3-(3'-carboxypropyl)xanthine.

If the increase in absorbance per 30 s, $\Delta A/\Delta t$, of the sample is greater than that of the highest standard, dilute the sample further with Tris buffer (1) and re-assay.

Calibration curve: correct all readings for blank (zero calibrator), yielding $\Delta A/\Delta t$. Plot values of $\Delta A/\Delta t$ of the standards *versus* the corresponding caffeine mass con-

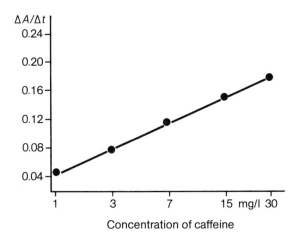

Fig. 1. Typical calibration curve for the assay of caffeine.

centrations, ρ (mg/l). By use of special graph paper* matched with the reagents, a linear calibration curve is obtained (Fig. 1). The construction of this paper is based on the logit-log function. So far the log-logit model (cf. Vol. I, p. 243) has proved to be useful to fit calibration curves of this EMIT assay [32].

Calculation: for the mean $\Delta A/\Delta t$ of the sample take the caffeine mass concentration, ρ (mg/l), from the calibration curve. For conversion to substance concentration, c (µmol/l), multiply by 5.1495 (molecular weight is 194.19).

Validation of Method

Precision, accuracy, detection limit and sensitivity: in order to obtain an adequate precision with this assay the use of partially or fully mechanized analytical systems is recommended [28]. With appropriate mechanization, between-days RSD varied between 2.7 and 5.6% at caffeine concentrations between 1 and 20 mg/l [28].

The recovery of caffeine added to drug-free pooled human serum was 99.5% at a concentration of 11 mg/l [28]. The accuracy has been proved by comparison with a highly specific HPLC method: [25]. The results obtained by EMIT with samples from adult patients containing caffeine agreed well with those determined by HPLC. (EMIT *vs* HPLC: $y = 1.06\,x + 0.24$ mg/l, $r = 0.99$, $n = 69$) [28]. The higher value of caffeine obtained with EMIT is presumably due to significant cross-reactivity with the major caffeine metabolite, paraxanthine [28].

Data on the detection limit according to the definition of *Kaiser* [33] apparently have not been established so far. The lower limit of the working range is at about 0.3 mg/l [28]. Sensitivity depends on the shape of the calibration curve and is highest on the steepest part of the curve, plotted on linear graph paper.

Sources of error: sera from grossly haemolyzed, icteric and lipaemic blood give erroneous results.

Specificity: the specificity depends on the quality of the antiserum. A common method of assessing specificity is to add to drug-free serum potentially cross-reacting substances (e.g. for caffeine its main metabolite paraxanthine). With reagents obtained from *Syva*, significant cross-reactivity was obtained with paraxanthine, whereas theophylline and theobromine cross-reactivities were insignificant (Table 1) [28].

Clinically insignificant cross-reactivity is observed with theophylline, theobromine, diprophylline and most of the major metabolites of the drug. The 1,7-dimethylxanthine shows a cross-reactivity of ca. 25% at the midpoint of the assay. However, it is of no importance in the assay since this compound is not a major metabolite in neonates [15].

---

* Supplied by *Syva*; for using this graph paper multiply each value of $\Delta A/\Delta t$ by 2.667.

Table 1. Cross-reactivity of caffeine metabolites in the Emit caffeine assay*. Aqueous solutions containing 2.5, 5, 10, 20, 40 and 80 µmol paraxanthine, theobromine or theophylline, respectively, per litre were assayed for relative interference in the caffeine assay. The relative interference of each metabolite over the concentration range tested was constant.

	Concentration range of metabolite tested µmol/l	% of metabolite concentration measured as caffeine $\bar{x} \pm SD$
Paraxanthine	2.5 – 80	27.8 ± 2.5
Theobromine	2.5 – 80	2.7 ± 1.7
Theophylline	2.5 – 80	1.3 ± 0.8

* Reprinted with permission from ref. [28].

Since paraxanthine is not found in newborns [15], the present enzyme-immunoassay is primarily recommended for the measurement of caffeine in neonates. In our institution the test has also been validated in the discrimination of adults with liver disease from normals. The results indicate that in most cases, the Emit assay may be used for this purpose, although the concentrations tended to be higher than those obtained by an HPLC method [28].

Therapeutic ranges: in neonates with apnoea, therapeutic levels of 5 – 20 mg/l have been suggested [4]. In young patients (8 – 18 years old) a bronchodilator effect at peak caffeine concentrations of 13.5 mg/l has been demonstrated [5]. In saliva, caffeine concentrations are approximately 30% lower than in plasma [9, 16].

Appendix

G6P-DH-3-(3'-carboxypropyl)-1,7-dimethylxanthine conjugate preparation: the N-hydroxysuccinimide ester method leads to a satisfactory conjugate [34].

- Dissolve 2.8 mg (≥1.82 kU, 27 nmol) G6P-DH, 10 mg (27 µmol) G-6-P, monosodium salt, 20 mg NADH, disodium salt and 150 µl dimethylformamide (DMF) in 1 ml carbonate buffer (100 mmol/l, pH 9.0) at 4°C;

- add with stirring in seven 10 µl portions the N-hydroxysuccinimide ester of 3-(3'-carboxypropyl)-1,7-dimethylxanthine (0.36 mol/l), previously prepared in N,N-dimethylformamide; this process takes about 2 h,

- then dialyze this solution five times at 4°C against Tris buffer (1).

The enzyme-drug conjugate obtained reportedly retains 60% of the original enzyme activity. A maximum inhibition of 90 – 100% of the remaining activity was observed, when antibodies to 3-(3'-carboxypropyl)-1,7-dimethylxanthine were added. The amount of enzyme conjugate used in the assay is based upon optimization procedures which match this reagent with the antibody/substrate solution (2) (cf. p. 26).

References

[1] D. M. Graham, Caffeine – Its Identity, Dietary Sources, Intake and Biological Effects, Nutr. Rev. 36, 97 – 102 (1978).

[2] T. W. Rall, The Xanthines, in: A. G. Gilman, L. S. Goodman, A. Gilman (eds.), The Pharmacological Basis of Therapeutics, Macmillan Publishing Co., New York 1980, pp. 592 – 607.

[3] Anonymous, Coffee and Cardiovascular Disease, Medical Letter 19, 65 – 67 (1977).

[4] J. V. Aranda, C. E. Cook, W. Gorman, J. M. Collinge, P. M. Loughnan, E. W. Outerbridge, A. Aldridge, A. H. Neims, Pharmacokinetic Profile of Caffeine in the Premature Newborn Infant with Apnea, J. Pediatr. 94, 663 – 668 (1979).

[5] A. B. Becker, K. J. Simons, C. A. Gillespie, F. E. R. Simons, The Bronchodilator Effects and Pharmacokinetics of Caffeine in Asthma, N. Engl. J. Med. 310, 743 – 746 (1984).

[6] R. W. Butcher, E. W. Sutherland, Adenosine 3',5'-phosphate in Biological Materials, J. Biol. Chem. 237, 1244 – 1250 (1962).

[7] D. D.-S. Tang-Liu, R. L. Williams, S. Riegelman, Disposition of Caffeine and Its Metabolites in Man, J. Pharmacol. Exp. Ther. 224, 180 – 185 (1983).

[8] M. Bonati, R. Latini, F. Galletti, J. F. Young, G. Tognoni, S. Garattini, Caffeine Disposition after Oral Doses, Clin. Pharmacol. Ther. 32, 98 – 106 (1982).

[9] R. Newton, L. J. Broughton, M. J. Lind, P. J. Morrison, H. J. Rogers, I. D. Bradbrook, Plasma and Salivary Pharmacokinetics of Caffeine in Man, Eur. J. Clin. Pharmacol. 21, 45 – 52 (1981).

[10] E. Renner, A. Wahlländer, P. Huguenin, H. Wietholtz, R. Preisig, Coffein – ein ubiquitärer Indikator der Leberfunktion, Schweiz. Med. Wochenschr. 113, 1074 – 1081 (1983).

[11] E. Renner, H. Wietholtz, P. Huguenin, M. J. Arnaud, R. Preisig, Caffeine – a Model Compound for Measuring Liver Function, Hepatology 4, 38 – 46 (1984).

[12] D. C. May, C. H. Jarboe, A. B. Van Bakel, W. M. Williams, Effects of Cimetidine on Caffeine Disposition in Smokers and Nonsmokers, Clin. Pharmacol. Ther. 31, 656 – 661 (1982).

[13] E. C. Rietveld, M. M. M. Broekman, J. J. G. Houben, T. K. A. B. Eskes, J. M. van Rossum, Rapid Onset of an Increase in Caffeine Residence Time in Young Women Due to Oral Contraceptive Steroids, Eur. J. Clin. Pharmacol. 26, 371 – 373 (1984).

[14] G. Jost, U. v. Mandach, A. Wahlländer, T. Zysset, R. Preisig, A New and Noninvasive Approach to Quantify Liver Function: Overnight Caffeine Clearance Measured in Saliva (OCCS), J. Hepatology, suppl. I, 572 (1984).

[15] A. Aldrige, J. V. Aranda, A. H. Neims, Caffeine Metabolism in the Newborn, Clin. Pharmacol. Ther. 25, 447 – 453 (1979).

[16] E. Zylber-Katz, L. Granit, M. Levy, Relationship between Caffeine Concentrations in Plasma and Saliva, Clin. Pharmacol. Ther. 36, 133 – 137 (1984).

[17] J. Blanchard, Protein Binding of Caffeine in Young and Elderly Males, J. Pharm. Sci. 71, 1415 – 1418 (1982).

[18] W. Banner, P. A. Czajka, Acute Caffeine Overdose in the Neonate, Am. J. Dis. Child 134, 495 – 498 (1980).

[19] P. B. Kulkarni, R. D. Dorand, Caffeine Toxicity in a Neonate, Pediatrics 64, 254 – 255 (1979).

[20] C. Bory, P. Baltassat, M. Porthault, M. Bethenod, A. Frederick, J. V. Aranda, Metabolism of Theophylline to Caffeine in Premature Newborn Infants, J. Pediatr. 94, 988 – 993 (1979).

[21] K.-J. Morgan, V. J. Stults, M. E. Zabik, Amount and Dietary Sources of Caffeine and Saccharin Intake by Individuals Ages 5 to 18 Years, Regul. Toxicol. Pharmacol. 2, 296 – 307 (1982).

[22] Martindale, The Extra Pharmacopoeia, 28th edit., The Pharmaceutical Press, London 1982, p. XXIV.

[23] J. L. Cohen, C. Cheng, J. P. Henry, Y.-L. Chan, GLC Determination of Caffeine in Plasma Using Alkali Flame Detection, J. Pharm. Sci. 67, 1093 – 1095 (1978).

[24] C. E. Cook, C. R. Tallent, E. W. Emerson, M. W. Myers, J. A. Kepler, G. F. Taylor, H. D. Christensen, Caffeine in Plasma and Saliva by a Radioimmunoassay Procedure, J. Pharmacol. Exp. Ther. 199, 679 – 689 (1976).

[25] A. Wahlländer, E. Renner, G. Karlaganis, High-performance Chromatographic Determination of Dimethylxanthine Metabolites of Caffeine in Human Plasma, J. Chromatogr. 338, 369 – 372 (1985).

[26] *T. Foenander, D. J. Birkett, J. O. Miners, L. M. H. Wing,* The Simultaneous Determination of Theophylline, Theobromine and Caffeine in Plasma by High Performance Liquid Chromatography, Clin. Biochem. *13,* 132 – 134 (1980).

[27] *F. L. S. Tse, D. W. Szeto,* Reversed-phase High-performance Liquid Chromatographic Determination of Caffeine and Its N-Demethylated Metabolites in Dog Plasma, J. Chromatogr. *226,* 231 – 236 (1981).

[28] *T. Zysset, A. Wahlländer, R. Preisig,* Evaluation of Caffeine Plasma Levels by an Automated Enzyme Immunoassay (EMIT) in Comparison with a High Performance Liquid Chromatographic Method, Therap. Drug Monit. *6,* 348 – 354 (1984).

[29] *K. E. Rubenstein, R. S. Schneider, E. F. Ullman,* "Homogeneous" Enzyme Immunoassay. A New Immunochemical Technique, Biochem. Biophys. Res. Commun. *47,* 846 – 851 (1972).

[30] *M. W. Hu, P. Singh,* Antibodies for Caffeine and Their Preparation, U. S. Patent (pending).

[31] Emit® Caffeine Assay. (Package insert). *Syva,* Palo Alto, U.S.A. (1980).

[32] *R. Cook, D. Wellington,* Data Handling for EMIT® Assays, *Syva,* Palo Alto, U.S.A. (1980).

[33] *H. Kaiser,* Zum Problem der Nachweisgrenze, Fresenius Z. anal. Chem. *209,* 1 – 18 (1965).

[34] *G. W. Anderson, J. E. Zimmerman, F. M. Callahan,* The Use of N-Hydroxysuccinimide in Peptide Synthesis, J. Am. Chem. Soc. *86,* 1839 (1964).

1.4   Phenytoin

5,5-Diphenyl-2,4-imidazolidinedione

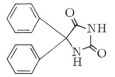

Leonello Leone and Emilio Perucca

General

Phenytoin was first synthesized in 1908 [1], but pharmacological studies with the compound were not reported until 3 decades later, when *Merritt & Putnam* [2] described its effectiveness against electroshock-induced convulsions in cats. In the same year, the drug was tested and found to be effective against generalized tonic-clonic seizures in patients. Today, phenytoin is still extensively used in the treatment of epilepsy, mainly in patients with partial seizures (with or without secondary generalization) and/or generalized convulsive seizures [3]. Phenytoin is ineffective in absence seizures and it is of no value in the prophylaxis of the recurrence of febrile convulsions. Besides epilepsy, phenytoin can be used for the treatment of various cardiac dysrhythmias, including those occurring after myocardial infarction or digitalis intoxication [3]. The therapeutic effects of phenytoin can generally be ascribed to widespread depression of excitable tissues. The mechanism underlying this action is not completely understood, but may be related to stimulation of the sodium pump, inhibition of passive sodium influx, interference with calcium movements or enhancement of GABA function [3].

Phenytoin is generally well absorbed from the gastro-intestinal tract, although with some pharmaceutical formulations the oral availability may be incomplete [4]. In serum, the drug is approximately 90% bound to serum proteins, mainly albumin. An increase in the unbound fraction is observed in hypoalbuminaemic states (e.g. in neonates), in certain diseases (e.g. uraemia) and in the presence of displacing agents [5]. The volume of distribution of the drug is $0.5 - 0.8$ l/kg. Less than 5% of the administered dose of phenytoin is excreted unchanged in urine. The remainder is extensively metabolized, mainly to 5-(4-hydroxyphenyl)-5-phenylhydantoin. The metabolic pathway may become saturated at therapeutic concentrations. This results in a non-linear relationship between drug concentration and dosage; when saturation is approached, small dosage increments can result in a disproportionate rise in serum drug concentrations at steady-state.

The degradation of phenytoin follows *Michaelis-Menten* kinetics. Elimination half-life calculated after administration of very small doses which avoid saturation of the enzyme system ranges from 10 to 20 h. Half-life values determined in this way over-estimate the actual elimination rate in patients stabilized on maintenance therapy with therapeutic doses. After administering tracer doses of radiolabelled phenytoin, half-life values in these patients are in the range of 15 to 70 h, with extremes up to 140 h in intoxicated patients [4].

Patients receiving the same phenytoin dose may show a very wide variability in steady-state serum concentrations, due to inter-patient differences in compliance and in drug absorption, distribution and, especially, metabolic elimination. Factors affecting phenytoin kinetics include genetic background, age (metabolism of the drug may be slow in newborns, while in children it is usually faster than in adults), pregnancy, associated disease(s) and interactions with concurrently administered drugs [4]. Examples of interactions include the displacement of phenytoin from plasma protein binding sites by valproic acid and phenylbutazone, and the inhibition of phenytoin metabolism by isoniazid, chloramphenicol and sulthiame [6].

It has been shown that the pharmacological effects of phenytoin correlate better with the serum concentration of the drug than with the prescribed daily dose [7]. The unusual and variable kinetics of phenytoin, coupled with its narrow therapeutic range and the difficulty of obtaining a direct estimate of drug effect, makes the monitoring of serum concentrations extremely useful in clinical practice. Measurement of serum phenytoin concentrations is especially valuable in the following situations [8]:

- when a patient has been stabilized on an optimal dosage; knowledge of the serum concentration on that dosage may be useful at a later stage in assessing the cause of an unexpected change in response (e.g. recurrence of seizures due to poor compliance or to a drug interaction causing a decrease in serum phenytoin concentration);

− in deciding the magnitude of a dosage adjustment; due to the occurrence of saturation kinetics, the same dosage increment may produce different changes in serum concentration depending on the initial serum concentration;

− when a patient is likely to show a variable kinetic pattern (e.g. neonates, infants, children, pregnant women, elderly patients, patients with unstable disease states, before and after addition of a possibly interacting drug);

− when a patient has a disease potentially affecting phenytoin kinetics;

− when a patient shows a pharmacological response unexpected at the prescribed dosage;

− in overdose cases, or when toxicity is suspected; some of the toxic effects of phenytoin may not be easily identifiable clinically: for example, toxic concentrations may exacerbate seizures;

− in the presence of combined drug therapy, when the response is inadequate; monitoring serum concentrations in this situation will help in identifying the drug whose dosage needs to be modified;

− when poor compliance is suspected.

Although approximately 90% of phenytoin present in serum is bound to proteins, only the free drug is pharmacologically active and correlates with clinical effect [5]. Since in most patients the inter-individual variation in unbound fraction is relatively small, measurement of total concentrations is usually adequate in providing an indirect estimate of effect. In patients with reduced serum protein binding, however, total phenytoin concentrations may provide a misleading estimate of the amount of drug which is present in free, pharmacologically active form [9]. Conditions associated with reduced protein binding of phenytoin include uraemia (even at normal albumin concentrations), marked hyperbilirubinaemia, hypoalbuminaemia (e.g. in neonates, late pregnancy, burns, chronic disease, etc.) and treatment with displacing agents (e.g. valproic acid, phenylbutazone, tolbutamide, salicylic acid, etc.) [5]. The possibility of an alteration in drug binding capacity should be considered when interpreting total serum phenytoin concentrations in these cases. In selected situations, monitoring of free (unbound) serum phenytoin concentrations may avoid the interpretative problems related to variation in serum binding capacity [9].

Application of method: in clinical chemistry, pharmacology and toxicology.

Substance properties relevant in analysis: phenytoin (molecular weight 252.26) is a weak organic acid, with a pK_a of 8.3. Phenytoin as the free acid is very poorly soluble in water but it dissolves in alkali and many organic solvents. The solubility in water at pH 7.4 (about 80% ionized) is 20.5 mg/l at 25°C. Phenytoin sodium (molecular weight 274.25) is not recommended for the preparation of analytical standards because of its variable water content and partial conversion to the free acid on exposure to carbon dioxide.

Methods of determination: currently used methods for measuring total serum phenytoin concentration include chromatographic procedures (gas-liquid chromatography and high-performance liquid chromatography) and homogeneous immunoassays (enzyme-multiplied immunoasssay, substrate-labelled fluorescent immunoassay, fluorescence-polarization immunoassay, nephelometric inhibition immunoassay).

Recently, heterologous solid-phase enzyme-immunoassays for phenytoin have been described, in which β-galactosidase-monosaccharide derivatives are used as conjugates [9a].

Chromatographic procedures [10, 11] are employed as a reference to assess the accuracy of other methods in comparative studies. Their main advantage is in allowing the simultaneous determination of metabolites and/or other concurrently administered drugs. However, chromatographic techniques are time-consuming and homogeneous immunoassays are usually preferred for the routine monitoring of serum phenytoin concentration. These immunoassays have proved to be as precise and usually as accurate as chromatographic procedures. Occasionally, however, some cross-reactivity with structurally related substances may be encountered. If this is suspected, chromatographic procedures should be chosen.

There are two possible approaches to the measurement of free (non-protein-bound) phenytoin. One is to determine the drug in protein-free fluids containing phenytoin at concentrations theoretically identical to the free concentration in plasma. These fluids are represented by cerebrospinal fluid (CSF) and saliva. However, the measurement of drug concentrations in the CSF is restricted by the invasive nature of the method. The measurement of phenytoin in saliva is discussed on p. 40 and is also not entirely satisfactory for a number of reasons. The second approach is to determine the free concentration of the drug in plasma or serum. This approach requires a preliminary step to separate the free from the protein-bound drug: the separation can be accomplished by ultrafiltration, ultracentrifugation or equilibrium dialysis. Each of these techniques has advantages and disadvantages, a discussion of which is beyond the scope of this article (for details cf. refs. [5, 12, 13]). It should be mentioned, however, that different techniques or different procedures within the same technique may not yield identical concentrations due to variation in important factors such as temperature, pH, sample dilution and *Donnan's* effect. These methodological variables may adversely affect the comparability of data (e.g. optimal concentration ranges) from different laboratories. However, comparable results can be obtained when adequate methods are used (cf. [43]).

Once the free drug has been separated, analysis can be performed by using any of the assay methods described for total phenytoin; however, the sensitivity required for the measurement of free phenytoin is about ten-fold greater than required for total phenytoin [5].

International reference method and standards: no standardization at the international level is known. Reference standard material for phenytoin, at three different concentrations, can be obtained from the *National Bureau of Standards*, U.S.A. (Antiepileptic SRM 900 material).

1.4.1   Total Phenytoin

1.4.1.1   Determination with Enzyme-multiplied Immunoassay Technique

Assay

Method Design

The method has been described by *Rubenstein et al.* [14], and was first used for the assay of phenytoin by *Pippenger et al.* [15].

Principle: as described in detail on pp. 7, 8.

A constant amount of an enzyme-phenytoin conjugate (E-L) competes with phenytoin to be assayed (L) for the binding sites of a specific antibody present in limited concentration. When the conjugate E-L is coupled with the antibody Ab the activity of the enzyme is reduced by about 75% [16], probably due to steric or allosteric hindrance [17]. As a consequence, the marker behaves differently in its "bound" and "free" forms. High enzymatic activity is related to a high concentration of the enzyme-phenytoin conjugate not bound to the antibody, and hence to a high concentration of the competing analyte in the sample (cf. Vol. I, chapter 2.7, p. 244). The catalytic activity of G6P-DH is measured as the increase in absorbance per unit time, $\Delta A/\Delta t$.

Selection of assay conditions and adaptation to the individual characteristics of the reagents: the assay is designed for rapid analysis (results of samples from patients available within 15 min) and can be mechanized on various analytical intruments.

Specific anti-phenytoin antibodies have been produced by immunizing sheep with a bovine γ-globulin conjugate of a closely related phenytoin derivative*. Enzyme-drug conjugate is obtained by coupling the phenytoin derivative to G6P-DH from *Leuconostoc mesenteroides*. Endogenous G6P-DH does not interfere because only the bacterial enzyme is able to convert the coenzyme NAD.

---

* The exact nature of the phenytoin derivative used in raising antibody and in coupling to G6P-DH has apparently not been published so far.

The ratio between the concentration of the antibody and that of the conjugate must be optimized in order to provide the maximum discriminatory power between phenytoin calibrators, with the minimum cross-reactivity. The optimization is carried out as described in detail on pp. 8, 9.

Each matched set of reagents has to contain concentrations of enzyme-drug conjugate and antibody adjusted to give an assay range of 2.5 to 30.0 mg/l, with an optimum sensitivity at about 10 mg/l.

The immunological and enzymatic reactions take place simultaneously and the resulting overall reaction approaches zero order. However, a slight decrease in reaction rate has been observed [16, 18, 19], and this should be taken into account when adapting the method to different instruments.

Equipment

A spectrophotometer or spectral-line photometer capable of measuring accurately the absorbance at 339 nm or Hg 334 nm with a thermostatted flow-cell which consistently maintains $30.0 \pm 0.1\,^{\circ}C$ throughout the working day. The use of a time-printer (or equivalent) is recommended in order to measure accurately and print out the absorbance readings at fixed time after the start of the reaction. Work rack and disposable 2 ml beakers with conical bottoms. Pipetter-diluter capable of delivering $50 \pm 1\,\mu l$ (sample and reagent) and $250 \pm 5\,\mu l$ (buffer) with an imprecision less than 0.25%. The delivery force must be sufficient to ensure adequate mixing of the reagents. Laboratory centrifuge.

Reagents and Solutions

Purity of reagents: G6P-DH should be of the best available purity (cf. p. 27). Chemicals are of analytical grade.

Preparation of solutions* (for 100 determinations): all solutions in re-purified water (cf. Vol. II, chapter 2.1.3.2)

1. Tris buffer (Tris, 55 mmol/l; Triton X-100, 0.1 ml/l; pH 8.0):

   dissolve 1.996 g Tris base in 140 to 160 ml water; add 0.03 ml Triton X-100; adjust to pH 8.0 with HCl, 1.0 mol/l; make up with water to 300 ml.

   Alternatively, dilute concentrated buffer provided with each kit to 150 or 200 ml (please check newest instruction!) with water, according to the manufacturer's instructions.

---

* Reagents, calibrators and buffer are commercially available from *Syva Co.,* Palo Alto, U.S.A. The solutions contain sodium azide as a preservative.

2. Antibody/substrate solution (γ-globulin*, ca. 50 mg/l; G-6-P, 66 mmol/l; NAD, 40 mmol/l; Tris, 55 mmol/l; pH 5.0):

   reconstitute available lyophilized preparation with 6.0 ml water.

3. Conjugate solution (G6P-DH coupled to phenytoin, ca. 1 kU/l; Tris, 55 mmol/l; pH 8.0):

   reconstitute available lyophilized preparation with 6.0 ml water. This reagent must be standardized to match the antibody/substrate solution (2) (cf. p. 38).

4. Phenytoin standard solutions (2.5 to 30 mg/l):

   dissolve 60 mg phenytoin in 100 ml Tris buffer (1); dilute 0.5 ml with phenytoin-free human serum containing sodium azide (5 g/l) and Thimerosal (0.5 g/l) to 10 ml. Dilute this solution (30 mg/l) 1 + 11, 1 + 5, 1 + 2, 1 + 0.5 with phenytoin-free human serum, yielding the following concentrations of phenytoin: 2.5, 5, 10, 20 mg/l. Take phenytoin-free human serum as zero calibrator.

   Alternatively, reconstitute each of the commercially available lyophilized standard preparations with 3 ml water.

Stability of solutions: when reconstituted from lyophilized material, the antibody/substrate solution (2) and the conjugate solution (3) must be stored at room temperature (20 °C to 25 °C) for at least 8 h before use. Standards (4) need a minimum reconstitution time of one hour. The reagents are stable for 12 weeks if stored at 2 °C to 8 °C. However, frequent warming to room temperature, necessary to perform the test, may unpredictably shorten the reagent life. The buffer solution (1) can be used for 12 weeks when stored at room temperature. Do not use plastic (e.g. polyethylene) bottles since proteins in the solutions may be adsorbed onto the container walls. Do not freeze the solutions. Do not keep the solutions at temperatures above 32 °C for prolonged periods of time.

Procedure

Collection and treatment of specimen: the specimen should be collected at a standardized time after drug administration, preferably in the morning before the first daily dose. In patients receiving phenytoin in 2 or more divided daily doses, diurnal fluctuations in serum phenytoin concentrations are usually relatively small and therefore the influence of time of sampling on the measured value will be limited. If exceptions are made for special situations (e.g. status epilepticus, suspected toxicity, etc.) samples

---

* Preparation from immunized sheep.

should be drawn at steady state, i.e. preferably at least 2 to 3 weeks after the last dosage adjustment.

Plasma or serum can be used in this assay. Blood cannot be used. The anticoagulants heparin and EDTA do not interfere, but it is advisable to collect the samples in a uniform manner (always serum, or plasma anticoagulated with the same anticoagulant). Sera from haemolyzed (up to 8 g haemoglobin per litre) or icteric (up to 450 μmol bilirubin per litre) blood may be assayed without clinically significant interference [20]. Results on lipaemic sera with triglycerides over 6.8 mmol/l should be re-checked on a fresh clear sample [20].

Saliva can also be used in this assay; however, since the salivary concentration is on average about one-tenth the total plasma concentration, a calibration range, a dilution procedure, and an optimal concentration range identical to those described for the assay of free phenytoin in serum or plasma ultrafiltrates should be used (cf. p. 50). In fact, salivary phenytoin concentrations have been proposed as a measure of free drug concentration. Salivary phenytoin concentrations appear to be independent of salivary flow or salivary pH. If mixed saliva is used, appreciable differences in concentration can occur between the sediment and the supernatant; it is preferable to perform measurements on the supernatant. The presence of sputum, food, food contaminants, drug residues in the mouth and protein-rich exsudates (e.g. in patients with gingivitis!) can result in clinically misleading concentration values: for these and other reasons, monitoring of phenytoin therapy by measuring salivary concentrations has not been recommended [5].

Stability of the substance in the sample: samples should be stored at 2 °C to 8 °C upon collection. Sera can be stored at 4 °C for up to four weeks, or at − 20 °C for five months without appreciable loss.

Assay conditions: perform measurements of samples and zero calibrator in duplicate. Single determinations can be made of the remaining calibrators. In the authors' experience it is critical to keep the time elapsing between the last addition (enzyme-drug conjugate) and the start of the incubation in the spectrophotometer cell absolutely constant.

A pre-dilution of samples and standards is required: mix one part (0.05 ml) sample or standard with five parts (0.25 ml) Tris buffer (1).

Wavelength 339 nm or Hg334 nm; light path 10 mm; temperature 30.0 ± 0.1 °C; total incubation time 45 s; measurement during the last 30 s against water; total volume of the assay mixture 0.90 ml.

Establish a calibration curve under exactly the same conditions (within the series). The zero calibrator serves as a reagent blank: all measurements are corrected for this value.

Measurement

Pipette into a disposable beaker:			concentration in the assay mixture
pre-diluted samples			
or standards (4)		0.05 ml	phenytoin up to 278 µg/l
Tris buffer	(1)	0.25 ml	Tris 52 mmol/l
antibody/substrate			
solution	(2)	0.05 ml	sheep serum γ-globulin*
			ca. 2.8 mg/l
			G-6-P 3.7 mmol/l
			NAD 2.2 mmol/l
Tris buffer	(1)	0.25 ml	
conjugate solution	(3)	0.05 ml	G6P-DH (coupled to
			phenytoin) ca. 56 U/l
Tris buffer	(1)	0.25 ml	
mix well; immediately aspirate into the clean spectrophotometer flow-cell; after a 15 s delay read absorbance over a period of 30 s.			

* There are also new lots containing monoclonal antibodies, amount is titre-dependent.

If $\Delta A/\Delta t$ of a sample is greater than that of the highest standard, dilute the sample further with Tris buffer and re-assay.

Calibration curve: plot the corrected $\Delta A/\Delta t$ values of the standards *versus* the corresponding phenytoin concentrations (mg/l). The calibration curve can be linearized by use of a graph paper matched with the reagents and based on a log-logit model [17, 19] (Fig. 1). To use this graph paper, multiply each $\Delta A/\Delta t$ value by 2.667. If good linearity cannot be obtained, check instrumental conditions and expiry date of reagents.

Calculation: read from the calibration curve the phenytoin mass concentration, ρ (mg/l), corresponding to the mean corrected $\Delta A/\Delta t$ readings of the sample. For conversion from mass concentration to substance concentration, c (µmol/l), multiply by 3.9641.

Validation of Method

Precision, accuracy, detection limit and sensitivity: the precision is highly dependent on the level of mechanization of the analytical system. At a phenytoin concentration

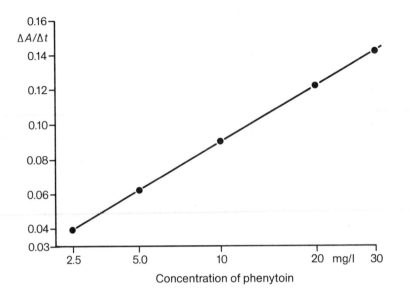

Fig. 1. Typical calibration curve for the assay of phenytoin in serum.

of 15 mg/l the imprecision between days is less than 10% [22] and ranges between 3.8 and 5.9% with mechanized analyzers [18, 20, 23 – 25]. The overall imprecision between laboratories was 10.8% at a mean concentration of 15.1 mg/l, while in the same report GLC methods gave imprecisions of 9.8 – 10% [26].

The recovery of phenytoin added to drug-free pooled human serum has been found to range from 97 to 95% at a concentration of 15 mg/l [23], and from 109.5 to 101.0% (with *National Bureau of Standards* (NBS) Antiepileptic SRM 900 material) at concentrations between 4.2 and 60.7 mg/l [27].

In a large European quality control survey [26] a mean difference between measured and added drug concentration of −9.8% was found at a concentration of 15.1 mg/l. The results obtained by EMIT in random patients' sera agree well with those determined by GLC (EMIT *vs* GLC: y = 1.09x − 0.90 mg/l, r = 0.98, n = 197 [28]; y = 1.10x + 1.12 mg/l, r = 0.98, n = 35 [29]).

The lower limit of the working range is 2.5 mg/l. Sensitivity is higher in the range 10 – 20 mg/l (steepest part of the calibration curve) and decreases consistently with increasing concentrations. Values as low as 0.5 mg/l have been measured with reasonable precision (within-series RSD = 15.9%, n = 10) [20].

Sources of error: sera from highly lipaemic blood (above 8 mmol triglycerides per litre) give falsely decreased phenytoin concentrations [20].

Specificity: the specificity depends on the quality of the antiserum. It has been assessed by adding known concentrations of potentially cross-reacting metabolites or other drugs to the 20 mg/l standard. The degree of cross-reactivity was expressed as the concentration of the added substance that produces a quantitation error of $+30\%$ [22]. With commercially available EMIT® reagents, no significant cross-reactivity was found with other commonly used antiepileptic drugs nor with various benzodiazepines, tricyclic antidepressants, or other frequently administered drugs [18]. The main phenytoin metabolite, 5-(4-hydroxyphenyl)-5-phenylhydantoin, and its glucuronide exhibit a distinct cross-reactivity [20, 30 – 32], variable between antiserum preparations [31]. In patients with normal renal function the interference is negligible because the concentration of metabolites in serum is low. In uraemic patients, however, the glucuronide may accumulate markedly, and the EMIT results in these samples can be falsely elevated [20, 30]. It is unclear whether the observed cross-reactivity is due entirely to 5-(4-hydroxyphenyl)-5-phenylhydantoin or whether other metabolites contribute to the interference [20]. With the use of a recently developed mouse monoclonal antibody [32] these interferences seem to have been overcome.

Therapeutic ranges: in many patients with epilepsy, optimal seizure control is achieved at total serum phenytoin concentrations between 10 and 20 mg/l. However, an appreciable number of patients may be completely controlled at concentrations lower than these [7, 8]. Side-effects (ataxia, nystagmus, tremor, incoordination, dysarthria, sedation, mental and behavioural disturbances) are frequently observed at concentrations above 20 – 25 mg/l. The therapeutic range is usually lower in newborns and in infants up to 2 months of age and in other patients showing an increased free fraction of the drug in plasma. In these patients, the lowering of the therapeutic range is assumed to be roughly proportional to the increase in free fraction.

When phenytoin is used as maintenance therapy for suppression of cardiac ventricular dysrhythmias, optimal effects are reportedly achieved at concentrations between 10 and 18 mg/l.

In all situations, drug concentrations must be interpreted in the context of all other information available to the clinician, including the clinical symptoms and the results of other tests.

1.4.1.2   Determination with
Substrate-labelled Fluorescent Immunoassay

Assay

Method Design

The method was developed by *Wong et al.* [33] and is based on the principle of the substrate-labelled fluorescent immunoassay (SLFIA) proposed by *Burd et al.* [34].

Principle: as described in detail on pp. 15, 16.

Phenytoin to be assayed (L) and a substrate-phenytoin conjugate (S-L) compete for a limited amount of an anti-phenytoin antibody (Ab). Under the assay conditions the substrate-phenytoin conjugate is non-fluorescent; it can be hydrolyzed by β-galactosidase*, to the fluorescent product (P-L), but the enzyme is ineffective when the substrate-phenytoin conjugate is bound to the antibody. Therefore, the increase in fluorescence intensity after a fixed time, $\Delta F / \Delta t$, depends on the concentration of the unbound conjugate (S-L) which, in turn, depends on the concentration of phenytoin (L).

Selection of assay conditions and adaptation to the individual characteristics of the reagents: the assay is designed for rapid analysis and requires no separation step (results of samples from patients are available within about 35 min).

Anti-phenytoin antibodies are produced by immunization of rabbits with 5-[2-(4-carboxybutoxy)-phenyl]-5-phenylhydantoin coupled to bovine serum albumin (8.1 moles per mole albumin) [33].

The substrate-phenytoin conjugate (β-galactosyl-umbelliferyl-phenytoin**) is prepared by coupling 5-(4-aminobutoxyphenyl)-5-phenylhydantoin to 7-β-galactosylcoumarin-3-carboxylic acid [33].

The umbelliferyl moiety in the umbelliferyl-phenytoin conjugate is fluorescent at pH 8.3, with excitation and emission maxima at 400 and 450 nm, respectively. The choice of pH is a compromise designed to obtain an adequate fluorescence intensity of the product of hydrolysis and a reasonable enzymatic activity, which at pH 8.3 is about 0.6 times the maximum activity (pH 7.2 to 7.4).

Shorter or longer incubation times than 20 min can be selected in the range of 5 – 90 min, without significant loss in precision and with only a slight decrease in sensitivity [33]. A sample blank is normally not required (cf. Appendix, p. 49). The immunological and enzymatic reactions take place simultaneously.

The ratio between the concentration of antibody and that of the substrate-phenytoin conjugate has to be adjusted in order to decrease the production of fluorescence in the absence of phenytoin to 10% of that observed without addition of antibody. The assay range is from 5 to 30 mg/l.

Equipment

A fluorimeter capable of measuring fluorescence with excitation and emission wavelengths of 400 nm and 450 nm, respectively. Cuvettes suitable for the fluorimeter used. Disposable polystyrene cuvettes are useful for large batch of samples. Glass test tubes (12 mm × 75 mm) for specimen and calibrators dilution. Timer. Pipetter-diluter capable of delivering 50 µl sample and reagents and 500 µl or 2.5 ml buffer. Laboratory centrifuge.

---

* β-D-Galactoside galactohydrolase, EC 3.2.1.23.
** 5-[4-(7-β-galactosylcoumarin-3-carboxamido) butoxy]-5-phenylhydantoin.

Check periodically the linearity of the fluorimeter response, using a solution of 7-hydroxycoumarin-3-[*N*-(2-hydroxyethyl)]carboxamide* in Bicine buffer (50 mmol/l, pH 8.3) at concentrations of 0, 1, 5, 10 and 15 nmol/l.

Reagents and Solutions

Purity of reagents: β-galactosidase (from *Escherichia coli*) of the best available purity. Chemicals are of analytical grade.

Preparation of solutions** (for 100 determinations): all solutions in re-purified water (cf. Vol. II, chapter 2.1.3.2).

1. *N,N*-bis(2-Hydroxyethyl)glycine (Bicine) buffer (Bicine, 1.0 mol/l; NaN$_3$, 308 mmol/l; pH 8.3):

   dissolve 4.08 g Bicine and 0.5 g sodium azide in 20 ml water; adjust to pH 8.3 (25 °C) with NaOH, 12.5 mol/l; make up with water to 25 ml; filter to remove particles. 25 ml of this buffer is provided with each kit.

2. Diluted Bicine buffer (Bicine, 50 mmol/l; NaN$_3$, 15.4 mmol/l):

   add one part solution (1) to 19 parts water.

3. Antibody/enzyme solution (γ-globulin***; β-galactosidase, 0.56 kU†/l; NaN$_3$, 15.4 mmol/l; Bicine, 50 mmol/l; pH 8.3):

   5 ml of this solution is provided with each kit.

4. Substrate-phenytoin conjugate solution (β-galactosyl-umbelliferyl-phenytoin, 1.08 μmol/l; sodium formate, 5 mmol/l, pH 3.5):

   5 ml of this solution is provided with each kit.

---

  * Provided by *Ames*.
 ** Commercially available reagents for this assay are provided as a matched set (*Ames* TDA® Phenytoin, *Ames Division, Miles Laboratories,* Elkhart, U.S.A.). Reagents, calibrators and buffer contain sodium azide.
*** Standardized preparation, cf. p. 44; amount is titre-dependent.
  † Enzyme activity is defined as in *Wong et al.* [33]: one unit of enzyme activity hydrolyzes 1.0 μmol 2-nitrophenyl-β-D-galactoside per min at 25 °C in Bicine buffer, 50 mmol/l, pH 8.2, containing 2-nitrophenyl-β-D-galactoside, 3 mmol/l, and sodium azide, 15.4 mmol/l. The molar absorption coefficient for the product of the reaction (2-nitrophenol) is 4.27×10^2 l$\times$ mol$^{-1}\times$ mm^{-1} at 415 nm.

5. Phenytoin standard solutions:

dissolve 60 mg phenytoin in 100 ml Bicine buffer (2); dilute 0.5 ml with 9.5 ml phenytoin-free human serum; dilute this solution (30 mg/l) $1+0.5, 1+2, 1+5$ with phenytoin-free human serum, yielding the following concentrations of phenytoin: 20, 10, 5 mg/l. Take phenytoin-free serum as zero calibrator.

Standard solutions containing the indicated phenytoin concentrations are provided with each kit.

Stability of solutions: reagents and calibrators are stable for 12 months at 2°C to 8°C. The stability is 6 months for diluted Bicine buffer (2) at 2°C to 23°C, and 2 weeks for diluted standards (5) at 2°C to 8°C. The substrate-phenytoin conjugate (4) must be protected from light.

Procedure

Collection and treatment of specimens: each assay requires 50 µl serum, or plasma anticoagulated with EDTA or heparin. Sera from slightly haemolyzed, lipaemic or icteric blood can be assayed without interference but results on sera from severely haemolyzed or lipaemic blood should be re-checked on fresh clear samples. If frozen specimens are used, thaw and mix thoroughly before testing and clarify by centrifugation.

For suitable collection times, cf. p. 39.

Stability of the substance in the sample: cf. p. 40.

Assay conditions: standards and samples should be assayed in duplicate under identical conditions. When analyzing a batch of samples, initiate the reaction at 30 s intervals. Since the incubation time is 20 min, do not analyze batches containing more than 35 samples. For larger batches, increase the incubation time accordingly up to 90 min. All solutions must be at room temperature before performing the assay.

A pre-dilution of samples and standard is required: mix one part (0.05 ml) sample or standard solutions (5) with 50 parts (2.5 ml) Bicine buffer (2). Use only glass test tubes for this step.

Room temperature (15°C to 30°C). Excitation wavelength 400 nm; emission wavelength 450 nm; total volume of the assay mixture 1.65 ml.

Establish a calibration curve under exactly the same conditions (within series).

Measurement

Pipette into a reaction cuvette:			concentration in the assay mixture	
antibody/enzyme solution	(3)	0.05 ml	β-galactosidase rabbit sera	17 U/l
			antibody	variable*
			NaN₃	14.9 mmol/l
diluted Bicine buffer	(2)	0.50 ml	Bicine	48.5 mmol/l
pre-diluted sample or standard (5)		0.05 ml	phenytoin	up to 17.8 µg/l
diluted Bicine buffer	(2)	0.50 ml		
substrate-phenytoin conjugate solution	(4)	0.05 ml	β-galactosyl-umbelliferyl- phenytoin	32.8 nmol/l
			formate	152 µmol/l
diluted Bicine buffer	(2)	0.50 ml		
mix well after each addition of Bicine buffer; incubate for exactly 20 min; measure fluorescence**.				

*   The final antibody concentration should be sufficient to decrease the fluorescence to 10% of that observed in the absence of antibody.

** Adjust fluorimeter reading to zero with a cuvette containing diluted Bicine buffer (2). Then adjust to 90% of full scale with the highest calibrator cuvette incubated for about 19 min.

Since setting the full-scale detection of the fluorimeter is the first operation (after zeroing), the highest standard should be the first sample assayed in the run: avoid carry-over contamination when assaying the other calibrators.

If $\Delta F/\Delta t$ of a sample is greater than that of the highest standard, dilute the pre-diluted sample further with Bicine buffer (2) and assay again. The result must then be multiplied by the dilution factor.

Calibration curve: plot mean $\Delta F/\Delta t$ of duplicates *versus* the corresponding phenytoin concentrations (mg/l) on linear graph paper (Fig. 2).

Calculation: read from the calibration curve the phenytoin mass concentration, ρ (mg/l), corresponding to the mean fluorescence value of each sample. For conversion from mass concentration to substance concentration, c (µmol/l), multiply by 3.9641.

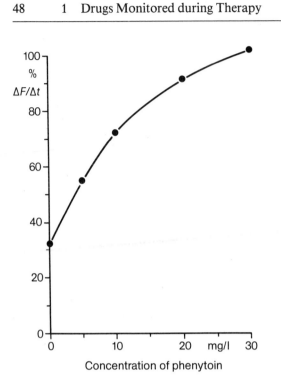

Fig. 2. Typical calibration curve for the assay of phenytoin in serum.

Validation of Method

Precision, accuracy, detection limit and sensitivity: at concentrations in the range of 5 to 25 mg/l the between-day RSD is in the range of 4.4 to 12.7% [33]. Mechanized analytical systems may provide better precision, with overall RSD in the range of 2.4 to 4.6% [20, 35 – 37].

The recovery of phenytoin added to reference sera ranged from 84 to 112% at concentrations between 2.5 and 30.0 mg/l [20, 36, 37].

The accuracy has been assessed by comparison with GLC [37], in clinical samples (SLFIA *vs* GLC: $y = 0.97x - 0.35$ mg/l, $r = 0.99$, $n = 50$).

Data on detection limit according to the definition of *Kaiser* [38] apparently have not been established so far. The concentration of phenytoin that can be distinguished from zero with 95% confidence is 1 mg/l [39], or 0.8 mg/l with mechanized analytical systems [35]. With manual methods, however, low recovery (84%) and relatively high imprecision (RSD = 11.2%) have been reported at a concentration of 2.5 mg/l [20].

Sensitivity depends on the shape of calibration curve, and is highest on the steepest part of the curve, plotted on linear graph paper.

Sources of error: highly fluorescent compounds, such as triamterene, may cause a false elevation of phenytoin concentration (cf. Appendix, below).

Interference by β-galactosidase from human serum has not been observed [34].

Sera from highly haemolyzed, lipaemic or icteric blood may adversely affect the results. However concentrations of haemoglobin up to 4.6 g/l, triglycerides up to 6.6 mmol/l, or bilirubin up to 222 µmol/l did not appreciably affect the assay of phenytoin at concentrations of 16.6 – 21.6 mg/l [35].

Specificity: the specificity depends on the quality of the antibody. No clinically relevant cross-reactions have been observed with other commonly used anticonvulsant drugs. The major metabolite 5-(4-hydroxyphenyl)-5-phenylhydantoin exhibits a distinct cross-reactivity [20, 33]; its glucuronide is also likely to cross-react, and this could explain the elevated results obtained in uraemic samples [20]. This problem has been overcome by the use of monoclonal antibody [40].

Therapeutic range: cf. p. 43.

Appendix

Correction for the presence of fluorescent compounds: when the sample is suspected to contain inherently fluorescent compound (e.g., triamterene or other highly fluorescent drugs) it is advisable to prepare a sample blank (SB) and a zero calibrator blank (ZB) by adding 0.05 ml pre-diluted sample and zero calibrator, respectively, to 1.6 ml Bicine buffer (2). Incubate for 20 min and measure increase in fluorescent intensity per unit time of the sample blank, $(\Delta F/\Delta t)_{SB}$, and of the zero calibrator blank $(\Delta F/\Delta t)_{ZB}$. Assay the suspected sample, and read $(\Delta F/\Delta t)_{SS}$; the corrected value will be

$$\Delta F/\Delta t = (\Delta F/\Delta t)_{SS} - (\Delta F/\Delta t)_{SB} + (\Delta F/\Delta t)_{ZB} \ .$$

With the resulting $\Delta F/\Delta t$ value, determine the phenytoin concentration from the calibration curve.

1.4.2   Free Phenytoin

Determination by Ultrafiltration and
Enzyme-multiplied Immunoassay Technique

Assay

Method Design

The method consists of ultrafiltering the sample to separate the free from protein-bound phenytoin and analyzing the ultrafiltrate by enzyme-multiplied immunoassay (cf. p. 37).

Principle: in the ultrafiltration step, the sample is placed in the reservoir tube of a suitable ultrafiltration device (e. g. *Amicon* Centrifree Micropartition System), which is then capped and centrifuged. During centrifugation, the serum or plasma is forced against the filter membrane. Proteins, protein-bound drug and a constant concentration of free drug remain in the reservoir tube. Protein-free ultrafiltrate containing the same free drug concentration is collected in the filter cup; the drug concentration in the ultrafiltrate can be analyzed by the EMIT assay. In a system in which there were no protein-protein interactions and no specific binding of the drug to the filter membrane, the free drug concentration in the ultrafiltrate would remain identical to the free concentration in the sample retained in the filter reservoir throughout the ultrafiltration process. This equilibrium can be maintained under appropriate experimental conditions even when the volume of ultrafiltrate represents a relatively large proportion of the total original sample. Therefore, analysis of the ultrafiltrate yields a direct measurement of free drug concentration in the ultrafiltered sample.

The EMIT determination of the phenytoin concentration in the ultrafiltrate is performed according to the principle outlined for total phenytoin (p. 37).

Optimized conditions for measurement: the desired assay range in ultrafiltrates is 0.25 to 5 mg/l. Since this range is much lower than the desired range for total phenytoin concentrations, the sensitivity required for the determination of the drug in the ultrafiltrate is greater than that required for the assay of total phenytoin in unfiltered samples. This increase in sensitivity is achieved by omitting the initial dilution step preceeding the assay of unfiltered samples.

Equipment

Suitable ultrafiltration device (e.g. *Amicon* Centrifree Micropartition System), which consists of a reservoir tube, a hydrophilic selectively-permeable membrane with low non-specific binding properties for phenytoin, a removable collection cup and a cap that covers the reservoir tube and prevents a change in pH during centrifugation. Another cap is provided that covers the collection cup.

Centrifuge. The ultrafiltration requires a thermostatted centrifuge equipped with an angle-head rotor capable of operating at 1000 to 2000 g with wells to accomodate the ultrafiltration devices (e.g. wells for 17 mm × 100 mm test tubes). The centrifuge must also be capable of maintaining an interior temperature of 25 °C with less than ± 3 °C fluctuation during ultrafiltration. Temperature control is essential because temperature affects phenytoin binding to proteins. Although an angle-head rotor is preferred, swing-out rotors may be used but entail slow flow-rates [41].

Spectrophotometer, work rack, disposable beakers and pipetter-diluter as described for the EMIT assay of total phenytoin (p. 38).

Reagents and Solutions

Reagents and buffer solutions are identical to those described for the EMIT assay of total phenytoin (pp. 45, 46). EMIT®-Free Level Calibrators differ from those used for the total phenytoin assay and contain phenytoin at a concentration of 0, 0.25, 0.75, 2.0 and 5.0 mg/l in Tris/HCl buffer, 55 mmol/l, pH 7.4, with a preservative (sodium azide).

Stability of materials, reagents and solutions: when reconstituted from lyophilized material, the antibody/substrate solution and the conjugate solution must be stored at room temperature (20 °C to 25 °C) for at least 8 h before use. Calibrators (provided in liquid form) and reagents are stable for 12 weeks at 2 °C to 8 °C. However, frequent warming to room temperature, necessary to perform the test, may unpredictably shorten the reagent life. The buffer solution can be used for 12 weeks when stored at room temperature. Do not use plastic (e.g. polyethylene) bottles since proteins in the solutions may become adsorbed to the wall of the container. Do not freeze the solutions. Do not keep the solutions at temperatures above 32 °C for prolonged periods. The filters should be stored at 15 °C to 30 °C and used before the expiry date.

Procedure

Collection and treatment of specimen: the optimal time for sample collection is identical to that previously described for total phenytoin (p. 39). Serum or plasma can be used in this assay. Blood cannot be used. Heparin is an acceptable anticoagulant, but citrate or oxalate should not be used [42]. A least 0.5 ml (preferably 1 ml) are required for the ultrafiltration and at least 200 µl ultrafiltrate should be collected.

Stability of the substance in the sample: samples should be filtered as soon as possible after collection. However, unfiltered serum or plasma can be refrigerated at 2 °C to 8 °C for up to 24 h. Unfiltered samples may also be stored at − 20 °C for one week. The sample should be brought to 25 °C before filtering.

Filtered samples can be stored capped at 2 °C to 8 °C for up to two weeks. The ultrafiltrate should be brought to room temperature before assaying.

Assay conditions: protein retention by the ultrafiltration device, measured in serum and heparinized plasma, should be greater than 99.9%. The adsorption of phenytoin to the filter should be negligible under normal working conditions [43]. The flow-rate measured for 1 ml samples filtered for 20 min in an angle-head centrifuge at 25 °C and 2000 g should be greater than 10 µl/min.

Ultrafiltration: place the sample (0.5 to 1.2 ml serum or plasma) in the reservoir tube of the ultrafiltration device (avoid formation of bubbles), cap and insert the tube in a centrifuge chamber equilibrated at 25 ± 3 °C. Using a thermocouple or a thermistor, ensure that the sample temperature is 25 ± 3 °C. Cap the filter again and centrifuge at 25 °C and 1000 to 2000 g for 10 to 20 min. At least 200 µl ultrafiltrate must be collected. Ensure that the sample temperature is still 25 ± 3 °C. The cap can be used as a handle to remove the filter from the centrifuge at the end of the centrifugation. When centrifugation is completed, the cap covering the reservoir tube should be left in place to prevent spilling when the device is discarded. The second cap provided can be used to cover the collection cup containing the ultrafiltrate. Single ultrafiltrations should be performed. The calibrators and control should not be ultrafiltered.

Measurement: determine ultrafiltrates and calibrators in duplicate, with the results averaged.

Assay the ultrafiltrates, calibrators and control exactly as described for EMIT of total phenytoin in serum or plasma (cf. p. 41), except that the pre-dilution step is omitted.

If $\Delta A/\Delta t$ of a sample is greater than that of the highest calibrator, dilute the specimen further with Tris buffer and re-assay.

Calibration curve: plot the corrected $\Delta A/\Delta t$ readings of each standard *versus* concentration (mg/l) on two-cycle semi-logarithmic graph paper. The curve should be constructed by fitting a smooth curve through the points.

Calculation: read from the calibration curve the phenytoin mass concentration, ρ (mg/l), corresponding to the mean corrected $\Delta A/\Delta t$ readings of the sample. For conversion from mass concentration to substance concentration, c (µmol/l), multiply by 3.9641.

Validation of Method

Precision, accuracy, detection limit and sensitivity: in order to attain an adequate precision with the assay the use of a partially or fully mechanized analytical system is recommended. The between-days RSD for a control containing 1.5 mg/l was 3.3% ($n = 15$). The accuracy of the assay was proved by comparison with a specific GLC method. Results obtained with the two methods were in good agreement (EMIT *vs* GLC: y = 0.99x + 0.04 mg/l, r = 0.99, n = 20) [41]; y = 1.04x − 0.05 mg/l, r = 0.96, n = 39 [20].

The lower limit of the working range is 0.25 mg/l.

Filter-to-filter precision was determined by filtering and assaying human serum. Within series RSD of 2.5% and 4% at phenytoin concentrations of 0.96 and 1.85 mg/l were observed in two laboratories with an ultrafiltration device which was previously commercially available [41]. Between-days RSD at concentrations of 0.9 and 1.3 mg/l have been reported to be 11.7% [43].

The performance was evaluated by comparing ultrafiltration, equilibrium dialysis and ultracentrifugation separation methods in about 40 samples from non-uraemic patients. Results for the free phenytoin fraction by these methods were in good agreement (ultrafiltration *vs* ultracentrifugation: y = 0.94x + 0.60%, standard error of the residuals 0.72; ultrafiltration *vs* equilibrium dialysis: y = 1.02x − 0.60%, standard error of the residuals 1.25) [43].

Sources of error: appropriate ultrafiltration devices retain haemoglobin, lipoproteins and protein-bound bilirubin. Therefore, no interference is expected in the assay from samples from haemolytic, lipaemic or icteric blood. However, since the free drug concentration in blood differs from that in plasma or serum, samples from haemolyzed blood can result in artefactual alteration of the free phenytoin concentration in serum or plasma. Quantitation of free phenytoin by EMIT in specimens from uraemic patients may yield falsely elevated values due to cross-reactivity with metabolites such as 5-(4-hydroxyphenyl)-5-phenylhydantoin glucuronide, which shows comparatively lower protein binding.

Specificity: cross-reactivity with 5-(4-hydroxyphenyl)-5-phenylhydantoin glucuronide may give falsely high results in uraemic samples (cf. above).

Therapeutic ranges: although a therapeutic range of free phenytoin concentrations has not been clearly established, this is likely to depend on the procedure used to separate the free from protein-bound drug. With the method described above, the free phenytoin therapeutic range is expected to be about one tenth the range reported for concentrations of total phenytoin. Therefore, optimal seizure control should be achieved at concentrations of free phenytoin below 2 mg/l [9]. Since associated disease may alter the individual sensitivity to phenytoin, both total and free drug concentrations must be interpreted flexibly and any decision about a change in dosage must always take primarily into account the patient's clinical response.

References

[1] *H. Biltz,* Über die Konstitution der Einwirkungsprodukte von substituierten Harnstoffen auf Benzil und über einige neue Methoden zur Darstellung der 5,5-Diphenylhydantoine, Ber. Dtsch. Chem. Ges. *41,* 1379 (1908).

[2] *H. H. Merritt, T. J. Putnam,* A New Series of Anticonvulsant Drugs Tested by Experiments on Animals, Arch. Neurol. Psychiatry *39,* 1003 – 1015 (1983).

[3] *G. L. Jones, G. H. Wimbish,* Hydantoins, in: *H. H. Frey, D. Janz* (eds.), Antiepileptic Drugs, Handbook of Experimental Pharmacology, Vol. *74,* Springer-Verlag, Berlin 1985, pp. 351 – 419.

[4] *A. Richens,* Clinical Pharmacokinetics of Phenytoin, Clin. Pharmacokinet. *4,* 153 – 169 (1977).

[5] *E. Perucca,* Plasma Protein Binding of Phenytoin in Health and Disease: Relevance to Therapeutic Drug Monitoring, Ther. Drug Monit. *2,* 331 – 344 (1980).

[6] *E. Perucca,* Pharmacokinetic Interactions with Antiepileptic Drugs, Clinical Pharmacokinetics *7,* 57 – 84 (1982).

[7] *L. Lund,* Anticonvulsant Effect of Diphenylhydantoin Relative to Plasma Levels, Arch. Neurol. *31,* 289 – 294 (1974).

[8] *E. Perucca, A. Richens,* Antiepileptic Drugs: Clinical Aspects, in: *A. Richens, V. Marks* (eds.), Therapeutic Drug Monitoring, Churchill-Livingstone, Edinburgh 1981, pp. 320 – 348.

[9] *E. Perucca,* Free Level Monitoring of Antiepileptic Drugs: Clinical Usefulness and Case Studies, Clinical Pharmacokinetics *9,* (suppl. 1), 71 – 78 (1984).

[9a] *J. A. Hinds, C. F. Pincombe, S. Smith, P. Duffy,* The Use of a Monosaccharide Linkage Group in a Heterologous Solid-phase Enzyme Immunoassay for Phenytoin, J. Immunol. Methods *80,* 239 – 253 (1985).

[10] *B. Rambeck, J. W. A. Meijer,* Gas-Chromatographic Methods for the Determination of Antiepileptic Drugs: a Systematic Rewiev, Ther. Drug Monit. *2,* 385 – 396 (1980).

[11] *M. Riedmann, B. Rambeck, J. W. A. Meijer,* Quantitative Simultaneous Determination of Eight Common Antiepileptic Drugs and Metabolites by Liquid Chromatography, Ther. Drug Monit. *3,* 397 – 413 (1981).

[12] *W. F. Bowers, S. Fulton, J. Thompson,* Ultrafiltration *vs.* Equilibrium Dialysis for Determination of Free Fraction, Clinical Pharmacokinetics *9,* (suppl. 1), 49 – 60 (1984).

[13] *E. Perucca, A. Richens,* Interpretation of Drug Levels: Relevance of Plasma Protein Binding, in: Drug Concentrations in Neuropsychiatry, Excerpta Medica, Amsterdam, pp. 51 – 68 (1980).

[14] *K. E. Rubenstein, R. S. Schneider, E. F. Ullman,* "Homogeneous" Enzyme Immunoassay. A New Immunochemical Technique, Biochem. Biophys. Res. Commun. *47,* 846 – 851 (1972).

[15] *C. E. Pippenger, D. L. Sichler, L. Lichtblau,* Preliminary Clinical Evaluation of an Experimental Enzyme Immunoassay System for Diphenylhydantoin and Phenobarbital in Serum, Clin. Chem. *20,* 869 (1974).

[16] *F. D. Lasky, K. K. Ahuja, A. Karmen,* Enzyme Immunoassays with the Miniature Centrifugal Fast Analyzer, Clin. Chem. *23,* 1444 – 1448 (1977).

[17] *A. H. W. M. Schuurs, B. K. van Weemen,* Enzyme Immunoassay, Clin. Chim. Acta *81,* 1 – 40 (1977).

[18] *D. N. Dietzler, C. R. Hoelting, M. P. Leckie, C. H. Smith, V. L. Tieber,* EMIT Assays for Five Major Anticonvulsant Drugs, Am. J. Clin. Pathol. *74,* 41 – 50 (1980).

[19] *D. N. Dietzler, M. P. Leckie, C. R. Hoelting, S. E. Porter, C. H. Smith, V. L. Tieber,* Logit-Log Calibrations Curves for EMIT Assays, Clin. Chim. Acta *127,* 239 – 250 (1983).

[20] *W. R. Kulpmann, S. Gey, M. Beneking, B. Kohl, M. Oellerich,* Determination of Total and Free Phenytoin in Serum by Non-Isotopic Immunoassays and Gas Chromatography, J. Clin. Chem. Biochem. *22,* 773 – 779 (1984).

[21] *D. Rodbard, S. W. McClean,* Automated Computer Analysis for Enzyme-Multiplied Immunological Techniques, Clin. Chem. *23,* 112 – 115 (1977).

[22] Phenytoin Assay Package Insert, *Syva Co.,* Palo Alto, U.S.A. (1977).

[23] *W. Shaw, J. McHan,* Adaptation of EMIT Procedures for Maximum Cost Effectiveness to Two Different Centrifugal Analyzer System, Ther. Drug Monit. *3,* 185 – 191 (1981).

[24] *N. Urquhart, W. Godolphin, D. J. Campbell,* Evaluation of Automated Enzyme Immunoassays for Five Anticonvulsants and Theophylline Adapted to a Centrifugal Analyzer, Clin. Chem. *25,* 785 – 787 (1979).

[25] *R. Becker, W. Klonizchii, S. A. Leeder, K. Schulkamp, D. M. Ward,* Homogeneous Enzyme Immunoassay (EMIT) Protocol for the Cobas Bio Centrifugal Analyzer, Clin. Chem. *30,* 103 (1984).

[26] *J. F. Wilson, R. W. Marshall, J. Williams, A. Richens,* Comparison of Assay Methods Used to Measure Antiepileptic Drugs in Plasma, Ther. Drug Monit. *5,* 449 – 460 (1983).

[27] *D. Studts, G. T. Haven, E. J. Kiser,* Adaptation of Microvolume EMIT Assays for Theophylline, Phenobarbital, Phenytoin, Carbamazepine, Primidone, Ethosuximide and Gentamicin to a CentrifiChem. Chemistry Analyzer, Ther. Drug Monit. *5,* 335 – 340 (1983).

[28] *H. E. Booker, B. A. Darcey,* Enzymatic Immunoassay *vs* Gas/Liquid Chromatography for Determination of Phenobarbital and Diphenylhydantoin in Serum, Clin. Chem. *21,* 1766 – 1768 (1975).

[29] *J. W. A. Meijer, B. Rambeck, M. Riedmann,* Antiepileptic Drug Monitoring by Chromatographic Methods and Immunotechniques – Comparison of Analytical Performance, Practicability, and Economy, Ther. Drug Monit. *5,* 39 – 53 (1983).

[30] *A. K. N. Nandedkar, R. Williamson, H. Kutt, G. F. Fairclough Jr.,* A Comparison of Plasma Phenytoin Level Determinations by EMIT and Gas-Liquid Chromatography in Patients with Renal Insufficiency, Ther. Drug Monit. *2,* 427 – 430 (1980).

[31] *L. Aldwin, D. S. Kabakoff,* Metabolite Interference in Homogeneous Enzyme Immunoassay of Phenytoin, Clin. Chem. *27,* 770 – 771 (1981).

[32] *D. Ngo, K. K. Mundy, D. E. Petrich, J. M. Centofanti, M. Fong, T. D. Kempe, G. L. Rowley,* Phenytoin Monoclonal Antibody for EMIT Quantitation in Uremic Patients Sera, Clin. Chem. *30,* 1022 (1984) (abstract).

[33] *R. C. Wong, J. F. Burd, R. J. Carrico, R. T. Buckler, J. Thoma, R. C. Boguslaski,* Substrate-labelled Immunoassay for Phenytoin in Human Serum, Clin. Chem. *25,* 686 – 691 (1979).

[34] *J. F. Burd, R. C. Wong, J. E. Feeney, R. J. Carrico, R. C. Boguslaski,* Homogeneous Reactant-labelled Fluorescent Immunoassay for Therapeutic Drugs Exemplified by Gentamycin Determination in Human Serum, Clin. Chem. *23,* 1402 – 1408 (1977).

[35] *T. M. Li, S. P. Robertson, T. H. Crouch, E. E. Pahuski, G. A. Bush, S. J. Hydo,* Automated Fluorometer/Photometer System for Homogeneous Immunoassay, Clin. Chem. *29,* 1628 – 1634 (1983).

[36] *M. Sheehan, G. Caron,* Evaluation of an Automated System (Optimate) for Substrate-labelled Fluorescent Immunoassay, Ther. Drug Monit. *7,* 108 – 114 (1985).

[37] *U. Klotz,* Performance of a New Automated Substrate-labeled Fluorescence Immunoassay System Evaluated by Comparative Therapeutic Monitoring of Five Drugs, Ther. Drug Monit. *6,* 355 – 359 (1984).

[38] *H. Kaiser,* Zum Problem der Nachweisgrenze, Fresenius Z. anal. Chem. *209,* 1 – 18 (1965).

[39] *Ames* TDA Phenytoin, Package Insert, *Ames Division, Miles Laboratories,* Elkhart (1983).

[40] *E. E. Pahuski, B. L. Halmo, L. A. Lewis,* Elimination of Immunoassay High Bias for Phenytoin in Uremic Specimens Using a Substrate-labelled Fluorescent Immunoassay with Monoclonal Antibody, Clin. Chem. *30,* 1021 (1984) (abstract).

[41] Package Insert, EMIT Free Level System I, *Syva Co.,* Palo Alto, California, U.S.A., 1982.

[42] *W. Godolphin, J. Trepanier, K. Farrel,* Serum and Plasma and Free Anticonvulsant Drug Analyses: Effects on EMIT Assays and Ultrafiltration Devices, Ther. Drug Monit. *5,* 319 – 323 (1983).

[43] *M. Oellerich, H. Müller-Vahl,* The EMIT Free Level Ultrafiltration Technique Compared with Equilibrium Dialysis and Ultracentrifugation to Determine Protein Binding of Phenytoin, Clinical Pharmacokinetics *9* (suppl. 1), 61 – 70 (1984).

1.5   Primidone

5-Ethyldihydro-5-phenyl-4,6-(1*H*, 5*H*)-pyrimidinedione

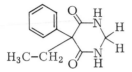

Leonello Leone and Emilio Perucca

General

Primidone, the 2-deoxy analogue of phenobarbital, is effective in the treatment of patients with generalized convulsive seizures and/or partial seizures (with or without secondary generalization) [1]. In animal models, the spectrum of anticonvulsant activity of primidone differs from that of its metabolite phenobarbital. In particular, compared with phenobarbital, primidone shows similar potency against maximal electroshock-induced convulsions but it is much weaker in antagonizing pentylentetrazole-induced seizures. The mode of action of the drug is unknown but may be related to enhancement of GABA function [2]. Much of the pharmacological activity of primidone can in any case be ascribed to metabolically derived phenobarbital. The other metabolite, phenylethylmalonamide (PEMA), also shows anticonvulsant activity in animal models but its potential contribution to clinical efficacy is questionable [2].

Primidone is well absorbed from the gastro-intestinal tract, peak serum concentrations being attained between 0.5 and 7 h after a single oral dose [3, 4]. The binding of the drug to serum proteins has been estimated to be around 35%, but in other studies no significant binding has been detected [5]. The apparent volume of distribution is of the order of 0.6 l/kg. Primidone is partly excreted unchanged in urine and partly metabolized to PEMA and phenobarbital; the latter, in turn, is partly converted to 4-hydroxyphenobarbital. The fraction of the primidone dose that is converted to PEMA varies between 15 and 65%, while the proportion converted to phenobarbital is much lower (less than 8%) [6]. The half-life of primidone is about 7 h (4 to 22 h); half-life values are generally shorter in patients receiving concurrent treatment with phenytoin or carbamazepine than in patients not receiving these drugs [6]. The half-life of phenobarbital is much longer (50 to 170 h in adults) [5].

Although only a minor proportion of primidone is transformed to phenobarbital, the latter is eliminated very slowly and therefore it accumulates in serum during chronic treatment at concentrations higher than those of the parent drug [7]. The concentrations of the other metabolite, PEMA, are comparable to those of primidone [8].

Although both unchanged primidone and PEMA show anticonvulsant activity in animal models, the presence of phenobarbital makes it difficult to establish to what extent (if any) the parent drug or PEMA contribute to the therapeutic activity in epileptic patients. In spite of the fact that some patients appear to do better on primidone than on phenobarbital, there is little doubt that in many clinical situations phenobarbital can account for most (if not all) of the observed therapeutic effect [1, 2, 9]. Unchanged primidone can be responsible for manifestations of toxicity in some cases [1].

In most patients treated with primidone, therapy can be adequately monitored by measuring the serum concentrations of the metabolite phenobarbital [10]. Measuring primidone concentrations may be useful in selected cases; for example, in overdose patients, in patients showing signs of toxicity at low phenobarbital concentrations, and in those with associated renal or hepatic disease.

Application of method: in clinical chemistry, pharmacology and toxicology.

Substance properties relevant in analysis: primidone (molecular weight 218.25) is poorly soluble in water (0.6 g/l at 37 °C) and slightly more soluble in methanol and ethanol (about 6 g/l) and nearly insoluble in most organic solvents.

Methods of determination: methods currently used for measuring primidone concentration include chromatographic procedures (gas-liquid chromatography and high-performance liquid chromatography) and homogeneous immunoassays (enzyme-multiplied immunoassay, substrate-labelled fluorescent immunoassay, fluorescence polarization immunoassay, nephelometric inhibition immunoassay).

Chromatographic procedures [11, 12] are employed as a reference to assess the accuracy of other methods in comparative studies. Their main advantage is in allowing the simultaneous determination of metabolites and/or other concurrently administered drugs. However, chromatographic techniques are time-consuming, and for the routine monitoring of serum primidone concentration homogeneous immunoassays are usually preferred. These immunoassays have proved to be as precise and usually as accurate as chromatographic procedures. Occasionally, however, some cross-reactivity with structurally related substances may be encountered. If this is suspected, chromatographic procedures should be chosen.

International reference method and standards: no standardization at the international level is known. Reference standard material for primidone at three different concentrations, can be obtained from the *National Bureau of Standards, U.S.A.* (Antiepileptic SRM 900 material).

1.5.1 Determination with Enzyme-multiplied Immunoassay Technique

Assay

Method Design

The method was first described by *Rubenstein et al.* [13].

Principle: as described in detail on pp. 7, 8. It is identical to that described for the assay of total phenytoin (cf. p. 37).

Selection of assay conditions and adaptation to the individual characteristics of the reagents: the assay is designed for rapid analysis (results of samples from patients available within 15 min) and can be mechanized on various analytical instruments.

Specific anti-primidone antibodies have been produced by immunizing sheep with a bovine γ-globulin conjugate of a closely related primidone derivative*. Enzyme-drug conjugate is obtained by coupling the primidone derivative to G6P-DH from *Leuconostoc mesenteroides*. Endogenous G6P-DH does not interfere because only the bacterial enzyme is able to convert the coenzyme NAD.

The ratio between the concentration of the antibody and that of the conjugate has to be optimized in order to provide the maximum discriminatory power between primidone standards, with the minimum cross-reactivity. The optimization is carried out as described in detail on pp. 8, 9. The desired assay range is from 2.5 to 20.0 mg/l and an optimum response should be obtained where the assay sensitivity is most desirable (e.g. at 10 mg/l).

The immunological and enzymatic reactions take place simultaneously and the resulting overall reaction approaches zero order. However, a slight decrease in reaction rate has been observed [14], and this should be taken into account when adapting the method to different instruments.

Equipment

The equipment is identical to that required for the EMIT assay of total phenytoin (cf. p. 38).

Reagents and Solutions

Purity of reagents: glucose-6-phosphate dehydrogenase should be of the best available purity (cf. p. 27). Chemicals are of analytical grade.

---

* The exact nature of the primidone derivative used in raising antibody and in coupling to G6P-DH has apparently not been published so far.

Preparation of solutions* (for 50 determinations): all solutions in re-purified water (cf. Vol. II, chapter 2.1.3.2).

1. Tris buffer (Tris, 55 mmol/l; Triton X-100, 0.1 ml/l; pH 8.0):

    dissolve 1.996 g Tris base in 140 to 160 ml water; add 0.03 ml Triton X-100; adjust to pH 8.0 with HCl, 1.0 mol/l; make up with water to 300 ml.

    Alternatively, dilute concentrated buffer provided with each kit to 150 ml with water.

2. Antibody/substrate solution (γ-globulin**, ca. 50 mg/l; G6P, 66 mmol/l; NAD, 40 mmol/l; Tris, 55 mmol/l; pH 5.0):

    reconstitute available lyophilized preparation with 3.0 ml water.

3. Conjugate solution (G6P-DH coupled to primidone, ca. 1 kU/l; Tris, 55 mmol/l; pH 8.0):

    reconstitute available lyophilized preparation with 3.0 ml water. This reagent must be standardized to match the antibody/substrate solution (2) (cf. p. 58).

4. Primidone standard solutions (2.5 to 20 mg/l):

    dissolve 80 mg primidone in 400 ml Tris buffer (1); dilute 1 ml with primidone-free human serum containing sodium azide (5 g/l) and Thimerosal (0.5 g/l) to 10 ml (20 mg/l). Dilute this solution $3+1$, $1+1$, $1+3$ and $1+7$ with primidone-free human serum, yielding the following concentration of primidone: 15.0, 10.0, 5.0, 2.5 mg/l. Take primidone-free human serum as zero calibrator.

    Alternatively, reconstitute each of the available lyophilized standard preparations with 3 ml water.

Stability of solutions: cf. p. 39.

Procedure

Collection and treatment of specimen: the specimen must be collected at a standardized time after drug administration. Because of the short half-life of primidone, serum primidone concentrations may fluctuate considerably during a dosing interval; a trough sample (just before the morning dose) is mainly collected but determination of peak concentration (usually 2 to 4 h after each dose) may also be useful, depending on the clinical situation. Steady-state serum primidone concentrations are attained within 2 to 4 days of starting therapy; however, the concentrations of the active metabolite phenobarbital, require about three weeks to reach a steady-state plateau.

---

 * Commercially available reagents are provided as a matched set (EMIT® Primidone Assay, *Syva Co.,* Palo Alto, CA, U.S.A.). Reagents, calibrators and buffer contain sodium azide.

** Preparation from immunized sheep.

Plasma or serum can be used in this assay. Blood cannot be used. The anticoagulants heparin and EDTA do not interfere, but it is advisable to collect the specimens in an uniform manner (always serum, or plasma anticoagulated with the same anticoagulant). Sera from haemolyzed (up to 8 g hemoglobin per litre), lipaemic (up to 11.3 mmol triglycerides per litre) or icteric (up to 513 μmol bilirubin per litre) blood may be assayed without clinically significant interference [15].

Stability of the substance in the sample: samples should be stored at 2 °C to 8 °C upon collection; sera can be stored at 4 °C for up to four weeks, or at − 20 °C for five months without appreciable loss.

Assay conditions: perform measurements of samples and zero calibrator in duplicate. Single determinations can be made of the remaining calibrators. In the authors' experience it is critical to keep the delay between the last addition (drug-enzyme conjugate) and the start of the incubation in the spectrophotometer cell absolutely constant.

A pre-dilution of samples and standards is required: mix one part (0.05 ml) sample or standard (4) with five parts (0.25 ml) Tris buffer (1).

Wavelength 339 nm or Hg 334 nm; light path 10 mm; temperature 30.0 ± 0.1 °C; total incubation time 45 s; measurement during the last 30 s against water; total volume of the assay mixture 0.90 ml.

Establish a calibration curve under exactly the same conditions (within series). The zero calibrator assay serves as a reagent blank; all measurements are corrected for this value.

Measurement

Pipette into a disposable beaker:			concentration in the assay mixture	
pre-diluted samples				
or calibrators (4)		0.05 ml	primidone	up to 185 μg/l
Tris buffer	(1)	0.25 ml	Tris	52 mmol/l
antibody/substrate solution	(2)	0.05 ml	sheep serum	variable
			γ-globulin	ca. 2.8 mg/l
			G-6-P	3.7 mmol/l
			NAD	2.2 mmol/l
Tris buffer	(1)	0.25 ml		
conjugate solution	(3)	0.05 ml	G6P-DH*	variable
				ca. 56 U/l
Tris buffer	(1)	0.25 ml		
mix well; immediately aspirate into the clean spectrophotometer flow-cell; after a 15 s delay read absorbance over a period of 30 s.				

* Coupled to primidone.

If $\Delta A/\Delta t$ of a sample is greater than that of the highest standard, dilute the sample further with Tris buffer (1) and re-assay.

Calibration curve: plot the corrected $\Delta A/\Delta t$ values of the standards *versus* the corresponding primidone concentrations (mg/l). The calibration curve can be linearized by use of a graph paper matched with the reagents and based on a log-logit model [14, 16] (Fig. 1). To use this graph paper, multiply each $\Delta A/\Delta t$ value by 2.667. If a good linearity cannot be obtained, check instrumental conditions and expiry date of reagents.

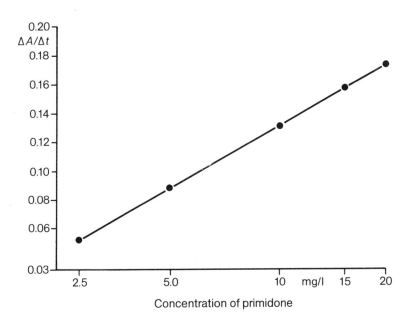

Concentration of primidone

Fig. 1. Typical calibration curve for the assay of primidone in serum.

Calculation: read from the calibration curve the primidone mass concentration, ρ (mg/l), corresponding to the mean corrected $\Delta A/\Delta t$ readings of the sample. For conversion from mass concentration to substance concentration, c (μmol/l), multiply by 4.5819.

Validation of Method

Precision, accuracy, detection limit and sensitivity: the precision is highly dependent on the level of mechanization of the analytical system. At a primidone concentration of 12 mg/l the between-days RSD is less than 10% [15] and ranges from 4.1 to 6% with centrifugal analyzers [17–19]. The RSD between laboratories was 9.4% at a mean concentration of 8.7 mg/l; this is lower than that reported for GLC methods (11.2 to 14.5%) [20].

The recovery of primidone added to drug-free pooled human serum ranges from 98 to 100% at a concentration of 12 mg/l [17] and from 106 to 109% (with Antiepileptic SRM 900 material from the *National Bureau of Standards*, NBS) at concentrations between 3.6 and 18.6 mg/l [21].

The results obtained by EMIT in patients' sera agree well with those determined by GLC method (EMIT *vs* GLC: y = 0.92x + 0.81 mg/l, n = 30 [19]; y = 0.93x + 1.47 mg/l, r = 0.96, n = 19 [22]).

In a large European quality control survey [20] the mean difference between measured and added drug concentrations was − 0.8% at a concentration of 8.7 mg/l.

Data on detection limit according to the definition of *Kaiser* [23] have apparently not been established so far. The lower limit of the working range is 2.5 mg/l. Sensitivity is highest in the range 5 to 10 mg/l (steepest part of the calibration curve plotted on linear graph paper) and decreases with increasing concentrations.

Sources of error: sera from highly haemolyzed (above 8 g haemoglobin per litre), lipaemic (triglycerides above 11.3 mmol/l) or icteric (bilirubin above 513 µmol/l) blood may adversely affect the results [15].

Specificity: the specificity depends on the quality of the antiserum. It has been assessed by adding to the standard with 10 mg/l a known concentration of potentially cross-reacting metabolites or other drugs. The degree of cross-reactivity was expressed as the concentration of the added substance that produces a quantitation error of + 30% [15].

With commercially available EMIT reagents, no significant cross-reactivity was found with the main metabolites, phenobarbital and PEMA, nor with other commonly used antiepileptic drugs, nor with various benzodiazepines, tricyclic antidepressants or other frequently administered drugs [19].

Therapeutic ranges: patients exposed for the first time to primidone may show toxic symptoms (drowsiness, malaise, vertigo, nausea, vomiting) at very low serum primidone concentrations. After a few days, tolerance to the drug improves considerably. In patients receiving maintenance therapy, optimal seizure control is often seen at serum primidone concentrations between 5 and 15 mg/l [1]. At higher concentrations adverse effects are more likely to be seen. In any case, serum primidone concentrations must be interpreted very flexibly, in the context of the clinical findings. The concentrations of the metabolite, phenobarbital, also need to be considered [10].

1.5.2 Determination with Substrate-labelled Fluorescent Immunoassay

Assay

Method Design

The method was developed by *Johnson et al.* [24] and is based on the principle of the substrate-labelled fluorescent immunoassay (SLFIA) proposed by *Burd et al.* [25].

Principle: as described in detail on pp. 15, 16. It is identical to that described for the SLFIA of total phenytoin (cf. p. 43).

Selection of assay conditions and adaptation to the individual characteristics of the reagents: the assay is designed for rapid analysis and requires no separation step (results on samples from patients are available within about 35 min).

Anti-primidone antibodies are obtained in goats or rabbits by immunization with a bovine serum albumin conjugate of primidone [26].

After enzymatic hydrolysis of the substrate-primidone conjugate* the umbelliferyl moiety at pH 8.5 exhibits an excitation maximum at 405 nm and a fluorescence maximum at 450 nm.

Shorter or longer incubation times than 20 min can be selected in the range of 5 to 90 min, without significant loss in precision and with only a slight decrease in sensitivity. A sample blank is usually required only in the presence of interfering fluorescent compounds (cf. p. 49). The immunological and enzymatic reactions take place simultaneously.

The ratio between the concentration of antibody and that of the substrate-primidone conjugate has to be adjusted in order to decrease the production of fluorescence in the absence of primidone to 10% of that observed without addition of antibody. The assay range is from 5 to 20 mg/l.

Equipment

The equipment is identical to that required for the SLFIA of total phenytoin (cf. p. 44).

---

* β-Galactosyl-umbelliferyl-primidone. The method for its synthesis has apparently not been published so far.

Reagents and Solutions

Purity of reagents: β-galactosidase from *Escherichia coli* of the best available purity. Chemicals are of analytical grade.

Preparation of solutions* (for 100 determinations): all solutions in re-purified water (cf. Vol. II, chapter 2.1.3.2).

1. *N,N*-bis(2-Hydroxyethyl)glycine (Bicine) buffer (Bicine, 1.0 mol/l; NaN_3, 308 mmol/l; pH 8.5):

   dissolve 4.08 g Bicine and 0.5 g sodium azide in 20 ml water; adjust to pH 8.5 (25 °C) with NaOH, 12.5 mol/l; make up with water to 25 ml; filter to remove particles. 25 ml of this buffer is provided with each kit.

2. Diluted Bicine buffer (Bicine, 50 mmol/l; NaN_3, 15.4 mmol/l; pH 8.5):

   add one part buffer (1) to 19 parts water.

3. Antibody/enzyme solution (γ-globulin**; β-galactosidase, ca. 1 kU***/l; NaN_3, 15.4 mmol/l; Bicine, 50 mmol/l; pH 8.5):

   5 ml of this solution is provided with each kit.

4. Substrate-primidone conjugate solution (β-galactosyl-umbelliferyl-primidone, ca. 0.7 μmol/l; sodium formate, 5 mmol/l; pH 3.5):

   5 ml of this solution is provided with each kit.

5. Primidone standard solutions (5 to 20 mg/l):

   dissolve 80 mg primidone in 400 ml Bicine buffer (2); dilute 1 ml with 9 ml primidone-free human serum (20 mg/l). Dilute this solution 3+1, 1+1, 1+3, with primidone-free human serum, yielding the following concentrations of primidone: 15, 10, 5 mg/l. Take primidone-free serum as zero calibrator.

   Standard solutions containing the indicated primidone concentrations are provided with each kit.

Stability of solutions: reagents and standards are stable for 12 months at 2 °C to 8 °C. The stability is 6 months for diluted Bicine buffer (2) at 2 °C to 23 °C, and 2 weeks for diluted calibrators at 2 °C to 8 °C. The substrate-primidone conjugate (4) must be protected from light [26].

---

* Commercially available reagents are provided as a matched set (*Ames* TDA Primidone, *Ames Division, Miles Laboratories,* Elkhart, U.S.A.). Reagents, calibrators and buffer contain sodium azide.
** Standardized preparation (cf. p. 63), amount is titre-dependent.
*** For definition of units (U), cf. p. 17, footnote.

Procedure

Collection and treatment of specimens: each assay requires 50 µl serum, or plasma anticoagulated with EDTA or heparin. Sera from slightly haemolyzed, lipaemic or icteric blood can be assayed without interference, but results on sera from severely haemolyzed or lipaemic blood should be re-checked on fresh clear samples. If frozen samples are used, thaw and mix thoroughly before testing and clarify by centrifugation.

For suitable collection times, cf. p. 59.

Stability of the substance in the sample: cf. p. 60.

Assay conditions: standards and samples should be assayed in duplicate under identical conditions. When analyzing a batch of samples, initiate the reaction at 30 s intervals. Since the incubation time is 20 min, do not analyze batches containing more than 40 samples. All solutions must be at room temperature before performing the assay.

A pre-dilution of samples and standard is required: mix one part (0.05 ml) sample or standard solutions (5) with 50 parts (2.5 ml) Bicine buffer (2). Use only glass test tubes for this step.

Room temperature (15 °C to 30 °C); excitation wavelength 405 nm; emission wavelength 450 nm; total volume of the assay mixture 1.65 ml.

Establish a calibration curve under exactly the same conditions (within series).

Since setting the full-scale deflection of the fluorimeter is the first operation (after zeroing), the highest standard should be the first sample assayed within the run: avoid carry-over contamination when assaying the other standards.

If $\Delta F/\Delta t$ of a sample is greater than that of the highest standard, dilute the pre-diluted sample further with Bicine buffer (2) and assay again. The result must then be multiplied by the dilution factor.

Calibration curve: plot mean $\Delta F/\Delta t$ values of duplicates *versus* the corresponding primidone concentrations (mg/l) on linear graph paper (Fig. 2, p. 66).

Calculation: read from the calibration curve the primidone mass concentration, p (mg/l), corresponding to the mean $\Delta F/\Delta t$ value of each sample. For conversion from mass concentration to substance concentration, c (µmol/l), multiply by 4.5819.

Measurement

Pipette into a reaction cuvette:			concentration in the assay mixture
antibody/enzyme solution	(3)	0.05 ml	β-galactosidase ca. 30 U/l rabbit or goat serum variable*
diluted Bicine buffer	(2)	0.50 ml	NaN₃ 14.9 mmol/l Bicine 48.5 mmol/l
pre-diluted sample or standard	(5)	0.05 ml	primidone up to 12 µg/l
diluted Bicine buffer	(2)	0.50 ml	

substrate-primidone
   conjugate
	(4)	0.05 ml	β-galactosyl-
			   umbelliferyl-
			   primidone ca. 21 nmol/l
			   formate 152 µmol/l

diluted Bicine buffer	(2)	0.50 ml	
mix well after each addition of Bicine buffer; incubate for exactly 20 min; measure fluorescence**.			

   * The final antibody concentration should be sufficient to decrease the fluorescence to 10% of that observed in the absence of antibody.
** Adjust fluorimeter reading to zero with a cuvette containing diluted Bicine buffer (2). Then adjust to 90% of full scale with the highest calibrator cuvette incubated for about 19 min.

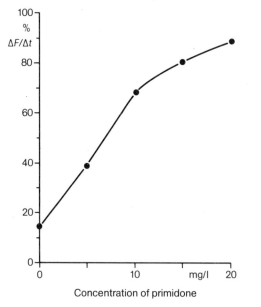

Fig. 2. Typical calibration curve for the assay of primidone in serum.

Validation of Method

Precision, accuracy, detection limit and sensitivity: at concentrations in the range 4 to 16 mg/l the between-days RSD is less than 6% [24]. Mechanized instruments may achieve better precision, with overall RSD in the range of 3.2 to 5.8% at primidone concentrations between 4 to 16 mg/l [28, 29].

The recovery of primidone added to reference sera ranges from 92.5 to 105% at concentrations of 4 to 16 mg/l [29, 30]. The accuracy has been assessed by comparison with HPLC [28], in samples from patients (SLFIA *vs* HPLC: $y = 1.06 x + 0.04$ mg/l, $r = 0.97$, $n = 90$).

Data on detection limit according to *Kaiser* [23] have apparently not been established so far. The concentration of primidone that can be distinguished from zero with 95% confidence is 1 mg/l [26], although with mechanized instruments values as low as 0.5 mg/l have been measured [28].

Sensitivity depends on the shape of the calibration curve, and is highest on the steepest part of the curve, plotted on linear graph paper.

Sources of error: highly fluorescent compounds such as triamterene may cause a false elevation of primidone concentration (correction for the presence of interfering fluorescent compounds, cf. chapter Phenytoin, p. 49). An interference of β-galactosidase from human serum has not been observed [25].

Serum from highly haemolyzed, lipaemic or icteric blood may adversely affect the results. However, concentrations of haemoglobin up to 4.6 g/l or bilirubin up to 222 μmol/l did not appreciably affect the results at primidone concentrations of 11.1 to 12.8 mg/l [28].

Specificity: the specificity depends on the quality of the antibody. No clinically relevant cross-reactions have been observed with other commonly used anticonvulsants, nor with the main metabolites (phenobarbital and PEMA).

Therapeutic ranges: cf. p. 62.

References

[1] *R. W. Fincham, D. D. Schottelius,* Primidone. Relation of Plasma Concentration to Seizure Control, in: *D. M. Woodbury, J. K. Penry, C. E. Pippenger* (eds.), Antiepileptic Drugs, Raven Press, New York 1982, pp. 429 – 440.

[2] *H. H. Frey,* Primidone, in: *H. H. Frey, D. Janz* (eds.), Antiepileptic Drugs, Handbook of Experimental Pharmacology, Vol. *74,* Springer-Verlag, Berlin 1985, pp. 449 – 477.

[3] *J. C. Cloyd, K. W. Miller, I. E. Leppik,* Primidone Kinetics: Effects of Concurrent Drugs and Duration of Therapy, Clin. Pharmacol. Ther. *29,* 402 – 407 (1981).

[4] *B. B. Gallagher, I. P. Baumel, R. H. Mattson,* Metabolic Disposition of Primidone and Its Metabolites in Epileptic Subjects after Single and Repeated Administration, Neurology *22,* 1186 – 1192 (1972).

[5] *E. Perucca, A. Richens,* Clinical Pharmacokinetics of Antiepileptic Drugs, in: *H. H. Frey, D. Janz* (eds.), Handbook of Experimental Pharmacology, Vol. *74,* Springer-Verlag, Berlin 1985, pp. 661 – 723.

[6] *R. E. Kauffman, R. Habersang, L. Lansky,* Kinetics of Primidone Metabolism and Excretion in Children, Clin. Pharmacol. Ther. *22,* 200 – 205 (1977).

[7] *E. H. Reynolds, G. Fenton, P. Fenwick, A. C. Johnson, M. Laundy,* Interaction of Phenytoin and Primidone, Br. Med. J. *2,* 594 – 595 (1975).

[8] *D. Haidukewych, E. A. Rodin,* Monitoring of 2-Ethyl-3-Phenylmalonamide in Serum by Gas-Liquid Chromatography: Application to Retrospective Study in Epileptic Patients Dosed with Primidone, Clin. Chem. *26,* 1537 – 1539 (1980).

[9] *O. V. Olesen, M. Dam,* The Metabolic Conversion of Primidone to Phenobarbitone in Patients under Long-Term Treatment, Acta Neurol. Scand. *43,* 348 – 356 (1967).

[10] *E. Perucca, A. Richens,* Antiepileptic Drugs: Clinical Aspects, in: *A. Richens, V. Marks* (eds.), Therapeutic Drug Monitoring, Churchill-Livingstone, Edinburgh 1981, pp. 320 – 348.

[11] *B. Rambeck, J. W. A. Meijer,* Gas-Chromatographic Methods for the Determination of Antiepileptic Drugs: a Systematic Review, Ther. Drug Monit. *2,* 385 – 396 (1980).

[12] *M. Riedmann, B. Rambeck, J. W. A. Meijer,* Quantitative Simultaneous Determination of Eight Common Antiepileptic Drugs and Metabolites by Liquid Chromatography, Ther. Drug Monit. *3,* 397 – 413 (1981).

[13] *K. E. Rubenstein, R. S. Schneider, E. F. Ullman,* "Homogeneous" Enzyme Immunoassay. A New Immunochemical Technique, Biochem. Biophys. Res. Commun. *47,* 846 – 851 (1972).

[14] *D. N. Dietzler, M. P. Leckie, C. R. Hoelting, S. E. Porter, C. H. Smith, V. L. Tieber,* Logit-Log Calibration Curves for EMIT Assays, Clin. Chim. Acta *127,* 239 – 250 (1983).

[15] *Syva* Primidone Assay Package Insert, *Syva Co.,* Palo Alto, U.S.A. 1977.

[16] *A. H. W. M. Schuurs, B. K. Van Weemen,* Enzyme Immunoassay, Clin. Chim. Acta *81,* 1 – 40 (1977).

[17] *W. Shaw, J. McHan,* Adaptation of EMIT Procedures for Maximum Cost Effectiveness to Two Different Centrifugal Analyzer System, Ther. Drug Monit. *3,* 185 – 191 (1981).

[18] *N. Urquhart, W. Godolphin, D. J. Campbell,* Evaluation of Automated Enzyme Immunoassays for Five Anticonvulsants and Theophylline Adapted to a Centrifugal Analyzer, Clin. Chem. *25,* 785 – 787 (1979).

[19] *D. N. Dietzler, C. R. Hoelting, M. P. Leckie, C. H. Smith, V. L. Tieber,* EMIT Assays for Five Major Anticonvulsant Drugs, Am. J. Clin. Pathol. *74,* 41 – 50 (1980).

[20] *J. F. Wilson, R. W. Marshall, J. Williams, A. Richens,* Comparison of Assay Methods Used to Measure Antiepileptic Drugs in Plasma, Ther. Drug Monit. *5,* 449 – 460 (1983).

[21] *D. Studts, G. T. Haven, E. J. Kiser,* Adaptation of Microvolume EMIT Assays for Theophylline, Phenobarbital, Phenytoin, Carbamazepine, Primidone, Ethosuximide and Gentamicin to a Centrifichem Chemistry Analyzer, Ther. Drug Monit. *5,* 335 – 340 (1983).

[22] *J. W. A. Meijer, B. Rambeck, M. Riedmann,* Antiepileptic Drug Monitoring by Chromatographic Methods and Immunotechniques – Comparison of Analytical Performance, Practicability, and Economy, Ther. Drug Monit. *5,* 39 – 53 (1983).

[23] *H. Kaiser,* Zum Problem der Nachweisgrenze, Fresenius Z. anal. Chem. *209,* 1 – 18 (1965).

[24] *P. K. Johnson, L. J. Messenger, L. M. Krausz, R. T. Buckler, J. F. Burd,* Substrate-Labeled Fluorescent Immunoassay for Primidone, Clin. Chem. *27,* 1093 (1981) (Abstract).

[25] *J. F. Burd, R. C. Wong, J. E. Feeney, R. J. Carrico, R. C. Boguslaski,* Homogeneous Reactant-Labeled Fluorescent Immunoassay for Therapeutic Drugs Exemplified by Gentamicin Determination in Human Serum, Clin. Chem. *23,* 1402 – 1408 (1977).

[26] AMES TDA Primidone Package Insert, *Ames Division, Miles Laboratories,* Elkhart, 1983, revised 3/84.

[27] *L. M. Krausz, J. B. Hitz, R. T. Buckler, J. F. Burd,* Substrate-Labeled Fluorescent Immunoassay for Phenobarbital, Ther. Drug Monit. *2,* 261 – 272 (1980).

[28] *T. M. Li, S. P. Robertson, T. H. Crouch, E. E. Pahuski, G. A. Bush, S. J. Hydo,* Automated Fluorometer/Photometer System for Homogeneous Immunoassay, Clin. Chem. *29,* 1628 – 1634 (1983).

[29] *M. Sheehan, G. Caron,* Evaluation of an Automated System (Optimate) for Substrate-Labeled Fluorescent Immunoassay, Ther. Drug Monit. *7,* 108 – 114 (1985).

[30] *P. A. Toseland, J. F. C. Wicks, R. G. Newall,* Application of Substrate-Labelled Fluorescent Immunoassay to the Measurement of Anticonvulsant and Antiasthmatic Drug Levels in Plasma and Serum, Ther. Drug Monit. *5,* 501 – 505 (1983).

1.6   Phenobarbital

5-Ethyl-5-phenyl-2,4,6-(1*H*,3*H*,5*H*)-pyrimidinetrione

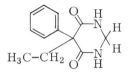

Leonello Leone and Emilio Perucca

General

Phenobarbital, introduced into clinical practice in 1912, is still widely used for the treatment of patients with generalized convulsive seizures (with or without secondary generalization). The drug is also effective in the prophylaxis of the recurrence of febrile convulsions, but it is of no value for the treatment of petit mal seizures [1, 2]. Phenobarbital has been prescribed in the past as a hypnotic/sedative drug, but this use is no longer indicated. The mechanism of action of phenobarbital is only partly known. The drug increases the threshold of neuronal excitability to various electrical and chemical stimuli. Since this effect is seen in both normal and pathological tissues, phenobarbital may suppress both the activity of the epileptogenic focus and the spread of the electrical discharge to adjacent neurones.

Phenobarbital is well absorbed from the gastro-intestinal tract, peak serum concentrations being attained 1 – 3 h after oral intake [3]. The drug is about 50% bound to serum proteins and its volume of distribution is about 0.6 l/kg. Approximately 20 to 40% of the administered dose is excreted unchanged in urine, while the remainder is metabolized in the liver. The major metabolic pathway is oxidation to 4-hydroxy-phenobarbital, which is excreted in urine partly as glucuronide [4]. The elimination half-life of phenobarbital in adults ranges from 50 to 170 h [3, 5, 6]. The half-life is prolonged in patients with severe liver disease [6], in patients receiving concurrent therapy with valproic acid [7] and in newborns, especially those born prematurely [5]. Conversely, a faster rate of phenobarbital elimination is observed in children. Due to the long elimination half-life, steady-state serum concentrations are not achieved until at least 2 – 3 weeks have elapsed after an adjustment in dosage [5].

Because of the wide inter-patient pharmacokinetic variation, patients receiving the same dosage show wide differences in serum drug concentrations at steady-state. Monitoring of phenobarbital concentrations may be valuable in individualizing therapy in epileptic patients and children treated for the prophylaxis of the recurrence of febrile convulsions [8]. The measurement of serum drug concentrations is especially useful in newborns and children, in pregnant women, in the elderly, in cases of overdose, in patients showing an inadequate pharmacological response, in patients with suspected toxicity or poor compliance, and in those with associated renal or hepatic disease. In patients on multiple drug therapy, the determination of serum phenobarbital may be helpful in assessing which drug is more likely to be responsible for an inadequate response (e.g. toxicity or poor seizure-control).

Application of method: in clinical chemistry, pharmacology and toxicology.

Substance properties relevant in analysis: phenobarbital is a white crystalline compound with a molecular weight of 232.23 and pK_a of 7.3. The free acid is only sparingly soluble in water while the sodium salt is freely soluble. Phenobarbital is soluble in many organic solvents, including diethyl ether, chloroform and ethanol. The partition coefficient between chloroform and water is 4.2 at pH 3.4.

Methods of determination: current methods for measuring phenobarbital concentration include chromatographic procedures (gas-liquid chromatography and high-performance liquid chromatography) and homogeneous immunoassays (enzyme-multiplied immunoassay, substrate-labelled fluorescent immunoassay, fluorescence polarization immunoassay, nephelometric inhibition immunoassay).

Chromatographic procedures [9, 10] are employed as a reference to assess the accuracy of other methods in comparative studies. Their main advantage is in allowing the simultaneous determination of metabolites and/or other concurrently administered drugs. However, chromatographic techniques are time-consuming, and homogeneous immunoassays are usually preferred for the routine monitoring of serum phenobarbital concentration. These immunoassays have proved to be as precise and usually as accurate as chromatographic procedures. Occasionally, however, some cross-reactivity with structurally similar substances may be encountered. If this is suspected, chromatographic procedures should be chosen.

International reference method and standards: no standardization at the international level is known. Reference standard material for phenobarbital, at three different concentrations, can be obtained from the *National Bureau of Standards,* U.S.A. (Antiepileptic SRM 900 material).

1.6.1   Determination with Enzyme-multiplied Immunoassay Technique

Assay

Method Design

The method was first described by *Rubenstein et al.* [11] and used to assay phenobarbital by *Pippenger et al.* [12].

Principle: as described in detail on pp. 7, 8. It is identical to that described for the assay of total phenytoin (cf. p. 37).

Selection of assay conditions and adaptation to the individual characteristics of the reagents: the assay is designed for rapid analysis (results of samples from patients available within 15 min) and can be mechanized on various analytical instruments.

Specific anti-phenobarbital antibodies have been produced by immunizing sheep with a bovine γ-globulin conjugate of a closely related phenobarbital derivative*. Enzyme-drug conjugate is obtained by coupling the phenobarbital derivative to G6P-DH from *Leuconostoc mesenteroides*. Endogenous G6P-DH does not interfere because only the bacterial enzyme is able to convert the coenzyme NAD.

The ratio between the concentration of the antibody and that of the conjugate must be optimized in order to provide the maximum discrimination between phenobarbital calibrators, with the minimum cross-reactivity. The optimization procedure is carried out as described in detail on pp. 8, 9. The desired assay range is from 5 to 80 mg/l; an optimum response should be obtained where the assay sensitivity is most desirable (e.g. at 20 mg/l).

The immunological and enzymatic reactions take place simultaneously and the resulting overall reaction approaches zero order. However, a slight decrease in reaction rate has been observed [13 – 15], and this should be taken into account when adapting the method to different instruments.

Equipment

The equipment is identical to that required for the EMIT assay of total phenytoin (cf. p. 38).

Reagents and Solutions

Purity of reagents: glucose-6-phosphate dehydrogenase should be of the best available purity (cf. p. 27). Chemicals are of analytical grade.

Preparation of solutions** (for 100 determinations): all solutions in re-purified water (cf. Vol. II, chapter 2.1.3.2).

---

* The exact nature of the phenobarbital derivative used in raising antibody and in coupling to G6P-DH has not apparently been published so far.

** Commercially available reagents are provided as a matched set (EMIT® Phenobarbital Assay, *Syva Co.,* Palo Alto, CA, U.S.A.). Reagents, calibrators and buffer contain sodium azide.

1. Tris buffer (Tris, 55 mmol/l; Triton X-100, 0.1 ml/l; pH 8.0):

   dissolve 1.996 g Tris base in 140 – 160 ml water; add 0.03 ml Triton X-100; adjust to pH 8.0 with HCl, 1.0 mol/l; make up with water to 300 ml.

   Alternatively, dilute concentrated buffer provided with each kit to 150 or 200 ml with water, according to the instructions of the manufacturer.

2. Antibody/substrate solution (γ-globulin*, ca. 50 mg/l; G-6-P, 66 mmol/l; NAD, 40 mmol/l; Tris, 55 mmol/l; pH 5.0):

   reconstitute available lyophilized preparation with 6.0 ml water.

3. Conjugate solution (G6P-DH coupled to phenobarbital, ca. 1 kU/l; Tris, 55 mmol/l; pH 8.0):

   reconstitute available lyophilized preparation with 6.0 ml water. This reagent must be standardized to match the antibody/substrate solution (2) (cf. p. 71).

4. Phenobarbital standard solutions (5 to 80 mg/l):

   dissolve 160 mg phenobarbital in 100 ml Tris buffer (1); dilute 0.5 ml with pheno-barbital-free human serum containing sodium azide (5 g/l) and Thimerosal (0.5 g/l) to 10 ml (80 mg/l). Dilute this solution 1 + 1, 1 + 3, 1 + 7 and 1 + 15 with phenobarbital-free human serum, yielding the following concentrations of pheno-barbital: 40, 20, 10, 5 mg/l. Take phenobarbital-free human serum as zero calibrator.

   Alternatively, reconstitute each of the available lyophilized standard preparations with 3 ml water.

Stability of solutions: these are identical to those described for total phenytoin (cf. p. 39).

Procedure

Collection and treatment of specimen: the specimen should be collected at a standardized time after drug administration, preferably in the morning before the first daily dose. Due to the long half-life of phenobarbital, however, diurnal fluctuations in serum phenobarbital concentrations are relatively small and therefore the time of sampling is usually unimportant. If exception is made for special situations (e.g. status epilepticus, suspected toxicity, etc.) samples should be drawn at steady-state, i.e. at least 3 weeks after the last dosage adjustment. Plasma or serum can be used in this assay. Blood cannot be used. The anticoagulants heparin and EDTA do not interfere,

---

* Preparation from immunized sheep.

but it is advisable to collect the samples in a uniform manner (always serum, or plasma anticoagulated with the same anticoagulant). Sera from haemolyzed (up to 8 g haemoglobin per litre), lipaemic (triglycerides up to 11.3 mmol/l) or icteric (up to 513 µmol bilirubin per litre) blood may be assayed without clinically significant interference [16].

Stability of the substance in the sample: samples should be stored at 2 °C to 8 °C upon collection. Sera can be stored at 4 °C for up to four weeks, or at − 20 °C for five months without appreciable loss.

Assay conditions: perform measurements of samples and zero calibrator in duplicate. Single determinations can be made of the remaining standards. In the authors' experience it is critical that the time elapsing between the last addition (enzyme-drug conjugate) and the start of the incubation in the spectrophotometer cell is kept absolutely constant.

A pre-dilution of samples and standards is required: mix one part (0.05 ml) sample or standard with five parts (0.25 ml) Tris buffer (1).

Wavelength 339 nm or Hg 334 nm; light path 10 mm; temperature 30.0 ± 0.1 °C; total incubation time 45 s; measurement during the last 30 s against water; total volume of the assay mixture 0.90 ml.

Establish a calibration curve under exactly the same conditions (within the series). The zero calibrator assay serves as a reagent blank: all measurements are corrected for this value.

Measurement

Pipette into a disposable beaker:			concentration in the assay mixture	
pre-diluted samples or calibrators (4)		0.05 ml	phenobarbital	up to 740 µg/l
Tris buffer	(1)	0.25 ml	Tris	52 mmol/l
antibody/substrate solution	(2)	0.05 ml	sheep serum	variable
			γ-globulin	ca. 2.8 mg/l
			G-6-P	3.7 mmol/l
			NAD	2.2 mmol/l
Tris buffer	(1)	0.25 ml		
conjugate solution	(3)	0.05 ml	G6P-DH*	variable
				ca. 56 U/l
Tris buffer	(1)	0.25 ml		
mix well; immediately aspirate into the clean spectrophotometer flow-cell; after a 15 s delay read absorbance over a period of 30 s.				

* Coupled to phenobarbital.

If $\Delta A/\Delta t$ of a sample is greater than that of the highest standard, dilute the sample further with Tris buffer (1) and re-assay.

Calibration curve: plot the corrected $\Delta A/\Delta t$ values of the standards *versus* the corresponding phenobarbital concentrations (mg/l). The calibration curve can be linearized by use of a graph paper matched with the reagents and based on a log-logit model [15, 17] (Fig. 1). To use this graph paper, multiply each $\Delta A/\Delta t$ value by 2.667. If a good linearity cannot be obtained, check instrumental conditions and expiry date of reagents.

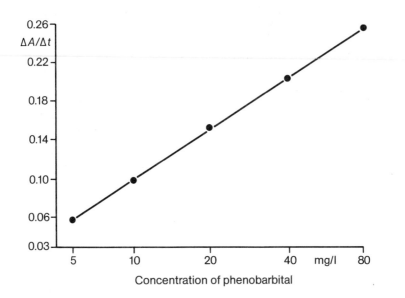

Fig. 1. Typical calibration curve for the assay of phenobarbital in serum.

Calculation: read from the calibration curve the phenobarbital mass concentration, ρ (mg/l), corresponding to the mean corrected $\Delta A/\Delta t$ readings of the sample. To convert from mass concentration to substance concentration, c (μmol/l), multiply by 4.306.

Validation of Method

Precision, accuracy, detection limit and sensitivity: the precision is highly dependent on the degree of mechanization of the analytical system. At a phenobarbital concentration of 30 mg/l the between-days RSD is less than 10% [16] and ranges between 4.1

and 6% with centrifugal analyzers [18 – 20]. The RSD between laboratories was 10% at a mean concentration of 27.8 mg/l; this is lower than that reported for GLC methods (11.2 to 12.9%) [21].

The recovery of phenobarbital added to drug-free pooled human serum has been found to range from 96 to 98% at a concentration of 30 mg/l [18], and from 113.2 to 103.3% (with *National Bureau of Standards* (NBS) Antiepileptic SRM 900 material) at concentrations between 5.3 and 103.6 mg/l [22].

The results obtained by EMIT in patients' sera agree well with those determined by GLC (EMIT *vs* GLC: $y = 0.79x + 2.15$ mg/l, $r = 0.97$, $n = 202$ [23]; $y = 0.91x + 2.31$ mg/l, $r = 0.95$, $n = 50$ [24]).

In a large European quality control survey [21] a mean difference between measured and added drug concentrations of -3.7% was found at a concentration of 27.8 mg/l. Data on detection limit according to the definition of *Kaiser* [25] have apparently not been established so far. The lower limit of the working range is 5 mg/l. Sensitivity is highest in the range 10 to 30 mg/l (steepest part of the calibration curve) and decreases with increasing concentrations.

Sources of error: sera from highly haemolyzed (haemoglobin above 8 g/l), lipaemic (triglycerides above 11.3 mmol/l) or icteric (bilirubin above 513 µmol/l) blood may adversely affect the results [16].

Specificity: the specificity depends on the quality of the antiserum. It has been assessed by adding to the 20 mg/l standard known concentrations of potentially cross-reacting metabolites and other drugs. The dregree of cross-reactivity was expressed as the concentration of the added substance that produces a quantitation error of $+30\%$ [16].

With commercially available EMIT reagents, no significant cross-reactivity was found with other commonly used antiepileptic drugs nor with various benzodiazepines, tricyclic antidepressants or other frequently administered drugs [14]. Other barbiturates such as methylphenobarbital and cyclobarbital show a marked cross-reactivity, thereby preventing the use of this assay for measuring phenobarbital concentrations in the serum of patients receiving these medications also [25a].

Therapeutic ranges: in patients with epilepsy, optimal seizure control is often achieved at serum concentrations between 15 and 40 mg/l, but there is a wide variation in the individual response at a given serum concentration [7]. In children receiving prophylactic treatment against the recurrence of febrile convulsions, serum phenobarbital concentrations between 15 and 20 mg/l are required [2]. Although toxic effects (sedation, drowsiness, ataxia, mental and behavioural disturbances) are more common above 40 mg/l, some patients receiving chronic therapy may tolerate well serum concentrations higher than these. In part, this is due to the development of tolerance within the CNS, as demonstrated by the fact that chronically treated patients may not show any toxic symptoms at serum phenobarbital concentrations that cause coma in cases of acute poisoning [8].

1.6.2   Determination with Substrate-labelled Fluorescent Immunoassay

Assay

Method Design

The method was developed by *Krausz et al.* [26] and is based on the principle of the substrate-labelled fluorescent immunoassay (SLFIA) proposed by *Burd et al.* [27].

Principle: as described in detail on pp. 15, 16.

5-(4-Aminobutyl)-5-phenyl barbituric acid is labelled with a fluorigenic enzyme substrate, 7-β-galactosylcoumarin-3-carboxylic acid. This conjugate (S-L) is non-fluorescent and competes with phenobarbital (L) in the sample for a limited amount of antibody (Ab). When the substrate-phenobarbital conjugate (S-L) is bound by the antibody, it cannot be hydrolyzed by β-galactosidase to a fluorescent product (P-L). The amount of conjugate available for reaction with the enzyme and the resulting increase in fluorescence intensity per unit time, $\Delta F/\Delta t$, are related to the concentration of phenobarbital in the sample (cf. Vol. I, chapter 2.7, p. 244).

Selection of assay conditions and adaptation to the individual characteristics of the reagents: the assay is designed for rapid analysis and requires no separation step (results of samples from patients are available within about 35 min).

Anti-phenobarbital antibodies are obtained in both New Zealand white and Dutch belted rabbits by immunization with a bovine serum albumin conjugate of phenobarbital [28]. Recently a specific monoclonal antibody has been made commercially available [29].

The substrate-phenobarbital conjugate (5-[4-(7-β-galactosylcoumarin-3-carbox-amido)-butyl-5-phenyl barbituric acid) is prepared by coupling 5-(4-aminobutyl)-5-phenyl barbituric acid to 7-β-galactosylcoumarin-3-carboxylic acid. After enzymatic hydrolysis, the umbelliferyl moiety at pH 8.5 exhibits an excitation maximum at 405 nm and a fluorescence maximum at 450 nm.

The ratio between the concentration of the antibody and that of the substrate-phenobarbital conjugate must be adjusted in order to decrease the production of fluorescence in the absence of phenobarbital to 5% of that observed without addition of antibody [26]. The assay range of concentrations is from 10 to 60 mg/l.

Shorter or longer incubation times than 20 min, in the range of 5 to 90 min provide acceptable calibration curves [26]. The immunological and enzymatic reactions take place simultaneously.

Equipment

The equipment is identical to that required for the SLFIA of total phenytoin (cf. p. 44).

Reagents and Solutions

Purity of reagents: β-galactosidase from *Escherichia coli* of the best available purity. Chemicals are of analytical grade.

Preparation of solutions* (for 100 determinations): all solutions in re-purified water (cf. Vol. II, chapter 2.1.3.2).

1. *N,N*-bis(2-Hydroxyethyl)glycine (Bicine) buffer (Bicine, 1.0 mol/l; NaN_3, 308 mmol/l; pH 8.5):

   dissolve 4.08 g Bicine and 0.5 g sodium azide in 20 ml water; adjust to pH 8.5 (25 °C) with NaOH, 12.5 mol/l; make up with water to 25 ml; filter to remove particles. 25 ml of this buffer is provided with each kit.

2. Diluted Bicine buffer (Bicine, 50 mmol/l; NaN_3, 15.4 mmol/l; pH 8.5):

   add one part buffer (1) to 19 parts water.

3. Antibody/enzyme solution (γ-globulin**; β-galactosidase, 1.5 kU***/l; NaN_3, 15.4 mmol/l; Bicine, 50 mmol/l; pH 8.5):

   5 ml of this solution is provided with each kit.

4. Substrate-phenobarbital conjugate solution (β-galactosyl-umbelliferyl-phenobarbital, 0.7 µmol/l; sodium formate, 5 mmol/l; pH 3.5):

   5 ml of this solution is provided with each kit.

5. Phenobarbital standard solutions (10 to 60 mg/l):

   dissolve 120 mg phenobarbital in 100 ml Bicine buffer (2); dilute 0.5 ml with 9.5 ml phenobarbital-free human serum (60 mg/l). Dilute this solution 1 + 0.5, 1 + 2, 1 + 5 with phenobarbital-free human serum, yielding the following concentrations of phenobarbital: 40, 20, 10 mg/l. Take phenobarbital-free serum as zero calibrator.

   Standard solutions containing the indicated phenobarbital concentrations are provided with each kit.

---

   * Commercially available reagents are provided as a matched set (*Ames* TDA Phenobarbital, *Ames Division, Miles Laboratories,* Elkhart, U.S.A.). Reagents, calibrators and buffer contain sodium azide.
  ** Standardized preparation (cf. p. 76), amount is titre-dependent.
 *** For definition of units (U), cf. p. 17.

Stability of solutions: reagents and standards are stable for 12 months at 2 °C to 8 °C. The stability is 6 months for diluted Bicine buffer (2) at 2 °C to 23 °C, and 2 weeks for diluted calibrators at 2 °C to 8 °C. The substrate-phenobarbital conjugate (4) must be protected from light.

Procedure

Collection and treatment of specimens: each assay requires 50 μl serum, or plasma anticoagulated with EDTA or heparin. Sera from slightly haemolyzed, lipaemic or icteric blood can be assayed without interference, but results on sera from severely haemolyzed or lipaemic blood should be re-checked on new clear samples. If frozen samples are used, thaw and mix thoroughly before testing and clarify by centrifugation. For the appropriate collection time, cf. p. 72.

Stability of the substance in the sample: cf. p. 73.

Assay conditions: standards and samples should be assayed in duplicate under identical conditions. When analyzing a batch of samples, initiate the reaction at 30 s intervals. Since the incubation time is 20 min, do not analyze batches containing more than 40 samples. For larger batches, increase the incubation time accordingly up to 90 min. All solutions must be at room temperature before performing the assay.

A pre-dilution of samples and standard is required: mix one part (0.05 ml) sample or standard solutions (5) with 50 parts (2.5 ml) Bicine buffer (2). Use only glass test tubes for this step.

Room temperature (15 °C to 30 °C); excitation wavelength 400 nm; emission wavelength 450 nm; total volume of the assay mixture 1.65 ml.

Establish a calibration curve under exactly the same conditions (within series).

Since setting the full-scale deflection of the fluorimeter is the first operation (after zeroing), the highest standard should be the first sample assayed within the run: avoid carry-over contamination when assaying the other standards.

If $\Delta F/\Delta t$ of a sample is greater than that of the highest standard, dilute the pre-diluted sample further with Bicine buffer (2) and assay again. The result must then be multiplied by the dilution factor.

Calibration curve: plot mean $\Delta F/\Delta t$ values of duplicates *versus* the corresponding phenobarbital concentrations (mg/l) on linear graph paper (Fig. 2).

Calculation: read from the calibration curve the phenobarbital mass concentration, ρ (mg/l), corresponding to the mean $\Delta F/\Delta t$ value of each sample. To convert from mass concentration to substance concentration, c (μmol/l), multiply by 4.306.

Measurement

Pipette into a reaction cuvette:			concentration in the assay mixture
antibody/enzyme solution (3)		0.05 ml	β-galactosidase 45 U/l rabbit or goat serum variable*
diluted Bicine buffer (2)		0.50 ml	NaN₃ 14.9 mmol/l Bicine 48.5 mmol/l
pre-diluted sample or standard (5)		0.05 ml	drug up to 36 µg/l
diluted Bicine buffer (2)		0.50 ml	
substrate-phenobarbital conjugate (4)		0.05 ml	β-galactosyl-umbelliferyl- phenobarbital 21 nmol/l formate 152 µmol/l
diluted Bicine buffer (2)		0.50 ml	
mix well after each addition of Bicine buffer; incubate for exactly 20 min; measure fluorescence**.			

* The final antibody concentration should be sufficient to decrease the fluorescence to 10% of that observed in the absence of antibody.

** Adjust fluorimeter reading to zero with a cuvette containing diluted Bicine buffer (2). Then adjust to 90% of full scale with the highest calibrator cuvette incubated for about 19 min.

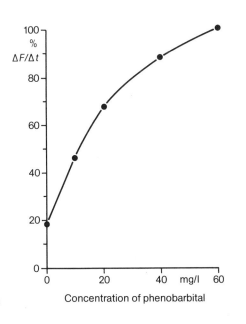

Concentration of phenobarbital

Fig. 2. Typical calibration curve for the assay of phenobarbital in serum.

Validation of Method

Precision, accuracy, detection limit and sensitivity: at concentrations in the range of 15 to 50 mg/l the between-days RSD is in the range of 6.0 to 7.3% [26]. A better precision may be achieved with mechanized instruments (RSD in the range of 2.4 to 4.6%) [30 – 32].

The recovery of phenobarbital added to sera ranges from 92.5 to 108% at concentrations between 8.1 and 32.2 mg/l [30 – 32].

The accuracy has been assessed by comparison with GLC [32] in samples from patients (SLFIA *vs* GLC: y = 0.99x + 0.75 mg/l, r = 0.98, *n* = 45).

Data on the detection limit according to the definition of *Kaiser* [25] have apparently not been established so far. The concentration of phenobarbital that can be distinguished from zero with 95% confidence is 3 mg/l with rabbit or goat antiscra [33], 2 mg/l with monoclonal antibody [29], and 1.1 mg/l with mechanized instruments [30].

Sensitivity depends on the shape of calibration curve, and is highest on the steepest part of the curve, plotted on linear graph paper.

Sources of error: highly fluorescent compounds such as triamterene may cause a false elevation of phenobarbital concentration (correction for the presence of fluorescent compounds, cf. Phenytoin, chapter 1.4.1.2, p. 49). Interference by β-galactosidase in human serum has not been observed [27].

Sera from highly haemolyzed, lipaemic or icteric blood may adversely affect the results. However, concentration of haemoglobin up to 4.6 g/l, triglycerides up to 6.6 mmol/l, or bilirubin up to 222 μmol/l did not appreciably affect the assay of phenobarbital at concentrations of 21.3 to 29.5 mg/l [30].

Specificity: the specificity depends on the quality of the antibody. With the monoclonal antibody currently used in the *Ames* TDA Phenobarbital kit, no clinically relevant cross-reactions have been observed with the other commonly used anticonvulsants, nor with 4-hydroxy-phenobarbital (a major metabolite), hexabarbital, amobarbital, butabarbital and methylphenobarbital. However, the last three barbiturates exhibited a marked cross-reactivity with the antisera previously employed.

Therapeutic ranges: cf. p. 75.

References

[1] *M. J. Eadie,* Clinical Use of Antiepileptic Drugs, in: *H. H. Frey, D. Janz* (eds.), Antiepileptic Drugs, Handbook of Experimental Pharmacology, Vol. *74,* Springer-Verlag, Berlin 1985, pp. 765 – 790.

[2] *M. A. Fishman,* Febrile seizures: The Treatment Controversy, J. Pediatr. *94,* 177 – 184 (1979).

[3] *A. J. Wilensky, P. N. Friel, R. H. Levy, C. P. Comfort, S. P. Kaluzny,* Kinetics of Phenobarbital in Normal Subjects and Epileptic Patients, Eur. J. Clin. Pharmacol. *23,* 87 – 92 (1982).

[4] *N. Kallberg, S. Agurell, O. Ericsson, E. Bucht, B. Jalling, L. O. Boreus,* Quantitation of Phenobarbital and Its Main Metabolites in Human Urine, Eur. J. Clin. Pharmacol. *9*, 161 – 168 (1975).

[5] *E. Perucca, A. Richens,* Clinical Pharmacokinetics of Antiepileptic Drugs, in: *H. H. Frey, D. Janz* (eds.), Antiepileptic Drugs, Handbook of Experimental Pharmacology, Vol. *74*, Springer-Verlag, Berlin 1985, pp. 661 – 723.

[6] *J. Alvin, T. McHorse, A. Hoyumpa, M. T. Bush, S. Schencker,* The Effect of Liver Disease in Man on the Disposition of Phenobarbital, J. Pharmacol. Exp. Ther. *192*, 224 – 235 (1975).

[7] *E. Perucca,* Pharmacokinetic Interactions with Antiepileptic Drugs, Clin. Pharmacokinet. *7*, 57 – 84 (1982).

[8] *E. Perucca, A. Richens,* Antiepileptic Drugs: Clinical Aspects, in: *A. Richens, V. Marks* (eds.), Therapeutic Drug Monitoring, Churchill-Livingstone, Edinburgh 1981, pp. 320 – 348.

[9] *B. Rambeck, J. A. A. Meijer,* Gas-Chromatographic Methods for the Determination of Antiepileptic Drugs: a Systematic Review, Ther. Drug Monit. *2*, 385 – 396 (1980).

[10] *M. Riedmann, B. Rambeck, J. W. A. Meijer,* Quantitative Simultaneous Determination of Eight Common Antiepileptic Drugs and Metabolites by Liquid Chromatography, Ther. Drug Monit. *3*, 397 – 413 (1981).

[11] *K. E. Rubenstein, R. S. Schneider, E. F. Ullman,* "Homogeneous" Enzyme Immunoassay. A New Immunochemical Technique, Biochem. Biophys. Res. Commun. *47*, 846 – 851 (1972).

[12] *C. E. Pippenger, D. L. Sichler, L. Lichtblau,* Preliminary Clinical Evaluation of an Experimental Enzyme Immunoassay System for Diphenylhydantoin and Phenobarbital in Serum, Clin. Chem. *20*, 869 (1974).

[13] *F. D. Lasky, K. K. Ahuja, A. Karmen,* Enzyme Immunoassays with the Miniature Centrifugal Fast Analyzer, Clin. Chem. *23*, 1444 – 1448 (1977).

[14] *D. N. Dietzler, C. R. Hoelting, M. P. Leckie, C. H. Smith, V. L. Tieber,* EMIT Assays for Five Major Anticonvulsant Drugs, Am. J. Clin. Pathol. *74*, 41 – 50 (1980).

[15] *D. N. Dietzler, M. P. Leckie, C. R. Hoelting, S. E. Porter, C. H. Smith, V. L. Tieber,* Logit-log Calibration Curves for EMIT Assays, Clin. Chim. Acta *127*, 239 – 250 (1983).

[16] *Syva* Phenobarbital Assay Package Insert, *Syva Co.,* Palo Alto, CA, U.S.A. 1977.

[17] *A. H. W. M. Schuurs, B. K. Van Weemen,* Enzyme Immunoassay, Clin. Chim. Acta *81*, 1 – 40 (1977).

[18] *W. Shaw, J. McHan,* Adaptation of EMIT Procedures for Maximum Cost Effectiveness to Two Different Centrifugal Analyzer System, Ther. Drug Monit. *3*, 185 – 191 (1981).

[19] *N. Urquhart, W. Godolphin, D. J. Campbell,* Evaluation of Automated Enzyme Immunoassays for Five Anticonvulsants and Theophylline Adapted to a Centrifugal Analyzer, Clin. Chem. *25*, 785 – 787 (1979).

[20] *R. Becker, W. Klonizchii, S. A. Leeder, K. Schulkamp, D. M. Ward,* Homogeneous Enzyme Immunoassay (EMIT) Protocol for the Cobas Bio Centrifugal Analyzer, Clin. Chem, *30*, 1033 (1984).

[21] *J. F. Wilson, R. W. Marshall, J. Williams, A. Richens,* Comparison of Assay Methods Used to Measure Antiepileptic Drugs in Plasma, Ther. Drug Monit. *5*, 449 – 460 (1983).

[22] *D. Studts, G. T. Haven, E. J. Kiser,* Adaptation of Microvolume EMIT Assays for Theophylline, Phenobarbital, Phenytoin, Carbamazepin, Primidone, Ethosuximide and Gentamicin to a Centrifi-Chem Chemistry Analyzer, Ther. Drug Monit. *5*, 335 – 340 (1983).

[23] *H. E. Booker, B. A. Darcey,* Enzymatic Immunoassay *vs* Gas/Liquid Chromatography for Determination of Phenobarbital and Diphenylhydantoin in Serum, Clin. Chem. *21*, 1766 – 1768 (1975).

[24] *J. W. A. Meijer, B. Rambeck, M. Riedmann,* Antiepileptic Drug Monitoring by Chromatographic Methods and Immunotechniques – Comparison of Analytical Performance, Practicability, and Economy, Ther. Drug Monit. *5*, 39 – 53 (1983).

[25] *H. Kaiser,* Zum Problem der Nachweisgrenze, Fresenius Z. anal. Chem. *209*, 1 – 18 (1965).

[25a] *M. Oellerich, W. R. Külpmann, R. Haeckel, R. Heyer,* Determination of Phenobarbital and Phenytoin in Serum by a Mechanized Enzyme-immunoassay (EMIT) in Comparison with a Gas-Liquid Chromatographic Method, J. Clin. Chem. Clin. Biochem. *15*, 353 – 358 (1977).

[26] *L. M. Krausz, J. B. Hitz, R. T. Buckler, J. F. Burd,* Substrate-Labeled Fluorescent Immunoassay for Phenobarbital, Ther. Drug. Monit. *2*, 261 – 272 (1980).

[27] *J. F. Burd, R. C. Wong, J. E. Feeney, R. J. Carrico, R. C. Boguslaski,* Homogeneous Reactant-Labeled Fluorescent Immunoassay for Therapeutic Drugs Exemplified by Gentamicin Determination in Human Serum, Clin. Chem. *23*, 1402 – 1408 (1977).

[28] *C. E. Cook, H. D. Christensen, E. W. Amerson, J. A. Kepler, C. R. Tallent, G. F. Taylor,* Radioimmunoassay of Anticonvulsant Drugs: Phenytoin, Phenobarbital, and Primidone, in: *P. Keller, I. Petersen* (eds.), "Quantitative" Analytical Studies in Epilepsy, Raven Press, New York 1976, pp. 39 – 58.

[29] *Ames* TDA Phenobarbital Package Insert, *Ames, Division of Miles Laboratories,* Elkhart 1984, revised 12/84.

[30] *T. M. Li, S. P. Robertson, T. H. Crouch, E. E. Pahuski, G. A. Bush, S. J. Hydo,* Automated Fluorometer/Photometer System for Homogeneous Immunoassay, Clin. Chem. *29*, 1628 – 1634 (1983).

[31] *M. Sheehan, G. Caron,* Evaluation of an Automated System (Optimate) for Substrate-Labeled Fluorescent Immunoassay, Ther. Drug Monit. *7*, 108 – 114 (1985).

[32] *U. Klotz,* Performance of a New Automated Substrate-Labeled Fluorescence Immunoassay System Evaluated by Comparative Therapeutic Monitoring of Five Drugs, Ther. Drug Monit. *6*, 355 – 359 (1984).

[33] *Ames* TDA Phenobarbital Package Insert, *Ames Division, Miles Laboratories,* Elkhart 1983.

1.7   Ethosuximide

3-Ethyl-3-methyl-2,5-pyrrolidinedione

Leonello Leone and Emilio Perucca

General

The antiepileptic properties of ethosuximide in animal models have been described by *Chen et al.* [1], who remarked on the much greater effectiveness of this drug against pentylentetrazole-induced seizures as compared to electrically-induced convulsions. In the clinical setting, ethosuximide is primarily used for the treatment of absence seizures and myoclonic seizures.

Ethosuximide is well absorbed from the gastro-intestinal tract, peak serum level being achieved within 2 to 7 hours after oral intake. The drug is not bound to serum proteins and has an apparent volume of distribution of approximately 0.7 l/kg. About 20% of the administered dose of ethosuximide is excreted unchanged in urine while another 60% is excreted as the isomers of 2-(1-hydroxyethyl)-2-methylsuccinimide, partly in glucuronide form [2]. Several other metabolites have been identified [3]. The half-life of ethosuximide in adults ranges from 25 to 75 h (average 50 h); shorter values (average 30 h) have been reported in children [4].

Due to a wide inter-individual variability in elimination kinetics, patients receiving the same ethosuximide dosage show wide differences in serum drug concentrations at steady state. In many patients with absence seizures (usually children), individualization of dosage can be achieved on purely clinical grounds by direct measurement of clinical response. It may be helpful to monitor serum ethosuximide concentrations in selected cases, especially in overdose patients, in patients with associated disease(s), and in patients showing an inadequate therapeutic response despite an apparently adequate prescribed dosage [5, 6]. An unsatisfactory response to ethosuximide is often caused by poor compliance.

Application of the method: in clinical chemistry, pharmacology and toxicology.

Substance properties relevant in analysis: ethosuximide (molecular weight 141.17) is highly soluble in water (190 mg/ml at 25°C). Its low melting point (64°C to 65°C) and its high volatility may result in losses during the concentration of organic solvent extracts by evaporation.

Methods of determination: methods currently used for measuring ethosuximide concentration include chromatographic procedures (gas-liquid chromatography and high-performance liquid chromatography) and homogeneous immunoassays (enzyme-multiplied immunoassay, substrate-labelled fluorescent immunoassay, fluorescence polarization immunoassay, nephelometric inhibition immunoassay).

Chromatographic procedures [7, 8] are employed as a reference to assess the accuracy of other methods in comparative studies. Their main advantage is in allowing the simultaneous determination of metabolites and/or other concurrently administered drugs. However, chromatographic techniques are time-consuming, and homogeneous immunoassays are usually preferred for the routine monitoring of serum ethosuximide concentration. These immunoassays have proved to be as precise and usually as accurate as chromatographic procedures. Occasionally, however, some cross-reactivity with structurally similar substances may be encountered. If this is suspected, chromatographic procedures should be chosen.

International reference method and standards: no standardization at the international level is known. Reference standard material for ethosuximide at three different concentrations, can be obtained from the *National Bureau of Standards,* U.S.A. (Anti-epileptic SRM 900 material).

1.7.1   Determination with Enzyme-multiplied Immunoassay Technique

Assay

Method Design

The method was described by *Rubenstein et al.* [9] and first used to assay ethosuximide by *Pippenger et al.* [10].

Principle: as described in detail on pp. 7, 8. It is identical to that described for the assay of total phenytoin (cf. p. 37).

Selection of assay conditions and adaptation to the individual characteristics of the reagents: the assay is designed for rapid analysis (results of samples from patients available within 15 min) and can be mechanized on various analytical instruments.

Specific anti-ethosuximide antibodies have been produced by immunizing sheep with a bovine γ-globulin conjugate of a closely related ethosuximide derivative*. Enzyme-drug conjugate is obtained by coupling the ethosuximide derivative to G6P-DH from *Leuconostoc mesenteroides*. Endogenous G6P-DH does not interfere because only the bacterial enzyme is able to convert the coenzyme NAD.

The ratio between the concentration of the antibody and that of the conjugate has to be optimized in order to provide the maximum discriminatory power between ethosuximide standards, with the minimum cross-reactivity. The optimization procedure is carried out as described in detail on pp. 8, 9. The desired assay range is from 10 to 150 mg/l and an optimum response should be obtained where the assay sensitivity is most important (e.g. at 50 mg/l).

The immunological and enzymatic reactions take place simultaneously and the resulting overall reaction approaches zero order. However, a slight decrease in reaction rate has been observed [11, 12], and this should be taken into account when adapting the method to different instruments.

Equipment

The equipment is identical to that required for the EMIT assay of total phenytoin (cf. p. 44).

---

* The exact nature of the ethosuximide derivative used in raising antibody and in coupling to G6P-DH has apparently not been published so far.

Reagents and Solutions

Purity of reagents: glucose-6-phosphate dehydrogenase should be of the best available purity (cf. p. 27). Chemicals are of analytical grade.

Preparation of solutions* (for 50 determinations): all solutions in re-purified water (cf. Vol. II, chapter 2.1.3.2).

1. Tris buffer (Tris, 55 mmol/l; Triton X-100, 0.1 ml/l; pH 8.0):

   dissolve 1.996 g Tris base in 140 to 160 ml water; add 0.03 ml Triton X-100; adjust to pH 8.0 with HCl, 1.0 mol/l; make up with water to 300 ml.

   Alternatively, dilute concentrated buffer provided with each kit (10 ml) to 150 ml with water.

2. Antibody/substrate solution (γ-globulin**, ca. 50 mg/l; G-6-P, 66 mmol/l; NAD, 40 mmol/l; Tris, 55 mmol/l; pH 5.0):

   reconstitute available lyophilized preparation with 3.0 ml water.

3. Conjugate solution (G6P-DH coupled to ethosuximide, ca. 1 kU/l; Tris, 55 mmol/l; pH 8.0):

   reconstitute available lyophilized preparation with 3.0 ml water. This reagent must be standardized to match the antibody/substrate solution (2) (cf. p. 84).

4. Ethosuximide standard solutions (10 to 150 mg/l):

   dissolve 300 mg ethosuximide in 100 ml Tris buffer (1); dilute 0.5 ml with ethosuximide-free human serum containing sodium azide (5 g/l) and Thimerosal (0.5 g/l) to 10 ml (150 mg/l). Dilute this solution 2 + 1, 1 + 2, 1 + 5 and 1 + 14 with ethosuximide-free human serum, yielding the following concentrations of ethosuximide: 100, 50, 25, 10 mg/l. Take ethosuximide-free human serum as zero calibrator.

   Alternatively, reconstitute each of the available lyophilized standard preparations with 3 ml water.

Stability of solutions: cf. p. 39.

---

 * Commercially available reagents are provided as a matched set (EMIT® Ethosuximide Assay, *Syva Co.,* Palo Alto, CA, U.S.A.). Reagents, calibrators and buffer contain sodium azide.
** Preparation from immunized sheep.

Procedure

Collection and treatment of specimen: the specimen should be collected at a standardized time after drug administration, preferably in the morning before the first daily dose. In patients receiving ethosuximide in 2 or more divided daily doses, diurnal fluctuations in serum drug concentrations are usually relatively small and therefore the influence of time of sampling on the measured value is limited. If exception is made for special situations (e.g. status epilepticus, suspected toxicity, etc.), samples should be drawn at steady state, i.e. at least 10 to 12 days after the last dosage adjustment. Plasma or serum can be used in this assay. Blood cannot be used. The anticoagulants heparin and EDTA do not interfere, but it is advisable to collect the samples in an uniform manner (always serum, or plasma anticoagulated with the same anticoagulant). Sera from haemolyzed (up to 8 g haemoglobin per litre), lipaemic (triglycerides up to 11.3 mmol/l) or icteric (up to 513 µmol bilirubin per litre) blood may be assayed without clinically significant interference [13].

Stability of the substance in the sample: samples should be stored at 2°C to 8°C upon collection. Sera can be stored at 4°C for up to four weeks, or at −20°C for five months without appreciable loss.

Assay conditions: perform measurements of samples and zero calibrator in duplicate. Single determinations can be made of the remaining calibrators. In the authors' experience it is critical to keep the time elapsing between the last addition (enzyme-drug conjugate) and the start of the incubation in the spectrophotometer cell absolutely constant.

A pre-dilution of samples and standards is required: mix one part (0.05 ml) sample or standard (4) with five parts (0.25 ml) Tris buffer (1).

Wavelength 339 nm or Hg 334 nm; light path 10 mm; temperature 30.0 ± 0.1°C; total incubation time 45 s; measurement during the last 30 s against water; total volume of the assay mixture 0.90 ml.

Establish a calibration curve under exactly the same conditions (within series). The zero calibrator assay serves as a reagent blank; all measurements are corrected for this value.

If $\Delta A/\Delta t$ of a sample is greater than that of the highest standard, dilute the sample further with Tris buffer and re-assay.

Calibration curve: plot the corrected $\Delta A/\Delta t$ values of the standards *versus* the corresponding ethosuximide concentrations (mg/l). The calibration curve can be linearized by use of a graph paper matched with the reagents and based on a log-logit model [11, 14] (Fig. 1). To use this graph paper, multiply each $\Delta A/\Delta t$ value by 2.667. If a good linearity cannot be obtained, check instrumental conditions and expiry date of reagents.

Measurement

Pipette into a disposable beaker:			concentration in the assay mixture	
pre-diluted samples or calibrators (4)		0.05 ml	ethosuximide up to 1388 µg/l	
Tris buffer	(1)	0.25 ml	Tris	52 mmol/l
antibody/substrate solution	(2)	0.05 ml	sheep serum	variable
			γ-globulin	ca. 2.8 mg/l
			G-6-P	3.7 mmol/l
			NAD	2.2 mmol/l
Tris buffer	(1)	0.25 ml		
conjugate solution	(3)	0.05 ml	G6P-DH*	variable
				ca. 56 U/l
Tris buffer	(1)	0.25 ml		
mix well; immediately aspirate into the clean spectrophotometer flow-cell; after a 15 s delay read absorbance over a period of 30 s.				

* Coupled to ethosuximide.

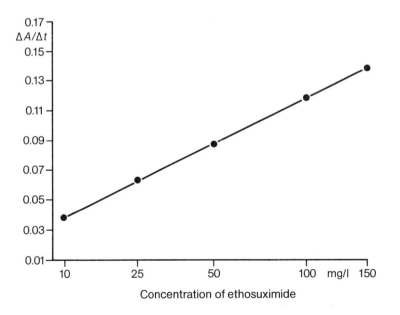

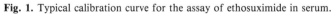

Fig. 1. Typical calibration curve for the assay of ethosuximide in serum.

Calculation: read from the calibration curve the ethosuximide mass concentration, ρ (mg/l), corresponding to the mean corrected $\Delta A/\Delta t$ readings of the sample. To convert from mass concentration to substance concentration, c (μmol/l), multiply by 7.0836.

Validation of Method

Precision, accuracy, detection limit and sensitivity: the precision is highly dependent on the level of mechanization of the analytical system. At an ethosuximide concentration of 75 mg/l the between-days RSD is less than 10% [13] and ranges from 3.1 to 6.7% with centrifugal analyzers [12, 15]. The overall imprecision between laboratories is 14.1% at a mean concentration of 70.6 mg/l; this is greater than that reported for GLC (11.8 to 13.0%) and HPLC methods (8.3%) [16].

The recovery of ethosuximide added to reference serum (*National Bureau of Standards*, Antiepileptic SRM 900 material) ranges from 100.5 to 112% at concentrations between 11.8 and 174.7 mg/l [17].

In a large European quality control survey [16] the mean difference between measured and added drug concentrations was +2.6% at a concentration of 70.6 mg/l.

The results obtained by EMIT in patients' sera agree well with those determined by GLC (EMIT *vs* GLC: y = 0.99x +0.34 mg/l, n = 50) [12].

Data on detection limit according to the definition of *Kaiser* [18] have apparently not been established so far. The lower limit of the working range is 10 mg/l. Sensitivity is highest in the range 25 to 50 mg/l (steepest part of the calibration curve) and decreases consistently with increasing concentrations.

Sources of error: sera from highly haemolyzed (above 8 g haemoglobin per litre), lipaemic (triglycerides above 11.3 mmol/l) or icteric (bilirubin above 513 μmol/l) blood may adversely affect the results [13].

Specificity: the specificity depends on the quality of the antiserum. It has been assessed by adding to the 50 mg/l standard known concentrations of potentially cross-reacting metabolites or other drugs. The degree of cross-reactivity was expressed as the concentration of the added substance that produces a quantitation error of +30% [13]. With commercially available EMIT reagents, no significant cross-reactivity was found with other commonly used antiepileptic drugs nor with various benzodiazepines, tricyclic antidepressants, or other frequently administered drugs [12]. A relevant cross-reactivity is exhibited by N-desmethylmethsuximide, a metabolite of methsuximide. As a consequence, ethosuximide cannot be determined by the EMIT procedure in sera from patients also receiving methsuximide.

Therapeutic ranges: serum ethosuximide concentrations ranging between 40 and 100 mg/l are considered optimal for effective seizure control. Very occasionally, concentrations up to 150 mg/l may be required [5, 6].

1.7.2   Determination with Substrate-labelled Fluorescent Immunoassay

Assay

Method Design

The method was developed by *Pahuski et al.* [19] and is based on the principle of the substrate-labelled fluorescent immunoassay (SLFIA) proposed by *Burd et al.* [20].

Principle: as described in detail on pp. 15, 16. It is identical to that described for the SLFIA of total phenytoin (cf. pp. 43, 44).

Selection of assay conditions and adaptation to the individual characteristics of the reagents: the assay is designed for rapid analysis and requires no separation step (results of samples from patients are available within about 35 min).

Anti-ethosuximide antibodies are obtained in goats by immunization with a bovine serum albumin conjugate of ethosuximide [21].

After enzymatic hydrolysis of the substrate-ethosuximide conjugate* the umbelliferyl moiety at pH 8.5 exhibits an excitation maximum at 405 nm and a fluorescence maximum at 450 nm.

The ratio between the concentration of antibody and that of the substrate-ethosuximide conjugate has to be adjusted in order to decrease the production of fluorescence in the absence of ethosuximide to 10% of that observed without addition of antibody. The assay range is from 20 to 150 mg/l.

Shorter or longer incubation times than 20 min can be selected in the range of 5 to 90 min, without significant loss in precision and with only a slight decrease in sensitivity. The immunological and enzymatic reactions take place simultaneously.

Equipment

The equipment is identical to that required for the SLFIA of total phenytoin (cf. p. 44).

Reagents and Solutions

Purity of reagents: β-galactosidase from *Escherichia coli* of the best available purity. Chemicals are of analytical grade.

---

* β-Galactosyl-umbelliferyl-ethosuximide. The method for its synthesis has apparently not been published so far.

Preparation of solutions* (for 100 determinations): all solutions in re-purified water (cf. Vol. II, chapter 2.1.3.2).

1. *N,N*-bis(2-Hydroxyethyl)glycine (Bicine) buffer (Bicine, 1.0 mol/l; NaN$_3$, 308 mmol/l; pH 8.5):

   dissolve 4.08 g Bicine and 0.5 g sodium azide in 20 ml water; adjust to pH 8.5 (25 °C) with NaOH, 12.5 mol/l; make up with water to 25 ml; filter to remove particles. 25 ml of this buffer is provided with each kit.

2. Diluted Bicine buffer (Bicine, 50 mmol/l; NaN$_3$, 15.4 mmol/l; pH 8.5):

   add one part buffer (1) to 19 parts water.

3. Antibody/enzyme solution (γ-globulin**; β-galactosidase, ca. 1 kU***/l; NaN$_3$, 15.4 mmol/l; Bicine, 50 mmol/l; pH 8.5):

   5 ml of this solution is provided with each kit.

4. Substrate-ethosuximide conjugate solution (β-galactosyl-umbelliferyl-ethosuximide, ca. 0.7 µmol/l; sodium formate, 5 mmol/l; pH 3.5):

   5 ml of this solution is provided with each kit.

5. Ethosuximide standard solutions (20 to 150 mg/l):

   dissolve 300 mg ethosuximide in 100 ml Bicine buffer (2); dilute 0.5 ml with 9.5 ml ethosuximide-free human serum (150 mg/l). Dilute this solution 2+1, 1+2, 1+6.5, with ethosuximide-free human serum, yielding the following concentrations of ethosuximide: 100, 50, 20 mg/l. Take ethosuximide-free serum as zero calibrator.

   Standard solutions containing the indicated ethosuximide concentrations are provided with each kit.

Stability of solutions: reagents and standards are stable for 12 months at 2 °C to 8 °C. The stability is 6 months for diluted Bicine buffer (2) at 2 °C to 23 °C, and 2 weeks for diluted calibrators at 2 °C to 8 °C. The substrate-ethosuximide conjugate (4) must be protected from light [23].

Procedure

Collection and treatment of specimens: each assay requires 50 µl serum, or plasma anticoagulated with EDTA or heparin. Sera from slightly haemolyzed, lipaemic or

---

* Commercially available reagents are provided as a matched set (*Ames, Division of Miles Laboratories,* Elkhart, U.S.A.).

** Standardized preparation (cf. p. 89), amount is titre-dependent.

*** For definition of units (U), cf. p. 17 and [22].

icteric blood can be assayed without interference, but results on sera from severely haemolyzed or lipaemic blood should be re-checked on fresh clear samples. If frozen samples are used, thaw and mix thoroughly before testing and clarify by centrifugation. For the suitable collection time, cf. p. 86.

Stability of the substance in the sample: cf. p. 86.

Assay conditions: standards and samples should be assayed in duplicate under identical conditions. When analyzing a batch of samples, initiate the reaction at 30 s intervals. Since the incubation time is 20 min, do not analyze batches containing more than 40 samples. All solutions must be at room temperature before performing the assay.

A pre-dilution of samples and standard is required: mix one part (0.05 ml) sample or standard solutions (5) with 50 parts (2.5 ml) Bicine buffer (2). Use only glass test tubes for this step.

Room temperature (15 °C to 30 °C); excitation wavelength 405 nm; emission wavelength 450 nm; total volume of the assay mixture 1.65 ml.

Establish a calibration curve under exactly the same conditions (within series).

Measurement

Pipette into a reaction cuvette:			concentration in the assay mixture	
antibody/enzyme solution	(3)	0.05 ml	β-galactosidase ca. 30 U/l rabbit or	
			goat serum	variable*
diluted Bicine buffer	(2)	0.50 ml	NaN$_3$	14.9 mmol/l
			Bicine	48.5 mmol/l
pre-diluted sample or standard	(5)	0.05 ml	ethosuximide	up to 89 µg/l
diluted Bicine buffer	(2)	0.50 ml		
substrate-ethosuximide				
conjugate	(4)	0.05 ml	β-galactosyl-umbelliferyl-etho-	
			suximide	ca. 21 nmol/l
			formate	152 µmol/l
diluted Bicine buffer	(2)	0.50 ml		
mix well after each addition of Bicine buffer; incubate for exactly 20 min; measure fluorescence**.				

* The final antibody concentration should be sufficient to decrease the fluorescence to 10% of that observed in the absence of antibody.

** Adjust fluorimeter reading to zero with a cuvette containing diluted Bicine buffer (2). Then adjust to 90% of full scale with the highest calibrator cuvette incubated for about 19 min.

Since setting the full scale deflection of the fluorimeter is the first operation (after zeroing), the highest standard should be the first sample assayed within the run: avoid carry-over contamination when assaying the other calibrators.

If $\Delta F/\Delta t$ of a sample is greater than that of the highest standard, perform a further dilution of the pre-diluted sample with Bicine buffer (2) and assay again. The result must then be multiplied by the dilution factor.

Calibration curve: plot mean $\Delta F/\Delta t$ values of duplicates *versus* the corresponding ethosuximide concentrations (mg/l) on linear graph paper (Fig. 2).

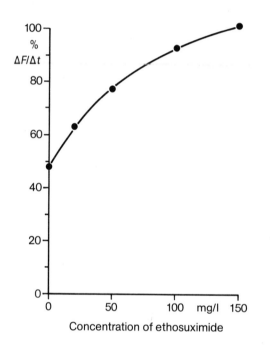

Fig. 2. Typical calibration curve for the assay of ethosuximide in serum.

Calculation: read from the calibration curve the ethosuximide mass concentration, ρ (mg/l), corresponding to the mean $\Delta F/\Delta t$ value of each sample. To convert from mass concentration to substance concentration, c (μmol/l), multiply by 7.0836.

Validation of Method

Precision, accuracy, detection limit and sensitivity: at concentrations in the range of 40 to 120 mg/l, the between-days RSD is less than 4.2% [21]. A good precision (between-days RSD of 4 to 6.2%) is also obtained with mechanized instruments [24].

The recovery of ethosuximide added to reference sera ranges from 94.5 to 98.5% at concentrations between 40 and 120 mg/l [21].

The accuracy has been assessed by comparison with GLC [21] in samples from patients (SLFIA vs. GLC: y = 0.97x + 0.01 mg/l, r = 0.95, n = 80).

Data on detection limit according to *Kaiser* [18] have apparently not been established so far. The concentration of ethosuximide that can be distinguished from zero with 95% confidence is 10 mg/l [23].

Sensitivity depends on the shape of calibration curve, and is highest on the steepest part of the curve, plotted on linear graph paper.

Sources of error: highly fluorescent compounds such as triamterene may cause a false elevation of ethosuximide concentration (correction for the presence of interfering fluorescent compounds, cf. Phenytoin, p. 49). Interference by β-galactosidase from human serum has not been observed [20]. Sera from highly haemolyzed, lipaemic or icteric blood may adversely affect the results.

Specificity: the specificity depends on the quality of the antibody. No clinically relevant cross-reactions have been observed with many other commonly used anticonvulsant drugs [19, 23]. However, trimethadione and paramethadione exhibit a relevant cross-reactivity at concentrations that can be found in patients receiving these medications [23]. The related drug methsuximide does not interfere [23] but no data have been reported on the cross-reactivity of its metabolite N-desmethylmethsuximide, which interferes in the EMIT procedure.

Therapeutic range: cf. p. 88.

References

[1] G. Chen, J. K. Weston, A. C. Bratton, Anticonvulsant Activity and Toxicity of Phensuximide, Methsuximide and Ethosuximide, Epilepsia 4, 66–76 (1963).

[2] J. R. Goulet, A. W. Kinkel, T. C. Smith, Metabolism of Ethosuximide, Clin. Pharmacol. Ther. 20, 213–218 (1976).

[3] E. Perucca, A. Richens, Clinical Pharmacokinetics of Antiepileptic Drugs, in: H. H. Frey, D. Janz (eds.), Antiepileptic Drugs, Handbook of Experimental Pharmacology, Vol. 74, Springer-Verlag, Berlin 1985, pp. 661–723.

[4] R. A. Buchanan, L. Fernandez, A. W. Kinkel, Absorption and Elimination of Ethosuximide in Children, J. Clin. Pharmacol. 9, 393–398 (1969).

[5] A. L. Sherwin, J. P. Robb, M. Lechter, Improved Control of Epilepsy by Monitoring Plasma Ethosuximide, Arch. Neurol. 28, 178–181 (1973).

[6] T. R. Browne, F. E. Dreifuss, P. R. Dyken, D. J. Goode, J. K. Penry, R. J. Porter, B. G. White, P. T. White, Ethosuximide in the Treatment of Absence (Petit Mal) Seizures, Neurology 25, 515–524 (1975).

[7] B. Rambeck, J. W. A. Meijer, Gas-Chromatographic Methods for the Determination of Antiepileptic Drugs: a Systematic Review, Ther. Drug Monit. 2, 385–396 (1980).

[8] M. Riedmann, B. Rambeck, J. W. A. Meijer, Quantitative Simultaneous Determination of Eight Common Antiepileptic Drugs and Metabolites by Liquid Chromatography, Ther. Drug Monit. 3, 397–413 (1981).

[9] *K. E. Rubenstein, R. S. Schneider, E. F. Ullman,* "Homogeneous" Enzyme Immunoassay. A New Immunochemical Technique, Biochem. Biophys. Res. Commun. *47,* 846 – 851 (1972).

[10] *C. E. Pippenger, R. J. Bastiani, R. S. Schneider,* Preliminary Report on the Analysis of Carbamazepine and Ethosuximide Using the EMIT System, in: *G. Gardner-Thorpe, D. Janz, H. Meinardi* (eds.), Clinical Pharmacology of Antiepileptic Drugs, Pitman Medical Publ. Co., London 1977, pp. 3 – 6.

[11] *D. N. Dietzler, M. P. Leckie, C. R. Hoelting, S. E. Porter, C. H. Smith, V. L. Tieber,* Logit-Log Calibration Curves for EMIT Assay, Clin. Chim. Acta *127,* 239 – 250 (1983).

[12] *D. N. Dietzler, C. R. Hoelting, M. P. Leckie, C. H. Smith, V. L. Tieber,* EMIT Assays for Five Major Anticonvulsant Drugs, Am. J. Clin. Pathol. *74,* 41 – 50 (1980).

[13] Ethosuximide Assay Package Insert, *Syva Co.,* Palo Alto, CA, U.S.A. 1977.

[14] *A. H. W. M. Schuurs, B. K. Van Weemen,* Enzyme Immunoassay, Clin. Chim. Acta *81,* 1 – 40 (1977).

[15] *N. Urquhart, W. Godolphin, D. J. Campbell,* Evaluation of Automated Enzyme Immunoassays for Five Anticonvulsants and Theophylline Adapted to a Centrifugal Analyzer, Clin. Chem. *25,* 785 – 787 (1979).

[16] *J. F. Wilson, R. W. Marshall, J. Williams, A. Richens,* Comparison of Assay Methods Used to Measure Antiepileptic Drugs in Plasma, Ther. Drug Monit. *5,* 449 – 460 (1983).

[17] *D. Studts, G. T. Haven, E. J. Kiser,* Adaptation of Microvolume EMIT Assays for Theophylline, Phenobarbital, Phenytoin, Carbamazepine, Primidone, Ethosuximide and Gentamicin to a CentrifiChem Chemistry Analyzer, Ther. Drug Monit. *5,* 335 – 340 (1983).

[18] *H. Kaiser,* Zum Problem der Nachweisgrenze, Fresenius Z. anal. Chem. *209,* 1 – 18 (1965).

[19] *E. E. Pahuski, C. S. Hixson, A. K. Petrozolin, R. P. Hatch, T. M. Li,* Homogeneous Substrate-Labeled Fluorescent Immunoassay (SLFIA) for Ethosuximide in Human Serum, Clin. Chem. *28,* 1664 (1982) (abstract).

[20] *J. F. Burd, R. C. Wong, J. E. Feeney, R. J. Carrico, R. C. Boguslaski,* Homogeneous Reactant-Labeled Fluorescent Immunoassay for Therapeutic Drugs Exemplified by Gentamicin Determination in Human Serum, Clin. Chem. *23,* 1402 – 1408 (1977).

[21] *S. M. Walker, R. E. Hill,* Substrate-Labeled Fluorescent Immunoassay for Ethosuximide Evaluated, Clin. Chem. *29,* 1567 – 1568 (1983).

[22] *L. M. Krausz, J. B. Hitz, R. T. Buckler, J. F. Burd,* Substrate-Labeled Fluorescent Immunoassay for Phenobarbital, Ther. Drug Monit. *2,* 261 – 272 (1980).

[23] *Ames* TDA Ethosuximide Package Insert, *Ames Division, Miles Laboratories,* Elkhart, 1983, revised 3/84.

[24] *E. E. Pahuski, D. M. Manzuk, J. L. Maurer, T. M. Li,* Total Automation of the SLFIAs for Ethosuximide and Valproic Acid, Clin. Chem. *29,* 1236 (1983) (abstract).

1.8   Carbamazepine

5*H*-Dibenz[*b*,*f*]azepine-5-carboxamide

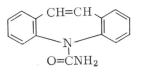

Thomas M. Li

General

Carbamazepine was first synthesized by *Schindler* [1] in the *Geigy* Laboratories in Basel in 1960. The use of carbamazepine as an anti-epileptic agent was reported in 1964 [2]. Carbamazepine is a tricylic compound that is effective against psychomotor and grand mal seizures. It also alleviates symptoms of trigeminal neuralgia.

Absorption of carbamazepine is slow. The time required to attain peak plasma levels after oral ingestion of a dose is 6 to 12 hours [3]. After absorption, carbamazepine is distributed to all tissues, with the brain concentration equal to or higher than that of plasma [4]. Carbamazepine is metabolized by the hepatic mixed-function oxidase system, first to its 10,11-epoxide, which has been shown to be pharmacologically active in animals and then to the 10,11-dihydroxide. Another metabolite of carbamazepine is iminostilbene which is formed by cleaving off the carbamyl group [5]. One percent of carbamazepine is eliminated in urine as the parent compound, 2% as the 10,11-epoxide, 20% as the 10,11-dihydroxide and less than 1% as iminostilbene [6]. Seventy-five percent of carbamazepine in serum is protein-bound. The concentration of unbound drug in serum is in agreement with cerebrospinal fluid and saliva levels [6]. The apparent serum half-life ranges from about 8 to 60 h. In patients who are also receiving other anti-epileptic drugs, such as phenytoin, phenobarbital, or primidone, the half-life of carbamazepine is considerably shorter than in patients taking carbamazepine alone. The time required for a carbamazepine dose to reach steady-state serum concentration is 2 to 4 days.

The margin of safety between therapeutic and toxic serum concentrations of carbamazepine is small. Serious adverse effects, including fatal cases, were reported as early as 1965 by *Donaldson & Graham* [7]. Because of inter-individual variability in pharmacokinetics, the optimal drug dose in mg per kg body weight can vary to a great extent among patients treated with similar doses. Furthermore, the half-life of carbamazepine is shorter during long-term carbamazepine treatment. Also, the serum concentration of carbamazepine decreases when phenytoin or phenobarbital is co-administered.

Adequate control of seizures is provided when the concentration of carbamazepine in serum is maintained in the range of 4 to 10 mg/l [8]. Toxic side effects including nystagmus, ataxia, unsteadiness and diplopia may be seen at varying serum carbamazepine levels. Concentrations above 20 mg/l may result in lethargy, severe sedation and even increased seizure frequency. Thus, to ensure adequate carbamazepine therapy without toxicity, determination of serum or plasma carbamazepine is necessary, especially during chronic treatment.

Application of method: in clinical chemistry, in pharmacology and in toxicology.

Substance properties relevant in analysis: carbamazepine has a molecular weight of 236.26. Carbamazepine appears as white crystals with a melting point of 190°C to 193°C. It is practically insoluble in water but dissolves in alcohol, acetone or propylene glycol.

Methods of determination: methods include gas-liquid chromatography [9], high-pressure liquid chromatography [10], enzyme-multiplied immunoassay technique (EMIT®) [11], substrate-labelled fluorescent immunoassay (SLFIA) [12] and fluorescence polarization immunoassay [13].

International reference method and standards: no international reference method or standard is established so far. However, for calibrators, material can be obtained from *United States Pharmacopeia-National Formulary Reference Standards* (Rockville, MD, U.S.A.).

Assay

Method Design

The method was developed by *Li et al.* [12] based on the principle of substrate-labelled fluorescent immunoassay [14]. The SLFIA for carbamazepine makes use of competitive protein-binding reactions.

Principle: as described in detail on pp. 15, 16.
A derivative of carbamazepine is covalently bound to a fluorigenic enzyme substrate, β-D-galactosyl-umbelliferone. This conjugate* is non-fluorescent until β-galactosidase** catalyzes the hydrolysis of the galactosyl portion of the molecule and produces a fluorescent product. When the conjugate is bound to antibody to carbamazepine, it becomes inactive as an enzyme substrate. Carbamazepine in a clinical sample

---

* β-Galactosyl-umbelliferyl-carbamazepine.
** β-D-Galactoside galactohydrolase, EC 3.2.1.23.

competes with the conjugate for antibody binding sites. The competitive-binding reactions are set up using a constant amount of conjugate and a limiting amount of antibody. Unbound conjugate is hydrolyzed by the enzyme, and the magnitude of increase in fluorescence intensity per unit time, $\Delta F/\Delta t$, is related to the carbamazepine concentration in the sample or calibrator.

Selection of assay conditions and adaptation to the individual characteristics of the reagents: the assay is designed for rapid determination of carbamazepine in serum or plasma with no sample pre-treatment or separation step.

Antiserum to carbamazepine was obtained from rabbits that had been injected with a derivative of carbamazepine covalently coupled to bovine serum albumin as the carrier protein [12].

The reaction mixture should contain sufficient antibody to decrease the production of fluorescence in the absence of carbamazepine to 10% of that observed without addition of antibody. The desired assay range is from 1 to 20 mg/l.

Equipment

Fluorimeter capable of measuring fluorescence intensity with excitation and emission wavelengths of 400 and 450 nm, respectively. Polystyrene disposable cuvettes suitable for the fluorimeter. Glass test tubes 12 mm × 75 mm for specimen and calibrator dilutions; stopwatch; pipetter-diluter capable of delivering 50 µl sample and 500 µl or 2.5 ml buffer. Laboratory centrifuge.

The fluorimeter should be checked periodically for linearity of the relationship of readings with concentrations. Use a solution of 7-hydroxycoumarin-3-[N-(2-hydroxyethyl)] carboxamide* in Bicine buffer (50 mmol/l, pH 8.5) at concentrations of 0.0, 1.0, 5.0, 10.0 and 15.0 nmol/l.

Reagents and Solutions

Purity of reagents: β-galactosidase from *E. coli* should be of the best available purity. Carbamazepine used for preparation of standard solution is from *United States Pharmacopeia – National Formulary Reference Standards,* Rockville, MD, U.S.A. Chemicals are of A.R. grade.

Preparation of solutions**: use only re-purified water (cf. Vol. II, chapter 2.1.3.2).

---

* Provided by *Ames*.
** Reagents, calibrators and buffer are commercially available from *Ames Division, Miles Laboratories,* Elkhart, IN, U.S.A.

1. Bicine buffer (Bicine, 50 mmol/l; NaN$_3$, 15.4 mmol/l; pH 8.5):

   dissolve 8.16 g *N,N*-bis-(2-hydroxyethyl)glycine and 1 g NaN$_3$ in approximately 900 ml water, adjust to pH 8.5 with NaOH, ca. 12 mmol/l; make up to 1 litre with water.

2. Antibody/enzyme solution (γ-globulin*, β-galactosidase**, 1 kU/l; Bicine, 50 mmol/l; NaN$_3$, 15.4 mmol/l; pH 8.5):

   5 ml of this solution is provided with each kit.

3. Substrate-carbamazepine conjugate solution (β-galactosyl-umbelliferyl-carb-amazepine, 0.55 µmol/l; formate, 30 mmol/l; Tween 20, 2 ml/l; pH 3.5):

   5 ml of this solution is provided with each kit.

4. Carbamazepine standard solutions:

   use vials with 0, 5, 10, 15 and 20 mg/l in normal carbamazepine-free serum with NaN$_3$, 15.4 mmol/l (provided with each kit).

Stability of solutions: reagents and calibrators are stable for 12 months if stored at 2°C to 8°C. Bicine buffer (1) is stable for 6 months at 2°C to 23°C, the antibody/enzyme solution (2) is stable for 12 months stored at 4°C. The substrate-carbamazepine conjugate solution (3) can be used for 12 months if stored at 4°C protected from light. Do not use diluted standard solutions (4) for longer than 2 weeks, if they are stored at 2°C to 8°C.

Procedure

Collection and treatment of specimen: during long-term oral therapy blood specimens should be taken in the "steady state". Draw the specimen immediately before administration of the next dose. 50 µl serum or plasma is required per assay. Acceptable anticoagulants are EDTA and heparin. Fresh serum samples are preferred. Thaw frozen samples and mix well before analysis. There is no interference in the assay when samples from slightly haemolyzed, icteric or lipaemic blood are used.

Stability of the analyte in the sample: samples can be stored at 2°C to 8°C upon collection at 2°C to 8°C for several weeks without appreciable loss.

Assay conditions: measure standards and samples in duplicate under identical conditions (within series). When analyzing series of samples, initiate the reaction at 20-s intervals. The maximum number of cuvettes in a run is 60 using the 20-min incubation time and 20-second intervals. All solutions must be allowed to reach room temperature before performing the assay.

---

* Antiserum to carbamazepine, standardized preparation; concentration is titre-dependent.
** International units, U, at 25°C and pH 8.5 with 2-NPgal as the substrate [15].

A pre-dilution of samples and standard solutions is required: mix one part (0.05 ml) sample or standard solution (4) with 50 parts (2.5 ml) Bicine buffer (1).

Incubation time 20 min; ambient temperature; reaction volume 1.65 ml; excitation wavelength 400 nm; emission wavelength 450 nm.

Establish a calibration curve under exactly the same conditions (within series).

Measurement

Pipette into reaction cuvettes:			concentration in assay mixture	
pre-diluted sample or standard (4)		0.05 ml	carbamaze- pine	up to 12 µg/l
Bicine buffer	(1)	0.50 ml	Bicine NaN$_3$	48.5 mmol/l 14.9 mmol/l
antibody/enzyme solution	(2)	0.05 ml	antibody galactosidase	variable* 30 U/l
Bicine buffer	(1)	0.50 ml		
substrate/carb- amazepine conjugate	(3)	0.05 ml	conjugate formate	16.7 nmol/l 0.9 mmol/l
Bicine buffer	(1)	0.50 ml		
mix well, incubate for exactly 20 min; measure increase in fluorescence intensity**, ΔF.				

* Amount of antibody sufficient to decrease the fluorescence to 10% of that observed in the absence of antibody.

** Adjust fluorimeter reading to zero by use of a cuvette containing Bicine buffer, and to 90% of full scale with the highest calibrator cuvette incubated for about 19 min after addition of the substrate-carbamazepine conjugate (3).

If ΔF per 20 min of the sample is greater than that of the highest calibrator, dilute the sample further with Bicine buffer (1) and re-assay.

Calibration curve: plot mean increase in fluorescence per 20 min, $\Delta F/\Delta t$, of duplicates *versus* the corresponding carbamazepine concentrations (mg/l) on linear graph paper. Fig. 1 shows a typical calibration curve.

Carbamazepine calibration curves can be fitted by using a four-parameter logistic model [16].

Calculation: read or calculate carbamazepine mass concentrations, ρ (mg/l), from the calibration curve using the mean $\Delta F/\Delta t$ value of each sample. To convert from mass concentration to substance concentration, c (µmol/l), multiply by 4.2319.

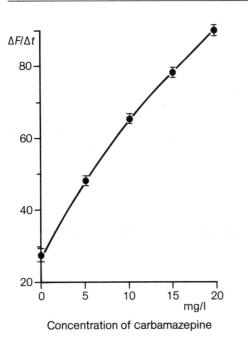

Concentration of carbamazepine

Fig. 1. Typical calibration curve for the assay of carbamazepine in serum.

Validation of Method

Precision, accuracy, detection limit and sensitivity: the imprecision of the carbamazepine SLFIA indicates that the between-days relative standard deviations for the manual and mechanized procedure are all less than 6% at concentrations between 4 and 16 mg/l [12, 16].

The accuracy has been proven by comparison with HPLC, GC, and EMIT [12, 16]. Good correlation was obtained with patient samples:

SLFIA *vs.* HPLC: y = 0.96x + 0.57 mg/l,   r = 0.97,   *n* = 60
SLFIA *vs.* GC: y = 1.01x − 0.30 mg/l,   r = 0.95,   *n* = 56
SLFIA *vs.* EMIT®:   y = 1.07x − 0.82 mg/l,   r = 0.98,   *n* = 53

The assay has a detection limit* of 1 mg/l. Sensitivity is highest on the steepest part of the calibration curve.

---

* Detection limit is defined as mean value plus 3 standard deviations obtained with the zero calibrator in a series of 20 determinations.

Sources of error: sera from highly lipaemic blood give falsely elevated carbamazepine concentrations. Certain fluorescent compounds such as triamterene may cause false elevations of carbamazepine concentrations. For samples containing triamterene (Dyrenium®, Dyazide®) which is fluorescent, subtraction of sample fluorescence background is recommended (cf. p. 49). Interference by β-galactosidase from human serum was not observed [15].

Specificity: the specificity of the carbamazepine SLFIA is assessed [12] by measuring the amount of potential cross-reactant required to compete with the conjugate for 50% of the carbamazepine antibody binding sites. Table 1 indicates that the 10,11-epoxide metabolite of carbamazepine is recognized to some extent (6.5%) by the antibody used in the carbamazepine SLFIA assay. The other tricyclic compounds do not show significant cross-reactivity. No other anti-epileptic drugs cross-react. The concentration of the 10,11-epoxide necessary to increase a carbamazepine control of 8 mg/l by 30% over its value determined in the absence of 10,11-epoxide is found to be > 20 mg/l.

Therapeutic ranges: usually, serum (plasma) carbamazepine concentrations between 4 and 10 mg/l (17 to 42 μmol/l) provide adequate seizure-control [8].

Table 1. Cross-reactivity of carbamazepine antibody at 50% competition

Drug	50% Competition mol/l	Percentage* cross-reactivity
Carbamazepine	1.95×10^{-4}	100.0
Carbamazepine-10,11-epoxide	3.00×10^{-3}	6.5
Iminostilbene	5.20×10^{-3}	3.8
Protriptyline	5.20×10^{-3}	3.8
Amitriptyline	8.50×10^{-3}	2.3
Nortriptyline	8.70×10^{-3}	2.2
Chlorpromazine	9.00×10^{-3}	2.1
Desipramine	1.10×10^{-2}	1.8
Doxepin	1.55×10^{-2}	1.3
Imipramine	1.90×10^{-2}	1.0
Phenylbutazone	3.00×10^{-2}	0.7
Methaqualone	4.95×10^{-1}	0.04
Procyclidine	$>8.9 \times 10^{-1}$	<0.02
Diazepam	$\geqslant 2.1 \times 10^{-1}$	$\ll 0.09$
Indomethacin	$\geqslant 2.8 \times 10^{-1}$	$\ll 0.07$
Phenobarbital	$\geqslant 4.3 \times 10^{-1}$	$\ll 0.05$
Phenytoin	$\geqslant 4.0 \times 10^{-1}$	$\ll 0.05$
Primidone	$\geqslant 4.6 \times 10^{-1}$	$\ll 0.04$
Caffeine	$\geqslant 5.1 \times 10^{-1}$	$\ll 0.04$
Ethosuximide	$\geqslant 7.1 \times 10^{-1}$	$\ll 0.03$

* (mol carbamazepine per litre at 50% competition)/(mol compound per litre at 50% competition) × 100.

Appendix

Preparation of β-galactosyl-umbelliferyl-carbamazepine conjugate: use the procedure described on p. 21 for the theophylline conjugate: however, use a carbamazepine derivative instead of 8-(3-aminopropyl) theophylline.

References

[1] *W. Schindler,* 5H-Dibenz[b,f]azepines, U.S. Patent 2.948.718 (1960).

[2] *M. Bonduelle,* Clinical Experiments with the Antiepileptic G 32.883 (Tegretol), Result of 100 Cases Observed over a Three-Year Period, Rev. Neurol. *110,* 200 – 225 (1964).

[3] *S. I. Johannessen,* Antiepileptic Drugs: Pharmacokinetic and Clinical Aspects, Therapeutic Drug Monitoring *3,* 17 – 37 (1981).

[4] *P. L. Morselli,* Carbamazepine: Absorption, Distribution and Excretion, in: *J. K. Penry, D. D. Daly* (eds.), Complex Partial Seizures and Their Treatment, Advances in Neurology, Vol. *11,* Raven Press, New York 1975, pp. 279 – 293.

[5] *P. L. Morselli, A. Frigeris,* Metabolism and Pharmacokinetics of Carbamazepine, Drug Metab. Rev. *4,* 97 – 113 (1975).

[6] *H. Kutt,* Clinical Pharmacology of Carbamazepine, in: *C. E. Pippenger, J. K. Penry, H. Kutt* (eds.), Antiepileptic Drugs: Quantitative Analysis and Interpretation, Raven Press, New York 1978, pp. 297 – 305.

[7] *G. W. K. Donaldson, J. G. Graham,* Aplastic Anaemia Following the Administration of Tegretol, Br. J. Clin. Pract. *19,* 699 – 702 (1965).

[8] *H. Kutt,* Clinical Pharmacology of Carbamazepine, in: *C. E. Pippenger,* Antiepileptic Drugs: Quantitative Analysis and Interpretation, Raven Press, New York 1978.

[9] *H. Kutt,* Carbamazepine: Chemistry and Methods of Determination, in: *J. K. Penry, D. D. Daly* (eds.), Complex Partial Seizures and Their Treatment, Advances in Neurology, Vol. *11,* Raven Press, New York 1975, pp. 249 – 261.

[10] *J. J. MacKichan,* Simultaneous Liquid Chromatographic Analysis for Carbamazepine and Carbamazepine 10,11-epoxide in Plasma and Saliva by Use of Double Internal Standardization, J. Chromatogr. *181,* 373 – 383 (1980).

[11] *D. D. Schottelius,* Homogeneous Immunoassay System (EMIT) for Quantitation of Antiepileptic Drugs in Biological Fluids, in: *C. E. Pippenger, J. K. Penry, H. Kutt* (eds.), Antiepileptic Drugs: Quantitation Analysis and Interpretation, Raven Press, New York 1978, pp. 95 – 108.

[12] *T. M. Li, J. E. Miller, F. E. Ward, J. F. Burd,* Homogeneous Substrate-Labeled Fluorescent Immunoassay for Carbamazepine, Epilepsia *23,* 391 – 398 (1982).

[13] *D. W. Forman, J. Dudek, J. E. O'Brien, M. W. Fordice, T. E. Worthy,* Therapeutic Drug Monitoring by Fluorescence Polarization: Performance of Immunoassays for Carbamazepine and Primidone, Clin. Chem. *28,* 1664 (1982).

[14] *J. F. Burd, R. C. Wong, J. E. Feeney, R. J. Carrico, R. C. Boguslaski,* Homogeneous Reactant-Labelled Fluorescent Immunoassay for Therapeutic Drugs Exemplified by Gentamicin Determination in Human Serum, Clin. Chem. *23,* 1402 – 1408 (1977).

[15] *R. C. Wong, J. F. Burd, R. J. Carrico, R. T. Buckler, J. Thoma, R. C. Boguslaski,* Substrate-Labeled Fluorescent Immunoassay for Phenytoin in Human Serum, Clin. Chem. *25,* 686 – 691 (1979).

[16] *T. M. Li, S. P. Robertson, T. H. Crouch, E. E. Pahuski, G. A. Bush, S. J. Hydo,* Automated Fluorometer/Photometer System for Homogeneous Immunoassays, Clin. Chem. *29,* 1628 – 1634 (1983).

[17] *T. M. Li, J. L. Benovic, R. T. Buckler, J. F. Burd,* Homogeneous Substrate-Labeled Fluorescent Immunoassay for Theophylline in Serum, Clin. Chem. *27,* 22 – 26 (1981).

1.9   Valproic Acid

2-Propylpentanoic acid

$$H_3C-CH_2-CH_2$$
$$\setminus$$
$$CH-COOH$$
$$/$$
$$H_3C-CH_2-CH_2$$

Hans Reinauer

General

The anticonvulsant properties of valproic acid (VPA) were first described in 1963 [1]. The drug has been used since 1977 as antiepileptic agent [2] and was approved by the U.S. Food and Drug Administration in 1978 [3] for the treatment of simple and complex absence seizures, including petit mal epilepsy. Due to its chemical structure, which is different from that of other antiepileptic drugs, VPA shows unique therapeutic properties [4].

The bio-availability is complete: sodium valproate is rapidly and fully absorbed (within 0.5 – 4 h). Therefore, the drug is administered almost exclusively by the oral route [1, 4, 5 – 8].

The distribution volume of VPA was calculated as 0.15 ± 0.04 l/kg, the serum clearance being 12.7 ± 3.7 ml/min [7, 9]. The biological half-life of valproate was reported as 6 to 15 h.

It is generally considered that the half-life in children is shorter than in adults. Assuming a mean half-life of 11 h the steady state would be reached after 2 to 3 days [5, 7, 10].

The correlation is poor between therapeutic dosage and the resulting concentrations in serum on the one side, and the therapeutic effect on the other. The therapeutic range of VPA in serum is considered to lie between 50 – 100 mg/l. There is disagreement in the literature concerning the relationship between serum levels of VPA and therapeutic effect: the differing kinds of epilepsy may be the reason for the discrepancies [11]. Adverse effects are seen at serum levels of VPA above 100 mg/l [12].

The free fraction of VPA depends not only on the serum protein concentration, but also on the concentration of additional drugs and of free fatty acids. Generally, the higher the total VPA concentration, the larger the free fraction of VPA. The protein binding of VPA varies from 84 to 95% [5, 13 – 15]. Interactions with the protein binding of VPA are found with salicylates, phenytoin and free fatty acids, increased concentrations of which increase the free fraction of VPA [13 – 17].

The concentration of free VPA can be determined after separation of the protein fraction from the ultrafiltrate by Centriflo (CF 50) cones (*Amicon Corp.,* Danvers, MA 01923).

Elevated free fatty acid concentrations affecting the free fraction of VPA were seen during exercise, high-fat diet and fasting [5, 18 – 20].

Only a small part of the drug is excreted unchanged into the urine. The biotransformation of VPA follows at least three different pathways.

1. About 40 to 60% of VPA is coupled with glucuronic acid and renally excreted.
2. About 30 to 40% undergoes β-oxidation leading to 2-propyl-3-oxopentanoic acid [21].
3. A small part is transformed by ω-oxidation in the microsomal monooxygenase-system leading to 2-propyl-4-hydroxypentanoic acid or 2-propylglutaric acid [21].

Of the metabolites tested, 5-hydroxyvalproic acid and 2-propylglutaric acid are nontoxic. However, aberrant pathways of VPA may be present and may be responsible for hepatoxicity. In experiments with ^{14}C-labelled valproate, almost 95% of the dose was recovered and the breakdown of the molecule to CO_2 was negligible.

It is assumed that valproic acid develops its cerebral effect by increasing the level of the inhibitory transmitter, γ-aminobutyric acid (GABA). The accumulation of GABA (Fig. 1) is achieved by inhibition of the degradative enzymes aminobutyrate aminotransferase and succinate-semialdehyde dehydrogenase [22]. Another and even

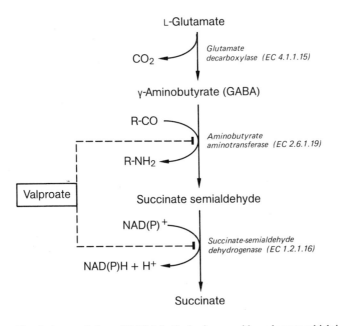

Fig. 1. Accumulation of GABA in the brain caused by valproate which inhibits the enzymes aminobutyrate aminotransferase and succinate-semialdehyde dehydrogenase [23].

stronger correlation exists between the anticonvulsant potency and the reduction of cerebral aspartate levels [23].

Studies with valproic acid, derivatives and analogues suggest a physicochemical action on membranes rather than direct actions on a specific receptor or recognition site, or active site of an enzyme. This physicochemical action may secondarily modify the catalytic properties of carriers or enzymes [7, 24].

The most frequently reported side-effects are gastro-intestinal complaints from patients at the beginning of therapy. Other side-effects – ataxia, dizziness, blurred vision – occur infrequently and do not appear to be dose-related. Valproate therapy of pregnant women seems to promote neural tube defects of the foetus [25].

Other side effects, including hepatotoxicity, pancreatitis and hyperammonaemia have been observed. However, it is not clear at present why a minority of patients develop hepatic and pancreatic disorders during VPA therapy. It is assumed that unknown metabolic pre-disposition may complicate therapy with VPA.

Therefore, the Commission on Drugs of the American Academy of Pediatrics stated that VPA is not the drug of first choice for most patients with seizures. After starting valproate therapy the activities of aspartate aminotransferase, amylase and lipase in serum, as well as the concentrations of ammonia and urea, should be monitored regularly in the patients.

Application of method: in clinical chemistry, laboratory medicine, pharmacology, toxicology.

Substance properties relevant in analysis: valproic acid (molecular weight 144.21) is a volatile weak acid. The pK_a of valproic acid is 4.95. Its solubility in water is low, 1.3 g/l (9 mmol/l), but it is easily soluble in organic solvents. The partition coefficient is not yet available. Salts: sodium salt, hygroscopic and freely soluble in water; magnesium salt.

Methods of determination: various methods have been developed for the determination of valproic acid in serum: gas chromatography, gas chromatography combined with mass spectrometry, isotope dilution-gas chromatography and mass spectrometry, homogeneous immunoassays (enzyme-multiplied immunoassay technique, substrate-labelled fluorescent immunoassay, fluorescence polarization immunoassay). In routine analysis gas chromatography and the homogeneous immunoassay are mainly used.

Definitive methods have been developed by use of tetradeutero isomers of valproic acid as the internal standard. After isolation from body fluids the acid is methylated with diazomethane. The *McLafferty* fragments of the methyl esters are used for the mass-spectrometric quantitation. RSD is below 1% [26, 27].

International reference method and standards: no internationally approved reference method exists. However, gas chromatography or the combination of isotope dilution with gas chromatography and mass spectrometry could be regarded as preliminary reference methods. The existence of standard reference material is not known so far.

Assay

Method Design

The routine analytical method has not yet been published in detail but is mainly based on the enzyme-multiplied immunoassay technique (EMIT®) of *Rubenstein et al.* [28].

Principle: as described in detail on pp. 7, 8.

Selection of assay conditions and adaptation to the individual characteristics of the reagents: the assay is designed for rapid analysis and requires no separation step. The antibody has been raised in sheep; the conditions of immunization have not been published (*Syva*).

Equipment

Spectrophotometer or spectral-line photometer (at 339 or Hg 334 nm); the flow-cell of the photometer must consistently maintain $30.0 \pm 0.1\,°C$ throughout the analytical procedure. Pipetter-diluter with an imprecision of RSD $< 0.25\%$. Laboratory centrifuge. Data processer: delay 15 s, measuring time 30 s.

Reagents and Solutions

Purity of reagents: G6P-DH from *Leuconostoc mesenteroides* should be of the best available purity (cf. p. 27). All chemicals are of analytical grade. Valproic acid from *Sigma* is of sufficient purity.

Preparation of solutions* (for 100 determinations): all solutions in re-purified water (cf. Vol. II, chapter 2.1.3.2).

1. Tris buffer (Tris, 55 mmol/l; Triton X-100, 0.1 ml/l; pH 8.0):

   dissolve 1.996 g Tris base in about 150 ml water; add 0.03 ml Triton X-100; adjust pH to 8.0 with HCl, 1 mol/l, add water to 300 ml.

   Alternatively, dilute the concentrated buffer provided with each kit with water to 150 ml.

---

* Reagents, calibrators and buffer are commercially available from *Syva Co.,* Palo Alto, U.S.A. The solutions cotain sodium azide. The single components and their concentrations are not declared – even on repeated request.

2. Antibody/substrate solution (γ-globulin*, ca. 50 mg/l; G-6-P, 66 mmol/l; NAD, 40 mmol/l; Tris, 55 mmol/l; pH 5.2):

reconstitute available lyophilized preparation with 6.0 ml water.

3. Conjugate solution (G6P-DH, ca. 1 kU/l; Tris, 55 mmol/l; pH 8.0):

reconstitute available lyophilized preparation with 6.0 ml water. This reagent must be standardized to match the antibody/substrate reagent.

4. Valproic acid standard solutions (10 mg/l to 150 mg/l):

dissolve 100 mg valproic acid in 100 ml Tris buffer (1); dilute 3.0 ml of this stock solution with VPA-free human serum**, containing 5 g sodium azide per litre, to 10 ml. Dilute this solution $1+1$, $1+2$, $1+5$, $1+11$, $1+29$ with VPA-free human serum. The final concentrations of VPA in the different preparations are 150, 100, 50, 25, 10 mg/l, corresponding to 1040, 693.4, 346.7, 173.4, 69.3 μmol/l. Use VPA-free serum as zero calibrator.

Alternatively, reconstitute each of the commercially available standard preparations with 1 ml water.

5. VPA control (valproic acid, 75 mg/l):

sheep serum to which 75 mg valproic acid per litre has been added is provided by the manufacturer for internal quality control (precision). Reconstitute the lyophilized preparation with 3.0 ml water. The control can be prepared from the stock solution of the calibrator.

Stability of solutions: if stored at 2°C to 8°C, the reconstituted reagents should not be frozen or exposed to temperatures above 32°C, they are stable for about 12 weeks.

Procedure

Collection and treatment of specimen: collect the specimen at an appropriate time following the administration of the drug. During long-term oral therapy draw blood specimens under steady-state conditions, i.e. after treatment with a constant dose over at least 4 half-lives of the drug (1 to 4 days). The peak serum levels are reached 1 to 4 h after oral administration. Take a sample before the next scheduled dose.

Serum or plasma can be used in this assay. EDTA, citrate and oxalate as anticoagulants induce a high bias in quantitation of valproic acid. Therefore, use only heparin plasma or serum for analysis. Furthermore, samples from severely haemolyzed, lipaemic or icteric blood may cause poor reproducibility and lower results.

---

* Standardized preparation from immunized sheep.

** Standard solutions and zero control are prepared from non-sterile sheep serum. Therefore, the same safety precautions should be employed as when handling other potentially infectious biological material.

Stability of the drug in the sample: if stored in sealed tubes at 2°C to 8°C (temperature should also not exceed 8°C during transport of specimens), the analyte is stable for at least a week. Serum can be stored at −20°C in sealed tubes for 12 weeks or more without appreciable loss.

Assay conditions: measure samples, standards and zero calibrator in duplicate under identical conditions, average readings. Perform single measurements for the remaining standards.

A pre-dilution of samples, control and standards is required: mix 0.05 ml sample or standard solution (4) and 0.25 ml buffer (1) in a 2.0 ml beaker.

Incubation time 45 s (in the spectrometer flow-cell); 0.90 ml; 30°C; measurement during the last 30 s against water; wavelength 339 nm or Hg 334 nm; light path 10 mm.

All reagents, samples and materials must be brought to room temperature. Equilibrate at room temperature (20°C to 25°C) for 1 h before use.

Establish a calibration curve under exactly the same conditions (within the series). The zero calibrator assay serves as a reagent blank; all measurements are corrected for this value. A new calibration curve should be prepared whenever a new set of reagents is used. If a new bottle of buffer is used the system may be validated by running the VPA control (5) only.

Measurement

Pipette into a disposable beaker:			concentration in the assay mixture	
pre-diluted sample or standard solutions (4)		0.05 ml		up to 1.4 mg/l
Tris buffer	(1)	0.25 ml	Tris	52 mmol/l
antibody/substrate solution	(2)	0.05 ml	sheep serum	variable
			γ-globulin	ca. 2.8 mg/l
			G-6-P	3.7 mmol/l
			NAD	2.2 mmol/l
Tris buffer	(1)	0.25 ml		
conjugate solution	(3)	0.05 ml	G6P-DH	variable
				ca. 56 U/l
Tris buffer	(3)	0.25 ml		
mix well; immediately aspirate into the clean spectrophotometer flow-cell; after 15 s delay read ΔA for a further incubation time of 30 s.				

If ΔA per 30 s of a sample is greater than that of the highest calibrator, dilute the sample further with Tris buffer (1) and re-assay.

Calibration curve: plot the corrected $\Delta A/\Delta t$ readings of the standards *versus* the corresponding valproic acid concentration (mg/l). A linear calibration curve is obtained by use of a special graph paper* (Fig. 2). The difference between zero calibrator and that with 10 mg/l should be greater than $\Delta A/\Delta t = 0.015$, the difference between the calibrators with 10 mg/l and 150 mg/l should be $\Delta A/\Delta t > 0.100$.

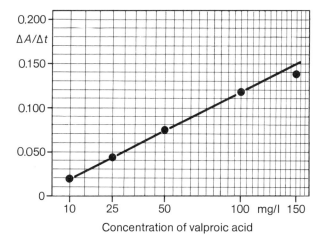

Fig. 2. Typical calibration curve for the assay of valproic acid in serum

Calculation: read the mass concentration, ρ (mg/l), corresponding to the mean corrected $\Delta A/\Delta t$ readings from the calibration curve. For conversion from mass concentration to substance concentration, c (µmol/l), multiply the mass concentration by 6.9343 (molecular weight of valproic acid is 144.21).

Validation of Method

Precision, accuracy, detection limit and sensitivity: the within-run imprecision (concentration of valproic acid 50 mg/l) is between RSD 2.5 to 6%. The between-run imprecision should not exceed a RSD of 7%.

The accuracy of the assay was assessed by analysis of a set of samples with added VPA. The samples were prepared in sheep serum. The correlation coefficient between theoretical values and found values was $r = 0.997$, with a regression line of $y = 0.94x - 0.57$ mg/l ($n = 9$). Different groups have compared gas-chromatographic analysis

* Supplied by *Syva*; for using this graph paper multiply each value of $\Delta A/\Delta t$ by 2.667.

with the enzyme-immunoassay technique. A high correlation between the two methods was achieved over a wide range of concentrations: $r = 0.999$, $y = 1.06x - 3.90$ mg/l [16, 29 – 33].

The detection limit, defined as the mean value plus 3 standard deviations of the zero calibrator in a series of 20 determinations, was found to be 0.9 mg/l ([16] and *Susanto*, unpublished).

Specificity: the enzyme-immunoassay measures the total valproic acid concentration in serum or heparinized plasma. Since valproic acid is usually administered together with other antiepileptic drugs, studies on cross-reactivity were concentrated on these additional drugs (cf. Table 1). No cross-reactivity was observed with phenytoin, primidone, carbamazepine or ethosuximide. The unsaturated metabolites of valproic acid (2-propyl-2-pentenoic acid and 2-propyl-4-pentenoic acid) cross-react in this assay, though concentration of the metabolite 2-propyl-2-pentenoic acid averaged only 9.5% of the corresponding serum valproic acid concentration. The metabolite 2-propyl-4-pentenoic acid and other interfering metabolites could not be found in serum samples.

Table 1. Cross-reactivities in the valproic acid assay with EMIT®. Cross-reactivity is defined as the minimum concentration of the compound needed to produce an error of 30% in a sample containing 50 mg VPA per litre (to an apparent concentration of 65 mg/l).

Compounds	Concentrations producing a 30% error (mg/l)
Diazepam	>100
Clonazepam	>100
2-n-Propyl-3-hydroxypentanoic acid	>100
2-n-Propyl-4-hydroxypentanoic acid	>100
2-n-Propyl-5-hydroxypentanoic acid	>100
2-n-Propyl-3-oxopentanoic acid	>100
Phenobarbital	>750
2-Propylglutaric acid	>500

Sources of error: no clinically significant interferences from endogeneous substances were observed, with haemoglobin up to 3 g/l, triglycerides up to 2 g/l, and bilirubin up to 300 mg/l.

Therapeutic ranges: the therapeutic range of VPA in plasma is between 50 to 100 mg/l.

References

[1] *H. Meunier, G. Carraz, Y. Meunier, P. Eymard, M. Aimard,* Proprietés pharmacodynamiques de l'acide n-dipropylacétique, Therapie *18*, 435 – 438 (1963).
[2] *W. Fröscher, H.-U. Schulz, R. Gugler,* Valproinsäure in der Behandlung der Epilepsien unter besonderer Berücksichtigung des Serumspiegels, Fortschr. Neurol. Psychiatr. *46*, 327 – 341 (1978).

[3] Valproic Acid and Sodium Valproate: Approved for Use in Epilepsy, FDA Drug Bulletin *8*, 14 – 15 (1978).

[4] *R. M. Pinder, R. N. Brogden, T. M. Speight, G. S. Avery,* Sodium Valproate: a Review of Its Pharmacological and Properties and Therapeutic Efficacy in Epilepsy, Drugs *13*, 18 – 123 (1977).

[5] *R. Gugler, G. E. von Unruh,* Clinical Pharmacokinetics of Valproic Acid, Clin. Pharmacokin. *5*, 67 – 83 (1981).

[6] *R. Gugler, A. Schell, M. Eichelbaum, W. Fröscher, H. U. Schulz,* Disposition of Valproic Acid in Man, Eur. J. Clin. Pharmacol. *12*, 125 – 132 (1977).

[7] *U. Klotz, K. H. Antouin,* Pharmacokinetics and Bioavailability of Valproate, Clin. Pharmacol. Ther. *21*, 736 – 743 (1977).

[8] *C. A. May,* Prediction of Steady-state Concentrations of Valproic Acid as Determined from Single Plasma Concentrations after the First Dose, Clin. Pharm. *2*, 143 – 147 (1983).

[9] *F. Hoffmann, G. E. von Unruh, B. C. Jancik,* Valproic Acid Disposition in Epileptic Patients during Combined Antiepileptic Maintenance Therapy, Eur. J. Clin. Pharmacol. *19*, 383 – 385 (1981).

[10] *E. Perucca, G. Gatti, G. M. Frigo, A. Crema, S. Calzetti, D. Visintini,* Disposition of Valproate in Epileptic Patients, Br. J. Clin. Pharmacol. *5*, 495 – 499 (1978).

[11] *F. Schobben, E. van der Kleijn, T. B. Vree,* Therapeutic Monitoring of Valproic Acid, Ther. Drug Monit. *2*, 61 – 71 (1980).

[12] *J. B. Isom,* On the Toxicity of Valproic Acid, Am. J. Dis. Child. *138*, 901 – 903 (1984).

[13] *J. A. Cramer, D. M. Bennett, R. H. Mattson,* Free and Bound Valproic Acid Separated by two Methods of Ultrafiltration, Clin. Chem. *29*, 1441 – 1442 (1983).

[14] *J. A. Cramer, K. A. McCarthy, R. H. Mattson,* Free Fraction of Valproic Acid is the Same in Serum and Plasma, Clin. Chem. *29*, 1449 (1983).

[15] *D. Haidukewych,* Fluorescence Polarization Immunoassay and Enzyme-immunoassay Compared for Free Valproic Acid in Serum Ultrafiltrates from Epileptic Patient, Clin. Chem. *31*, 156 (1985).

[16] *D. Kuschak, D. Roffhack, A. Jacobs-Sturm, K. V. Toyka, H. Reinauer,* Analysis of Total and Free Valproic Acid in Serum/Plasma by Gaschromatography and Enzyme-immunoassay, J. Clin. Chem. Clin. Biochem. *19*, 743 (1981).

[17] *R. Riva,* Valproic Acid Free Fraction in Epileptic Children under Chronic Monotherapy, Ther. Drug Monit. *5*, 197 – 200 (1983).

[18] *T. A. Bowdle, I. H. Patel, A. J. Wilensky, C. Comfort,* Hepatic Failure from Valproic Acid, N. Engl. J. Med. *301*, 435 – 436 (1979).

[19] *D. M. Turnbull,* Plasma Concentrations of Sodium Valproate: Their Clinical Value, Ann. Neurol. *14*, 38 – 42 (1983).

[20] *H. J. Zimmermann, K. G. Ishak,* Valproate-induced Hepatic Injury: Analysis of 23 Fatal Cases, Hepatology *2*, 591 – 597 (1982).

[21] *E. Kingsley, P. Gray, K. G. Tolman, R. Tweedale,* The Toxicity of Metabolites of Sodium Valproate in Cultured Hepatocytes, J. Clin. Pharmacol. *23*, 178 – 185 (1983).

[22] *R. L. Boeckx,* Valproic Acid: Physical and Pharmacological Properties, in: *T. P. Moyer, R. L. Boeckx* (eds.), Applied Therapeutic Drug Monitoring, Vol. *II*, Am. Assoc. Clin. Chem., Washington 1984, pp. 79 – 81.

[23] *A. G. Chapman, B. S. Meldrum, E. Mendes,* Acute Anticonvulsant Activity of Structural Analogues of Valproic Acid and Changes in Brain GABA and Aspartate Content, Life Sci. *32*, 2023 – 2031 (1983).

[24] *C.-M. Becker, R. A. Harris,* Influence of Valproic Acid on Hepatic Carbohydrate and Lipid Metabolism, Arch. Biochem. Biophys. *223*, 381 – 392 (1983).

[25] *P. M. Jeavons,* Sodium Valproate and Neural Tube Defects, Lancet *II*, 1282 – 1283 (1982).

[26] *M. Eichelbaum, G. E. von Unruh, A. Somogyi,* Application of Stable Labelled Drugs in Clinical Pharmacokinetic Investigations, Clin. Pharmacokinet. *7*, 490 – 507 (1982).

[27] *D. M. Turnbull, A. J. Bone, K. Bartlett, P. P. Koundakjian, H. S. A. Sheratt,* The Effects of Valproate on Intermediary Metabolism in Isolated Rat Hepatocytes and Intact Rats, Biochem. Pharmacol. *32*, 1887 – 1892 (1983).

[28] *K. E. Rubenstein, R. S. Schneider, E. F. Ullman,* "Homogeneous" Enzyme-immunoassay, A New Immunochemical Technique, Biochem. Biophys. Res. Commun. *47*, 846 – 851 (1972).

[29] *S. L. Braun, A. Tausch, W. Vogt, K. Jacob, M. Knedel,* Evaluation of a New Valproic Acid Enzyme-immunoassay and Comparison with Capillary Gas-chromatographic Method, Clin. Chem. *27,* 169 – 172 (1981).

[30] *H. Grote, L. Herbertz, D. Kuschak, M. Meiertoberens, R. Reinauer,* Quantitative Bestimmung von Valproinsäure aus Serum mit Hilfe der Dünnfilmglaskapillar-Gaschromatographie, J. Clin. Chem. Clin. Biochem. *17,* 157 (1979).

[31] *C. J. Jensen, R. Gugler,* Sensitive Gas-liquid Chromatographic Method for Determination of Valproic Acid in Biological Fluids, J. Chromatogr. *137,* 188 – 193 (1977).

[32] *H. J. Kupferberg,* Gas-liquid Chromatographic Quantitation of Valproic Acid, in: *C.E. Pippeneger, J. K. Penry, H. Kutt* (eds.), Antiepileptic Drugs: Quantitative Analysis and Interpretation, Raven Press, New York 1978, pp. 147 – 151.

[33] *M. Puuka, R. Puuka, M. Reunane,* A Rapid and Simple Gas-liquid Chromatographic Determination of Valproic Acid (α-Propyl Valeric Acid) in Serum, J. Clin. Chem. Clin. Biochem. *18,* 497 – 499 (1980).

1.10   Digoxin, Digitoxin

(3β,5β,12β)-3-[(*O*-2,6-Dideoxy-β-D-*ribo*-hexopyranosyl-(1→ 4)-*O*-2,6-dideoxy-β-D-*ribo*-hexopyranosyl-(1→ 4)-2,6-dideoxy-β-D-*ribo*-hexopyranosyl)oxy]-12,14-dihydroxycard-20(22)-enolide

(3β,5β)-3-[(*O*-2,6-Dideoxy-β-D-*ribo*-hexopyranosyl-(1→ 4)-*O*-2,6-dideoxy-β-D-*ribo*-hexopyranosyl-(1→ 4)-2,6-dideoxy-β-D-*ribo*-hexopyranosyl)oxy]-14-hydroxycard-20(22)-enolide

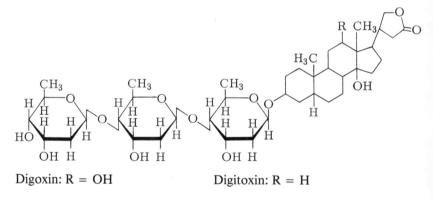

Digoxin: R = OH   Digitoxin: R = H

Gerd Kleinhammer and Hans Mattersberger

General

Digitalis preparations have been used in the treatment of "dropsy" for the first time 200 years ago, in 1785, by William Withering [1]. Today, digoxin and digitoxin are the most frequently used cardiac glycosides. Both glycosides consist of an aglycone or genin structure to which three sugars are attached. The pharmacological activity of increasing the force of myocardial contraction resides in the steroid nucleus. The sugars modify the water solubility, cell penetrability and potency of the resulting glycoside.

Cardiac glycosides are still irreplaceable in treatment of chronic cardiac insufficiency, though they are extremely poisonous. Intoxication with digoxin or digitoxin and their derivatives severely affects not only cardiac function but also the central nervous system. Before the correlation between serum digitalis concentration and these side-effects had been established, signs of intoxication could be observed in 20% of hospitalized patients under treatment with cardiac glycosides [2].

Digoxin is hepatically metabolized to the extent of only ca. 30% giving cardioactive cleavage products (digoxigenin, digoxigenin mono- and bis-digitoxoside) and the cardio-inactive derivative dihydrodigoxin. Up to 70 – 80% of digoxin is eliminated by renal excretion, whereas digitoxin is degraded in the liver mainly to inactive genins.

All cardiac glycosides act in the same way. They differ in strength and pharmacokinetic characteristics. The distinctly different pharmacokinetic behaviour of digoxin and digitoxin is due to the additional OH-group of digoxin, which makes it less lipophilic. Today it is generally accepted that about 24% of digoxin and 90 – 97% of digitoxin in serum is bound to albumin. Albumin is the only protein of importance in binding digoxin and digitoxin. The apparent association constants at 37 °C, as reported by *Lukas et al.* [3], are: $9 \times 10^2 \, 1 \times \text{mol}^{-1}$ for digoxin and $9.62 \times 10^4 1 \times \text{mol}^{-1}$ for digitoxin. This clear difference is reflected, e.g., in the half-lives of these compounds in man: 2 days for digoxin but 4 to 6 days for digitoxin.

Peak serum concentrations after oral administration of digoxin or digitoxin can be measured after 1 to 4 hours and 4 to 6 hours, respectively. A considerable variation has been observed from patient to patient in the peak level achieved and the rate of increase.

Cardiac glycosides are present in serum to only a minor extent and are stored mainly in different body tissues in varying degrees. It is assumed that inter-individual variations in the respective equilibria of partition are responsible for these observations. When interpreting serum digoxin concentrations certain factors have to be taken into consideration. The pharmacokinetics of digoxin may be altered due to renal impairment or drug interactions (particularly the interaction of digoxin with quinidine) [4]. Moreover there are other factors which affect the pharmacological effect of digoxin, such as disturbances of electrolyte balance (particularly hypokalaemia) and abnormalities of thyroid function [4]. For example, a usually therapeutic digoxin serum concentration can be associated with toxic symptoms, if the serum potassium concentration of the patient is below the reference range. Therefore the whole clinical picture of a patient receiving cardiac glycosides has to be considered when digoxin serum concentrations are interpreted. Indications for monitoring digoxin serum concentrations are: suspected overdosage, lack of therapeutic effect, medication in patients with altered pharmacokinetics, unknown pre-medication of cardiac glycosides (particularly in unconscious patients) and ECG alterations which do not allow the detection of digitalis effects (e.g. in patients with pacemakers).

Application of method: in clinical chemistry, pharmacology and toxicology.

Substance properties relevant in analysis

Digoxin: $M_r = 780.92$
Digitoxin: $M_r = 764.92$

Digoxin and digitoxin differ in position C12 of the aglycone part, in which digoxin carries an hydroxyl-group whereas digitoxin is unsubstituted. Both substances are very poorly soluble in water but soluble in diluted alcohol and chloroform.

Methods of determination: before radioimmunoassay procedures were introduced no simple, sensitive and rapid methods were available for routine measurement of serum digitalis concentrations. Earlier methods were based on the effect that cardiac glycosides inhibit the uptake of ^{86}Rb by erythrocytes [5]. The radioimmunoassay for digoxin was introduced in 1969 by *Smith et al.* [6]. Comprehensive reviews of the different radioimmunoassay approaches for digoxin and digitoxin appeared in [7] and [8]. The first enzyme-immunoassays for digoxin were presented in 1975 (homogeneous [9]) and in 1976 (heterogeneous [10]). Further digoxin enzyme-immunoassays have been described by *Monji et al.* [11] and *Healey et al.* [12]. A new immunochemical technique has been introduced recently: fluorescence polarization immunoassay [13]. After depletion of proteins in order to eliminate disturbances by quenching effects, serum samples can conveniently be assayed for digoxin with good reliability with this method. In 1984 *Litchfield et al.* [14] described a homogeneous liposome-mediated immunoassay.

However, enzyme-immunoassays have not been widely applied to digitoxin assay. Based on the observation that digitoxin may cross-react significantly with anti-digoxin antiserum, the possibility of measuring digitoxin-containing samples with the use of digoxin assay reagents was investigated [15]. Also, the successful development of a fluoroimmunoassay for digitoxin was reported [16]. Recently a newly developed heterogeneous enzyme-immunoassay for digitoxin was presented [17].

Digoxin has been determined by gas chromatography and by a combination of HPLC and radioimmunoassay. The latter technique has proved to be of special value in the quantitative analysis of digoxin metabolites in human serum.

International reference method and standards: no reference method is known so far. The only attempt at standardization at the international level may be seen in the availability of the standard reference materials for digoxin and digitoxin (cf. Vol. II, p. 409, 410 for details).

1.10.1   Determination with Enzyme-linked Immunosorbent Assay

Assay

Method Design

The assays are based on the methods published by *Kleinhammer et al.* in 1976 [10] and *Mattersberger et al.* [17].

Principle

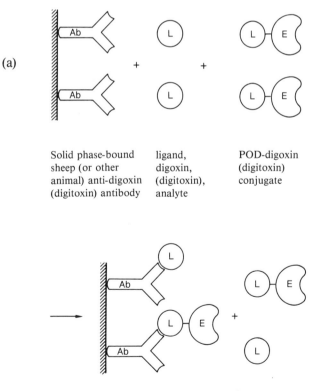

(a)

| Solid phase-bound sheep (or other animal) anti-digoxin (digitoxin) antibody | ligand, digoxin, (digitoxin), analyte | POD-digoxin (digitoxin) conjugate |

POD-labelled ligand-Ab complex

(b)

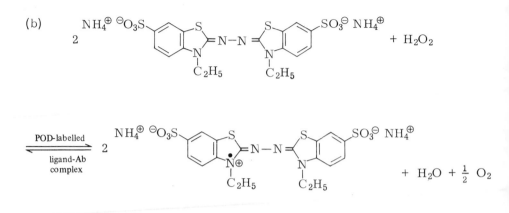

The assays described in this section follow the principle of competition between labelled and unlabelled ligand for a limited number of antibody binding sites. Polystyrene tubes with adsorptively bound antibodies provide the solid phase for separation of free from antibody-bound conjugate, after an appropriate period of simultaneous incubation with labelled and unlabelled ligand. The enzymatic activity in the antibody-bound fraction is quantitated after one washing step. Enzyme activity measured as increase in absorbance per unit time, $\Delta A/\Delta t$, and ligand concentration are inversely related in this assay.

Horseradish peroxidase (POD)* is used as the label. The conjugated enzyme catalyzes the oxidation of ABTS®** by H_2O_2.

Selection of assay conditions and adaptation to the individual characteristics of the reagents: *Butler & Chen* [18] were the first successfully to generate antibodies against digoxin. Their basic approach was to use the terminal digitoxose residue for derivatization and subsequent coupling to bovine serum albumin. To be useful, an anti-digoxin antiserum has to exhibit its highest specificity against that region of the digoxin molecule which is formed by C-atoms 12 to 23. Most of the practically relevant cross-reacting substances in digoxin immunoassays are distinguishable by that epitope: digitoxin, canrenoate and dihydrodigoxin.

Antisera have been raised in sheep against immunogens in which digoxin or digitoxin is covalently linked at the terminal digitoxose *via* a glutaryl residue as spacer to primary amino groups of the protein. Coupling to POD is achieved by the same chemical reactions.

For the digoxin or digitoxin EIA, polystyrene tubes, 10 mm × 40 mm, are used as the solid phase. The antiserum dilution used for coating is adjusted to give a detection limit in the digoxin assay of ca. 0.3 µg/l, or 3 µg/l in the digitoxin assay, and a good sensitivity in the respective upper therapeutic ranges (cf. p. 124). A useful guideline is

---

* Peroxidase, donor: hydrogen peroxide oxidoreductase, EC 1.11.1.7.
** ABTS®, diammonium 2,2'-azinobis-[3-ethyl-2,3-dihydrobenzothiazole-6-sulphonate].

that analyte concentration which diminishes the signal for the zero standard by 50%. This value should lie between 2.0 and 2.5 µg/l for digoxin or 15 to 25 µg/l for digitoxin, respectively, in order to fulfil the above requirements.

The enzyme-conjugate concentration is optimized to give an enzyme activity corresponding to $\Delta A/\Delta t$ at 405 nm in the absence of an analyte of approximately 1.5 against the substrate solution. In order to make the assay rapid and simple the time for the immunoreaction as well as for the enzymatic reaction was fixed arbitrarily at 30 minutes at ambient temperature. The appropriate range of conjugate concentration for these conditions is 24 to 40 µg/l.

The dependence of the immunoreaction, measured as the binding of the POD-labelled ligand in the absence of analyte, on the pH of the reaction mixture is shown in Fig. 1. In view of this relationship the incubation buffer has been adjusted to a neutral pH.

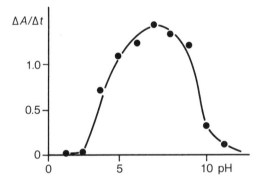

Fig. 1. Influence of pH on binding of digoxin-POD conjugate to solid-phase anti-digoxin antibody in the absence of unlabelled digoxin.

The POD-catalyzed oxidation of ABTS® in the presence of H_2O_2 is strongly pH-dependent with a sharp optimum at pH 4.0. At this pH occasional precipitation of bovine serum albumin remaining from the tube-preparation process occurred. Therefore, a slightly more alkaline pH of 4.4 had to be chosen. The rate of the reaction further depends on the concentrations of H_2O_2 and ABTS®. The concentrations of both components were finally fixed considering not only maximal reaction rate, but also the practical requirement for a low blank value of the reaction mixture at 405 nm against water. Increase in absorbance at Hg 405 nm or 420 nm, $\Delta A/\Delta t$, is determined photometrically by a two-point kinetic measurement in order to quantitate the activity of the enzyme.

Both the immuno- and enzymatic reactions are terminated far from the equilibrium state or end-point, respectively. Incubation times for all determinations included in one assay have to be kept constant. The assay system is temperature dependent. If the immuno- and enzymatic reactions are carried out at 37 °C, $\Delta A/\Delta t$ is approximately twofold higher than at 25 °C for both incubations. The assay is optimized for the range between 20 °C and 25 °C.

Equipment

Spectrophotometer or spectral-line photometer capable of exact measurement at 420 nm or Hg 405 nm; semi-micro cuvette with 10 mm light path, flow-through and suction cuvettes have proved to be especially useful; vacuum source; *Pasteur* pipette or a similar device for aspiration of the incubation mixture; plunger-type pipettes with disposable tips of 1000, 100 µl and 20 µl volume; wash-bottle as container for the washing water with 1 litre volume; stopwatch.

Reagents and Solutions

Purity of reagents: horseradish peroxidase should be of grade I, RZ $\triangleq$ 3, 1000 U/mg (25 °C, ABTS® and H_2O_2 as substrates).

Chemicals are of analytical grade. Digoxin should be of USP XVIII and digitoxin USP XX quality.

Preparation of solutions* (for about 100 determinations): all solutions in re-purified water (cf. Vol. II, chapter 2.1.3.2).

1. Incubation buffer (phosphate, 40 mmol/l; BSA, 2 g/l; Merthiolate, 0.1 g/l; pH 6.8):

   dissolve 426 mg Na_2HPO_4, 153 mg $NaH_2PO_4 \cdot 2 H_2O$, 200 mg bovine serum albumin and 10 mg Merthiolate in 100 ml water.

2. Conjugate solutions (POD, 20 U/l; phosphate, 40 mmol/l; BSA, 2 g/l; Merthiolate, 0.1 g/l; pH 6.8):

   use the preparations according to Appendix, p. 131. Dilute the stock solutions with incubation buffer (1) to the concentrations resulting from the experimental optimization processes.

3. Substrate solution (phosphate, 60 mmol/l; citrate, 40 mmol/l; $NaBO_2 \cdot H_2O_2$, 3.25 mmol/l; ABTS®, 1.7 mmol/l; pH 4.4):

   dissolve 841 mg citric acid, monohydrate, 1068 mg $Na_2HPO_4 \cdot 2 H_2O$, 50 mg $NaBO_2 \cdot H_2O_2 \cdot 3 H_2O$ and 95 mg ABTS® in 100 ml water.

4. Digoxin standard solutions (0.75 to 5.0 µg/l):

   dissolve 5 mg digoxin USP XVIII in 10 ml ethanol.

   Dilute this stock solution to a concentration of 100 µg/l with buffer (1). Dilute the resulting solution with digoxin-free human serum (14) to yield digoxin concentrations of 5.0, 3.0, 1.5 and 0.75 µg/l. Digoxin-free serum serves as zero calibrator.

---

* Reagents are commercially available, e.g. from *Boehringer Mannheim*.

Do not use NaN_3 as preservative for serum because this substance strongly inhibits the enzyme activity of POD.

5. Digitoxin standard solutions (7.5 to 60.0 µg/l):

dissolve 40 mg digitoxin USP XX with ethanol/water 80:20 (v/v) in a 1 l volumetric flask. Dilute this solution further to a digitoxin concentration of 4 mg/l with ethanol/water 20:80 (v/v). Dilute this stock solution with buffer (1) to a concentration of 0.4 mg/l. Prepare the working solutions by serial dilution of the stock solution with digitoxin-free human serum (14) to yield concentrations of 7.5, 15.0, 30.0 and 60.0 µg/l. Digitoxin-free human serum is used as the zero calibrator.

6. Phosphate buffer (40 mmol/l, pH 7.2):

dissolve 1.08 g $Na_2HPO_4 \cdot 12\ H_2O$ and 160 mg $NaH_2PO_4 \cdot 2\ H_2O$ in 100 ml water.

7. Anti-digoxin antibody solution (IgG, 2 mg/l):

use preparation according to Appendix, p. 131. Dilute stock solution with buffer (5) according to the result of the experimental optimization.

8. NaCl/BSA solution (NaCl, 0.15 mol/l; BSA, 3 g/l):

dissolve 1.35 g NaCl and 0.45 g bovine serum albumin in 150 ml water.

9. Coated tubes:

prepare according to Appendix, p. 131.

10. Phosphate buffer (0.1 mol/l, pH 8.0):

dissolve 0.5 g KH_2PO_4 and 17.15 g $Na_2HPO_4 \cdot 2\ H_2O$ in 1 l water.

11. Tris/HCl buffer (40 mmol/l, pH 7.5):

dissolve 4.84 g tris(hydroxymethyl)aminomethane in 800 ml water; adjust pH to 7.5 with HCl, 1.0 mol/l; make up with water to 1 litre.

12. Dialysis buffer (phosphate, 15 mmol/l; NaCl, 50 mmol/l; pH 7.0):

dissolve 1.63 g $Na_2HPO_4 \cdot 2\ H_2O$, 0.81 g $NaH_2PO_4 \cdot H_2O$ and 2.92 g NaCl in 1 l water.

13. K_2CO_3 solution (0.1 mol/l):

dissolve 1.38 g K_2CO_3 in 100 ml water.

14. Standard matrix (fresh human serum; Merthiolate, 0.1 g/l):

dissolve 10 mg Merthiolate in 100 ml fresh human serum from individuals not treated with digoxin or digitoxin.

Stability of reagents and solutions: store all solutions in the refrigerator between 0 °C and 4 °C when not in use. Solutions (2) and (7) are used on the day of preparation. Coated tubes are stable for at least one year if kept at 4 °C. All other solutions are stable for at least 3 months if growth of micro-organisms is avoided.

Procedure

Collection and treatment of specimen: serum should be used in this assay; plasma samples show falsely elevated results. Digoxin in urine is present in a hundredfold higher concentration than in plasma. Urine therefore requires an appropriate pre-dilution. For therapeutic drug monitoring blood should not be taken earlier than 8 to 12 h after the last glycoside administration.

Stability of the substance in the sample: digoxin and digitoxin are stable in serum for 2 weeks at room temperature. At 4 °C samples are stable for 3 months and for more than 6 months if kept at − 20 °C.

Assay conditions: standards and samples are measured in duplicate under identical conditions (within series).

Immunoreaction: incubation for 30 min between 20 °C and 25 °C. During the incubation period the temperature should remain constant and the same for all assay tubes.

Enzymatic indicator reaction: incubation for 30 min at 25 °C. Avoid direct exposure to sunlight. Wavelength 405 nm; light path 10 mm; measurements against substrate solution (3).

If $\Delta A/\Delta t$ of the sample is lower than $\Delta A/\Delta t$ of the highest standard dilute with physiological saline and repeat the determination.

Calibration curve: calculate and plot mean $\Delta A/\Delta t$ readings of the standard measurements (ordinate) against the respective analyte concentrations (abscissa) on linear graph paper. Draw the calibration curve through the points. A typical example is given in Fig. 2.
   A list of suitable mathematical curve-fitting methods can be found in Vol. I, p. 243. The use of the spline approximation is recommended.

Calculation: for the mean $\Delta A/\Delta t$ readings of the samples take the analyte mass concentration, ρ (µg/l), from the calibration curve. For conversion from mass concentration to the substance concentration, c (µmol/l), multiply by 1.2805 for digoxin and 1.3073 for digitoxin.

Measurement

Pipette successively into an antibody-coated tube (9):			concentration in incubation/ assay mixture	
sample or standard solution (4)* conjugate solution** (2)	0.1 ml 1.0 ml		digoxin POD phosphate BSA	up to 0.5 µg/l 5 to 30 U/l 36 mmol/l 1.8 g/l
cover the tubes, incubate for 30 min; aspirate vial contents, discard and rinse vials once with 2 ml water; remove the water within 5 min as completely as possible;				
substrate solution** (3)	1.0 ml		H_2O_2 ABTS® phosphate citrate	3.25 mmol/l 1.9 mmol/l 60 mmol/l 40 mmol/l
pipette at constant time intervals between successive tubes, e.g. every 15 s; cover the tubes, mix, incubate for 30 min; zero the photometer against substrate solution (3); mix again thoroughly; read absorbance at the same time intervals as have been used for addition of substrate solution***.				

  * Or 0.02 ml digitoxin standard solution (5), up to 1.2 µg/l.
 ** In equilibrium with the laboratory temperature!
*** For sake of convenience the use of NaN_3 as stopping reagent has successfully been applied [19].

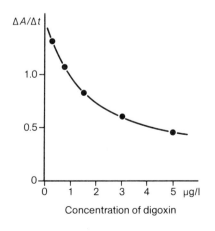

Fig. 2. Typical calibration curve for the assay of digoxin in serum.

Validation of Method

Precision, accuracy, detection limit and sensitivity: the between-days relative standard deviations (RSD) in the range of digoxin concentrations from 0.84 µg/l to 3.78 µg/l were between 6.3% for a concentration of 3.69 µg/l and 17.8% for 0.84 µg/l [20]. The accuracy has been proved by recovery experiments on human sera which were spiked with digoxin in the range from 1 to 2 µg/l. Recovery rates between 99 and 106% have been obtained. An acceptable correlation with well established digoxin radioimmunoassays has been demonstrated, e.g., y = 1.09x − 0.13 µg/l (y: digoxin-EIA, x: digoxin-RIA [21]).

The detection limit of the digoxin assay determined according to *Kaiser* [22] as the mean value plus the 3-fold standard deviation of a series of analyte-free human sera was calculated to be 0.3 µg/l. The sensitivity (Δc corresponding to the within-series imprecision of $\Delta A / \Delta t$ readings at the same concentration) at the 0.95 confidence level is 0.25 µg/l at a digoxin concentration of 2 µg/l.

The between-days relative standard deviation (RSD) for the digitoxin assay was found between 7.3% at a concentration of 7 µg/l and 9.4% at a concentration of 36 µg/l.

Accuracy was tested by spiking human sera with digitoxin. Recoveries between 100 and 106% were found. The detection limit determined according to *Kaiser* [22] was calculated to be 1.2 µg/l. The sensitivity of the assay was determined as 2.8 µg/l at a digitoxin concentration of 24 µg/l.

Table 1. Specificity of sheep anti-digoxin antibody (lot 01).

Substance	% Cross-reaction*	Substance	% Cross-reaction
Digoxin	100	Digitoxin	32
Dihydrodigoxin	32	k-Strophanthin	30
β-Methyldigoxin	100	Spironolactone	0.01
α-Acetyldigoxin	97	Canrenone	< 0.001
β-Acetyldigoxin	105	Cortisol	< 0.001
Lanatoside C	92	Prednisone	< 0.001
Digoxigenin	86	Progesterone	< 0.001
Digoxigenin-mono-		Proscillaridin	18
digitoxoside	88	Methyl-	
Digoxigenin-bis-		proscillaridin	11
digitoxoside	92	Oestradiol	< 0.001
		Oestriol	< 0.001
		Testosterone	< 0.001

* Cross-reaction: ratio of the molar concentrations of digoxin and the test substance effecting 50% displacement of digoxin-POD-conjugate under routine assay conditions.

Sources of error: the assays are affected by haemolysis, but not by lipaemia, bilirubin and uric acid in physiological concentrations [23]. A disturbance by high albumin concentrations has not been observed, although occasionally reported for radioimmunoassays. Plasma is unsuitable as sample material, since erroneously elevated analyte concentrations are obtained in comparison with serum samples. Presumably fibrinogen is the factor responsible for this disturbance.

Specificity: the specificity of the antibodies utilized is illustrated by Tables 1 and 2, which list the measured cross-reactivities.

Table 2. Specificity of sheep anti-digitoxin antibody.

Substance	% Cross reaction*	Substance	% Cross-reaction
Digitoxin	100	Digitoxigenin-bis-digitoxoside	80.0
Digoxin	1.5	Prednisolone	< 0.02
β-Methyldigoxin	1.3	Spironolactone	< 0.02
α-Acetyldigoxin	1.3	Canrenone	0.03
β-Acetyldigoxin	1.2	Prednisone	< 0.02
Lanatoside C	1.0	Progesterone	< 0.02
Digoxigenin	1.5	Dihydrodigoxin	0.02
Digoxigenin-mono-digitoxoside	1.8	Proscillaridin	3.1
Digoxigenin-bis-digitoxoside	1.5	K-Strophantin-β	41.1
Digitoxigenin	86.8	Cymarin	52.9
Digitoxigenin-mono-digitoxoside	98.7	Cortisol	< 0.02
		Ouabain	0.6

* Cross-reaction: ratio of the molar concentrations of digoxin and the test substance effecting 50% displacement of POD-digoxin conjugate under routine assay conditions.

Table 3. Comparison of 2 anti-digoxin antibodies. Digoxin has been determined by the heterogeneous enzyme-immunoassay in serum samples collected from 6 digitalized patients 30 and 60 minutes after administration of 400 mg canrenoate i.v.

Time after canrenoate administration (min)	Canrenone mg/l	Canrenoic acid mg/l	Digoxin measured with			
			Antiserum I*		Antiserum II**	
			μg/l	% of basic value	μg/l	% of basic value
0	0	0	1.3	100	1.1	100
30	1.3	26.8	1.5	115	2.6	236
60	1.2	14.3	1.5	115	2.2	200

 * Cross-reaction with canrenone: ≪ 0.001%.
** Cross-reaction with canrenone: 0.025%.

Due to the cross-reactivity of the digoxin antibody with digitoxin it is not possible to quantitate digoxin in the presence of therapeutic levels of digitoxin. The antibody does not discriminate between digoxin, β-methyl-, or α- and β-acetyldigoxin. Hence it follows that these glycosides can be measured in the assay with digoxin standards.

Special attention has to be given to the cross-reactivity with spironolactone and its metabolites. Canrenone and canrenoate are the main ones of at least 9 metabolites, which are rapidly formed after intravenous application. After a single dose of 200 mg spironolactone the concentration of canrenone can exceed the digoxin concentration by a factor of 1000 on a molar basis. Falsely elevated digoxin concentrations will be measured in serum samples prepared from blood collections soon after administration of spironolactone or canrenoate, if the antibody does not discriminate specifically enough between digoxin and canrenone. Table 3 illustrates this by comparing two anti-digoxin antisera. Furthermore, an endogenous digoxin-like substance can cause falsely-elevated digoxin concentrations [24].

Therapeutic ranges: digoxin concentrations from 0.8 to 2.0 µg/l and digitoxin concentrations from 13 to 25 µg/l are considered to be in the therapeutic range. If the digoxin concentration exceeds 3.0 µg/l, more than 90% of patients show signs of intoxication.

1.10.2   Determination with Enzyme-multiplied Immunoassay Technique

Assay

Method Design

The assay follows the basic EMIT® (enzyme-multiplied immunoassay technique) principle first published by *Rubenstein* [25]. The enzyme-inhibition mechanism, however, is not the steric mechanism described by *Rubenstein* but the conformational mechanism described by *Rowley et al.* [26]. Adaptation to the determination of digoxin was published in 1975 by *Chang et al.* [9].

Principle: as described in detail on pp. 7, 8.

G6P-DH is labelled with digoxin. If anti-digoxin antibodies bind to the attached haptens, the specific activity of the enzyme is diminished. Digoxin in the assay mixture derived from the sample binds to antibodies according to its concentration. Antibody-digoxin complexes will not inhibit the enzyme activity. The G6P-DH activity measured as increase in absorbance per unit time, $\Delta A/\Delta t$, is related to the digoxin concentration in the sample.

Selection of assay conditions and adaptation to the individual characteristics of the reagents: the assay is designed for rapid analysis. No separation step is required. Results on samples from patients are available within about 45 min.

G6P-DH-digoxin conjugate – the key reagent of this assay – is obtained starting from the carboxymethyl oxime of digoxigenin which is coupled to G6P-DH *via* a mixed anhydride reaction using carbitol [2-(2-ethoxyethoxy)ethanol] (cf. Appendix, p. 132). Bacterial G6P-DH (from *Leuconostoc mesenteroides*) has been chosen as the enzyme label. Since this enzyme employs NAD as a cofactor, interference from the endogenous NADP-dependent human enzyme is eliminated.

The G6P-DH-digoxin conjugate has to be optimized with regard to the inhibitability of its enzyme activity upon antibody binding. This is checked *in situ* during the coupling reaction as the molar ratio of added hapten derivative to enzyme increases (cf. Appendix, p. 132 for details). Conjugates having an inhibitability in the presence of antibody excess of about 50% and a de-activation compared to the unconjugated enzyme in the absence of antibody of not more than ca. 35% give the most sensitive calibration curves.

Serum and plasma samples must be pre-treated with sodium hydroxide and a detergent in order to eliminate interferences from endogenous enzyme activity.

The determination comprises three reaction phases: a) denaturation of interfering substances of the sample; b) binding of sample digoxin to a defined amount of anti-digoxin antibody (immunoreaction I); c) binding of the antibody remaining uncomplexed during step b) to a defined amount of G6P-DH conjugate (immunoreaction II), and G6P-DH catalyzed formation of NADH (enzymatic indicator reaction). Thus, immunoreaction II, leading to G6P-DH inhibition, and the enzymatic reaction run at the same time. This explains the observed changes in rate of enzyme reaction for this type of homogeneous enzyme-immunoassay [27]. Absorbance measurements take place at the beginning and at the end of phase c). The basic procedure is easily adapted to common automated clinical analyzers. The recommended timing for the manual procedure is 5 minutes for step a), 15 minutes for step b) and 30 minutes for step c).

The enzyme activity is determined by measuring the increase in NADH absorbance at 334 or 339 nm per unit time, $\Delta A/\Delta t$.

The optimal quantities and concentrations of sample and reagents are dependent on the characteristics of the antibodies present in the antiserum.

The optimal enzyme conjugate concentration is based upon setting the maximum reaction rate at a value of $\Delta A/\Delta t = 0.900$ (339 nm, 30°C, $\Delta t = 30$ min). This value was arrived at because it was desirable to have the final reading not higher than $\Delta A/\Delta t = 1.200$ of the calibrator with 8.0 µg/l.

Loading studies (addition of increasing amounts of antibody to a constant amount of enzyme conjugate) are conducted using the digoxin standards with 0.5 µg/l and 2.0 µg/l. These samples are assayed according to the regular assay protocol (cf. p. 128). Initial readings are subtracted from their corresponding final readings. $\Delta A/\Delta t$ values are plotted *vs.* antibody concentration for both standards. The difference in $\Delta A/\Delta t$, or separation, between the standards with 0.5 µg/l and 2.0 µg/l at each level of antibody loading is a measure of sensitivity at that particular loading. Obviously, loadings which result in large separations between the two calibrators are most desirable.

Further analysis involves the generation of complete calibration curves using all six calibrators exactly according to the assay procedure. That level of antibody which gives rise to the steepest calibration curve (i.e. largest separations between calibrators) is chosen.

Equipment

Spectrophotometer or spectral-line photometer capable of exact measurement at $30 \pm 0.1\,°C$ at 339 nm or Hg 334 nm; semi-micro suction cuvette; stopwatch; water-bath or constant temperature block at $30 \pm 0.1\,°C$; repeating dispenser for 1.25 ml; pipetter-diluter capable of delivering 50 ± 1 µl sample and 500 ± 5 µl buffer with a relative standard deviation of less than 0.5%; mixing device for 12 mm × 75 mm reaction vials; 12 mm × 75 mm glass tubes; tube rack; micro-pipette with disposable tips for 200 µl; dispensing syringe (2.5 ml).

Reagents and Solutions

Purity of reagents: G6P-DH from *Leuconostoc mesenteroides* should be of the best available purity (ca. 550 U/mg at 25°C, G-6-P and NAD as substrates). Chemicals are of analytical grade.

Preparation of solutions* (for 70 determinations): all solutions in re-purified water (cf. Vol. II, chapter 2.1.3.2).

1. Tris buffer (Tris, 55 mmol/l; NaCl, 85.5 mmol/l; NaN$_3$, 0.5 g/l; Thimerosal**, 50 mg/l; Triton X-100, 5 ml/l; pH 7.4):

   dissolve 1.996 g Tris base, 1.5 g NaCl, 150 mg NaN$_3$ and 15 mg Thimerosal in 150 ml water; add 1.5 ml Triton X-100; adjust to pH 7.4 with HCl, 1.0 mol/l. Make up with water to 300 ml.
   Dilute buffer concentrate provided with each kit to 225 ml with water.

2. Antibody/substrate solution (γ-globulin***, ca. 500 mg/l; G-6-P, 130 mmol/l; NAD, 80 mmol/l; rabbit serum albumin, 10 g/l; Tris, 55 mmol/l; NaN$_3$, 0.5 g/l; Thimerosal, 50 mg/l; pH 5.0):

   reconstitute available lyophilized preparation with 4.0 ml water.

---

* Reagents are commercially available from *Syva Co.,* Palo Alto, U.S.A.
** Thiomersal, Merthiolate® : 2-ethylmercurithiobenzoic acid, sodium salt.
*** Standardized preparation from immunized sheep.

3. Conjugate solution (G6P-DH*, ca. 1 kU/l; Tris, 55 mmol/l; pH 7.9; rabbit serum albumin, 10 g/l; NaCl, 0.15 mol/l; NaN$_3$, 0.5 g/l; Thimerosal, 50 mg/l):

reconstitute available lyophilized preparation with 5.0 ml water. This reagent must be standardized to match the antibody/substrate solution (2) (cf. p. 125).

4. Digoxin standard solutions (0.5 µg/l to 8.0 µg/l):

dissolve 10 mg digoxin in 10 ml dimethylformamide. Dilute this solution 1 + 99 with dimethylformamide to give a digoxin stock solution of 10 mg/l. Dilute stock solution 1 + 99 with solution (1), in order to obtain a working dilution with a concentration of 100 µg/l. Dilute this solution further with digoxin-free human serum (containing 0.5 g NaN$_3$ per litre and 50 mg Thimerosal per litre) 1 + 11.5, 1 + 24, 1 + 49, 1 + 99 and 1 + 199, yielding the following digoxin concentrations: 8.0, 4.0, 2.0, 1.0, 0.5 µg/l. Take pure serum as zero calibrator. Reconstitute each of the available lyophilized standard preparations with 3 ml water.

5. Pre-treatment solution (NaOH, 0.5 mol/l):

5 ml of a ready-to-use solution are supplied with each kit.

6. Tris buffer (55 mmol/l, pH 8.1):

dissolve 1.996 g Tris base in 150 ml water; adjust pH to 8.1 with HCl, 1 mol/l. Make up with water to 300 ml.

Stability of solutions: after reconstitution the antibody/substrate solution (2), conjugate solution (3) and standards (4) must be stored at room temperature (20°C to 25°C) for at least one hour before use. The reagents are then stable for 12 weeks, if stored at 2°C to 8°C. Solutions (1), (5) and (6) can be used for 12 weeks when stored at room temperature.

Procedure

Collection and treatment of specimen: serum or plasma (anticoagulants: heparin, EDTA, oxalate) can be used as sample material in this assay. For further information about specimen handling cf. p. 120.

Stability of the substance in the sample: cf. p. 120.

Assay conditions: standards and samples are measured in duplicate under identical conditions (within series). The zero calibrator assay serves as a reagent blank; all measurements are corrected for this value.

---

* Coupled to 3-oxodigoxigenin carboxymethoxine.

Pre-treatment reaction: incubation for 5 min at ambient temperature.

First immunoreaction: incubation for 15 min at ambient temperature; volume 1.5 ml.

Second immunoreaction and (simultaneous) enzyme reaction: incubation for 30 minutes at 30°C ± 0.1°C; wavelength 339 nm or Hg 334 nm; light path 10 mm; volume 2.05 ml; measurements against water.

Measurement

Pipette successively into a 12 mm × 75 mm glass test tube:			concentration in incubation/ assay mixture	
sample or standard solution (4)		0.20 ml	digoxin	up to 6.4 µg/l
pre-treatment solution (5)		0.05 ml	NaOH	0.1 mol/l
mix well and incubate for 5 min at ambient temperature;				
antibody/substrate solution (2)		0.05 ml	sheep γ-globulin	16.7 mg/l
			G-6-P	4.3 mmol/l
			NAD	2.7 mmol/l
			rabbit serum albumin	0.33 g/l
Tris buffer* (1)		1.20 ml	Tris	46 mmol/l
			NaCl	68.4 mol/l
			NaN$_3$	0.42 g/l
mix well and incubate for 15 min at ambient temperature;				
conjugate solution (3)		0.05 ml	G6P-DH**	variable ca. 24 U/l
Tris buffer (1)		0.50 ml	Tris	48.3 mmol/l
			NaCl	75 mmol/l
			rabbit serum albumin	0.49 g/l
			NaN$_3$	0.44 g/l
mix well; transfer 1.0 ml to the cuvette; read absorbance A_1; discard solution; incubate test tube with residual assay mixture for exactly 30 min at 30°C ± 0.1°C; read absorbance A_2; $A_2 - A_1 = \Delta A$ per 30 min.				

 * Equilibrated with the laboratory temperature!
** Coupled to 3-oxodigoxigenin carboxymethoxime.

If $\Delta A/\Delta t$ of the sample is greater than that of the highest standard, dilute the sample further with Tris buffer (1) and re-assay.

Calibration curve: plot the corrected $\Delta A/\Delta t$ readings of the standards *versus* the corresponding digoxin concentrations (µg/l). By use of special graph paper* matched with the reagents, a linear calibration curve is obtained (Fig. 3). The construction of this curve is based on the logit-log function. So far, the log-logit model (cf. Vol. I, p. 243) has proved to be useful to fit calibration curves of this assay.

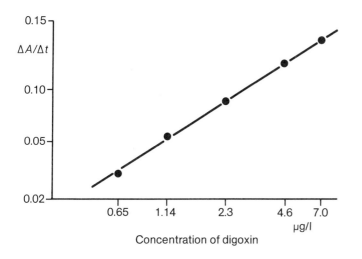

Fig. 3. Typical calibration curve for the assay of digoxin in serum.

Calculation: for the mean corrected $\Delta A/\Delta t$ readings of the sample take the digoxin mass concentration, p (µg/l), from the calibration curve. For conversion to substance concentration, c (µmol/l), multiply by 1.2805 (molecular weight is 780.92).

Validation of Method

Precision, accuracy, detection limit and sensitivity: significantly lower imprecisions have been reported for mechanized versions of this assay compared with its manual procedure. The between-days relative standard deviations for digoxin concentrations between 1.0 and 3.0 µg/l ranged from 13% to 8% without [28], but from 5% to 7% with, appropriate mechanization [29]. Recovery rates of digoxin between 94% and 104% have been obtained for human sera spiked with 0.5 to 4.0 µg/l [29].

---

* Supplied by *Syva*.

Data about the detection limit according to the definition in chapter 1.2.2, p. 20, have not been found in the literature. From reported imprecision values for low digoxin concentrations, however, the detection limit can be estimated to 0.5 µg/l for a mechanized procedure. The sensitivity at the 0.95 confidence level at a digoxin concentration of 1.9 µg/l was found to be 0.17 µg/l under mechanized conditions [30].

Sources of error: sera from highly haemolyzed, lipaemic and uraemic blood have been reported to give falsely decreased digoxin concentrations [29]. No clinically significant interference was observed with sera from haemolyzed and icteric blood at concentrations of haemoglobin and bilirubin up to 3 g/l and 513 µmol/l, respectively. Interference by a digoxin-like substance has been observed on patients with renal insufficiency [31].

Specificity: the cross-reactivity with digoxigenin of the antibody used has been found to be 67% and with digitoxin 10% [32]. Canrenone and canrenoate in concentrations up to 10 mg/l are without significant effect on the resulting digoxin concentrations [33].

Therapeutic range: cf. chapter 2.13.1, p. 124.

Appendix

Preparation of immunogen:

- dissolve 250 mg BSA in 10 ml K_2CO_3 solution (13), p. 119;
- add 100 mg digoxin or digitoxin carrying a reactive N-hydroxysuccinimide ester group at the terminal digitoxose (cf. p. 116);
- stir for 2 days at room temperature;
- dialyze against four 500 ml portions of phosphate buffer (6), p. 119, each time for 12 h at 4°C;
- determine protein concentration by the *Lowry* method (Vol. II, p. 88);
- store at −20°C.

Immunization: anti-digoxin and anti-digitoxin antibodies are elicited in sheep. Quantities are given per animal.

- Emulsify 1.0 ml BSA-digoxin or -digitoxin solution (see above) with a protein concentration of 0.5 g/l with 1.0 ml complete *Freund's* adjuvant;
- apply the immunogen mixture by intradermal injection into 40 to 50 sites;
- emulsify on days 7, 14, 30, 60, 90, 120 etc. after primary injection 0.5 ml BSA-digoxin or BSA-digitoxin solution with a protein concentration of 1.0 mg/ml with 0.5 ml complete *Freund's* adjuvant;
- immunize on days 7, 30, 90, etc., subcutaneously;
- immunize on days 14, 60, 120, etc., intramuscularly;

- collect serum after 6 months and check for titre and specificity;
- pool satisfactory antisera and store at $-20\,°C$.

Preparation of coated tubes: the solid phase is prepared by passive adsorption of IgG antibodies to the inner wall of polystyrene tubes.

- Fill tubes (12 mm in diameter, 40 mm in length) with 1.5 ml IgG solution (7) and incubate for 16 h at room temperature;
- remove the coating solution and fill tubes immediately with 1.5 ml BSA solution (8), incubate for $30-60$ min;
- remove solution and dry tubes overnight at room temperature.

Store tubes in a plastic bag in dry atmosphere at $4\,°C$.

Preparation of POD-digoxin conjugate*:

- dissolve 10 mg $\triangleq$ 10000 U POD (*Boehringer Mannheim*, grade I) in 6 ml phosphate buffer (10);
- dissolve 10 mg of a digoxin derivative carrying the reactive N-hydroxysuccinimid-ester group at the terminal digitoxose (cf. p. 116) in 0.7 ml absolute ethanol;
- add this solution dropwise under stirring to the POD solution;
- transfer the mixture to $4\,°C$ and allow to react for 3 days in the dark with stirring;
- dialyze the solution against 100 ml Tris/HCl buffer (11) for 24 h at $4\,°C$; change buffer four times;
- de-gas and fill into a 10 mm $\times$ 150 mm column 10 ml phenyl-Sepharose and equilibrate with Tris/HCl buffer (11);
- apply conjugate and start elution with Tris/HCl buffer (11);
- measure the absorbances ΔA_{280} and ΔA_{403};
- pool and evaluate the peak fractions using the desired assay conditions;
- dialyze the pool against 500 ml phosphate buffer (6) at $4\,°C$, change the buffer three times.

For storage, add bovine serum albumin (BSA) to a final concentration of 5 g/l, filter through 0.45 µm filter and add one granule of thymol. This material can be kept for at least 2 years at $4\,°C$.

Preparation of anti-digoxin antibodies: IgG antibodies from sheep for coating tubes are isolated by DEAE ion-exchange chromatography.

- Add slowly 2.7 g solid ammonium sulphate to 10 ml antiserum while stirring at room temperature; if necessary adjust pH to 7.0 with ammonia;
- stir slowly for about 2 h at room temperature; centrifuge for 30 min at 15000 g and discard the supernatant;
- wash the pellet in about 5 ml ammonium sulphate solution, 2 mol/l, centrifuge as before;
- re-suspend the pellet in about 2 ml dialysis buffer (12) and dialyze against the same buffer for 15 h at $4\,°C$; change the buffer four times.

---

* POD-digitoxin conjugate can also be prepared by this procedure.

For ion-exchange chromatography:

- weigh out 10 g pre-swollen DEAE-cellulose (*Whatman* DE 52) and equilibrate it by washing four times with dialysis buffer;
- de-gas the cellulose suspension, fill it into a column, allow to settle and wash with 60 – 80 ml dialysis buffer (12);
- free the protein dialysate from any possible precipitate by centrifugation;
- apply the solution to the column and elute with dialysis buffer (12); the IgG flows through almost without any delay;
- combine the fractions of the first protein peak and concentrate the solution if necessary to an IgG concentration of 10 – 20 g/l.

The preparation can be stored in this condition at $-20\,^\circ$C.

Preparation of G6P-DH-3-oxodigoxigenin conjugate: the digoxigenin derivative has to be prepared separately following the procedure described in [34]. In principle, digoxigenin is first oxidized with PtO_2 to 3-oxodigoxigenin; this substance is subsequently reacted with carboxymethoxylamine.

Dry the carboxymethoxime of digoxigenin overnight in an *Abderhalden* apparatus. Dry a 5 ml pear-shaped flask with a side arm and a 7 mm Teflon-coated stirring bar in an oven for 1 h at 105 °C. After reaching room temperature in a desiccator, place 23.05 mg (50 µmol) of the carboxymethoxime of digoxigenin in the flask. Then fit the flask with a serum stopper and a drying tube. Add a 250 µl aliquot of dry dimethyl-formamide and 7.1 µl (52 µmol) triethylamine (stored over KOH) through the serum stopper with a syringe while stirring at room temperature. Then cool the reaction flask to $-14\,^\circ$C or lower with a salt-ice bath. Subsequently add 9.3 µl carbitol chlorformate (50 µmol) below the surface of the DMF solution. Allow this mixture to stir at $-14\,^\circ$C for thirty minutes before proceeding with the conjugation. If necessary, it can be kept at $-14\,^\circ$C for several hours before use.

Put 2 ml G6P-DH (1 – 2 mg/ml in Tris buffer (6), p. 127), in a reaction vial fitted with a 7 mm stirring bar. Check the enzyme activity in a 5 µl aliquot diluted to 5 ml. The percent activity calculated during the rest of the conjugation will be based on this number.

To the 2 ml enzyme solution, add while stirring 20 mg G-6-P (disodium salt) and 40 mg NADH. Remove a 5 µl aliquot and check enzyme activity as above. Cool to 0 °C in an ice-bath. While stirring rapidly add 1080 µl carbitol slowly with a syringe below the surface of the enzyme. Check enzyme activity. Allow to stand for 30 minutes, remove a precipitate, if formed, and adjust pH of the enzyme solution to 9.0 ± 0.05 with NaOH, 1 mol/l. Check the enzyme activity again. The enzyme is now ready for conjugation.

Keep the mixed anhydride stirring at $-14\,^\circ$C and the enzyme stirring at 0 °C. Remove the mixed anhydride with a syringe through the serum stopper and add 1 µl at a time to the enzyme at a rate of 1 µl per minute. Make activity and inhibition checks after about every 10 µl mixed anhydride have been added. Determine the percent inhibition by adding 5 µl original antidigoxin antiserum having a high titre along with the enzyme and substrate.

It has been found that an inhibition of about 50% and a de-activation of 36% gave the most sensitive calibration curve. This usually occurs after adding 35 to 45 µl mixed anhydride.

When this point is reached, purify the enzyme conjugate by dialysis against Tris/ HCl buffer (6), p. 127 (with 0.5 g NaN$_3$ and 50 mg Thimerosal per litre as preservatives).

References

[1] *W. Withering,* An Account of the Foxglove and Some of Its Medical Uses, G. G. J. & J. Robinson, Birmingham 1785.

[2] *G. A. Beller, T. W. Smith, W. H. Abelmann, E. Haber, W. B. Hood,* Digitalis Intoxication: A Prospective Clinical Study with Serum Level Correlations, N. Engl. J. Med. *284,* 989 (1971).

[3] *D. S. Lukas, A. G. de Martino,* Binding of Digitoxin and Some Related Cardenolides to Human Plasma Proteins, J. Clin. Invest. *48,* 1041 – 1053 (1969).

[4] *J. K. Aronson,* Indications for the Measurement of Plasma Digoxin Concentrations, Drugs *26,* 230 – 242 (1983).

[5] *K. O. Haustein,* Methods for Determination of Cardiac Glycosides, Possibilities and Experiences, Z. med. Laboratoriumsdiagn. *20,* 99 – 110 (1979).

[6] *T. W. Smith, V. P. Butler, E. Haber,* Determination of Therapeutic and Toxic Serum Digoxin Concentrations by Radioimmunoassay, N. Engl. J. Med. *281,* 1212 – 1216 (1969).

[7] *H. Rutner, R. Rapun, N. Lewin,* Labelled 3-Acetyldigoxigenin Amino Acids for Digoxin Test from Digoxigenin-12-acetate-3-hemisuccinate and Amino Acids, US Patent 3 855 208 (1971).

[8] *K. Stellner,* Radioimmunologic Methods, in: *K. Greeff* (ed.), Handbook of Experimental Pharmacology, Vol. *56,* I, Springer-Verlag, Berlin 1981, pp. 57 – 81.

[9] *J. J. Chang, C. P. Crowl, R. S. Schneider,* Homogeneous Enzyme Immunoassay for Digoxin, Clin. Chem. *21,* 967 (1975).

[10] *G. Kleinhammer, H. Lenz, R. Linke, W. Gruber,* Enzyme Immunoassay for Determination of Serum Digoxin in Antibody Coated Tubes. (Abstract) 2nd European Congress on Clinical Chemistry, Prague 1976.

[11] *N. Monji, H. Ali, A. Castro,* Quantification of Digoxin by Enzyme Immunoassay: Synthesis of a Maleimide Derivative of Digoxigenine Succinate for Enzyme Coupling, Experientia *36,* 1141 – 1143 (1980).

[12] *K. Healey, H. M. Chandler, J. C. Cox, J. G. R. Hurrell,* A Rapid Semi Quantitative Capillary Enzyme Immunoassay for Digoxin, Clin. Chim. Acta *134,* 51 – 58 (1983).

[13] *K. Nelson, L. D. Bowers,* Evaluation of and Service Experience with a Fluorescence Polarization Immunoassay for Digoxin, Clin. Chem. *29,* 1175 (1983).

[14] *W. J. Litchfield, J. W. Freytag, M. Adamich,* Highly Sensitive Immunoassays Based on Liposomes without Complement, Clin. Chem. *30,* 1441 – 1445 (1984).

[15] *G. G. Belz, G. Belz,* Determination of Digitoxin Serum Concentration with a Solid Phase Enzyme Immunoassay, Med. Klin. *74,* 620 – 623 (1979).

[16] *M. H. H. Al-Hakiem, M. Simon, S. Mahmod, J. Landon,* Fluoroimmunoassay of Digitoxin in Serum, Clin. Chem. *28,* 1364 – 1366 (1982).

[17] *J. Mattersberger, H. Haug, H. G. Batz, G. Kleinhammer, R. Linke,* A Rapid Enzyme Immunoassay for Digitoxin Measurement, Clin. Chem. *30,* 1011 (1984).

[18] *V. P. Butler, J. P. Chen,* Digoxin-specific Antibodies, Proc. Natl. Acad. Sci. U.S.A. *57,* 71 – 78 (1967).

[19] *J. P. Persijn, K. M. Jonker,* A Terminating Reagent for the Peroxidase Labelled Immunoassay, J. Clin. Chem. Clin. Biochem. *16,* 531 – 532 (1978).

[20] *E. Munz, A. Kessler, P. U. Koller, E. W. Busch,* Experiences with an Enzyme Immuno Assay for Digoxin, Lab. med. *3,* 71 – 76 (1979).

[21] *M. Oellerich, H. Haindl, R. Haeckel,* Determination of Serum Digoxin with Enzyme-immuno Assays, Internist *19,* 188 – 190 (1978).

[22] *H. Kaiser,* About the Problem of the Detection Limit, Fresenius Z. anal. Chemie *209*, 1 – 18 (1965).

[23] *K. Borner, N. Rietbrock,* Determination of Digoxin in Serum. Comparison of Radioimmunoassay with Heterogeneous Enzyme Immuno Assay, J. Clin. Chem. Clin. Biochem. *16*, 335 – 342 (1978).

[24] *S. J. Soldin,* Digoxin-Issues and Controversies, Clin. Chem. *32*, 5 – 12 (1986).

[25] *K. E. Rubenstein, R. S. Schneider, E. F. Ullmann,* Homogeneous Enzyme Immunoassay. A New Immunochemical Technique, Biochem. Biophys. Res. Commun. *47*, 846 – 851 (1972).

[26] *G. L. Rowley, K. E. Rubenstein, J. Huisjen, E. F. Ullmann,* Mechanism by which Antibodies Inhibit Hapten-Malate Dehydrogenase Conjugates, J. Biol. Chem. *250*, 3759 – 3766 (1975).

[27] *H. M. Greenwood, J. Chandler,* Homogeneous Enzyme Immuno Assay and the Centrifugal Analyzer, Methods Lab. Med. *1*, 125 – 140 (1980).

[28] *L. Sun, V. Spiehler,* Radioimmunoassay and Enzyme Immunoassay Compared for Determination of Digoxin, Clin. Chem. *22*, 2029 – 2031 (1976).

[29] *J. Spitz, J. S. Braun, M. Schmidt,* Determination of Digoxin in Serum; Comparison of Radio- and Enzyme Immunoassay, Nucl. Med. *18*, 237 – 240 (1979).

[30] *P. B. Eriksen, O. Andersen,* Homogeneous Enzyme Immuno Assay of Serum Digoxin with Use of Bichromatic Analyzer, Clin. Chem. *25*, 169 – 171 (1978).

[31] *S. W. Graves, B. Brown, R. Valdes Jr.,* An Endogenous Digoxin-like Substance in Patients with Renal Impairment, Ann. Int. Med. *99*, 604 – 608 (1983).

[32] *M. Oellerich,* Enzyme Immuno Assay in Clinical Chemistry: Present Status and Trends, J. Clin. Chem. Clin. Biochem. *18*, 197 – 208 (1980).

[33] *W. Vogt, A. Tausch, K. Jacob, M. Knedel,* Determination of Digoxin in Serum with the EMIT-method by Means of a Rapid Analyzer, in: *W. Vogt* (ed.), Enzyme Immuno Assay; Basis and Practical Application, Thieme Verlag, Stuttgart 1978.

[34] *E. F. Ullmann, K. E. Rubenstein,* Cardiac Glycoside Enzyme Conjugates, US Patent 4039385 (1977).

1.11   Methotrexate

N-[4-[[(2,4-Diamino-6-pteridinyl)methyl]methyl-amino]benzoyl]-L-glutamic acid

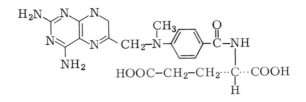

William E. Evans and Michael L. Christensen

General

Methotrexate (MTX), a folic acid antagonist, is used clinically as an anti-neoplastic agent in treatment of a variety of neoplasms. MTX is also used as an immunosuppressant in the treatment of severe psoriasis and rheumatoid arthritis. Initially, MTX was used in relatively low doses for the treatment of acute leukaemia and other diseases. However, subsequent studies have demonstrated that the use of high-dose methotrexate (HDMTX) with leucovorin rescue can successfully treat certain neoplasms previously found to be resistant to lower doses of the drug. HDMTX has proved useful in acute leukaemia, malignant lymphoma, carcinoma of the head and neck, and osteosarcoma [1, 2]. The efficacy of HDMTX for other neoplasms is currently being investigated.

MTX acts by inhibiting the enzyme dihydrofolate reductase, which converts dihydrofolate (FH_2) to tetrahydrofolate (FH_4). FH_4 is necessary for purine and pyrimidine synthesis and ultimately DNA synthesis. In HDMTX therapy, leucovorin is used to prevent the toxicity associated with high doses by replenishing intracellular pools of FH_4.

The oral absorption of MTX is incomplete, unpredictable and apparently dose-dependent, with low dosages of MTX (< 12 to 30 mg/m^2) reported to be absorbed better than higher doses [3 – 5]. Following intravenous administration, MTX has an initial volume of distribution of approximately 18% (0.18 l/kg) of body weight and has a steady-state volume of distribution of approximately 50 – 80% of body weight [3, 6]. In man, about 50% of MTX is bound to serum proteins, primarily albumin, at serum concentrations ranging from 0.1 to 1000 µmol/l [3, 4]. MTX disappears from the serum in a biphasic manner with an initial half-life of 1.5 to 3.5 h and a terminal phase half-life of 8 to 15 h [6, 7].

MTX undergoes oxidation to 7-hydroxy methotrexate (7-OH-MTX), a metabolite that has $\approx 1/200$th the cytotoxicity of MTX [8]. Following 4 to 6 h intravenous infusions of HDMTX, 1 to 11% of the dose is recovered in urine as the 7-OH metabolite. An even higher percentage of the drug can be accounted for as 7-OH-MTX in serum and urine following longer (i.e., 24 h) infusions of HDMTX [9]. 7-OH-MTX may pose analytical problems because concentrations of 7-OH-MTX commonly exceed MTX concentrations following HDMTX infusions. When utilizing immunoassays, the potential for metabolite cross-reactivity is particularly important at later times after the dose, even when cross-reactivity is relatively low (e.g., 2 to 4%), since 7-OH-MTX concentrations may be ten-fold higher than MTX [9, 10]. In these cirumstances, confirmation by chromatographic methods (i.e., HPLC) that allow quantitation of MTX and its metabolites, may be necessary.

MTX is also metabolized by intestinal bacteria to 4-amino-4-deoxy-N^{10}-methylpteroic acid (DAMPA) [11]. Although detection of this metabolite in serum and urine has been reported, it is usually undetectable and amounts to a very small percentage of the dose (i.e. < 1%).

MTX undergoes intracellular metabolism to MTX-polyglutamates, MTX-PG [12, 13]. MTX-PG may be more toxic to cells because they are retained by the cells for longer periods of time than MTX, are potent inhibitors of DHFR, and may have direct inhibitory effects on other enzymes (e.g., thymidylate synthase). Therefore, MTX-PG appear to have a major role in the cytotoxicity of MTX. Because MTX-PG are intracellular metabolites and not found outside the cell, they are not detected by routine clinical assay methodologies.

Low doses of MTX are not generally associated with significant systemic toxicity. In contrast, HDMTX may potentially result in life-threatening toxicity if not appropriately monitored and adequate leucovorin rescue administered. Toxicities associated with HDMTX include myelosuppression, stomatitis, nausea, vomiting, convulsions, and liver and renal abnormalities.

The cytotoxic effects of MTX are a function of both drug concentration and duration of exposure. MTX concentrations above 0.01 to 0.05 µmol/l are necessary to inhibit DNA synthesis [14]. Routine monitoring of MTX levels can aid in the early identification of patients at risk of developing serious toxicity. Depending on the high-dose treatment regimen, peak serum MTX concentrations may exceed 1000 µmol/l. Numerous relationships have been established between MTX concentrations in serum at selected times after drug administration and the occurrence of toxicity following a short (1 to 6 h) intravenous infusion. In general, patients with a serum concentration in excess of 10 µmol/l 24 h after initiation of the infusion, a concentration at 48 h greater than 0.5 to 1.0 µmol/l and a concentration at 72 h greater than 0.1 µmol/l are at increased risk for toxicity [15 – 17]. Altering leucovorin rescue in these patients has been shown to prevent MTX induced toxicity [16, 18].

Application of method: in clinical chemistry, pharmacology and toxicology.

Substance properties relevant in analysis: the molecular weight of MTX is 454.5. MTX is a weak acid with pK_a of 3.4, 4.7 and 5.7 for the α-COOH, γ-COOH and N-1 groups, respectively. MTX is practically insoluble in water, freely soluble in dilute alkaline solutions and slightly soluble in dilute hydrochloric acid.

Methods of determination: there are numerous methods by which MTX can be quantitated in biological fluids. These methods include homogeneous immunoassay (enzyme-immunoassay, fluorescence-polarization immunoassay), heterogeneous immunoassay (enzyme-linked immunosorbent assay, radioimmunoassay), microbiological, fluorimetric, competitive protein-binding assay, enzyme-inhibition assay, and high-pressure liquid chromatography (HPLC). These assays differ with regard to specificity, sensitivity, length of procedure, sample preparation, cost and ability to detect metabolites.

Comparison of enzyme-immunoassay, radioassay, radioimmunoassay and HPLC revealed no significant difference in results from patients' samples obtained during 24 h following a 6 h infusion of HDMTX (3500 to 5000 mg/m^2) [19]. MTX concentrations ranged from 0.1 to 10 µmol/l. Even samples in which HPLC analysis revealed that the 7-OH-MTX metabolite exceeded the parent compound, there was no statistical difference in results by any of the four methods. A more recent comparison of fluorescence-polarization immunoassay and enzyme-immunoassay with HPLC in patient samples following MTX administration, either 1000 mg/m^2 as a 24 h infusion or 12000 mg/m^2 as a 4 h infusion, demonstrated no difference between the two immunoassays, but results of both were significantly greater than those of HPLC [20]. However, in samples in which the 7-OH-MTX concentrations were less than the MTX concentrations, there was no significant difference between the 3 methods.

Further evaluations between immunoassays and HPLC are necessary to define the time following HDMTX at which the 7-OH-MTX metabolite concentration may be elevated to the extent that it may significantly alter results of immunoassays.

International reference method and standards: there is no reference standard. Analytical grade MTX is used to prepare standard concentrations. No reference method is currently known.

1.11.1   Determination with Enzyme-multiplied Immunoassay Technique

Assay

Method Design

The method was developed by *Gushaw et al.* [21] based on the enzyme-multiplied technique (EMIT®) first described by *Rubenstein et al.* [22]. The method described here follows the principle previously described (cf. Vol. I, chapter 2.7, p. 244).

Principle: as described in detail on pp. 7, 8.

When the enzyme-labelled MTX becomes bound to an antibody, the activity of the enzyme is reduced. MTX in a sample competes with the G6P-DH labelled MTX (E-L) for the binding sites of an antibody (Ab) present in a limited concentration. The activity of G6P-DH measured as increase in absorbance per unit time, $\Delta A/\Delta t$, is related to the MTX concentration in the sample (cf. Vol. I, chapter 2.7, p. 244). The method has been mechanized on various analytical systems [23, 24].

Selection of assay conditions and adaptation to the individual characteristics of the reagents: the assay is designed for rapid analysis and requires no separation step (results on samples from patients are available within 15 to 30 min).

Highly specific anti-methotrexate antibodies have been produced by immunizing sheep with a bovine γ-globulin conjugate of methotrexate. Antibodies are raised in sheep, which are first inoculated with an emulsion of 1 mg of this antigen in complete *Freund's* adjuvant and subsequently receive monthly inoculations of the same amount of antigen in incomplete *Freund's* adjuvant.

Enzyme-drug conjugate is obtained by coupling MTX to G6P-DH from *Leuconostoc mesenteroides*. Interference from serum G6P-DH is avoided by use of the coenzyme NAD which is converted only by the bacterial enzyme.

The optimal quantities and concentrations of sample and reagents depend on the characteristics of the antibodies present in the antiserum.

The assay range is from 0.2 to 2.0 µmol/l. Concentrations above this range can be assayed with the appropriate dilution prior to the assay. It is also possible to assay concentrations below this range by omitting the pre-dilution of the sample (samples with MTX concentrations as low as 0.02 µmol/l can be assayed) [25]. The sample volume required is usually 50 µl.

These parameters being fixed, the optimum ratio of antibody to enzyme conjugate in the presence of different MTX concentrations within the desired assay range must be determined. First, a prospective enzyme conjugate concentration is chosen, to which gradually increasing amounts of the corresponding antibody are added. Enzymic activity is increasingly inhibited as the amount of antibody is raised. A maximum inhibition of 90 to 100% of the enzymic activity can be achieved by addition of a sufficient amount of antibody. In order to assess the system's ability to discriminate between serum MTX concentrations likely to be encountered during MTX therapy, the following optimization procedure is used: calibrators containing 0.0, 0.2, 0.5, 1.0, 1.33 and 2.0 µmol/l are assayed according to the regular assay protocol (cf. p. 141) at a fixed enzyme-conjugate concentration and with different amounts of antibody. The zero calibrator serves as a reagent blank and all measurements are corrected for this value. The corrected readings of ΔA per unit time, $\Delta A/\Delta t$, obtained for each calibrator are plotted against the amount of antibody added. The antibody quantity is then selected according to the following considerations:

– an optimum response should be obtained at an analyte concentration within the assay range where assay sensitivity is most desirable (e.g. at 1.0 µmol MTX per litre).
– the assay response between MTX calibration values reflecting the discriminatory power of the assay configuration should attain a maximum possible value.

The first step of the assay procedure (addition of the antibody/substrate solution to the diluted sample) is carried out at room temperature. Immediately after the subsequent addition of the conjugate the reaction mixture is incubated for 45 s in the spectrophotometer flow-cell at 30 °C and pH 8.0. During this period the immuno- and enzymatic reactions take place simultaneously. Beginning 15 s after the start of the incubation, absorbance readings are made over a period of 30 s. The results are calculated from the change in absorbance over a 30 s measurement period.

Equipment

Spectrophotometer or spectral-line photometer capable of exact measurement at 339 or Hg 334 nm. It should be equipped with a thermostatted flow-cell which consistently maintains 30.0 ± 0.1 °C throughout the working day. Pipetter-diluter capable of delivering 50 ± 1 µl sample and 250 ± 5 µl buffer with a relative standard deviation less than 0.25%. Delivery must be of sufficient force to ensure adequate mixing of the components. The spectrophotometer should be connected to a suitable data handling device. Laboratory centrifuge.

Reagents and Solutions

Purity of reagents: G6P-DH from *Leuconostoc mesenteroides* should be of the best available purity (cf. p. 27). Chemicals are of analytical grade.

Preparation of solutions* (for 100 determinations): all solutions in re-purified water (cf. Vol. II, chapter 2.1.3.2).

1. Tris buffer (55 mmol/l; Triton X-100, 0.1 ml/l; pH 8.0):

   dissolve 19.96 g Tris base in 1400 – 1600 ml water; add 0.3 ml Triton X-100; adjust to pH 8.0 with HCl, 1.0 mol/l; make up with water to 3000 ml.

   Alternatively, dilute buffer concentrate provided with each kit to 150 ml with water.

2. Antibody/substrate solution (γ-globulin**, ca. 50 mg/l; G-6-P, 66 mmol/l; NAD, 40 mmol/l; Tris, 55 mmol/l; pH 5.0):

   reconstitute available lyophilized preparation with 6.0 ml water.

3. Conjugate solution (G6P-DH***, ca. 1 kU/l; Tris, 55 mmol/l; pH 8.0):

   reconstitute available lyophilized preparation with 6.0 ml water. This reagent must be standardized to match the antibody/substrate reagent. (The conjugate is prepared using the methodology described in Appendix, p. 143.)

4. MTX standard solution (0.2 to 2.0 µmol/l; 90.9 to 909 µg/l):

   dissolve 22.7 mg MTX in 1000 ml Tris buffer (1). Dilute 0.4 ml with MTX-free human serum containing sodium azide (5 g/l) and Thimerosal (0.5 g/l) to 10 ml (2.0 µmol/l). Dilute this solution 1 + 0.5, 1 + 1, 1 + 3, and 1 + 9 with MTX-free human serum, yielding the following concentrations of MTX: 2.0, 1.33, 1.0, 0.5, 0.2 µmol/l.

   Reconstitute each of the available lyophilized standard preparations with 1 ml water.

---

* Reagents, calibrators and buffers are commercially available from *Syva Co.,* Palo Alto, U.S.A. The solutions contain sodium azide as a preservative.
** Standardized preparation from immunized sheep.
*** Coupled to MTX.

Stability of reagents and solutions: after reconstitution, the antibody/substrate solution (2), conjugate solution (3) and standards (4) must be stored at room temperature (20 °C to 25 °C) for at least one hour before use. The reagents are then stable for 12 weeks if stored at 2 °C to 8 °C. The buffer solution (1) can be used for 12 weeks when stored at room temperature.

Procedure

Collection and treatment of specimen: it is important that the serum or plasma is collected at the appropriate time following the dose administration. Plasma, serum, urine, saliva or cerebrospinal fluid can be used in this assay. Blood cannot be used. 50 µl sample is required per assay. Acceptable anticoagulants are heparin, EDTA and oxalate. Sera from severely haemolyzed, lipaemic or icteric blood may cause poor reproducibility. If possible, a new clear sample should be obtained.

Stability of the analyte in the sample: samples should be stored at 2 °C to 8 °C upon collection and protected from light (MTX in solution is light-sensitive), or stored frozen (-20 °C or below) if not analyzed within 24 h. Samples can be stored at -20 °C for 6 months without appreciable loss.

Assay conditions: measurements of samples and zero calibrator (pure serum) in duplicate. Single determinations are made of the standards (4). A pre-dilution of samples and standard solution is required. Mix 0.05 ml sample or standard solution (4) with 0.25 ml Tris buffer (1).

Incubation time 45 s (in the spectrophotometer flow-cell); 0.90 ml; 30 °C; wavelength 339 or Hg 334 nm; light path 10 mm; measurement during the last 30 s against water.

Establish a calibration curve under the same conditions (within the series). The zero calibrator assay serves as a reagent blank; all measurements are corrected for this value.

Calibration curve: correct averaged readings of the calibrators for averaged blank readings, yielding $\Delta A/\Delta t$. Plot *versus* the corresponding methotrexate concentrations (µmol/l). By use of special graph paper* matched with the reagents, a linear calibration curve is obtained (Fig. 1). The construction of this paper is based on the logit-log function. So far the log-logit model (cf. Vol. I, p. 243) has proved to be useful to fit calibration curves of this EMIT [25].

---

* Supplied by *Syva*. For using this graph paper multiply each value of $\Delta A/\Delta t$ by 2.667.

Measurement

Pipette into a disposable beaker:			concentration in assay mixture	
pre-diluted sample or standard solution (4)		0.05 ml	MTX	up to 0.11 μmol/l 50 μg/l
Tris buffer antibody/substrate	(1)	0.25 ml	Tris	52 mmol/l
solution	(2)	0.05 ml	γ-globulin G-6-P NAD	ca. 2.8 mg/l 3.7 mmol/l 2.2 mmol/l
Tris buffer	(1)	0.25 ml		
conjugate solution	(3)	0.05 ml	G6P-DH	ca. 56 U/l
Tris buffer	(1)	0.25 ml		
mix well; immediately aspirate into the clean spectrophotometer flow-cell; after a 15 s delay, read ΔA over a period of 30 s.				

If reading of the sample is greater than that of the highest standard, dilute the sample further with Tris buffer (1) and re-assay.

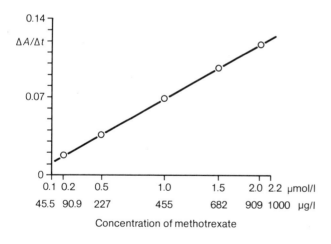

Fig. 1. Typical calibration curve for the assay of methotrexate in serum.

Calculation: read the methotrexate concentration, c (μmol/l), corresponding to the mean corrected $\Delta A/\Delta t$ readings of the sample from the calibration curve. If additional dilutions of the sample are required the actual MTX concentration of the sample is determined by multiplying the concentration obtained from the calibration curve by the additional dilution factor required. For conversion from μmol/l to μg/l multiply by molecular weight of 454.5.

Validation of Method

Precision, accuracy, detection limit and sensitivity: in order to attain an adequate precision with this assay, the use of partially or fully mechanized analytical systems is recommended. With the appropriate mechanized system, the between-days RSD ranged from 15.7% to 2.4% at MTX concentrations between 0.2 and 2.0 μmol/l [20]. The accuracy has been shown by comparison with either radioimmunoassay or an enzymatic technique based on the inhibition of dihydrofolate reductase. The results obtained by EMIT with samples from patients receiving MTX agreed well with those determined by two methods (EMIT *vs.* RIA, y = 0.90 x + 0.01 μmol/l, r = 0.970, n = 50 and EMIT *vs.* enzymatic method*, y = 0.97 x − 0.21 μmol/l, r = 0.997, n = 100) [23].

The detection limit, defined as the lowest concentration giving an absorbance significantly different from the zero standard, is about 0.1 μmol/l (with the modifications in the assay procedure [cf. p. 138] the detection limit is 0.02 μmol/l).

Sources of error: no pathophysiological influence is known to alter the measurement of MTX.

No clinically significant interference was observed with samples from haemolyzed, lipaemic and icteric blood at haemoglobin concentrations up to 8 g/l, triglycerides, 11 mmol/l, and bilirubin, 513 mmol/l.

Specificity: the specificity entirely relies on the quality of the antiserum. Compounds whose chemical structure or concurrent therapeutic use would suggest possible cross-reactivity have been evaluated. The concentrations of these compounds necessary to produce a maximum quantitation error of 30% in samples containing 1.0 μmol MTX per litre are shown in the following table [23].

Structurally unrelated compounds	Concentration (μmol/l)
Cyclophosphamide	> 1000
Doxorubicin (Adriamycin®)	> 1000
5-Fluorouracil	> 1000
Vinblastine	> 1000
Vincristine	> 1000
Allopurinol	> 1000

---

* Data on file *Syva Co.,* Palo Alto, U.S.A.

Structurally related compounds

Dihydrofolic acid	> 100
Folic acid	> 10
7-Hydroxymethotrexate	> 10
Leucovorin (Citrovorum factor)	> 100
Trimethoprim	> 100

The assay has approximately equal sensitivity to aminopterin, an anti-neoplastic agent not usually administered concurrently with MTX, and DAMPA – a minor metabolite of MTX.

Therapeutic range: no precise relationship between MTX concentration and anti-neoplastic efficacy has been established, although levels below 0.01 to 0.05 μmol/l are necessary for resumption of DNA synthesis. Levels following certain HDMTX dosages may exceed 1000 μmol/l. The potential for increased toxicity occurs when there is a delay in MTX clearance. Following short infusions of MTX, a patient with 24, 48 and 72 h serum MTX concentrations greater than 10 μmol/l, 1.0 μmol/l and 0.1 μmol/l, respectively, is at increased risk for MTX toxicity if standard low-dose leucovorin rescue is given [15 – 17].

Appendix

Preparation of G6P-DH-methotrexate conjugate: the *N*-hydroxysuccinimide ester method leads to a satisfactory conjugate [26].

Prepare the *N*-hydroxysuccinimide ester of methotrexate (0.1 mol/l), in dimethylformamide, with equimolar concentrations of *N*-hydroxysuccinimide (0.1 mol/l) and dicyclohexylcarbodiimide (0.1 mol/l).

For coupling the active ester of MTX with G6P-DH, the reaction mixture contains 2.8 mg (27 nmol/l; ca. 1.82 kU/l) G6P-DH, 10 mg (38 μmol/l) G-6-P, 20 mg (26 μmol/l) NADH, disodium salt, and 300 μl carbitol* in 1 ml sodium carbonate buffer (0.1 mol/l, pH 9) at 4 °C. Add to this reaction mixture the *N*-hydroxysuccinimide ester of MTX in seven 10 μl aliquots with stirring; this process takes approximately 2 h. Dialyze the solution against Tris buffer (55 mmol/l, pH 7.9) at 4 °C to remove low molecular-mass components.

The amount of enzyme conjugate used in the assay is based upon the optimization procedures that match this reagent with the antibody/substrate reagent (cf. p. 138).

---

* Diethyleneglycol monoethylether, 2-(2-ethoxyethoxy)ethanol.

1.11.2   Determination with Enzyme-linked Immunosorbent Assay

Assay

Method

The method was developed by *Ferrua* [27] based on the principle of enzyme-linked immunosorbent assay (ELISA).

Principle

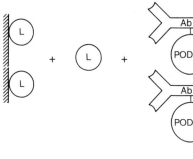

(a)

| Solid phase-bound methotrexate | ligand, analyte, methotrexate | POD-rabbit anti-methotrexate antibody conjugate |

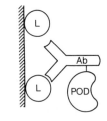

POD-labelled solid phase-bound ligand-Ab complex

(b) $2 \ \text{1,2-Phenylenediamine} + 2\,H_2O_2 \xrightarrow{} \text{2,2'-diaminoazobenzene} + 4\,H_2O$.

The method described here operates according to the immuno-enzymometric assay design (cf. Vol. I, chapter 2.7.2, p. 237). MTX in the sample competes with the MTX attached to the solid phase for the POD*-labelled anti-MTX antibody. All POD-labelled anti-MTX antibodies bound to MTX in the sample and unbound labelled antibody are washed out.

The quantity of enzyme-labelled antibody which has been bound onto the solid phase is measured by incubation with the appropriate substrate. The concentration of MTX in the sample is directly related to the enzyme activity present on the solid phase.

Selection of assay conditions and adaptation to the individual characteristics of the reagents: this is relatively slower than other methods described in this chapter (results available in 120 min).

Highly specific anti-MTX antisera have been produced by immunizing rabbits with bovine serum albumin (BSA) coupled to MTX through the carboxylic group by means of carbodiimide [28]. Antibodies are raised in rabbits which receive monthly injections of a solution containing 100 μg MTX, emulsified with an equal volume of complete *Freund's* adjuvant.

The optimal quantities and concentration of samples and reagents depend on the characteristic of the antibodies present in the antiserum.

The desired assay range is 0.1 to 100 μg/l. The serum sample is 100 μl.

These parameters being fixed, the concentration of MTX in the solution used to coat the spheres and the concentration of the conjugate are to be determined. A chessboard titration is made using different concentrations of the MTX preparation for coating and different conjugate dilutions. Each combination must be tested with at least a blank and a MTX concentration in the upper range of the desired calibration curve.

The selection of the optimal concentrations is made according to the following criteria.

a) The optimal dilution of the coating solution is estimated after plotting the enzyme activity in terms of increases in absorbance per unit time, $\Delta A/\Delta t$, *versus* MTX concentration when different immunogen (MTX-BSA) concentrations have been used for coating, for a given concentration of conjugate. The preferred concentration is the one which gives the maximal absorbance, or a concentration slightly below this (example in Fig. 2).

b) The optimal dilution of the conjugate is that which gives maximal $\Delta A/\Delta t$ in the range of 1.2 to 1.5 for the upper point of the desired assay range, together with a low blank.

The first step of the assay procedure (addition of POD-antibody, diluted sample and one MTX-coated sphere to each sample well) is carried out at room temperature with a one hour incubation. The sphere is removed and washed 3 times. The substrate solution is added and incubated in the dark at room temperature for 30 min. The reaction is stopped with HCl, 1 mol/l, and the absorbance is read directly through the tubes.

---

* Peroxidase, donor: hydrogen-peroxide oxidoreductase, EC 1.11.1.7.

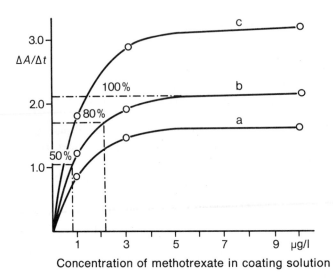

Fig. 2. Example. Optimization of the assay: polystyrene spheres were coated with different concentrations of immunogen (MTX-BSA) giving increasing amounts of solid phase-bound MTX. The different batches were incubated with 0.1 ml zero standard (10) and a) 65, b) 100 and c) 200 μg POD-labelled anti-MTX IgG in 0.1 ml conjugate buffer. Conjugate dilution was 1 : 500 from stock solution of conjugate, i.e., 1000 μg/l [27].

Equipment

Spectrophotometer capable of exact measurement at 492 nm; disposable trays, disposable polystyrene tubes 13 mm × 75 mm; a washing device capable of washing polystyrene balls is useful (Pentawash® from *Abbott*); laboratory centrifuge.

Reagents and Solutions

Purity of reagents: POD should be of the best available purity: grade I, RZ (A_{403}/A_{280}) = 3 or more; 250 U/mg at 25 °C with guaiacol and H_2O_2 as substrates (available from *Boehringer Mannheim Biochemicals,* Indianapolis, IN 46250, U.S.A). Chemicals are of analytical grade. Hydrogen peroxide, 30% (w/v) solution must be stored at 4 °C, in the dark. 1,2-Phenylenediamine should also be stored in the dark.

Preparation of solutions (for 100 determinations): all solutions in re-purified water (cf. Vol. II, chapter 2.1.3.2).

1. Phosphate-buffered saline, PBS (phosphate, 10 mmol/l; NaCl, 137 mmol/l; pH 7.2):

   dissolve 2.7 g $Na_2HPO_4 \cdot 12\ H_2O$, 0.4 g $NaH_2PO_4 \cdot 2\ H_2O$ and 8 g NaCl in 1 l water.

2. Phosphate buffer (0.1 mol/l, pH 7.2):

   dissolve 27 g $Na_2HPO_4 \cdot 12\ H_2O$ and 0.4 g $NaH_2PO_4 \cdot 2\ H_2O$ in 1 l water.

3. Carbonate buffer (10 mmol/l, pH 9.5):

   dissolve 530 mg $NaHCO_3$ and 700 mg $NaCO_3 \cdot 10\ H_2O$ in 1 l water.

4. $NaIO_4$ solution (0.1 mol/l):

   dissolve 420 mg $NaIO_4$ in 20 ml water.

5. Phosphate/citrate buffer (phosphate, 0.1 mol/l; citrate, 44 mmol/l; H_2O_2, 5.9 mmol/l; Thimerosal, 0.1 g/l; pH 5.5):

   dissolve 4.6 g citric acid, monohydrate, 10 g $Na_2HPO_4 \cdot 2\ H_2O$, 50 mg Thimerosal and 335 µl 30% (w/v) H_2O_2 in 500 ml water.

6. Substrate solution (phosphate, 0.1 mol/l; citrate, 44 mmol/l; H_2O_2, 5.9 mmol/l; 1,2-phenylenediamine, 16.5 mol/l; pH 5.0):

   dissolve 1.5 g 1,2-phenylenediamine, dihydrochloride, in 500 ml buffer (5).

7. Conjugate buffer (phosphate, 0.1 mol/l; pooled rabbit serum, 333 ml/l; Thiomersal, 0.1 g/l; pH 7.2):

   mix 200 ml phosphate buffer (2) and 100 ml pooled rabbit serum and 30 mg Thimerosal.

8. Conjugate solution (POD, 0.16 kU/l; phosphate, 0.1 mol/l; rabbit serum, 333 ml/l; Thimerosal, 0.1 g/l; BSA, 10 g/l; pH 7.2):

   use preparation according to Appendix, p. 151 (protein, ca. 10 g/l; POD, ca. 80 kU/l); dilute stock solution with buffer (7) containing 10 g BSA per litre according to optimal concentration determined experimentally, cf. p. 145.

9. MTX-coated polystyrene spheres:

   prepare according to Appendix, p. 152.

10. Standard and sample diluent (phosphate buffer, 10 mmol/l, pH 7.2; pooled chicken serum, 100 ml/l):

    dissolve 27 g $Na_2HPO_4 \cdot 12\ H_2O$ and 0.4 g $NaH_2PO_4 \cdot 2\ H_2O$ with water to 900 ml, add 100 ml pooled chicken serum.

11. Acetate buffer (1 mmol/l, pH 4.4):

dissolve 54.4 mg $CH_3COONa \cdot 3\ H_2O$ and 34 µl CH_3COOH in 1 l water.

12. Methotrexate standard solution (0.1 to 100 µg/l):

dissolve 25 mg MTX in 100 ml phosphate buffer (2). Dilute 0.4 ml with standard and sample diluent (10) to 1000 ml (100 µg/l). Dilute this solution, $1 + 9, 1 + 99$, $1 + 999$ with standard or sample diluent (10), yielding the following concentrations of MTX: 100.0, 10.0, 1.0 and 0.1 µg/l.

Stability of reagents and solutions: MTX-coated polystyrene spheres (9) are stable at least for 6 months when stored in liquid phase at 4°C. The undiluted conjugate solution (8) is stable for more than 4 years at $-20°C$. The MTX standards (10), the diluted conjugate (8), the phosphate/citrate buffer (5) and the incubation buffer (7) are prepared every week and kept at 4°C. Substrate solution (6) must be prepared and stored in a dark bottle and used within 1 h. Buffer solutions $(1) - (3)$ are stable for two months if microbial contamination is avoided.

Procedure

Collection and treatment of specimen: it is important to collect the serum or plasma sample at the appropriate time following administration of the dose. 100 µl serum or plasma is required per assay. Acceptable anticoagulants are heparin, EDTA and oxalate. Complete mixing of sample is necessary upon thawing. Interference by sera from grossly icteric blood is eliminated by blank readings.

Stability of analyte in the sample: the sample should be stored at 2°C to 8°C upon collection and stored frozen ($-20°C$ or below) if not analyzed within 24 h.

Assay conditions: all measurements of sample and standards in duplicate. A predilution of sample is required: mix one part (0.10 ml) sample with 9 parts (0.90 ml) phosphate-buffered saline (1) and then dilute this mixture; one part (1 ml) with one part (1 ml) standard and sample diluent (10).

Immunoreaction: incubation at room temperature (25°C) for 60 min.

Enzymatic indicator reaction: incubation at 25°C for 30 min in the dark; wavelength 492 nm; light path 10 mm. Measurements against substrate solution (6).

Establish a calibration curve under exactly the same conditions (within the series).

Measurement

Immunoreaction

Pipette or place into a disposable tray:			concentration in incubation mixture	
diluted sample or standard solution (12)		0.1 ml	MTX	up to 50 µg/l
			phosphate	55 mmol/l
			chicken serum	50 ml/l
diluted conjugate	(8)	0.1 ml	conjugate	1000 µg/l
			POD	80 U/l
			rabbit serum	167 ml/l
			BSA	5 g/l
coated sphere	(9)	1		
cover the tray and incubate for 60 min at 25 °C; wash* with 3 – 5 ml aliquots of water, transfer the sphere into a fresh reaction tube;				

* Preferably using a washing device (Pentawash® from *Abbott*).

Enzymatic indicator reaction

Pipette or place into a test tube:			concentration in assay mixture	
coated sphere after primary reaction		1		
substrate solution*	(6)	0.4 ml	phosphate	0.1 mol/l
			citrate	44 mmol/l
			H_2O_2	5.9 mmol/l
			1,2-phenylene- diamine	16.5 mmol/l
incubate in the dark for 30 min,				
HCl, 1 mol/l		2.0 ml	HCl	0.83 mol/l
reaction is stopped; read absorbance.				

* Freshly prepared!

If $\Delta A/\Delta t$ of the sample is greater than that of the highest standard, dilute the sample further with sample diluent (10) and re-assay.

Calibration curve: if $\Delta A_0/\Delta t$ corresponds to the enzyme activity at zero MTX concentration and $\Delta A/\Delta t$ to the antibody-labelled enzyme activity bound at a given MTX concentration, plot the ratios $(\Delta A/\Delta t)/(\Delta A_0/\Delta t) = \Delta A/\Delta A_0$ (ordinate) *versus* MTX concentrations (abscissa) on log-linear graph paper.

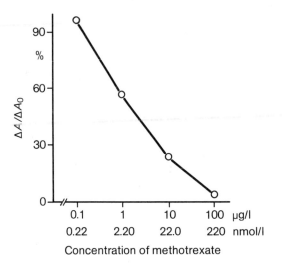

Fig. 3. Typical calibration curve for the assay of methotextrate.

Calculation: read the MTX mass concentration, ρ (μg/l), from the calibration curve using the $\Delta A/\Delta A_0$ value of each sample. For conversion from mass concentration to substance concentration, c (μmol/l), divide by 454.5.

Validation of Method

Precision, accuracy, detection limit and sensitivity: for concentrations within the major part of the calibration curve within-run RSD below 12.1% were observed [27].

The recovery of MTX added to drug-free human serum ranged from 90 to 96% at concentrations of 0.1 to 50 μg/l. The results obtained from ELISA agreed very well with HPLC method and RIA (ELISA *vs.* HPLC, y = 1.14x − 638 nmol/l, r = 0.978, n = 97 and ELISA *vs.* RIA, y = 1.34x − 351 nmol/l, r = 0.984, n = 83) [27].

The detection limit of the assay, defined as the lowest concentration of MTX giving $\Delta A/\Delta t$ values significantly different from the zero standard, was 10^{-10} mol/l or about 50 ng/l.

Sources of error: no pathophysiological influence is known to alter the measurement of MTX. The most common trouble originates from possible presence of anti-γ-globulin factor (rheumatoid factor) binding with the enzyme-antibody conjugate. It thus can simulate the antigen, leading to erroneously elevated values. In this assay, this drawback is prevented by the use of the incubation medium with an excess of pooled rabbit serum [29].

Specificity: the specificity depends on the quality of the antibody, which should be carefully checked. Cross-reactivity with folic acid, folinic acid, dihydrofolic acid, tetrahydrofolic acid and N_5-methyltetrahydrofolic acid is less than 0.05%; with 7-OH-MTX it is 0.9% and with MTX polyglutamate it is 100% [27].

Therapeutic range: cf. chapter 1.11.1, p. 143.

Appendix

Preparation of POD-IgG conjugate: the procedure of *Wilson & Nakane* [30] leads to a reproducible and satisfactory conjugate. The α-diol groups present in the carbohydrate groups of POD are oxidized by periodic acid. The newly formed aldehyde residues are allowed to react with the free amino groups of IgG to form *Schiff* bases. The unreacted aldehyde groups are reduced by sodium borohydride.

– Dissolve 4 mg POD (1 kU) (*Boehringer Mannheim,* grade I) in 1 ml water,
– add 0.2 ml freshly prepared $NaIO_4$ solution (6) to the POD solution;
– stir this mixture for 20 min at room temperature, then dialyze overnight at 4°C against 1 litre acetate buffer (11);
– adjust pH to 9 – 9.5 by addition of 20 μl carbonate buffer (3); add 8 g anti-MTX antisera (cf. p. 138) in 1 ml carbonate buffer (3) to the dialysate; incubate for 2 h at room temperature under gentle agitation;
– reduce unreacted aldehyde groups by addition of 0.1 ml anhydrous $NaBH_4$ solution (20 mg/ml) for 2 h at 4°C;
– add an equal volume of saturated aqueous ammonium sulphate to the conjugate solution for separation of unreacted POD from conjugated POD; stir for 15 min at room temperature;
– isolate the precipitate by centrifugation at 5000 g for 20 min and wash with neutral half-saturated ammonium sulphate solution; centrifuge the precipitate and wash similarly two additional times with the same half-saturated ammonium sulphate solution;
– after a final centrifugation, dissolve the precipitate with 8 ml buffered saline (2);
– measure the absorbances A_{280} and A_{403}. The ratio A_{280}/A_{403} should be between 1.5 and 2.5;
– add 100 to 200 ml serum/l of the same animal species which provided the antiserum. Alternatively, bovine or human serum albumin, 10 g/l, can be substituted for animal serum.

This stock solution of conjugate is filtered through 0.45 μ filters and can be stored frozen at − 20°C in small aliquots or can be lyophilized after addition of lactose to a final concentration of 10 g/l. It should be diluted before use to the appropriate concentration (cf. p. 145).

Coating of the solid phase: a variety of different solid phases may be used. Use polystyrene balls, 6.5 mm in diameter provided by *Precision Plastic Balls, Inc.,* Chicago, for the homogeneity of their surface properties and for the convenience of storage of the coated balls. Therefore, one can prepare a relatively high number of balls (i.e., several thousands) at the same time. They should be placed in a clean vessel (e.g., a disposable vessel) without touching by hands.

- Rinse several times with water;
- for coating, immerse polystyrene spheres for 1 h at room temperature in phosphate buffer (2) containing the diluted BSA-MTX solution ($0.8 - 2.2$ µg/l), cf. p. 138;
- incubate for 1 h;
- wash the balls four times with the same phosphate buffer and saturate for 24 h at 4°C with 10 g BSA per litre PBS (1).

Store the coated balls at 4°C in phosphate buffer (2) with Thimerosal, 0.1 g/l.

1.11.3   Determination with Enzyme-inhibition Assay

Assay

Method Design

The method was developed by *Falk* [31] and modified by *Brown* [32] based on the principle of enzyme-inhibition assay [33, 34].

Principle

$$\text{Dihydrofolate} + \text{NADPH} + \text{H}^+ \xrightarrow{\text{DHFR*}} \text{tetrahydrofolate} + \text{NADP}^+.$$

When MTX binds to the enzyme dihydrofolate reductase (DHFR), the enzyme is inhibited. The catalytic activity of DHFR measured as increase in absorbance per unit time, $\Delta A/\Delta t$, is inversely proportional to the MTX concentration of the sample. The method has been mechanized on various analytical systems [32, 35].

---

* 5,6,7,8-Tetrahydrofolate: NADP$^+$ oxidoreductase, EC 1.5.1.3.

Optimized conditions for measurement: the assay is designed for rapid analysis and requires no separation step (results on samples from patients are available within about 15 min, once the calibration curve is established).

The enzyme, DHFR, is obtained from either bovine liver or *Lactobacillus casei*. Although the bacterial enzyme is more stable, there is significant interference from trimethoprim, a potent inhibitor of bacterial, but not mammalian DHFR. (Trimethoprim is widely used as an antibiotic in combination with sulfamethoxazole.) Addition of albumin to the reaction mixture will prevent the enzyme from being adsorbed to the walls of the reaction vials and maintains the stability of the enzyme for over 24 h.

The assay range is from 0.02 to 0.30 µmol/l. Concentrations above this range can be assayed with the appropriate dilution prior to the assay. The serum sample required is 20 µl.

The first step of the assay procedure (incubation of NADPH and enzyme with the sample for 5 min) is carried out at room temperature. This has been shown to increase the linearity of the calibration curve [32]. Immediately after addition of dihydrofolate the enzymatic reaction runs. The reaction mixture is incubated in the spectrophotometer for 120 s at 30 °C, pH 7.5. The initial absorbance is read at 60 s after the start of the incubation and the final absorbance reading at 120 s. The results are calculated from the change in absorbance, ΔA, over a 60 s measurement period, $\Delta A / \Delta t$.

The optimal quantities and concentrations of sample and reagents depend on the characteristics of the enzyme, but in any case, dihydrofolate and NADPH concentrations should be at least 0.19 µmol/l and 16.3 mmol/l respectively, in the assay mixture.

Equipment

Spectrophotometer capable of exact measurement at 339 nm. It should be equipped with a thermostatted flow-cell which consistently maintains 30.0 ± 0.1 °C throughout the assay procedure. Work rack and disposable 2 ml reaction vials with conical bottoms. The spectrophotometer should be connected to a suitable data handling device. Laboratory centrifuge.

Reagents and Solutions

Purity of reagents: DHFR should be of the best available purity, e.g. preparation from bovine liver available from *Sigma*, 8 U/mg at 25 °C, 7,8 dihydrofolate and NADPH as substrates, pH 6.5. Chemicals are of analytical grade.

Preparation of solutions: all solutions in re-purified water (cf. Vol. II, chapter 2.1.3.2).

1. Tris buffer (1 mol/l, pH 7.5):

   dissolve 60.55 g Tris base in 250 to 300 ml water, dilute to 500 ml; adjust pH to 7.5 with HCl, 1.0 mol/l.

2. Ethylenediamine tetraacetate, EDTA (50 mmol/l):

   dissolve 1.86 g EDTA-Na$_2$H$_2$ · 2 H$_2$O in 100 ml water.

3. Reaction buffer (Tris 50 mmol/l; mercaptoethanol, 8.5 mmol/l; EDTA, 1 mmol/l; pH 7.5):

   mix 0.6 ml mercaptoethanol (sp. gr. 1.11), 50 ml Tris buffer (1) and 20 ml EDTA solution (2) and dilute to 1000 ml with water.

4. Dihydrofolate buffer (Tris buffer, 100 mmol/l; mercaptoethanol, 300 mmol/l):

   mix 25 ml Tris buffer and 5.3 ml mercaptoethanol (sp. gr. 1.11) and dilute to 250 ml with water.

5. Dihydrofolate solution (DHF, 10 mmol/l; mercaptoethanol, 14 mmol/l; Tris, 5 mmol/l; EDTA, 1 mmol/l; pH 7.5):

   transfer the contents of one ampoule (25 mg) dihydrofolic acid, 90% (from *Sigma*), corresponding to 22.5 mg DHF, quantitatively into a beaker with 10 ml of a solution containing mercaptoethanol, 14 mmol/l, and HCl, 5 mmol/l. Stir the suspension for 25 min. Dilute a 0.3 ml aliquot of this suspension (5.1 mmol/l) with 2.7 ml dihydrofolate buffer (4). Add 20 µl of this solution (510 µmol/l) to 1.0 ml reaction buffer (3). The absorbance of this suspension at 339 nm should be 0.075 ± 0.005. If necessary the concentration of the suspension is adjusted, altering the quantity of dihydrofolate buffer (4), to bring the absorbance within the proper range. Store this suspension in 0.3 ml aliquots at − 20°C.

6. Substrate solution (DHF, 1 mmol/l; mercaptoethanol, 300 mmol/l; Tris, 100 mmol/l; pH 7.5):

   prepare just before use by thawing a tube of solution (5) and dilute with 2.7 ml dihydrofolate buffer (4).

7. Dihydrofolate reductase, DHFR, (3.5 kU/l):

   individual tubes for assay can be prepared by diluting the contents of one vial DHFR (5 U, suspension in ammonium sulphate solution, 3.6 mol/l, pH 8) with 1.3 ml water. Place exactly 50 µl aliquote into plastic tubes and store at − 20°C until use.

8. Methotrexate standard solution (0.05 to 0.30 µmol/l):

   dissolve 34 mg MTX in 1000 ml Tris buffer (1). Dilute 0.4 ml with MTX-free human serum to 100 ml (0.30 µmol/l). Dilute this solution 1 + 0.2, 1 + 0.5,

1 + 1, 1 + 2, 1 + 5, with MTX free human serum, yielding the following concentration of MTX: 0.25, 0.2, 0.15, 0.10, 0.05 µmol/l.

9. Working solution (Tris buffer, 50 mmol/l; mercaptoethanol, 8.6 mmol/l; EDTA, 1 mmol/l; NADPH, 17 µmol/l; DHFR, 2.5 U/l):

rinse the contents of a 1 mg vial NADPH and of one tube of enzyme with 70 ml reaction buffer (3) into a 100 ml amber bottle, keep mixture on ice during use. Prepare fresh each day.

Stability of reagents and solutions: after preparation, the reaction buffer (3) and dihydrofolate buffer (4) are stable for two months at 4 °C. Dihydrofolic acid (5) is stable for two months at − 20 °C. The working solution (9) and substrate solution (6) must be discarded at the end of the day. The DHFR solution (7) is stable for 4 months at − 20 °C. Tris buffer (1) and EDTA solution (2) are stable for 2 months at 4 °C if microbial contamination is avoided.

Procedure

Collection and treatment of specimen: it is important to collect the serum or plasma sample at the appropriate time following administration of the dose. 50 µl serum or plasma is required per assay. Acceptable anticoagulants are heparin, EDTA and oxalate. Complete mixing of sample is necessary upon thawing.

Stability of substance in the sample: the sample should be stored at 2 °C to 8 °C upon collection and stored frozen (− 20 °C or less) if not analyzed within 24 h.

Assay conditions: samples and calibrators are each assayed once. Incubation time is 120 s (in the spectrophotometer flow-cell); 30 °C; wavelength 339 nm; light path 10 mm; measurement during last 60 s against water.

Establish a calibration curve under exactly the same conditions (within the series). A zero calibrator of water is included.

If $\Delta A/\Delta t$ of the sample is greater than that of the highest standard, dilute the specimen further with water and re-assay.

Calibration curve: plot the $\Delta A/\Delta t$ of the standards *versus* the corresponding MTX concentration, c (µmol/l), on linear graph paper.

Calculation: read or calculate the MTX substance concentration, c (µmol/l), from the calibration curve using the $\Delta A/\Delta t$ value of each sample. For conversion to mass concentration, ρ (µg/l), multiply by 454.5.

Measurement

Pipette into a disposable tube:			concentration in assay mixture
sample or standard solution (8)		0.02 ml	MTX up to 0.6 µmol/l
working solution (9)		1.00 ml	Tris 50 mmol/l
			mercaptoethanol 12.8 mmol/l
			EDTA 1 mmol/l
			DHFR 2.4 U/l
			NADPH 16 mmol/l
substrate solution (6)		0.02 ml	dihydrofolate 19 µmol/l
mix well; immediately aspirate into the clean spectrophotometer flow-cell; after a 60 s delay, read ΔA over a period of 60 s.			

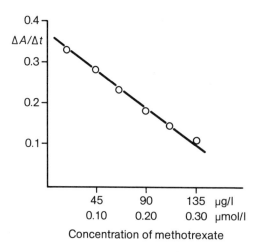

Fig. 4. Typical calibration curve for the assay of methotrexate in serum.

Validation of Method

Precision, accuracy, detection limit and sensitivity: in order to attain an adequate precision with this assay the use of a partially or fully automated analytical system is recommended [35]. With the appropriate mechanized system the within-day and between-day RSD range from 2.0 to 7.1% at MTX concentrations between 0.02 and 0.3 µmol/l [32].

The recovery of MTX added to samples containing drug ranged from 99 to 115% (average 105%) at concentrations ranging from 0.27 to 1.65 µmol/l. The accuracy has been shown by comparison with other methods (enzyme-inhibition assay *vs.* EMIT y = 0.93x + 0.49 µmol/l; r = 0.98, n = 36; enzyme-inhibition assay *vs.* RIA y = 1.00x + 0.04 µmol/l, r = 0.997, n = 62) [32, 35].

The lower limit of the working range is about 0.02 µmol/l as indicated by the precision at low MTX concentrations [32].

Sources of error: no interference by sera from lipaemic, haemolyzed or icteric blood with haemoglobin up to 2 g/l or bilirubin up to 171 µmol/l.

Specificity: inhibition of dihydrofolate reductase has been tested with many compounds whose chemical structures or potential co-administration with MTX could cause concern for potential interference with the assay. Potential MTX metabolites were also tested.

7-OH-MTX shows 1.0% inhibition by MTX; another potential minor metabolite, DAMPA, shows 17%; aminopterin, a structurally similar drug never co-administered with MTX, shows 100% inhibition. There was no noticeable inhibition by folic acid or folinic acid [36].

Trimethoprim is a potent inhibitor of bacterial DHFR and can cause significant interference with the assay. Trimethoprim does not inhibit the mammalian enzyme.

Therapeutic range: cf. chapter 1.11.1, p. 143.

References

[1] *E. Frei, N. Jaffe, M. Tattersal, S. Pittman, L. Parker,* New Approaches to Cancer Chemotherapy with Methotrexate, N. Engl. J. Med. *292,* 846 – 851 (1975).

[2] *R. L. Cappizi, R. C. DeConti, J. C. Marsh, J. R. Bertino,* Methotrexate Therapy of Head and Neck Cancer: Improvement in Therapeutic Index by the Use of Leucovorin Rescue, Cancer Res. *30,* 1782 – 1788 (1970).

[3] *E. S. Henderson, R. H. Adamson, V. T. Oliverio,* The Metabolic Fate of Tritiated Methotrexate II. Absorption and Excretion in Man, Cancer Res. *25,* 1018 – 1024 (1965).

[4] *S. H. Wan, D. H. Huffman, D. C. Azarnoff, R. Stephens, B. Hoogstraten,* Effect of Route of Administration and Effusion on Methotrexate Pharmacokinetics, Cancer Res. *34,* 3487 – 3491 (1974).

[5] *F. M. Balis, J. L. Savitch, W. A. Bleyer,* Pharmacokinetics of Oral Methotrexate in Children, Cancer Res. *43,* 2342 – 2345 (1983).

[6] *P. R. Leme, P. J. Creaven, L. M. Allen, M. Berman,* Kinetic Model for the Disposition and Metabolism of Moderate and High-Dose Methotrexate (NSC-740) in Man, Cancer Chemother. Rep. *59,* 811 – 817 (1975).

[7] *O. D. Shen, D. L. Azarnoff,* Clinical Pharmacokinetics of Methotrexate, Clin. Pharmacokinet. *3,* 1 – 13 (1978).

[8] *S. A. Jacobs, R. S. Stoller, B. A. Chabner, D. G. Johns,* 7-Hydroxymethotrexate as a Urinary Metabolite in Human Subjects and Rhesus Monkeys Receiving High Dose Methotrexate, J. Clinical Invest. *57,* 534 – 538 (1976).

[9] *W. E. Evans, C. F. Stewart, P. R. Hutson, D. A. Cairnes, W. P. Bowman, G. C. Yee, W. R. Crom,* Disposition of Intermediate-Dose Methotrexate in Children with Acute Lymphocytic Leukemia, Drug Intell. Clin. Pharm. *16,* 839 – 842 (1982).

[10] *D. A. Cairnes, W. E. Evans,* High-Performance Liquid Chromatographic Assay of Methotrexate, 7-Hydroxymethotrexate, 4-Deoxy-4-amino-N^{10}-methylpteroic Acid and Sulfamethoxazole in Serum, Urine and Cerebrospinal Fluid, J. Chromatogr. *231,* 103 – 110 (1982).

[11] *D. M. Valerino, D. G. John, D. S. Zaharko, V. T. Oliverio,* Studies of the Metabolism of Methotrexate by Intestinal Flora-I. Identification and Study of Biological Properties of the Metabolic 4-Amino-4-deoxy-*N*-methylpteroic Acid, Biochem. Pharmacol. *21,* 821 – 831 (1973).

[12] *D. S. Rosenblatt, V. M. Whitehead, N. Vera, N. V. Matiaszuk, A. Pottier, M. J. Vuchich, D. Beaulieu,* Differential Effects of Folinic Acid and Glycine, Adenosine, and Thymidine as Rescue Agents in Methotrexate-Treated Human Cells in Relaxation to the Accumulation of Methotrexate Polyglutamate, Mol. Pharmacol. *21,* 718 – 722 (1982).

[13] *J. Galivan,* Evidence for the Cytotoxic Activity of Polyglutamate Derivate of Methotrexate, Mol. Pharmacol. *17,* 105 – 110 (1980).

[14] *H. M. Pinedo, B. A. Chabner,* Role of Drug Concentration Duration of Exposure and Endogenous Metabolites in Determining Methotrexate Cytotoxicity, Cancer Treat. Rep. *61,* 709 – 715 (1977).

[15] *M. H. N. Tattersall, L. M. Parker, S. W. Pitman, E. Frei,* Clinical Pharmacology of High Dose Methotrexate, Cancer Treat. Rep. (part 3) *6,* 25 – 29 (1975).

[16] *W. E. Evans, C. B. Pratt, R. H. Taylor, L. F. Barker, W. R. Crom,* Pharmacokinetics Monitoring of High Dose Methotrexate: Early Recognition of High Risk Patients, Cancer Chemother. Pharmacol. *3,* 161 – 166 (1979).

[17] *A. Nirenberg, C. Mosende, B. M. Mehta, A. L. Gisolfi, G. Rosen,* High Dose Methotrexate Concentrations and Corrective Measures to Avert Toxicity, Cancer Treat. Rep. *61,* 779 – 783 (1977).

[18] *R. G. Stoller, K. R. Hande, S. A. Jacobs, S. A. Rosenberg, B. A. Chabner,* Use of Plasma Pharmacokinetics to Predict and Prevent Methotrexate Toxicity, N. Engl. J. Med. *297,* 630 – 634 (1977).

[19] *R. G. Buice, W. E. Evans, J. Karas, C. A. Nicholas, P. Sidhu, A. B. Straughn, M. C. Meyer, W. R. Crom,* Evaluation of Enzyme Immunoassay, Radioassay, and Radioimmunoassay of Serum Methotrexate, as Compared with Liquid Chromatography, Clin. Chem. *26,* 1902 – 1904 (1980).

[20] *W. R. Crom, E. T. Melton, R. K. Dodge, W. R. Evans,* Evaluation of Fluorescence Polarization Immunoassay and Enzyme Immunoassay of Plasma Methotrexate as Compared with Liquid Chromatography, Drug Intell. Clin. Pharmacol. *18,* 512 (1984).

[21] *J. B. Gushaw, J. G. Miller,* Homogeneous Enzyme Immunoassay for Methotrexate in Serum, Clin. Chem. *24,* 1032 (1978).

[22] *K. E. Rubenstein, R. S. Schneider, E. F. Ullman,* "Homogeneous" Enzyme Immunoassay. A New Immunochemical Technique, Biochem. Biophys. Res. Commun. *47,* 846 – 851 (1972).

[23] *M. Oellerich, P. Engelhardt, M. Schaadt, V. Diehl,* Determination of Methotrexate in Serum by a Rapid, Fully Mechanized Enzyme Immunoassay (EMIT), J. Clin. Chem. Clin. Biochem. *18,* 169 – 174 (1980).

[24] *P. R. Finley, R. J. Williams, F. Griffith, D. A. Lichti,* Adaptation of the Enzyme-Multiplied Immunoassay for Methotrexate to the Centrifugal Analyzer, Clin. Chem. *26,* 341 – 343 (1980).

[25] *L. Arnold, M. Cheng,* Modified *Syva* Emit® Procedure for Methotrexate (MTX) for Increased Sensitivity, Clin. Chem. *26,* 1001 (1980).

[26] *G. W. Anderson, J. E. Zimmerman, F. M. Callahan,* The Use of *N*-Hydroxysuccinimide in Peptide Synthesis, J. Am. Chem. Soc. *86,* 1836 – 1842 (1964).

[27] *B. Ferrua, G. Milano, B. Ly, J. Y. Guennec, R. F. Masseyeff,* An Enzyme Immunoassay Design Using Labelled Antibodies for the Determination of Haptens. Application to Methotrexate Assay, J. Immunol. Methods *60,* 257 – 268 (1983).

[28] *J. Hendel, L. J. Sarek,* Production of Methotrexate Antiserum in Rabbits: The Significance of Immunogen Solubility, Hapten Content, and Mode of Administration on the Antibody Response, Scand. J. Clin. Lab. Invest. *37,* 273 – 278 (1977).

[29] *R. Maiolini, R. Masseyeff,* A Sandwich Method of Enzyme Immunoassay. Application to Rat and Human Alpha-Fetoprotein, J. Immunol. Methods *8,* 223 – 234 (1975).

[30] *M. B. Wilson, P. K. Nakane,* Recent Developments in the Periodate Method of Conjugating Horse Radish Peroxidase to Antibodies, in: *W. Knapp, K. Holubar, G. Wick* (eds.), Immunofluorescence and Related Staining Techniques, Elsevier-North Holland, Biomedical Press, Amsterdam 1978, pp. 215 – 224.

[31] *L. C. Falk, D. R. Clark, S. M. Kalman, T. F. Long,* Enzymatic Assay for Methotrexate in Serum and Cerebrospinal Fluid, Clin. Chem. *22*, 785 – 788 (1976).

[32] *L. F. Brown, G. F. Johnson, D. L. Witte, R. D. Feld,* Enzymic Inhibition Assay for Methotrexate with a Discrete Analyzer, the ABA-100, Clin. Chem. *26*, 335 – 338 (1980).

[33] *W. C. Werkheiser, S. F. Zakrezewski, C. A. Nichol,* Assay for 4-Amino Folic Acid Analogues by Inhibition of Folic Acid Reductases, J. Pharmacol. Exp. Ther. *137*, 162 (1962).

[34] *J. R. Bertino, G. A. Fischer,* VI Techniques for Study of Resistance to Folic Acid Antagonists, Methods Med. Res. *10*, 297 – 307 (1964).

[35] *M. A. Pesce, S. H. Bodourian,* Enzyme Immunoassay and Enzyme Inhibition Assay of Methotrexate, with Use of the Centrifugal Analyzer, Clin. Chem. *27*, 380 – 384 (1981).

[36] *A. M. Imbert, T. Pignon, N. Lena,* Enzymatic Assay for Methotrexate with a Centrifugal Analyzer (Cobas Bio), Clin. Chem. *29*, 1317 – 1318 (1983).

1.12 Amikacin, Gentamicin, Netilmicin and Tobramycin

Determination with Enzyme-multiplied Immunoassay Technique

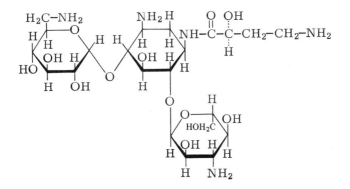

Amikacin

(S)-O-3-Amino-3-deoxy-α-D-glucopyranosyl-(1→ 6)-O-[6-amino-6-deoxy-α-D-glucopyranosyl-(1→ 4)]-N¹-(4-amino-2-hydroxy-1-oxobutyl)-2-deoxy-D-streptamine

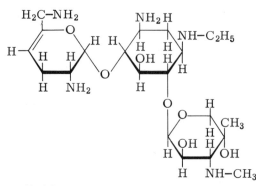

Netilmicin

O-3-Deoxy-4-*C*-methyl-3-(methylamino)-β-L-arabinopyranosyl-(1→6)-*O*-
[2,6-diamino-2,3,4,6-tetradeoxy-α-D-glycero-hex-4-enopyranosyl-(1→4)]-
2-deoxy-*N*¹-ethyl-D-streptamine

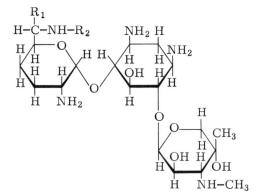

Gentamicin C₁ₐ $R_1 = R_2 = H$

O-3-Deoxy-4-*C*-methyl-3-(methylamino)-β-L-arabinopyranosyl-(1→6)-*O*-
[2,6-diamino-2,3,4,6-tetradeoxy-α-D-*erythro*-hexopyranosyl(1→4)]-2-
deoxy-D-streptamine

Gentamicin C₁ $R_1 = R_2 = CH_3$

O-2-Amino-2,3,4,6,7-pentadeoxy-6-(methylamino)-α-D-*ribo*-heptopyrano-
syl-(1→4)-*O*-[3-deoxy-4-*C*-methyl-3-(methylamino)-β-L-arabinopyrano-
syl-(1→6)]-2-deoxy-D-streptamine

Gentamicin C₂ $R_1 = CH_3, R_2 = H$

O-3-Deoxy-4-*C*-methyl-3-(methylamino)-β-L-arabinopyranosyl-(1→6)-*O*-
[2,6-diamino-2,3,4,6-pentadeoxy-α-D-*ribo*-heptopyranosy-(1→4)]-2-de-
oxy-D-streptamine

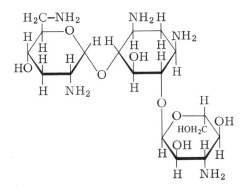

Tobramycin

O-3-Amino-3-deoxy-α-D-glucopyranosyl-(1→6)-*O*-[2,6-diamino-2,3,6-tri-deoxy-α-D-*ribo*-hexopyranosyl-(1→4)]-2-deoxy-D-streptamine

Markus Wenk

General

The chemical structure of the aminoglycoside antibiotics is characterized by the presence of aminosugars which are glycosidically linked to aminocyclitols (hence the name 'aminoglycoside'). The history of this group of drugs began with the discovery of streptomycin in 1944 [1]. Kanamycin, introduced in 1957 [2], was the first aminoglycoside containing a 2-deoxystreptamin, a property which is shared with all aminoglycosides currently in use. Today, amikacin, gentamicin, netilmicin, and tobramycin are the most widely used representatives of these antibiotics and therefore are discussed in this chapter.

All aminoglycosides possess similar antimicrobial and pharmacological properties. They have a broad spectrum of bactericidal activity, but they are used almost exclusively to treat severe infections by *Gram*-negative bacteria, including *Pseudomonas aeruginosa*. The bactericidal activity is due to inhibition of protein synthesis in susceptible organisms. Bacterial resistance to aminoglycosides is predominantly caused by R-factor-mediated aminoglycoside-inactivating enzymes which occur in various forms. Among such enzymes, three main groups have been characterized: phosphotransferases, nucleotidyltransferases, and acetyltransferases. Some of these enzymes have been isolated and can be used in radioenzymatic assays for the quantitative determination of aminoglycosides in biological fluids [3, 4].

Being very polar substances, the aminoglycosides are poorly absorbed from the gastro-intestinal tract and therefore must be administered parenterally. In patients with normal renal function the average dosage is 3 to 5 mg/kg per day, except for amikacin which is given at a dose of 10 to 20 mg/kg per day. The pharmacokinetics are very similar for all aminoglycosides and are most adequately described by a three-compartment open model [5, 6]. Using this model, a steady-state volume of distribution (Vd_{ss}) of 0.45 to 1.34 l/kg has been calculated. The rapid distribution phase with a half-life ($t_{1/2\alpha}$) of about 0.5 h is followed by a β-phase with half-life ($t_{1/2\beta}$) of about 2 h. A prolonged terminal elimination phase with a half-life ($t_{1/2\gamma}$) of about $40 - 100$ h reflects the elimination from deep tissue compartments which include the kidney and inner ear. No metabolism of the aminoglycosides is known so far; they are excreted unchanged by glomerular filtration. Therefore, renal function is the most important factor when dosage regimens are calculated. Several methods have been described for the prediction of concentrations of aminoglycosides in serum based on serum creatinine or creatinine clearance [7 – 9]. However, prediction of the serum aminoglycoside concentration is not always reliable, because of the wide inter-patient variability in distribution and elimination characteristics. The factors so far known to influence this variability are renal function, age, lean body mass, obesity, haematocrit, fever, and interaction with β-lactam antibiotics. In order to overcome these problems other dosing methods have been introduced which take into consideration individual patient pharmacokinetic parameters determined after a test dose [10], or which use the *Bayesian* feedback approach which also calculates individual kinetics based on routine concentration data [11].

The main drawback in the use of aminoglycoside antibiotics is a high incidence (5 to 15%) of clinically important side-effects, namely ototoxicity and nephrotoxicity. While nephrotoxicity is usually transient, ototoxicity often remains irreversible and thus may lead to permanent deafness. The controversy over whether peak or trough serum concentrations contribute more to toxic side-effects of the aminoglycosides is still unsettled. Because neither peak nor trough levels are necessarily representative of drug accumulation, the area under the serum concentration *versus* time curve (AUC) or the average serum concentration achieved within a dosing interval have been proposed as more reliable parameters to predict toxicity [12]. It is important to note, however, that even with the most careful drug monitoring procedure, aminoglycoside accumulation in tissue compartments and the consequent risk of side-effects cannot be avoided.

The rationale for monitoring aminoglycoside serum levels is to maintain effective serum concentrations with a minimal risk of toxic drug accumulation. Therefore, considering the wide inter-patient variability in pharmacokinetics discussed above, aminoglycoside measurements are indicated in the following situations.

1. Severe *Gram*-negative infections in patients with normal renal function in order to ensure adequate bactericidal antibiotic levels.
2. In patients with slight to moderate renal impairment, especially when unstable renal function and/or haemodynamics are present.
3. In patients with advanced renal failure to ensure therapeutically adequate drug levels without accumulation.
4. Neonates and small children as well as elderly patients.

Application of method: in clinical chemistry, pharmacology, and toxicology.

Substance properties relevant in analysis: amikacin (M_r = 585.62), gentamicin (C_1: M_r = 477.59; C_{1a}: M_r = 449.54; C_2: M_r = 463.57), netilmicin (M_r = 475.60), and tobramycin (M_r = 467.52) are water-soluble basic antibiotics. In pharmaceutical preparations all aminoglycosides are used in their sulphate form.

Methods of determination: a large number of methods have been developed for the quantitative determination of aminoglycoside concentrations: microbiological agar-diffusion assay (bioassay), chromatographic techniques (gas chromatography, high-performance liquid chromatography, thin-layer chromatography), spectrophotometric assays, radioenzymatic assays, urease assay, luciferase assay, immunological assays (radioimmunoassay, enzyme-multiplied immunoassay technique, fluorescence polarization immunoassay, haemagglutination inhibition assay, solid-phase peroxidase immunoassay, quenching fluoroimmunoassay, thermometric enzyme-linked immunosorbent assay, reactant-labelled fluorescent immunoassay, latex agglutination assay, particle-enhanced turbidimetric immunoassay, radial partition immunoassay).

Only a few of these many techniques are of practical importance. As with other antibiotics, bioassay is still widely used. However, because this technique is non-specific and time-consuming, it is being replaced by faster and more specific assays. Homogeneous enzyme-multiplied immunoassay technique, fluorescence polarization immunoassay, and radioimmunoassay are among the most widely employed advanced methods. High-performance liquid chromatography (HPLC) is the only technique which is capable of distinguishing between all three gentamicin components individually. Because it needs labour-intensive sample preparation and complex and expensive equipment, HPLC is used as a reference method and for research work rather than for routine drug monitoring of the aminoglycosides.

International reference method and standards: neither standardization at the international level nor the existence of reference standard materials is known so far.

Assay

Method Design

The enzyme-multiplied immunoassay technique (EMIT®) discussed here for all four aminoglycosides (amikacin, gentamicin, netilmicin, and tobramycin) has the same design as, and is based on, the method described by *Rubenstein et al.* [13].

Principle: as described in detail on pp. 7, 8.

In this assay an aminoglycoside which is covalently bound to glucose-6-phosphate dehydrogenase (G6P-DH) competes with the free drug in the patient's serum for an antibody directed against the aminoglycoside. When bound to the enzyme-labelled aminoglycoside, the antibody blocks the active site of the enzyme. However, the more drug there is in the patient's serum, the less enzyme-labelled drug is bound, so that more of the enzyme remains active.

Thus, the enzyme activity, measured as a change in absorbance at 339 nm per unit time ($\Delta A/\Delta t$), due to the conversion of NAD to NADH, correlates directly with the aminoglycoside concentration in the serum sample.

The assay can be mechanized by using either the EMIT® auto carousel* in connection with the EMIT® Lab 5000 System* or a centrifugal analyzer [14].

Selection of assay conditions and adaptation to the individual characteristics of the reagents: specific antibodies against the corresponding aminoglycoside were raised in sheep after conjugation of the drug with bovine serum albumin [15]. Initially sheep are immunized with a saline solution of 1 mg of this antigen in complete *Freund's* adjuvant and subsequently receive monthly injections of the same amount of antigen in complete *Freund's* adjuvant. The sheep are bled every month and the γ-globulin fraction is isolated with 50% ammonium sulphate.

Glucose-6-phosphate dehydrogenase from *Leuconostoc mesenteroides* is coupled to aminoglycoside in a complex three-step procedure (cf. ref. [15] for details). The aminoglycoside is first functionalized with a sulphydryl group on its deoxystreptamine ring. Then the enzyme is modified with a bromoacetyl group using bromoacetyl-glycine and, finally, is coupled to the sulphydryl-substituted aminoglycoside. In the case of gentamicin, only gentamicin C_1 was conjugated to the enzyme. However, no significant bias was observed in quantitation of the other gentamicins, C_{1a} and C_2. Interference from endogenous G6P-DH is avoided by use of the coenzyme NAD which is converted only by the bacterial enzyme.

The optimum quantities and concentrations of samples and reagents are dependent on the characteristics of the antibodies present in the antiserum.

The desired assay range is from 1 to 16 mg/l for gentamicin and tobramycin, 1 to 12 mg/l for netilmicin, and from 2.5 to 50 mg/l for amikacin. The assay needs 50 μl serum sample.

These parameters being fixed, the optimum ratio of antibody to enzyme conjugate in the presence of different aminoglycoside concentrations within the desired assay range must be determined.

First, a prospective enzyme conjugate concentration is chosen, to which gradually increasing amounts of the corresponding antibody are added. Enzymatic activity is increasingly inhibited as the amount of antibody is raised [15]. In the case of gentamicin, a ratio of antibody to enzyme was finally chosen at which the enzyme was 39% inhibited (5 μl sheep anti-gentamicin antiserum, 2.34×10^{-11} mol enzyme conjugate). In order to assess the system's ability to discriminate among serum aminoglycoside concentrations likely to be encountered during aminoglycoside therapy the following optimization procedure is used: calibrators with the same aminoglycoside concentrations which will be used for therapeutic monitoring are assayed according to the

---

* *Syva Corp.,* Palo Alto, CA, U.S.A.

regular assay protocol (cf. p. 167) at a fixed enzyme conjugate concentration and with different amounts of antibody. A zero calibrator serves as a reagent blank and all measurements are corrected for this value. The corrected $\Delta A/\Delta t$ readings of each calibrator thus obtained are plotted against the amount of antibody added. The antibody quantity is then selected according to the following considerations:

- an optimum response should be obtained at a point within the assay range at which assay sensitivity is most desirable (depending on the aminoglycoside used);
- the response of the assay between aminoglycoside calibration values reflecting the discriminatory power of the assay configuration should attain a maximum possible value.

Equipment

Spectrophotometer or spectral-line photometer for the exact determination of the absorbance at 339 or Hg 334 nm with a thermostatted micro-scale flow-cell, which maintains the temperature at $30.0 \pm 0.1\,°C$. The spectrophotometer should be connected to a data handling device which prints out the time-dependent readings of the photometer. A semi-automatic or automatic pipetter-diluter is highly recommended, although high-precision manual measuring systems may be used.

Reagents and Solutions

Purity of reagents: G6P-DH from *Leuconostoc mesenteroides* should be of the best available purity. Chemicals are of analytical grade.

Preparation of solutions* (for 100 determinations): all solutions in re-purified water (cf. Vol. II, chapter 2.1.3.2).

1. Tris buffer (Tris, 55 mmol/l; Triton X-100, 0.1 ml/l; pH 8.0):

   dissolve 1.996 g Tris base in 140 to 160 ml water; add 0.03 ml Triton X-100; adjust to pH 8.0 with HCl, 1.0 mol/l. Make up with water to 300 ml.

   Alternatively, dilute concentrated buffer provided with each kit with water according to the instructions of the manufacturer.

---

* Reagents, calibrators and buffer are commercially available from *Syva Co.,* Palo Alto, U.S.A. The solutions contain sodium azide as a preservative.

2. Antibody/substrate solution (γ-globulin*, ca. 110 mg/l; G-6-P, 66 mmol/l; NAD, 40 mmol/l; Tris, 55 mmol/l; pH 5.0):

reconstitute available lyophilized preparation with 6.0 ml water.

3. Conjugate solution (G6P-DH**, 0.7 to 2.7 kU/l; Tris, 55 mmol/l; pH 8.0):

reconstitute available lyophilized preparation with 6.0 ml water. This reagent must be standardized to match the antibody/substrate solution (cf. p. 164).

4. Aminoglycoside standard solutions (gentamicin, netilmicin, tobramicin, 1.0 to 16 mg/l; amikacin, 2.5 to 50 mg/l):

dissolve 40 mg gentamicin, netilmicin or tobramycin, or 125 mg amikacin in 100 ml Tris buffer (1). The exact amount of each aminoglycoside depends on the batch-specific potency. Dilute 0.4 ml with aminoglycoside-free human serum containing sodium azide (5 g/l) to 10 ml.

Dilute gentamicin, netilmicin and tobramycin standard solution (16*** mg/l) 1 + 1, 1 + 3, 1 + 7 and 1 + 15 with aminoglycoside-free human serum yielding the following concentrations: 8.0, 4.0, 2.0 and 1.0 mg/l.

For amikacin, dilute the standard solution (50 mg/l) 1 + 1.5, 1 + 4, 1 + 9 and 1 + 19 with aminoglycoside-free human serum, yielding the following concentrations: 20.0, 10.0, 5.0 and 2.5 mg/l. Take pure serum as zero calibrator.

Alternatively, reconstitute each of the available standard preparations with 1 ml water.

Stability of reagents and solutions: after reconstitution, the antibody/substrate solution (2), conjugate solution (3) and standards (4) must be stored at room temperature (20 °C to 25 °C) for at least one hour before use. The reagents are then stable for 12 weeks, if stored at 2 °C to 8 °C. The buffer solution (1) can be used for 12 weeks when stored at room temperature.

Procedure

Collection and treatment of specimen: draw blood for aminoglycoside peak serum levels 60 min after intramuscular injection or 30 to 60 min after the end of a intravenous infusion. Collect trough serum samples immediately before the next dose. During blood collection avoid any contamination with residues from drug application. Therefore, blood specimens should never be drawn through the same line

---

   * Standardized preparation from immunized sheep.
  ** Coupled to the aminoglycoside.
*** 12 mg/l in the commercially available netilmicin EMIT® standard preparation.

through which the dose has been given. Plasma can be used as an alternative to serum. Plasma can be obtained using oxalate or EDTA. Heparin is not recommended as it has been shown to interfere with aminoglycosides [16 – 18]. Biological specimens other than serum can also be measured with this assay (e.g. urine, tissue homogenate). However, the calibration curve must be made up with the appropriate medium and the method re-evaluated [19, 20].

Stability and storage of specimens: although the aminoglycosides are stable in solution at room temperature, samples should be stored frozen at $-20°C$ if they are not immediately analyzed (cf. Sources of error, p. 169).

Assay conditions: samples and zero calibrator are measured in duplicate. Single determinations are made for the remaining standards. Each assay requires 0.05 ml sample or standards (4) each of which is pre-diluted with 0.25 ml Tris buffer (1). The pre-diluted sample can be used for up to five determinations. Incubation time is 45 s (in the spectrophotometer flow-cell); 30°C; measurement during the last 30 s against water; wavelength 339 or Hg 334 nm; light path 10 mm.

Establish a calibration curve under exactly the same conditions (within the series). The zero calibrator serves as a reagent blank; all measurements are corrected for this value.

Measurement

Pipette into a disposable beaker:			concentration in assay mixture	
pre-diluted sample or standard solution (4)		0.05 ml	gentamicin, tobramycin, netilmicin amikacin	up to 148 µg/l up to 463 µg/l
Tris buffer	(1)	0.25 ml	Tris	52 mmol/l
antibody/substrate solution	(2)	0.05 ml	γ-globulin G-6-P NAD	ca. 6 mg/l 3.7 mmol/l 2.2 mmol/l
Tris buffer	(1)	0.25 ml		
conjugate solution	(3)	0.05 ml	G6P-DH*	variable ca. 40 – 150 U/l
Tris buffer	(1)	0.25 ml		
mix well; immediately aspirate into the spectro-photometer flow-cell; after a 15 s delay read ΔA over a period of 30 s.				

* Coupled to the aminoglycoside.

If ΔA per 30 s ($\Delta A/\Delta t$) of the sample is greater than that of the highest calibrator, dilute the sample further with Tris buffer (1) and re-assay.

Calibration curve: plot the corrected $\Delta A/\Delta t$ readings of the standards *versus* the corresponding aminoglycoside concentrations (mg/l). By use of special graph paper* matched with the reagents, a linear calibration curve is obtained (Fig. 1). The construction of this paper is based on the logit-log function. So far the log-logit model (cf. Vol. I, p. 243) has proved to be useful in fitting calibration curves of EMIT assays [21].

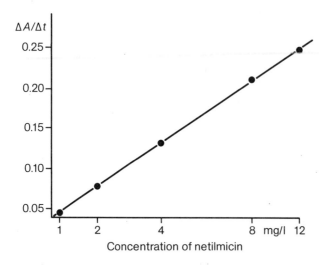

Fig. 1. Typical calibration curve for the assay of netilmicin in serum.

Calculation: read the aminoglycoside mass concentration, ρ (mg/l), corresponding to the mean corrected $\Delta A/\Delta t$ reading of the sample from the calibration curve. To convert from mass concentration to substance concentration, c (μmol/l), use a conversion factor of 1.71 for amikacin ($M_r = 585.6$), 2.16** for gentamicin (gentamicin-C_1: $M_r = 477.6$; gentamicin-C_{1a}: $M_r = 449.5$; gentamicin-C_2: $M_r = 463.6$), 2.11 for netilmicin ($M_r = 475.6$), and 2.14 for tobramycin ($M_r = 467.5$).

Validation of Method

Precision, accuracy, detection limit and sensitivity: the performance characteristics of this assay are very similar for all aminoglycosides [22 – 27]. For concentrations be-

---

* Supplied by *Syva Co.,* Palo Alto, CA, U.S.A. for using this graph paper, multiply each $\Delta A/\Delta t$ value by 2.667.
** Mean value.

tween 1.0 and 12.0 mg/l (gentamicin, netilmicin, tobramycin) and between 2.5 and 50 mg/l (amikacin), between-days RSD ranged from 1.9 to 5.3%. The recoveries of aminoglycosides added to drug-free pooled human serum were between 92 and 116%. The accuracy of the EMIT has been proved by comparison with RIA (amikacin, tobramycin), microbiological assay (gentamicin), and radioenzymatic assay (netilmicin). Linear regression analysis of results from patients' samples measured by both methods yielded correlation coefficients between 0.95 and 0.99. Detection limit (as defined in chapter 1.2.2, p. 20) is about 0.5 mg/l for gentamicin, netilmicin and tobramycin, and about 1 mg/l for amikacin. However, the detection limit for samples with low aminoglycoside concentrations can be improved by omitting the first sixfold predilution step [28]. Sensitivity depends on the shape of the calibration curve and is highest on the steepest part of the curve, plotted on linear graph paper.

Sources of error: heparin interferes with these assays if the heparin concentration exceeds 2×10^3 USP units/l (tobramycin, netilmicin) and 10^5 USP units/l (gentamicin, amikacin). Thus, heparin-coated tubes should be avoided [16–18].

High concentrations of β-lactam antibiotics, namely azlocillin, carbenicillin and ticarcillin, inactivate tobramycin, gentamicin and netilmicin in serum [29]. Amikacin is inactivated by cephalotin and moxalactam [30]. This reaction between the aminoglycoside and the β-lactam antibiotics depends on the concentrations of both components, temperature and time. Because the concomitant administration of β-lactam antibiotics is often unknown to the laboratory, all samples should be stored frozen at $-20°C$ until assayed.

Sera from severely lipaemic blood may cause poor reproducibility and questionable quantitation.

Specificity: the specificity depends on the quality of antiserum. Cross-reactions between some of the aminoglycosides have been described for all four assays. Amikacin assay: cross-reaction with tobramycin and gentamicin [22]. Gentamicin assay: cross-reaction with netilmicin and sisomicin [25]. Netilmicin assay: cross-reaction with sisomicin and gentamicin [26]. Tobramycin assay: cross-reaction with kanamycin and amikacin [27]. However, these cross-reactions have no clinical importance, as patients are not simultaneously treated with two aminoglycosides. Cross-reaction was also tested with a wide selection of penicillins and cephalosporins, chloramphenicol, tetracycline, sulfamethazole and trimethoprim. So far, no cross-reaction has been described with these agents.

Therapeutic ranges: peak serum levels of amikacin should be between 15 and 25 mg/l and trough levels below 5 mg/l. In patients treated with gentamicin, netilmicin or tobramycin, peak levels should be between 5 and 10 mg/l and trough levels below 2 mg/l. In patients with impaired renal function it is important to prevent toxic accumulation of the aminoglycoside antibiotics, but also to avoid under-treatment of the infection after dosage reduction or prolongation of the dosage interval [31].

References

[1] *A. Schatz, S. Bugie, S. A. Waksman,* Streptomycin a Substance Exhibiting Antibiotic Activity against *Gram*-positive and *Gram*-negative Bacteria, Proc. Soc. Exp. Biol. Med. *57*, 244 – 248 (1944).

[2] *H. Umezawa, M. Ueda, K. Maeda, K. Yagashita, S. Kondo, Y. Okami, R. Utahara, Y. Osato, K. Nitta, T. Takeuchi,* Production and Isolation of a New Antibiotic, Kanamycin, J. Antibiotics *10*, 181 – 189 (1957).

[3] *A. L. Smith, D. H. Smith,* Gentamicin: Adenine Mononucleotic Transferase: Partial Purification, Characterization, and Use in the Clinical Quantitation of Gentamicin, J. Infect. Dis. *129*, 391 – 401 (1974).

[4] *J. M. Broughall, D. S. Reeves,* The Acetyltransferase Enzyme Method for the Assay of Serum Gentamicin Concentrations and a Comparison with Other Methods, J. Clin. Pathol. *28*, 140 – 145 (1975).

[5] *J. J. Schentag, G. Lasezkay, T. J. Cumbo, M. E. Plaut, W. J. Jusko,* Accumulation Pharmacokinetics of Tobramycin, Antimicrob. Agents Chemother. *13*, 649 – 656 (1978).

[6] *M. Wenk, P. Spring, S. Vozeh, F. Follath,* Multicompartment Pharmacokinetics of Netilmicin, Eur. J. Clin. Pharmacol. *16*, 331 – 334 (1979).

[7] *L. C. Dettli,* Drug Dosage in Patients with Renal Disease, Clin. Pharmacol. Ther. *16*, 274 – 280 (1974).

[8] *R. A. Chan, E. J. Benner, R. D. Hoeprich,* Gentamicin Therapy in Renal Failure: A Nomogram for Dosage, Ann. Intern. Med. *76*, 775 – 778 (1978).

[9] *J. H. Hull, F. A. Sarubbi,* Gentamicin Serum Concentrations: Pharmacokinetic Predictions, Ann. Intern. Med. *85*, 183 – 189 (1976).

[10] *R. J. Sawchuk, D. E. Zaske, R. J. Cipolle, W. A. Wargin, R. G. Strate,* Kinetic Model for Gentamicin Dosing with the Use of Individual Patient Parameters, Clin. Pharmacol. Ther. *21*, 362 – 369 (1977).

[11] *M. E. Burton, C. Brater, P. S. Chen, R. B. Day, P. J. Huber, M. R. Vasko,* A *Bayesian* Feedback Method of Aminoglycoside Dosing, Clin. Pharmacol. Ther. *37*, 349 – 357 (1985).

[12] *F. Follath, M. Wenk, S. Vozeh,* Plasma Concentration Monitoring of Aminoglycosides, J. Antimicrob. Chemother. *8* (suppl. A), 37 – 43 (1981).

[13] *R. Rubenstein, R. S. Schneider, E. Ullman,* Homogeneous Enzyme Immunoassay, a New Immunochemical Technique, Biochem. Biophys. Res. Commun. *47*, 846 – 851 (1972).

[14] *M. A. Pesche, S. H. Bodourian,* Enzyme Immunoassay of Gentamicin with Use of a Centrifugal Analyzer, Clin. Chem. *27*, 1460 – 1462 (1981).

[15] *P. Singh, D. K. Leung, G. L. Rowley, C. Gagne, E. F. Ullman,* A Method for Controlled Coupling of Amino Compounds to Enzymes: A Homogeneous Enzyme Immunoassay for Gentamicin, Anal. Biochem. *104*, 51 – 58 (1980).

[16] *D. J. Krogstad, G. G. Granich, P. R. Murray, M. A. Pfaller, R. Valdes,* Heparin Interferes with the Radioenzymatic and Homogeneous Enzyme Immunoassays for Aminoglycosides, Clin. Chem. *28*, 1517 – 1521 (1982).

[17] *M. I. Walters, W. H. Roberts,* Gentamicin, Heparin Interactions: Effects on Two Immunoassays and on Protein Binding, Ther. Drug Monit. *6*, 199 – 302 (1984).

[18] *M. E. O'Connell, K. L. Heim, C. E. Hastenson, G. R. Matzke,* Analytical Accuracy of Determinations of Aminoglycoside Concentrations by Enzyme Multiplied Immunoassay, Fluorescence Polarization Immunoassay, and Radioimmunoassay in the Presence of Heparin, J. Clin. Microb. *20*, 1080 – 1082 (1984).

[19] *R. A. Giuliano, G. A. Verpooten, D. E. Pollet, L. Verbist, S. L. Scharpe, M. E. De Broe,* Improved Procedure for Extracting Aminoglycosides from Renal Cortical Tissue, Antimicrob. Agents Chemother. *25*, 783 – 784 (1984).

[20] *A. P. Provoost, W. P. Van Schalkwijk, O. Adejuyigbe, W. B. Van Leeuwen, J. H. T. Wagenvoort,* Determination of Aminoglycosides in Rat Renal Tissue by Enzyme Immunoassay, Antimicrob. Agents Chemother. *25*, 497 – 498 (1984).

[21] *P. Sandel, W. Vogt,* Rechnerunterstützte Ergebniswerterrechnung bei Enzymimmunoassays, in: *W. Vogt* (ed.), Enzymimmunoassay, Georg Thieme Verlag, Stuttgart 1978, pp. 67 – 72.

[22] *H. Fukuchi, M. Yoshida, S. Tsukiai, T. Kitaura, T. Konishi,* Comparison of Enzyme Immuno-assay, Radioimmunoassay, and Microbiologic Assay for Amikacin in Plasma, Am. J. Hosp. Pharm. *41*, 690 – 693 (1984).

[23] *J. M. Damien, R. J. Courcol, N. Houdret, M. Lhermitte, G. R. Martin,* Amikacin Assay: Correlation between Rapid Bioassay, Enzyme Immuno-assay (EMIT) and Fluoroimmuno-assay (FIA), Ann. Biol. Clin. *42*, 217 – 220 (1984).

[24] *Th. D. O'Leary, R. M. Ratcliff, T. D. Geary,* Evaluation of an Enzyme Immunoassay for Serum Gentamicin, Antimicrob. Agents Chemother. *17*, 776 – 778 (1980).

[25] *D. Phaneuf, E. Francke, H. C. Neu,* Rapid Reproducible Enzyme Immunoassay for Gentamicin, J. Clin. Microbiol. *11*, 266 – 269 (1980).

[26] *M. Wenk, R. Hemman, F. Follath,* Homogeneous Enzyme Immunoassay for Netilmicin, Antimicrob. Agents Chemother. *22*, 954 – 957 (1982).

[27] *E. L. Francke, S. Srinivasan, P. Labthavikul, H. C. Neu,* Rapid Reproducible Enzyme Immunoassay for Tobramycin, J. Clin. Microbiol. *13*, 93 – 96 (1981).

[28] *C. J. Voegeli, G. J. Burckart,* Improving the Sensitivity of Gentamicin Enzyme Immunoassay, Clin. Chem. *28*, 248 (1982).

[29] *L. J. Riff, J. L. Thomason,* Comparative Aminoglycoside Inactivation by Beta-lactam Antibiotics: Effect of a Cephalosporin and Six Penicillins on Five Aminoglycosides, J. Antibiot. *35*, 850 – 857 (1982).

[30] *J. H. Jorgensen, S. A. Crawford,* Selective Inactivation of Aminoglycosides by Newer Beta-lactam Antibiotics, Curr. Ther. Res. *32*, 25 – 35 (1982).

[31] *M. Wenk, S. Vozeh, F. Follath,* Serum Level Monitoring of Antibacterial Drugs: A Review, Clin. Pharmacokin. *9*, 475 – 492 (1984).

1.13   Gentamicin

Determination with Competitive Enzyme-linked Immunosorbent Assay

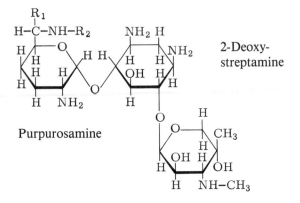

C₁   R₁ = R₂ = CH₃
C₂   R₁ = CH₃, R₂ = H
C₁ₐ   R₁ = R₂ = H

Jim Standefer

General

Gentamicin is one of the most widely used aminoglycoside antibiotics in hospitalized patients. Gentamicin was isolated and purified from the actinomycete *Micromono-spora purpurea* in 1963 [1]. Because gentamicin has a wider spectrum of activity compared to other aminoglycoside antibiotics, it has gained wide acceptance in the treatment of infections due to a variety of *Gram*-negative bacilli. In particular, gentamicin is especially effective against *Gram*-negative organisms such as *Proteus, Escherichia coli, Pseudomonas, Klebsiellae pneumoniae,* and *Serratia marcescens.* These organisms are associated with a number of severe diseases including bacteraemia, pneumonia, pyelonephritis, and osteomyelitis.

The antibiotic efficacy of gentamicin is directly related to its concentration within those bacteria which are sensitive to the drug.

Because the drug is actively transported across cell membranes, its concentration within the cell depends on the potency of the active transport system in the cell membrane. Assuming adequate active transport across the cell membrane, gentamicin accumulates inside the cell and combines with the 30-S ribosome to disrupt the transmission of genetic code and subsequently to disrupt protein synthesis. The rate of transport across cell membranes and therefore the rate of accumulation inside the cell varies with the organism and may account for a variable sensitivity to the drug. Since gentamicin is a polar compound (cf. formula), alterations in pH, osmolality, cellular respiration, and ionic composition of the local extracellular regions may account for variable membrane transport.

Resistance to gentamicin may develop as a result of low accumulation of the drug within cells (transportation failure), little or no binding of the drug to ribosomes (affinity failure), or increased drug metabolism which may occur when the organism produces enzymes that rapidly metabolize and degrade the aminoglycoside. The last process most likely accounts for resistance to gentamicin that may be acquired by the organism. Because of the specificity of enzymatic inactivation, an organism may lose its sensitivity to one aminoglycoside while remaining sensitive to another.

Therapeutic effectiveness of gentamicin has been linked directly to its concentration in serum. More particularly, the concentrations of gentamicin at two specific times during the dosage interval have been used to indicate optimum therapeutic effect. First, the maximum or "peak" gentamicin concentration, i.e., 30 to 60 min after completion of an i.v. infusion, correlates well with therapeutic response. In one study, 26 of 33 patients (79%) who achieved a peak serum gentamicin level of 5 mg/l responded to treatment within the first 72 hours [2]. Accordingly, peak gentamicin concentrations in the range of 5 to 10 mg/l have been recommended for effective treatment of serious *Gram*-negative infections [3]. This relatively narrow range must be closely regulated since toxicity has been associated with peak gentamicin concentrations maintained above 12 to 15 mg/l [4].

In addition to the peak gentamicin concentration, a second specimen collected at the time of the minimum or "trough" drug concentration, i.e., within 10 minutes prior to the next dose, has been used to indicate therapeutic effectiveness as well as tissue accumulation of gentamicin. When gentamicin concentrations in serum are below the minimum inhibitory concentration for prolonged periods, increases in bacteraemic "breakthroughs" occur [5]. In this respect, the dosing interval is particularly important, since serum concentrations may be lower than the minimum inhibitory concentration for considerable portions of long dosing intervals in patients in whom the elimination half-lives of gentamicin are short. Conversely, trough concentrations of gentamicin greater than 3 mg/l are associated with increased risk of ototoxicity and indicate excessive tissue accumulation of the drug [6, 7]. Thus, for optimum therapy, the concentration of gentamicin in serum should be regulated to maintain a peak concentration within the range of 5 to 10 mg/l and a trough concentration of less than 2 mg/l [6].

Gentamicin must be given to patients by either the intramuscular or the intravenous routes since it is not absorbed from the gastro-intestinal tract. The poor absorption from the intestinal tract is attributed to the polar, cationic nature of gentamicin at the pH of the intestine. Less than 1% of a dose is absorbed after oral or rectal administration; however, gentamicin is rapidly absorbed after topical application to serosal surfaces. Absorption from an intramuscular injection is usually complete in ambulatory patients; however, absorption from this route is unreliable in critically ill patients [8], in whom intravenous administration is preferred. While absorption from topical application sites can be appreciable during procedures such as wound irrigation [9], topical use of aminoglycosides has been discouraged in order to limit bacterial resistance induced by such chronic low-level exposure.

Aminoglycosides are distributed rapidly (5 to 15 min distribution half-life) into highly perfused organs and subsequently into a variety of extracellular fluids, including bile [10], synovial fluid [11], and lymph [12]. Following gentamicin's initial rapid distribution, a second slower distribution follows during which gentamicin slowly accumulates in tissues and is readily excreted into the urine [13]. A two-compartment pharmokinetic model has been proposed to account for the biphasic nature of gentamicin distribution and excretion [7]. The fact that about 50% of a parenteral dose of gentamicin is excreted unchanged in the urine during the first 24 hours is consistent with this model. Additionally, the concentration of gentamicin in urine decreases exponentially and traces of a single dose may be detected for up to 30 days [14]. During gentamicin therapy, urine concentrations of the drug may range between 50 to 200 mg/l [15]. Therapeutic concentrations of gentamicin are achieved during the phase of slower distribution and elimination in most extracellular fluids except cerebrospinal and vitreous fluids. Because of the lower penetration rate through the CSF barrier, an adequate cerebrospinal fluid concentration for treatment of meningitis is achieved by administering gentamicin by either intraventricular or intralumbar routes [16, 17].

Accumulation of gentamicin in specific fluids and tissues may account for its reported toxicity. For example, ototoxicity has been related to the accumulation of gentamicin in perilymph and endolymph leading directly to vestibular damage [11]. Furthermore, very high concentrations of gentamicin have been found at autopsy in kidneys of patients who had recieved gentamicin therapy [18], suggesting that such accumulation may account for the specific nephrotoxicity associated with aminoglycoside therapy.

Binding of gentamicin to serum proteins is variable (usually less than 10%) and is affected by ionic strength. Because of gentamicin's relatively low serum protein binding and its low penetration rate into cells, the volume of distribution of gentamicin is essentially equal to extracellular fluid volume, i.e., about 25% of lean body-weight [19]. The half-life of gentamicin ranges from 1 to 6 h (average 2.5 h) in healthy adult volunteers without sepsis, to 0.4 to 33 h in patients with *Gram*-negative sepsis but normal renal function [20 – 22].

Application of method: in clinical pharmacokinetics, clinical chemistry, pharmacology, and toxicology.

Substance properties relevant in analysis: gentamicin is soluble in water, methanol, ethanol and acetone, but is insoluble in benzene and chlorinated hydrocarbon solvents. Gentamicin, which has a pK_a of 8.2, forms a freely soluble salt in aqueous acidic media. Gentamicin available for therapy consists of three related compounds, C1, C2, and C1a, with molecular weights of 477.6, 463.6, and 449.5, respectively. As shown in the formula (cf. p. 172), gentamicin consists of two amino sugars, purpurosamine and garosamine, glycosidically linked to an aminocyclitol, deoxystreptamine, in 1 to 4 and 1 to 6 linkages.

Methods of determination: gentamicin has been assayed by a variety of methods including microbiological [23], radioenzymatic [24] and radiometric [25] assays, gas-liquid chromatography [26], high-performance liquid chromatography [27, 28], fluorescence-quench assay [29], bioassay [30], spectrophotometric assay [31], substrate-labelled fluorescent immunoassay [32, 33], radioimmunoassay [34, 35], latex agglutination immunoassay [36], bioluminescent immunoassay [37, 38], fluorescence polarization immunoassay [39, 40], competitive enzyme-immunoassay [41], and homogeneous enzyme-immunoassay [42]. The fluorescence-polarization and EMIT procedures are widely used because the analysis time is relatively short, the procedures are compatible with automation, less technical expertise is required, and the problems associated with disposal of radioactive wastes are avoided. The chromatographic methods have the advantage that each of the three main components of gentamicin, C1, C2, and C1a, can be quantitated.

International reference method and standards: no reference standards or reference method are known.

Assay

Method Design

Principle

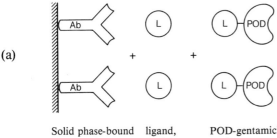

(a)

Solid phase-bound   ligand,   POD-gentamicin
rabbit anti-genta-   analyte,   conjugate
micin antibody   gentamicin

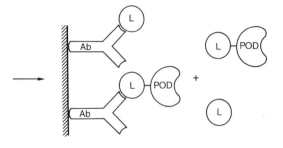

POD-labelled gentamicin-Ab complex

(b) $2 \text{ ABTS} + H_2O_2 \xrightarrow{\quad\quad} 2 \text{ ABTS-radical cations}^* + H_2O + 1/2\, O_2$.

Antibody to gentamicin that has been attached to a polystyrene solid phase reacts with a mixture of gentamicin conjugated to horseradish peroxidase** and gentamicin in the sample. After the reaction is complete, the solid phase is washed and a substrate for peroxidase is added to the reaction well. The increase in absorbance due to the enzyme reaction product is measured at 410 nm after 10 min of reaction ($\Delta A/\Delta t$) and is indirectly related to the gentamicin concentration.

Selection of assay conditions and adaptation to the individual characteristics of the reagents: a polystyrene microtitre plate is used for the reaction (96-well, flat bottom, *Linbro Chemical Co.,* New Haven, CT, 06065, U.S.A.). The wells of the plate are prepared for the immunological reaction by pre-coating each well with albumin followed by coating with antibody to gentamicin (cf. Appendix, p. 182). The pre-coating with albumin is necessary to limit non-specific binding of extraneous proteins to the reaction vessel walls. The amount of antibody, i.e. the dilution of antiserum, used to coat the reaction wells will vary according to the antibody titre and should be determined empirically. This may be accomplished by using a series of antiserum dilutions to coat the reaction wells. The optimum antibody dilution is selected after comparing the calibration curves generated for each dilution. Furthermore, the appropriate dilution of enzyme-conjugate is determined by assaying a series of standards (1 to 15 mg/l) using 1:100, 1:200, 1:500 and 1:1000 dilutions of the conjugate. Three calibration curves generated for dilutions of antibody of 250-, 500- and 1000-fold are shown in Fig. 1. While the 250-fold dilution is recommended for routine use, if greater sensitivity is required the 1000-fold dilution may be selected.

While the optimum pH for the immune reaction is approximately 7.0, the enzyme reaction is more rapid at pH 4.0.

No stabilizers are added during the reaction; however, the washing solution that is used to clean the reaction wells following the immune reaction contains a high concentration of sodium chloride as well as a detergent (Tween 80). These ingredients are necessary to limit non-specific binding of conjugate to the reaction wells.

---

* Cf. chapter 1.10, p. 116.
** POD, donor: hydrogen-peroxide oxidoreductase, EC 1.11.1.7.

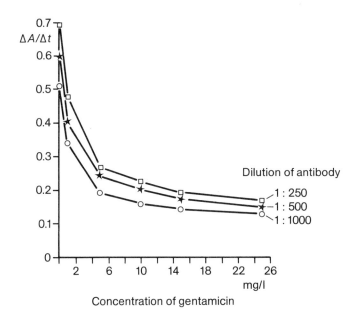

Fig. 1. Typical calibration curves for the assay of gentamicin in serum (three dilutions of antibody).

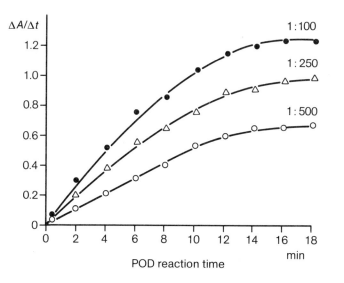

Fig. 2. The effect of conjugate dilution on linearity of the enzyme reaction. The peroxidase-gentamicin conjugate was diluted as indicated and linearity throughout 10 min reaction was obtained only with the 500-fold dilution.

The reaction times and temperatures of the immune and enzymatic reactions must be adjusted to ensure that the enzymatic reaction proceeds in a linear fashion. Higher temperatures and longer times will promote binding of the peroxidase-gentamicin conjugate to the antibody on the solid phase; and as more conjugate is bound, more substrate will be converted during the subsequent enzyme reaction. Consequently, more substrate will be required to maintain linear kinetics during the enzymatic reaction. It is important to note that non-linear conditions during the enzymatic reaction could lead to an inaccurate quantitation of the analyte.

As shown in Fig. 2, dilution of the conjugate used during the immune reaction influences the linearity of substrate conversion during the subsequent enzymatic reaction. Obviously, if the enzyme reaction interval is sufficiently long, substrate exhaustion may occur and non-zero order kinetics will lead to inaccurate quantitation.

Equipment

A spectrophotometer capable of exact measurement at 410 nm with a flow-cell requiring a total volume of less than 0.3 ml is required. The reaction mixture in the microtitre well may be diluted with a citrate buffer in order to stabilize the absorbance of the product prior to aspiration into the spectrophotometer cuvette; alternatively, the reaction mixture may be aspirated directly from the reaction wells.

Reagents and Solutions

Purity of reagents: all chemicals are reagent grade. Horseradish peroxidase, POD, should be of the highest available specific activity, e.g. *Sigma* type VI, RZ $\geqslant$ 3; specific activity 250 to 330 units*/mg.

Preparation of solutions (for 100 determinations): all solutions in re-purified water (cf. Vol. II, chapter 2.3.1.2).

1. Phosphate solution (0.5 mol/l):

   dissolve 116 g $K_2HPO_4 \cdot 3\,H_2O$ in 1 litre water.

2. Bovine serum albumin (2.4 g/l):

   add 0.8 ml bovine serum albumin solution (300 g/l, *Hyland Division Travenol Laboratories, Inc.*) to 100 ml water.

3. Citrate/phosphate buffer (citrate, 109 mmol/l; phosphate, 25 mmol/l; pH 4.0):

   dissolve 2.3 g citric acid, monohydrate, in 90 ml water. Adjust pH to 4.0 with 5 ml phosphate solution (1) and dilute to 100 ml with water.

---

* One unit will form 1 mg purpurogallin in 20 s from pyrogallol at pH 6.0 at 20 °C. This 20 s unit is equivalent to approximately 18 international enzyme units, U, at 25 °C.

4. Washing solution (NaCl, 0.15 mol/l; Tween 80, 5 ml/l; pH 7.6):

dissolve 9 g NaCl in water, add 5 ml Tween 80; adjust pH to 7.6 with NaOH, 0.1 mol/l; make up to 1 litre with water.

5. ABTS* stock solution (36 mmol/l):

dissolve 0.20 g ABTS (*Boehringer Mannheim Corp.*) in 10 ml water.

6. Substrate solution (ABTS, 2 mmol/l; H_2O_2, 1.5 mmol/l; citrate, 105 mmol/l; phosphate, 24 mmol/l; pH 4.0):

add 0.5 ml ABTS stock solution (5) and 0.1 ml hydrogen peroxide (5 g/l) to 9.6 ml citrate buffer (3).

7. Gentamicin standard solutions (1.88 to 30 mg/l):

dissolve 10 mg gentamicin sulphate (571 mg gentamicin per g powder; *Shering Corp.,* Kenilworth, NJ) in 10 ml water. Mix 1 ml of this solution with 18 ml gentamicin-free serum to obtain a standard of 30 mg/l. Dilute this standard 1 + 1, 1 + 3, 1 + 7, and 1 + 15 with gentamicin-free human serum to obtain gentamicin standards of 15, 7.5, 3.75, and 1.88 mg/l.

8. Conjugate solution (POD, ca. 10 U/l; phosphate, 50 mmol/l):

add 1 ml phosphate solution (1) to 9 ml water and mix this solution with 0.02 ml peroxidase-gentamicin conjugate (cf. pp. 182 – 184).

Stability of reagents and solutions: all solutions are stable at 4 °C for at least 4 months, except for the conjugate solution (8) which is stable for 1 week at 4 °C and the substrate solution (6) which is stable for 8 h at room temperature.

Procedure

Collection and treatment of specimen: specimens should be collected only at specific times during the course of therapy in order to facilitate interpretation of the results. Usually, two blood specimens are collected after the drug has reached a steady state, i.e., approximately 12 to 24 h after initiation of the drug therapy assuming a half-life of approximately 3 h. The first specimen is collected immediately prior to the next dose and a second specimen is collected 30 min after completion of an infused dose. Serum is separated. These two samples represent a "trough" or minimum serum concentration and a "peak" or maximum serum concentration. Based on these two values, an indication as to the tissue accumulation and potential toxicity of gentamicin (trough values greater than 2 mg/l) and as to the therapeutic efficacy (peak values between 5 and 10 mg/l) can be obtained.

---

* Diammonium 2,2′-azinobis-[3-ethyl-2,3-dihydrobenzothiazole-6-sulphonate].

Stability of the analyte in the sample: some precautions must be taken if specimens and samples are to be stored for longer than 3 to 4 h. Penicillin is usually given concomitantly with gentamicin to facilitate the therapeutic effect, and an interaction between gentamicin and two types of penicillin, carbenicillin and tricarcillin, has been described [43, 44]. If the gentamicin half-life is sufficiently prolonged, an *in vivo* reaction between gentamicin and carbinicillin can lead to the formation of a complex which has no antibiotic activity. In this circumstance, a 25 to 75% reduction in gentamicin half-life is observed. In addition to an *in vivo* inactivation, which generally is not significant when renal function is normal, an *in vitro* inactivation of gentamicin can occur in samples that are stored for longer than a few hours. Thus if any delay in analysis is anticipated, the serum samples must be frozen to ensure stability of gentamicin. Other than this potential loss of gentamicin, the drug is stable in serum samples stored at 4°C for at least 4 weeks.

Measurement

Pipette into a well of a coated microtitre plate:		concentration in assay mixture	
pre-diluted sample or standard solution (7) conjugate solution (8)	0.05 ml 0.05 ml	gentamicin POD phosphate NaCl	up to 682 μg/l ca. 5 units/l 25 mmol/l 68 mmol/l
incubate for 10 min at room temperature; wash* wells with five 2 ml portions of washing solution (4); tap to remove residual liquid;			
substrate solution (6)	0.10 ml	ABTS H_2O_2 citrate phosphate	2 mmol/l 1.5 mmol/l 105 mmol/l 24 mmol/l
incubate for 10 min at room temperature,			
HCl, 1.0 mol/l	0.10 ml	HCl	0.5 mol/l
transfer contents of reaction well to 1 ml citrate/ phosphate buffer (3); mix well; aspirate mixture into spectrophotometer flow-cell, read absorbance.			

* After the immunological reaction is complete and the excess reactants have been washed from the reaction well, vigorously tap the plate against an absorbent surface to ensure that all washing solution is removed from the reaction wells after each wash. This helps prevent non-specific carry-over of the conjugate and subsequent spurious conversion of substrate.

Assay conditions: pre-dilute all serum samples, standards and controls prior to the assay; mix 50 µl sample with 500 µl isotonic saline (0.15 mol/l) using an automatic pipette capable of delivering to within 1% of the stated volume. Analyze all samples, standards and controls in duplicate with an incubation time of 10 min at room temperature for both the immunological and enzymatic reactions (wavelength 410 nm).

The order of addition of substrate to the series of reaction wells should be the same as the order of aspiration of the sample into the spectrophotometer for the absorbance measurement. While the addition of HCl and mixing with citrate/phosphate buffer stabilizes the absorbance by slowing the enzyme reaction, small absorbance changes may occur. Therefore, the order of addition and aspiration should be the same and the measurement of absorbance should proceed without delay.

If the absorbance of the sample is greater than that of the highest standard, dilute the sample further with saline and re-assay.

Calibration curve: plot the average $\Delta A/\Delta t$ values of duplicate assays for each standard *versus* the corresponding gentamicin concentrations (mg/l) on logit-log paper. Alternatively, the average $\Delta A/\Delta t$ may be plotted on linear-log paper or used in a microcomputer-based programme for conversion of immunoassay data to logit-log plots.

Calculation: read the gentamicin mass concentration, ρ (mg/l), corresponding to the mean $\Delta A/\Delta t$ readings of the samples and controls from the calibration curve. For conversion from mass concentration to substance concentration, c (µmol/l), multiply by 2.1598 (molecular weight is 463*).

Validation of Method

Precision, accuracy, detection limit and sensitivity: the precision of the method is shown by between-days relative standard deviation (RSD) ranging from 9% to 16% for control samples with concentrations between 18.0 and 1.6 mg/l, respectively. The within-series RSD is 9% to 14% for control samples with concentrations between 19.0 and 1.5 mg/l, respectively.

The accuracy of the method was evaluated by comparison with a radioimmunoassay [33] (EIA *vs.* RIA: y = 0.94x + 0.28 mg/l). The detection limit, defined as the smallest amount of gentamicin that can be distinguished from no gentamicin in serum, was calculated to be 0.23 mg/l [45]. The sensitivity, defined in terms of the slope of the response curve, was calculated to be $\Delta A/\Delta t = 0.052$ per 1 mg/l [46] at a gentamicin concentration of 5 mg/l.

Sources of error: neither protein (up to 30-fold excess) nor triglycerides (up to 11 mmol/l) interferes with either the immunological or the enzymatic reactions.

---

* Mean value.

Specificity: the specificity of any immunoassay depends on the ability of the antibody to distinguish between compounds that are chemically similar and that may be present in the specimen. The antiserum used in this instance showed less than 1% cross-reactivity with other aminoglycosides, including amikacin, tobramycin, kanamycin, and penicillin, when these drugs were included in the gentamicin assay at serum concentrations 10-fold higher than their respective therapeutic levels.

Therapeutic ranges: serum (plasma) concentrations of gentamicin of less than 2 mg/l (trough) and 5 to 10 mg/l (peak) are associated with minimum toxicity and therapeutic success.

Appendix

Preparation of thyroglobulin-gentamicin conjugate: mix 10 mg thyroglobulin with 40 mg gentamicin in 1 ml PBS (phosphate, 10 mmol/l, containing sodium chloride, 0.15 mol/l, pH 7.5). To this solution add dropwise 1 ml PBS containing 400 mg ethyl carbodiimide. Stir the mixture gently for 4 h at room temperature and dialyze for 16 h against PBS.

Preparation of gentamicin antibodies: a thyroglobulin-gentamicin conjugate is used as an antigen for the production of gentamicin antibodies (cf. above). Obtain antibody to gentamicin by injecting rabbits subcutaneously with a homogenized mixture containing 0.5 ml dialyzed conjugate solution in 0.5 ml complete *Freund's* adjuvant. Inject animals once each week for three weeks and then monthly for three months. Thereafter give each one booster injection at six months. Collect serum and test for antibody titre after the first month following the initial injection. Affinity constant of the antibody is about 5×10^8 l/mol and the titre increases linearly during the first three months of injections.

Preparation of microtitre plates: soak each reaction well of a 96-well polystyrene plate with BSA solution (2) for 10 min, empty and wash the wells twice with washing solution (4), and dry the plate at room temperature. Add 50 µl gentamicin antiserum (diluted 500-fold) to each well and dry overnight at room temperature. Wash each well 5 times with washing solution (2) and shake dry.

The plate is now ready for use, or it may be stored at room temperature for at least one month. When only a portion of a plate is used, the remainder may be covered with plastic film (Microtitre plate sealers; *Cooke Laboratory Products,* 900 Slaters Lane, Alexandria, VA, 22314).

Preparation of peroxidase-gentamicin conjugate: the enzyme conjugate is prepared by a modification of the method of *Nakane* [47]. In this method, illustrated in Fig. 3, a reactive aldehyde is produced by controlled oxidation of the carbohydrate moieties associated with horseradish peroxidase. In a second step, the reactive aldehyde groups on the enzyme are coupled with the free amino groups of gentamicin. After reduction and elimination of excess aldehyde groups, a stable conjugate is produced.

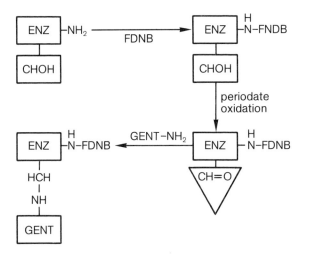

Fig. 3. Schematic reaction sequence for production of peroxidase-gentamicin conjugate. Free amino groups of the enzyme (ENZ) are protected by fluorodinitrobenzene (FDNB) followed by oxidation of carbohydrate moieties. The active aldehyde then is conjugated to the free amino groups of the gentamicin (GENT).

The molar ratio of gentamicin to peroxidase may be controlled by adjusting the relative concentrations of each component in the conjugation reaction in order to obtain the optimum ratio. Molar ratios of gentamycin to peroxidase greater than 2 may markedly decrease the enzymatic activity of the conjugate [48].

Preparation of peroxidase-aldehyde: dissolve 5 mg horseradish peroxidase, POD, in 1.0 ml sodium hydrogen carbonate, 0.3 mol/l. In order to limit internal conjugation of its own amino groups with newly formed aldehydes, add 0.1 ml 1% fluorodinitro-benzene in absolute ethanol to the solution containing the peroxidase and mix gently for 1 h at room temperature. (If the peroxidase is impure a precipitate may form and should be removed by centrifugation.) Initiate oxidation of the carbohydrate moieties by adding 1.0 ml sodium peroxidase solution, 0.08 mol/l. Mix gently for 30 min at room temperature. In order to limit the further oxidation of aldehyde groups, add 1.0 ml ethylene glycol, 0.16 mol/l, and mix gently for 1 h at room temperature. Dialyze this solution against sodium carbonate buffer, 0.01 mol/l, pH 9.5, for 1 day with several changes of buffer at 4°C. The reactive aldehyde produced by this procedure is stable for one month when stored at 4°C.

Conjugation: conjugate gentamicin to reactive peroxidase-aldehyde by adding 0.57 mg gentamicin sulphate to 2.5 ml aldehyde and mixing gently (no vigorous stirring) at room temperature for 3 h. Add 5 mg sodium borohydride and store at 4°C for

3 h. This treatment stabilizes the conjugate and limits further reactions of the activated aldehyde groups. Dialyze the reaction mixture in 1 l PBS for 24 h at 4 °C. (A small amount of precipitate may be formed inside the dialysis bag during the dialysis period and should be removed by centrifugation before proceeding to the purification step.)

Purification: purify the conjugate by passing the dialyzate through a 5 mm × 400 mm column filled with Sephadex G-100. Pool the fractions that elute in the first peak, sterilize by filtration through 0.45 µm filter and store in 1 ml aliquots at 4 °C. While this purification step is not invariably necessary, we have found that the sensitivity of the assay is improved and the stability of the conjugate is extended for purified, sterilized conjugates. The molar ratio of gentamicin to peroxidase in the conjugate is approximately 2 to 1.

References

[1] *J. P. Rosselot, J. Marquez, E. Meseck, A. Murawski, A. Hamdan, C. Joyner, R. Schmidt, D. Migliore, H. L. Herzog,* Isolation, Purification and Characterization of Gentamicin, in: *J. C. Sylvester* (ed.), Antimicrobial Agents and Chemotherapy, American Society of Microbiology, Ann Arbor, MI, 1964, pp. 14 – 16.

[2] *P. Noone, T. M. C. Parsons, J. R. Pattison, R. C. B. Slack, D. Garfield-Davies, K. Hughes,* Experience in Monitoring Gentamicin Therapy during Treatment of Serious *Gram*-Negative Sepsis, Br. Med. J. *1,* 477 – 481 (1974).

[3] *J. G. Dahlgren, E. Anderson, W. Hewitt,* Gentamicin Blood Levels, a Guide to Nephrotoxicity, Antimicrob. Agents Chemother. *8,* 58 – 62 (1975).

[4] *F. G. Falco, H. Smith, G. Arcieri,* Nephrotoxicity of Aminoglycosides in Gentamicin, J. Infect. Dis. *119,* 406 – 409 (1969).

[5] *E. T. Anderson, L. S. Young, W. L. Hewitt,* Simultaneous Antibiotic Levels in "Breakthrough" *Gram*-negative Rod Bacteremia, Am. J. Med. *61,* 493 – 497 (1976).

[6] *D. E. Zaske,* Aminoglycosides, in: *W. E. Evans, J. J. Schentag, W. J. Jusko* (eds.), Applied Pharmacokinetics, Principles of Therapeutic Drug Monitoring, Applied Therapeutics, Inc., San Francisco, CA, 1980, pp. 210 – 239.

[7] *J. J. Schentag, J. Jusko, M. Plaut, T. J. Cumbo, J. V. Vance, E. Abrutyn,* Tissue Persistence of Gentamicin in Man, J. Am. Med. Assoc. *238,* 327 – 329 (1977).

[8] *J. H. Hull, F. A. Sarubbi,* Gentamicin Serum Concentrations: Pharmacokinetic Predictions, Ann. Intern. Med. *85,* 183 – 189 (1976).

[9] *C. D. Ericsson, J. H. Duke, L. K. Pickering,* Clinical Pharmacology of Intravenous and Intraperitoneal Aminoglycoside Antibiotics, Ann. Surg. *188,* 66 – 70 (1978).

[10] *J. Mendelson, J. Portnoy, H. Sigman,* Pharmacology of Gentamicin in the Biliary Tract of Humans, Antimicrob. Agents Chemother. *4,* 538 – 541 (1973).

[11] *D. C. March, E. B. Matthew, R. H. Persellin,* Transport of Gentamicin into Synovial Fluid, J. Am. Med. Assoc. *228,* 607 (1974).

[12] *L. Nordstrom, G. Banck, S. Belfrage, I. Juhlin, O. Tjernstrom, N. G. Toremalm,* Prospective Study of the Ototoxicity of Gentamicin, Acta Pathol. Microbiol. Scand. *81* (suppl. 241), 58 – 61 (1973).

[13] *J. J. Schentag, W. J. Jusko, J. W. Vance, T. J. Cumbo, E. Abrutyn, M. DeLattre, L. M. Gergracht,* Gentamicin Disposition and Tissue Accumulation on Multiple Dosing, J. Pharmacokinet. Biopharm. *5,* 559 – 577 (1977).

[14] *J. J. Schentag, W. J. Jusko,* Renal Clearance and Tissue Accumulation of Gentamicin, Clin. Pharmacol. Ther. *22,* 364 – 370 (1977).

[15] *G. G. Jackson,* Present Status of Aminoglycoside Antibiotics and Their Safe Effective Use, Clin. Ther. *1,* 200 – 215 (1977).

[16] *A. B. Kaiser, Z. A. McGee,* Aminoglycoside Therapy of *Gram*-negative Meningitis, N. Eng. J. Med. *293,* 1215 – 1220 (1975).

[17] *J. J. Rahal, P. J. Hyams, M. S. Simberkoff, E. Rubinstein,* Combined Intramuscular and Intrathecal Gentamicin for *Gram*-negative Meningitis, N. Eng. J. Med. *290,* 1394 – 1398 (1974).

[18] *C. Q. Edwards, C. Smith, K. Baugham, J. F. Rogers, P. S. Lietman,* Concentrations of Gentamicin and Amikacin in Human Kidneys, Antimicrob. Agents Chemother. *9,* 925 – 927 (1976).

[19] *M. Barza, R. B. Brown, D. Shen, M. Gibaldi, L. Weinstein,* Predictability of Blood Levels of Gentamicin in Man, J. Infect. Dis. *132,* 165 – 174 (1975).

[20] *R. E. Cutler, A. M. Gyselynck, P. Fleet, A. W. Forrey,* Correlation of Serum Creatinine Concentration and Gentamicin Half-life, J. Am. Med. Assoc. *219,* 1037 – 1041 (1972).

[21] *M. C. McHenry, T. L. Gavan, R. W. Gifford, N. A. Geurkink, R. A. Van Ommen, M. A. Town, J. G. Wagner,* Gentamicin Dosage for Renal Insufficiency: Adjustment Based on Endogenous Creatinine Clearance and Serum Creatinine Concentration, Ann. Intern. Med. *74,* 192 – 197 (1971).

[22] *D. Kaye, M. E. Levinson, E. D. Labovitz,* The Unpredictability of Serum Concentrations of Gentamicin: Pharmacokinetics of Gentamicin in Patients with Normal and Abnormal Renal Function, J. Infect. Dis. *130,* 150 – 154 (1974).

[23] *M. E. Lund, D. J. Blazevic, J. M. Matsen,* Rapid Gentamicin Bioassay Using a Multiresistant Strain of *Klebsiella Pneumoniae,* Antimicrob. Agents Chemother. *4,* 569 – 573 (1973).

[24] *R. V. Case, L. M. Mezei,* An Enzymatic Radioassay for Gentamicin, Clin. Chem. *24,* 2145 – 2150 (1978).

[25] *B. A. Gunn, S. L. Brown, C. S. Otey, C. A. Gaydos, J. F. Keiser, F. A. Meeks, R. G. Trahan,* Serum Gentamicin Assay by a Radiometric Procedure, Am. J. Clin. Pathol. *73,* 259 – 262 (1980).

[26] *J. W. Mayhew, S. L. Gorbach,* Assay for Gentamicin and Tobramycin by Gas-Liquid Chromatography, Antimicrob. Agents Chemother. *14,* 851 – 855 (1978).

[27] *J. P. Anhalt, S. D. Brown,* High Performance Liquid Chromatographic Assay of Aminoglycoside Antibiotics, Clin. Chem. *24,* 1940 – 1947 (1978).

[28] *D. M. Barends, C. L. Zwaan, A. Hulshoff,* Improved Microdetermination of Gentamicin and Sisomicin in Serum by High-Performance Liquid Chromatography with Ultraviolet Detection, J. Chromatogr. *222,* 316 – 323 (1981).

[29] *E. J. Shaw, R. A. Watson, J. Landon, D. S. Smith,* Estimation of Serum Gentamicin by Quenching Fluoroimmunoassay, J. Clin. Pathol. *30,* 526 – 531 (1977).

[30] *L. Nilsson, H. Höjer, S. Anséhn, A. Thore,* A Rapid Semiautomated Bioassay of Gentamicin Based on Luciferase Assay of Bacterial Adenosine Triphosphate, Scand. J. Infect. Dis. *9,* 232 – 236 (1977).

[31] *J. W. Williams, J. S. Langer, D. B. Northrop,* A Spectrophotometric Assay for Gentamicin, J. Antibiot. (Tokyo) *28,* 982 – 987 (1975).

[32] *J. F. Burd, R. C. Wong, J. E. Feeney, R. J. Carrico, R. C. Goguslaski,* Homogeneous Reactant-labeled Fluorescent Immunoassay for Therapeutic Drugs Exemplified by Gentamicin Determination in Human Serum, Clin. Chem. *23,* 1402 – 1408 (1977).

[33] *A. H. Lau, E. Chow-Tung,* Comparison of a Fluorescent Immunoassay with an Enzyme Immunoassay and a Radioimmunoassay for Gentamicin, Am. J. Hosp. Pharm. *41,* 2647 – 2650 (1984).

[34] *A. Broughton, J. E. Strong,* Radioimmunoassay of Antibiotics and Chemotherapeutic Agents, Clin. Chem. *22,* 726 – 732 (1976).

[35] *W. Mahon, T. Wilson,* Radioimmunoassay for Measurement of Gentamicin in Blood, Antimicrob. Agents Chemother. *3,* 585 – 589 (1973).

[36] *H. C. Standiford, D. Bernstein, H. C. Nipper, E. Caplan, B. Tatum, J. S. Hall, J. Reynolds,* Latex Agglutination Inhibition Card Test for Gentamicin Assay: Clinical Evaluation and Comparison with Radioimmunoassay and Bioassay, Antimicrob. Agents Chemother. *19,* 620 – 624 (1981).

[37] *W. G. Wood, H. Fricke, L. von Klitzing, C. J. Strasburger, P. C. Scriba,* Solid Phase Antigen Luminescent Immunoassays (SPALT) for the Determination of Insulin, Insulin Antibodies and Gentamicin Levels in Human Serum, J. Clin. Chem. Clin. Biochem. *20,* 825 – 831 (1982).

[38] *L. Nilsson,* Correlation of Bioluminescent Assay of Gentamicin in Serum with Agar Diffusion Assay, Latex Agglutination Inhibition Card Test, Enzyme Immunoassay, and Fluorescence Immunoassay, J. Clin. Microbiol. *20,* 396 – 399 (1984).

[39] *R. A. Watson, J. Landon, E. J. Shaw, D. S. Smith,* Polarization Fluoroimmunoassay of Gentamicin, Clin. Chim. Acta *73*, 51 – 55 (1976).

[40] *M. E. Jolley, S. D. Stroupe, C. J. Wang, H. N. Panas, C. L. Keegan, R. L. Schmidt, K. S. Schwenzer,* Fluorescence Polarization Immunoassay 1. Monitoring Aminoglycoside Antibiotics in Serum and Plasma, Clin. Chem. *27*, 1190 – 1197 (1981).

[41] *J. C. Standefer, G. C. Saunders,* Enzyme Immunoassay for Gentamicin, Clin. Chem. *24*, 1903 – 1907 (1978).

[42] *K. E. Rubinstein, R. S. Schneider, E. F. Ullman,* Homogeneous Enzyme Immunoassay, a New Immunochemical Technique, Biochem. Biophys. Res. Commun. *42*, 846 – 851 (1972).

[43] *L. J. Riff, G. G. Jackson,* Laboratory and Clinical Conditions for Gentamicin Inactivation by Carbenicillin, Arch. Intern. Med. *130*, 887 – 891 (1972).

[44] *M. Davies, J. R. Morgan, C. Anand,* Interactions of Carbenicillin and Tlcarcillin with Gentamicin, Antimicrob. Agents Chemother. *7*, 431 – 434 (1975).

[45] *A. R. Midgely, Jr., G. Niswender, R. Rebar,* Principles for the Assessment for the Reliability of Radioassay Methods (Precision, Accuracy, Sensitivity, Specificity), Acta Endocrinol. *63* (suppl. 142), 163 – 184 (1969).

[46] *R. P. Eakins, G. Newman, J. O'Riordan,* Saturation Assays, in: *J. W. McArthur, T. Colton* (eds.), Statistics in Endocrinology, MIT Press, Cambridge, MA, 1970, pp. 345 – 378.

[47] *P. K. Nakane, A. Kawaoi,* Peroxidase-labeled Antibody: A New Method of Conjugation, J. Histochem. Cytochem. *22*, 1084 – 1091 (1974).

[48] *G. C. Saunders,* The Art of Solid-phase Enzyme Immunoassay Including Selected Protocols, in: *R. M. Nakamura, W. R. Dito, E. S. Tucker III* (eds.), Immunoassay in the Clinical Laboratory, Alan R. Liss, New York 1979, pp. 99 – 118.

1.14 Kanamycin A

3,5-Diamino-2-(3-amino-3-deoxy-β-D-glucosyloxy)-6-(6-amino-6-deoxy-β-D-glucosyloxy)-1-cyclohexanol

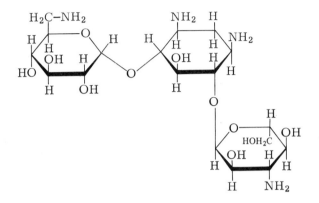

Tsunehiro Kitagawa and Kunio Fujiwara

General

Kanamycin (KM*), isolated from *Streptomyces kanamyceticus* [1], is a basic amino-glycosidic antibiotic, consisting of three components (6-amino-6-deoxy-D-glucose, 3-amino-3-deoxy-D-glucose and 2-deoxystreptamine in the case of kanamycin A) bonded by glycosides. KM exhibits strong antimicrobial activity, not only against a variety of aerobic bacteria, including *Mycobacteria, Staphylococci* and *Enterobacteria,* but also against drug-resistant mutants of *Staphylococci, Gonococci, Tuberculomyces* and genus *Shigella.* KM is also effective in septicaemia, and in urinary and intestinal infections caused by *Gram*-negative bacteria. KM is somewhat similar to streptomycin in its biological activities but resistance to KM may arise more slowly than to streptomycin.

KM is usually administered by intramuscular injection because its absorption in the gastro-intestinal tract is negligible [2]. Intravenous administration is reported to be a safe and effective alternative, especially also in neonates in whom i.m. injections are not possible [3,4]. The profiles of KM levels in serum following an i.m. dose show inter-individual variations. With 7.5 mg/kg in adults, peak serum levels occur be-

---

* Abbreviations:
  BKM   bekanamycin
  KM   kanamycin
  TOB   tobramycin

tween 45 min and 2 h after i.m. administration, ranging from 18 to 21 mg/l, and then progressively decrease with time, with an elimination half-life of ~ 2 h [3]. In children, the peak levels are attained after 30 min to 1 h, ranging from 10 to 20 mg/l with serum half-life of 4.3 h. Lower KM levels in serum were observed in diabetic children than in normal controls [5]. The apparent volumes of distribution at steady state average 30% of body weight [3]. Infants appear to have a slightly larger volume of distribution (0.47 l/kg) [4]. KM is not bound to serum proteins [3].

KM is excreted unchanged by glomerular filtration in the kidney into the urine at 94% of dose in 24 h [3]. An appreciable amount of KM (about 30%) might undergo tubular re-absorption after filtration [3]. A small amount is excreted in the bile and pancreatic juice [5]. No hepatic or pancreatic effect on metabolism of KM has been reported [6].

Recently, a variety of KM analogues such as tobramycin (TOB) [7], dibekacin [8], bekanamycin (BKM) [1] and amikacin [9] have been used clinically to a great extent because of their excellent therapeutic effects. These have somewhat similar pharmacodynamic effects to KM [10]. The most important adverse effects of aminoglycosidic antibiotics are oto- and nephrotoxicity. The factors that are alleged to increase the risk of ototoxicity in neonates are total dosage, duration of therapy, and peak serum concentrations that exceed approximately 30 mg KM per litre [4]. Great emphasis is being placed on determination of serum levels of aminoglycosides in the treatment of severe *Gram*-negative enteric infections, in order to adjust the dosage, route, and frequency of administration to obtain the same therapeutic effects with minimum side-effects.

Application of method: in clinical chemistry, pharmacology and toxicology.

Substance properties relevant in analysis: kanamycin A (M_r = 484.49) is freely soluble in water and practically insoluble in the common alcohols and non-polar solvents.

Methods of determination: the microbiological method using a strain of *Bacillus subtilis* as a test organism has been used most widely in clinical investigations [3 – 5, 11]. However, it may not afford the desired specificity and precision. A number of chromatographic procedures have also been reported for KM, among which are: thin-layer and paper chromatography [12], gas-liquid chromatography [13], ion-exchange liquid chromatography using anion-exchange resin [14 – 16] and micro-reticular anion-exchange resin in the hydroxyl form [15], and high-pressure liquid chromatography with cation-exchange resin [17]. Since kanamycin does not have suitable ultraviolet absorption, these liquid chromatographies were combined with refractive-index monitoring [15, 17], the ninhydrin test [14, 15] or a post-column fluorimetric detection system [17]. None of these methods of KM assay is ideal but each has its strong points; in particular, a highly specific high-pressure liquid-chromatographic procedure [17] appears to be very useful for rapid, precise analysis of kanamycins A and B in complex samples. Recently, a number of radioimmunoassays for aminoglycoside antibiotics excluding KM but including amikacin [18], gentamicin [19 – 21], TOB [22], netilmicin [23] and sisomicin [24] have appeared, among which a RIA for amikacin [20] cross-reacted strongly with KM, suggesting this RIA is capable of being used also as a system for KM measurement. A fluorescence polarization immunoassay for the determination of KM is commercially available. We have developed a heterologous enzyme-immunoassay (EIA) for KM-group antibiotics in which the reagents anti-KM serum and β-D-galactosidase*-labelled TOB as a tracer are used in combination. This EIA allows reproducible and precise determination of KM as well as bekanamycin (BKM), TOB and dibekacin at concentrations as low as 1 ng/tube [25].

For routine monitoring of serum KM concentrations, a high-pressure liquid-chromatographic procedure and heterologous enzyme-immunoassay appear so far to be reliable and most convenient. In particular, the latter method [25] would be an appropriate tool for comparison studies on the clinical pharmacology of KM, TOB, BKM and dibekacin.

International reference method and standards: neither standardization at the international level, nor the existence of reference standard materials is known so far.

Assay

Method Design

The method described here operates according to the well-known procedure of the double-antibody EIA, but using a heterologous enzyme label.

Principle

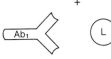

(a)

Rabbit-anti- kanamycin antibody	ligand analyte, kanamycin, TOB, or BKM	β-galactosidase- TOB (or BKM) conjugate

---

* β-D-Galactoside galactohydrolase, EC 3.2.1.23.

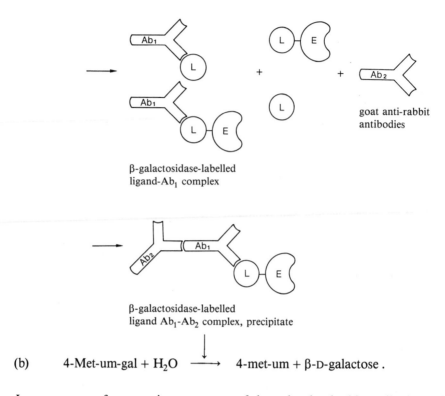

β-galactosidase-labelled
ligand-Ab₁ complex

goat anti-rabbit
antibodies

β-galactosidase-labelled
ligand Ab₁-Ab₂ complex, precipitate

(b)   4-Met-um-gal + H_2O $\longrightarrow$ 4-met-um + β-D-galactose .

In some cases of enzyme-immunoassay of drugs by the double-antibody method, a drug derivative or analogue with relatively low affinity for the antibody against the primary analyte drug to be assayed may act as a more favourable tracer when labelled by enzyme than the primary drug itself. This may occur because of the possibility that the conjugate formed between such a drug derivative or analogue and the enzyme binds less strongly to anti-drug antibody than the conjugate with the primary drug, and is more readily inhibited by low concentrations of the analyte drug in binding to anti-drug antibody, increasing the sensitivity of the heterologous assay for the drug.

This is true in the case of the EIA for KM described here, in which anti-KM antibody is used in combination with galactosidase-labelled TOB or BKM or KM analogues [25]. The galactosidase activity measured as increase in fluorescence intensity per unit time, $\Delta F/\Delta t$, is related to the kanamycin concentration of the sample.

Selection of assay conditions and adaptation to the individual characteristics of the reagents: for the preparation of KM immunogen the amino groups of KM are allowed to react with N-(3-maleimidobenzoyloxy)succinimide as a hetero-bifunctional cross-linking agent [26–29] to form an amide bond, introducing a maleimide residue into the KM molecule. The KM bearing the maleimide group is then coupled with thiol groups of dithiothreitol-reduced bovine serum albumin simultaneously prepared. The thiol groups result from the cleavage of the intramolecular disulphide bonds of bovine serum albumin (cf. p. 198).

Anti-KM antibodies are produced in rabbits, which are first inoculated with an emulsion of 1 mg of the antigen described above in complete *Freund's* adjuvant and on three subsequent occasions receive one-half the amount of the dose of the first inoculation at bi-weekly intervals.

Enzyme-drug conjugate is obtained by coupling KM, TOB or BKM to galactosidase from *Escherichia coli* (EC 3.2.1.23) by what is essentially the same principle as that used in the preparation for KM immunogen, using N-(γ-maleimidobutyryloxy) succinimide as a hetero-bifunctional cross-linking agent (cf. p. 198).

When using galactosidase-KM conjugate as a tracer in the EIA for KM (homologous EIA), the dose-response calibration curve slopes gently and lies within a wide range of KM concentrations ranging from 10 ng/l to 1 mg/l, so that even a subtle variation in the bound enzyme activity (cf. KM calibration curve) might represent a great change in KM concentration, thus offering insufficient accuracy and precision in KM EIA [25]. However, these disadvantages are overcome by replacing the galactosidase-KM conjugate with enzyme conjugated to a KM analogue in the EIA (heterologous EIA). A knowledge of cross-reactivity values of anti-KM antibody with KM analogues based on the homologous EIA procedure for KM might help in the selection of an appropriate drug as enzyme label. It has been shown that anti-KM anti-serum cross-reacts with KM analogues of amikacin, BKM, TOB, and dibekacin at values of 26.7, 10.2, 0.0667, and 0.0279%, respectively, when comparing the drug amount required to inhibit binding of galactosidase-KM to the antibody by 50% with that of KM [25]. Therefore, either TOB- or BKM-conjugate can be used in a heterologous EIA for KM.

The optimal quantities and concentrations of sample and reagents are dependent on the characteristics of the antibodies present in the antiserum.

The optimal assay conditions for KM detection at nanogram concentrations are established as follows: first, a prospective enzyme conjugate concentration is chosen. The advisable amount of galactosidase activity in the conjugate ranges from 40 to 50 µU* per tube. Anti-KM antiserum must be used after dilution to the level at which it is capable of binding about 30% of the added enzyme conjugate in control tubes containing unlabelled KM. The concentration of the second antibody (goat anti-rabbit IgG and normal rabbit serum), depending largely upon the concentration of the first antibody (rabbit IgG including anti-KM antibody), must be determined so that the antibody-bound fractions of KM and galactosidase-KM conjugate may each be completely separated from non-bound fractions by precipitation. The advisable concentration is a solution of 19 parts buffer to 1 part goat anti-rabbit IgG serum and a solution of 399 parts buffer to 1 part normal rabbit serum.

The first immunoreaction is carried out by mixing galactosidase-labelled TOB or BKM, specific amounts of KM or a sample, and anti-KM serum, followed by incubation at room temperature and pH 7.4 for 3 h. After addition of goat anti-rabbit IgG and normal rabbit serum the reaction mixture is further incubated at room temperature for 16 h. The resultant immune precipitate is measured for enzyme activity of galactosidase using a fluorigenic substrate at 30 °C and pH 7.0 for 1 h.

---

* 1 U is that enzyme activity which catalyzes the hydrolysis of 1 µmol substrate, 7-β-D-galactopyranosyl-oxy-4-methylcoumarin (4-methylumbelliferyl-β-D-galactopyranoside, 4-met-um-gal) per min at 30 °C.

Equipment

Spectrofluorimeter capable of measuring fluorescence emission at 448 nm with an excitation wavelength of 365 nm. Laboratory centrifuge, water-bath, stopwatch or timer, pH meter, glass tubes, pipettes.

Reagents and Solutions

Purity of reagents: KM, TOB, dibekacin and BKM are from the *Meiji Seika Co., Tokyo, Japan. Amikacin (Banyu Pharm. Ind., Ltd., Tokyo, Japan), N*-(3-maleimido-benzoyloxy)succinimide, MBS (*Sigma Co., U.S.A.*) and *N*-(γ-maleimidobutyryloxy)-succinimide, GMBS (*Dojin Chemical Co.,* Kumamoto, Japan) are commercial products with sufficient purity. β-Galactosidase from *Escherichia coli*, 195.9 U/mg lyophilized material (37 °C; 4-nitrophenyl β-galactoside as substrate) is from *Boehringer Mannheim,* W. Germany. Chemicals are of analytical grade.

Preparation of solutions (for 200 determinations): all solutions in re-purified water (cf. Vol. II, chapter 2.1.3.2).

1. Buffer A (phosphate, 20 mmol/l; NaCl, 0.1 mol/l; $MgCl_2$, 1 mmol/l; BSA, 1 g/l; NaN_3, 1 g/l; pH 7.0):

   dissolve 3.12 g $NaH_2PO_4 \cdot 2\ H_2O$, 5.85 g NaCl, 203 mg $MgCl_2 \cdot 6\ H_2O$, 1 g BSA and 1 g NaN_3 in water, adjust pH to 7.0 with NaOH, 1 mol/l, and make up to 1 l with water.

2. Buffer B (phosphate, 60 mmol/l; EDTA, 10 mmol/l; BSA 1 g/l; NaN_3, 1 g/l; pH 7.4):

   dissolve 9.36 g $NaH_2PO_4 \cdot 2\ H_2O$, 3.72 g $EDTA\text{-}Na_2H_2 \cdot 2\ H_2O$, 1 g BSA, and 1 g NaN_3 in water, adjust pH 7.4 with NaOH, 1 mol/l, and make up to 1 l with water.

3. Anti-KM antiserum solution (γ-globulin, ca. 0.40 mg/l; phosphate, 20 mmol/l; NaCl, 0.1 mmol/l; $MgCl_2$, 1 mmol/l; BSA, 1 g/l; NaN_3, 1 g/l; pH 7.0):

   prepare anti-KM antiserum according to the method described in Appendix, p. 197; dilute 0.4 µl serum to 25.0 ml with buffer A (1).

4. Galactosidase-labelled TOB solution (galactosidase, 800 mU/l; phosphate, 60 mmol/l; EDTA, 10 mmol/l; BSA, 1 g/l; NaN_3, 1 g/l; pH 7.4):

   use galactosidase-TOB conjugate as prepared according to Appendix, p. 198; dilute 0.2 ml conjugate to 10 ml with buffer B (2) to give 8.0 mU enzyme activity in 10 ml.

5. Galactosidase-labelled BKM solution (galactosidase, 800 mU/l; phosphate, 60 mmol/l; EDTA, 10 mmol/l; BSA, 1 g/l; NaN_3, 1 g/l; pH 7.4):

prepare galactosidase-BKM conjugate according to Appendix, p. 198; dilute 0.2 ml conjugate to 10 ml with buffer B to give 8.0 mU enzyme activity in 10 ml. This solution can be used alternatively to solution (4).

6. KM standards (10 ng/l to 1 mg/l):

   dissolve 10 mg KM in 10 ml buffer A (1); dilute 0.1 ml of it with KM-free human serum containing sodium azide (1 g/l) to 10 ml (10 mg/l); dilute this solution serially with KM-free human serum to yield concentrations from 1 mg/l to 10 ng/l. Take KM-free serum as zero calibrator.

7. Goat anti-rabbit IgG solution (γ-globulin, ca. 1.25 g/l; phosphate, 60 mmol/l; EDTA, 10 mmol/l; BSA, 1 g/l; NaN$_3$, 1 g/l; pH 7.4):

   dilute 0.5 ml goat anti-rabbit antiserum to 10 ml with buffer B (2).

8. Normal rabbit serum solution (serum, 2.5 ml/l; phosphate, 20 mmol/l; NaCl, 0.1 mmol/l; MgCl$_2$, 1 mmol/l; BSA, 1 g/l; NaN$_3$, 1 g/l; pH 7.0):

   dilute 0.025 ml normal rabbit serum to 10 ml with buffer A (1).

9. Substrate solution (4-met-um-gal, 0.1 mmol/l; phosphate, 12 mmol/l; EDTA, 2 mmol/l; BSA, 0.2 g/l; NaN$_3$, 0.2 g/l; pH 7.4):

   dissolve 3.4 mg 4-methylumbelliferylgalactoside in 80 ml water and mix with 20 ml buffer B (2).

10. Glycine buffer (glycine, 0.2 mol/l, pH 10.4):

    dissolve 15 g glycine in 700 ml water, adjust pH to 10.4 with NaOH, 1 mol/l, and make up to 1 l with water.

Stability of solutions: buffer solutions (1), (2) and (10) are stable at room temperature during storage as long as no bacterial growth is evident. Substrate solution (9) must be prepared freshly and can be kept in the dark for a week at 4 °C. Galactosidase-labelled solutions (4) and (5), and anti-KM antiserum solution (3) are stable for up to 3 months.

Procedure

Collection and treatment of specimen: collect specimens as described on p. 166. No special treatment of the sample (serum, plasma or urine) is necessary. Specimens should be collected at specific times during the course of therapy in order to facilitate interpretation of the results. Collect the first specimen immediately before the next dose ("through" concentration), and collect the second specimen 30 min after completion of an infused dose ("peak" concentration). Store samples at − 30 °C before use.

Assay conditions: standards and samples are measured in triplicate under identical conditions. A reference tube containing antibody solution (3) but no sample or standard, and a blank consisting of assay buffer (1) and galactosidase-labelled TOB solution (4) or BKM solution (5) are run simultaneously. Dilute the sample appropriately according to Fig. 1 (concentration range of ca. 3 to 100 µg/l).

Immunoreaction: incubation at room temperature for 3 h and for 16 h.

Enzymatic indicator reaction: incubation at 30°C for 60 min. Excitation wavelength 365 nm, emission wavelength 448 nm; light path 10 mm; measure against air at ambient temperature.

Establish a calibration curve with standard solution (6) instead of sample.

Measurement

Immunoreaction

Pipette successively into glass tubes:			concentration in incubation mixture	
sample or standard solution (6)		0.1 ml	kanamycin	up to 500 µg/l
galactosidase-labelled TOB (4)			β-galactosidase	200 mU/l
			phosphate	20 mmol/l
or				
galactosidase-labelled BKM (5)		0.05 ml	EDTA	2.5 mmol/l
			BSA	0.5 g/l
			NaN$_3$	0.5 g/l
anti-KM antiserum (3)		0.05 ml	γ-globulin	ca. 10 mg/l
			NaCl	25 mmol/l
			MgCl$_2$	0.25 mmol/l
mix well, let stand at 25°C for 3 h;				
goat anti-rabbit antiserum (7)		0.05 ml	goat serum	variable
normal rabbit serum (8)		0.05 ml	γ-globulin	ca. 200 mg/l
			phosphate	27 mmol/l
			NaCl	33 mmol/l
			MgCl$_2$	3.3 mmol/l
			BSA	0.67 g/l
			NaN$_3$	0.67 g/l
			EDTA	3.3 mmol/l
mix well, let stand for 16 h; centrifuge; discard supernatants by decantation; wash with 1 ml buffer A (1); centrifuge for 15 min at 800 *g*; wash the immune precipitate with buffer A (1).				

Enzymatic indicator reaction

Pipette into glass tubes containing sediment of immunoreaction:			concentration in assay mixture	
substrate solution	(9)	0.15 ml	4-met-um-gal phosphate EDTA BSA NaN$_3$	0.1 mmol/l 12 mmol/l 2 mmol/l 0.2 g/l 0.2 g/l
mix well, incubate exactly for 60 min at 30°C;				
glycine buffer	(10)	2.5 ml	glycine	0.19 mol/l
mix well, measure change in fluorescence ΔF^* per 60 min.				

* Adjust the fluorimeter to 100% of full scale with quinine sulphate, 10 μmol/l, in sulphuric acid, 0.1 mol/l.

If $\Delta F/\Delta t$ of the sample is greater than that of the highest calibrator, dilute the specimen with buffer A (1) and re-assay.

Calibration curve: correct $\Delta F/\Delta t$ readings of standards and reference tube for the blank. Calculate the ratio of $\Delta F/\Delta t$ at a given KM concentration and $(\Delta F/\Delta t)_0$ at zero concentration of KM. Plot the ratio, $\Delta F/\Delta F_0$, *versus* KM concentrations on semi-logarithmic graph paper. A typical calibration curve is given in Fig. 1.

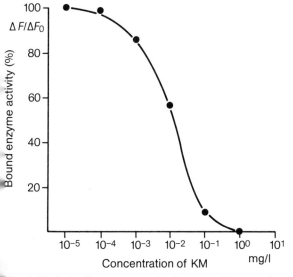

Fig. 1. Typical calibration curve for the assay of kanamycin.

Calculation: correct $\Delta F/\Delta t$ readings of sample for that of blank. Calculate the ratio $\Delta F/\Delta F_0$ of the sample. Read the corresponding KM mass concentrations, ρ (mg/l), from the calibration curve. To convert from mass concentration to substance concentration, c (μmol/l), multiply by 2.064 (molecular weight is 484.49).

Validation of Method

Precision, accuracy, detection limit and sensitivity: RSD of the assay has been determined to be between 5.4% and 16.6% ($n = 9$) within series 7.5% and 21.2% ($n = 5$) between-days in a range of 3.0 to 100 μg/l. The recovery values ranged from 98.5 to 103.6%. The lower limit of detection by the assay is at 3.0 μg KM per litre and the working range is from 3.0 to 100 μg/l [25]. No further data are available.

Sources of error: no pathophysiological influence is known to alter the measurement of KM. Disturbances in the results of the technique may be due to a subtle difference in assay conditions between samples and standards. It is important to prepare the calibration curve calculated for each subject in advance of treatment with KM.

Specificity: the specificity depends on the quality of the antiserum. A common method of assessing specificity is to add potentially cross-reacting substances (e.g. for KM, KM analogues and other antibiotics) to drug-free serum. The concentrations of these compounds which are necessary to produce 50% inhibition of galactosidase-TOB binding to anti-KM antibody are determined.

With our preparation of anti-KM antiserum in combination with galactosidase-TOB as a label, such KM analogues as TOB, dibekacin and BKM showed marked cross-reactions at concentrations of 16.0, 36.0 and 37.5 μg/l, respectively, almost comparable to KM at a concentration of 10.5 μg/l. However, the cross-reaction of amikacin was comparable only at a concentration of 850.0 μg/l [25]. Other antibiotics such as dihydrostreptomycin, ampicillin, and cephalexin showed no significant cross-reactivity at a concentration of 150 mg/l.

It thus appears that the heterologous EIA using anti-KM antiserum and galactosidase-TOB as immune reagents is an appropriate tool for measurement not only of KM, but also of TOB, BKM and dibekacin [25]. As these drugs, in general, are not used clinically in combination with each other, such cross-reactivities in the EIA do not offer any disadvantage to individual measurement of drugs unless during therapy one compound is replaced by another.

Therapeutic ranges: most of the pathogens sensitive to KM, BKM and amikacin have a minimum inhibitory concentration (MIC) of < 4 mg/l [10, 30]. Thus, to be effective, these three drugs would require serum levels of about 10 mg/l, keeping the side effects to a minimum. Acceptable peak serum concentrations are 15 to 25 mg/l and trough serum concentration < 5 mg/l.

Appendix

Preparation of immunogen for KM: the three-step reactions involved in the preparative process are shown in Fig. 2.

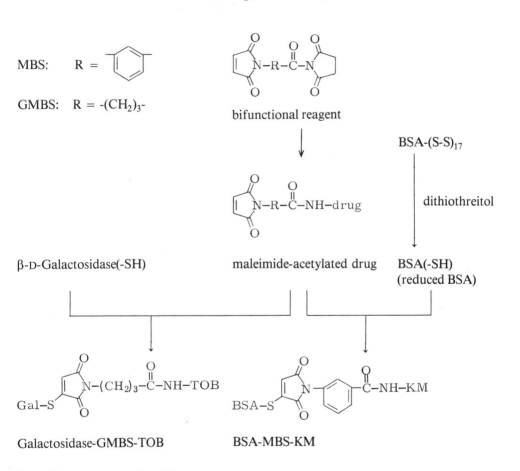

Fig. 2. Scheme for preparation of immunogen to KM and of galactosidase-TOB conjugate.

Step 1. MBS-acylated KM: add 3.14 mg (10 µmol) MBS in 0.5 ml tetrahydrofuran to 4.8 mg (10 µmol) KM in 1 ml phosphate buffer, 50 mmol/l, pH 7.0, and incubate at 30°C for 30 min with occasional stirring. After removing tetrahydrofuran by passing nitrogen through the reaction mixture, extract excess MBS three times with 5 ml portions of methylene chloride and use the aqueous layer in the reaction of step 3.

Step 2. Thiol-exchange reduction of disulphide bonds of BSA: incubate 10 mg (0.15 µmol) BSA in 2 ml Tris buffer, 0.2 mol/l, pH 8.6, containing urea, 6 mol/l, with 2.08 mg (13.5 µmol) dithiothreitol at room temperature for 1 h. Let the reduced BSA precipitate by addition of 2 ml trichloroacetic acid solution, 100 g/l. Wash the precipitate with 2 ml phosphate buffer, 50 mmol/l, pH 7.0, then re-dissolve it in 1 ml of the same phosphate buffer but containing urea, 6 mol/l. Use this solution in the reaction of step 3.

Step 3. Conjugation reaction: incubate the solution of reduced BSA with the aqueous layer of MBS-acylated KM at 25 °C for 2 h. Apply the mixture to a 25 mm × 570 mm column of Sephadex G-100 with an eluent of phosphate buffer, 0.1 mol/l, pH 7.0, containing urea, 3 mol/l. The conjugate appearing immediately after the void volume from the column is used as an immunogen for KM.

Preparation of β-galactosidase-labelled KM, TOB and BKM: KM, TOB or BKM can be coupled with β-galactosidase by what is essentially the same principle as that used in the preparation of the immunogen for KM, using GMBS as a hetero-bifunctional cross-linking agent. The preparation of all these conjugates can be performed similarly but is here represented only by the synthesis of galactosidase-TOB conjugate:

incubate 0.467 mg (1 µmol) TOB in 1 ml phosphate buffer, 20 mmol/l, pH 7.3, with 28 µg (0.1 µmol) GMBS in 30 µl tetrahydrofuran at 30 °C for 40 min. Add the reaction mixture to a solution of 28 µg galactosidase in 1 ml phosphate buffer, 0.1 mol/l, pH 7.0, and incubate at 30 °C for 2 h. Then apply the mixture to a 15 mm × 500 mm column of Sepharose 6B using buffer B (2) as an eluent. Collect 3.7 ml fractions and test for enzyme activity of the conjugate. The main peak (fractions 13 – 15) is used as a source of galactosidase-labelled TOB.

References

[1] *T. Takeuchi, T. Hikiji, K. Nitta, S. Yamazaki, S. Abe, H. Takeyama, H. Umezawa,* Biological Studies on Kanamycin, J. Antibiot. Japan Ser. A *10*, 107 – 114 (1957).
[2] *H. Umezawa,* Kanamycin: Its Discovery, Ann. N. Y. Acad. Sci. *76*, 20 – 26 (1958).
[3] *W. M. M. Kirby, J. T. Clarke, R. D. Libke, C. Regamey,* Clinical Pharmacology. Clinical Pharmacology of Amikacin and Kanamycin, J. Infect. Dis. *134*, S312 – S315 (1976).
[4] *G. H. McCracken, Jr., N. Threlkeld, M. L. Thomas,* Intravenous Administration of Kanamycin and Gentamycin in Newborn Infants, Pediatrics *60*, 463 – 466 (1977).
[5] *G. P. Gabriel, L. V. de Esther, T. S. Hugo,* Serum Levels and Urinary Concentrations of Kanamycin, Bekanamycin and Amikacin (BB-K8) in Diabetic Children and a Control Group, J. Int. Med. Res. *5*, 322 – 329 (1977).
[6] *L. P. Garrod, F. O'Grady,* Antibiotic and Chemotherapy, 2nd edit., The Williams and Wilkins Company, Baltimore 1968, pp. 106 – 114.
[7] *W. M. Stark, M. M. Hoehn, N. G. Knox,* Nebramycin, a New Broad-spectrum Antibiotic Complex. I. Detection and Biosynthesis, Antimicrob. Agents Chemother. *1*, 314 – 323 (1967).
[8] *H. Umezawa, S. Umezawa, T. Tsuchiya, Y. Okazaki,* 3',4'-Dideoxykanamycin B Active against Kanamycin Resistant *Escherichia coli* and *Pseudomonas aeruginosa,* J. Antibiot. (Tokyo) *24*, 485 – 487 (1971).

[9]   *H. Kawaguchi, T. Naito, S. Nakagawa, K. Fujisawa,* BBK 8 a New Semisynthetic Aminoglycoside Antibiotic, J. Antibiot. (Tokyo) *25*, 695 – 708 (1972).

[10]   *P. A. Uribe, S. H. Trujillo, M. Palacio, E. L. de Vidal,* Estudio de sensibilidad *in vitro* de 802 cepas de varios grupos bacterianos al nuevo aminoglycosido BBK 8 (Amikacin), A Folha Medica *69*, 339 – 345 (1974).

[11]   *A. Gourevitch, V. Z. Rossomano, T. A. Puglisi, J. M. Tynda, J. Lein,* Microbiological Studies with Kanamycin, Ann. N. Y. Acad. Sci. *76*, 31 – 41 (1958).

[12]   *E. G. C. Clark* (ed.), Isolation and Identification of Drugs in Pharmaceuticals, Body Fluids, and Post-mortem Materials, Pharmaceutical Press, London 1969.

[13]   *K. Tsuji, J. H. Robertoson,* Gas-liquid Chromatographic Determination of Amino Glycoside Antibiotics: Kanamycin and Paromomycin, Anal. Chem. *42*, 1661 – 1663 (1970).

[14]   *S. Inouye, H. Ogawa,* Separation and Determination of Amino Sugar Antibiotics and Their Degradation Products by Means of an Improved Method of Chromatography on Resin, J. Chromatogr. *13*, 536 – 541 (1964).

[15]   *T. Ottake, M. Yaguchi,* Liquid Chromatography, Applications Bulletin No. 5, Varian, Palo Alto, CA 1973.

[16]   *H. Umezawa, S. Kondo,* Ion-exchange Chromatography of Aminoglycoside Antibiotics, in: *J. H. Hash* (ed.), Methods in Enzymology, Vol. *XLIII*, Academic Press, New York 1975, pp. 263 – 278.

[17]   *D. L. Mays, R. J. V. Apeldoorn, R. G. Lauback,* High-performance Liquid Chromatographic Determination of Kanamycin, J. Chromatogr. *120*, 93 – 102 (1976).

[18]   *J. E. Lewis, J. C. Nelson, H. A. Elder,* Amikacin: A Rapid and Sensitive Radioimmunoassay, Antimicrob. Agents Chemother. *7*, 42 – 45 (1975).

[19]   *J. E. Lewis, J. C. Nelson, H. A. Elder,* Radioimmunoassay of an Antibiotic Gentamicin, Nat. New Biol. *239*, 214 – 216 (1972).

[20]   *W. A. Mahon, J. Ezer, T. W. Wilson,* Radioimmunoassay for Measurement of Gentamycin in Blood, Antimicrob. Agents Chemother. *3*, 585 – 589 (1973).

[21]   *A. Broughton, J. E. Strong,* Radioimmunoassay of Iodinated Gentamycin, Clin. Chim. Acta *66*, 125 – 129 (1976).

[22]   *A. Broughton, J. E. Strong, L. K. Pickering, G. P. Bodey,* Radioimmunoassay of Iodinated Tobramycin, Antimicrob. Agents Chemother. *10*, 652 – 656 (1976).

[23]   *P. Stevens, L. S. Young, W. L. Hewitt,* [125]I Radioimmunoassay of Netilmicin, Antimicrob. Agents Chemother. *11*, 768 – 770 (1977).

[24]   *A. Broughton, J. E. Strong, G. P. Bodey,* Radioimmunoassay of Sisomicin, Antimicrob. Agents Chemother. *9*, 247 – 250 (1976).

[25]   *T. Kitagawa, K. Fujiwara, S. Tomonoh, K. Takahashi, M. Koide,* Enzyme-immunoassay of Kanamycin Group Antibiotics with High Sensitivities Using Anti-Kanamycin as a Common Antiserum: Reasoning and Selection of a Heterologous Enzyme Label, J. Biochem. (Tokyo) *94*, 1165 – 1172 (1983).

[26]   *T. Kitagawa, T. Kanamura, H. Kato, S. Yano, Y. Asanuma,* Novel Enzyme Immunoassay of Three Antibiotics. New Method for Preparation of Antisera to the Antibiotics and for Enzyme Labelling Using a Combination of Two Hetero-bifunctional Reagents, in: *S. B. Pal* (ed.), Enzyme Labelled Immunoassay of Hormones and Drugs, Walter de Gruyter, Berlin 1978, pp. 59 – 66.

[27]   *K. Fujiwara, M. Yasuno, T. Kitagawa,* Enzyme Immunoassay for Pepleomycin, a New Bleomycin Analog, Cancer Res. *41*, 4121 – 4126 (1981).

[28]   *K. Fujiwara, H. Saikusa, M. Yasuno, T. Kitagawa,* Enzyme Immunoassay for the Quantification of Mitomycin C Using β-Galactosidase as a Label, Cancer Res. *42*, 1487 – 1491 (1982).

[29]   *K. Fujiwara, H. Nakamura, T. Kitagawa, N. Nakamura, A. Saito, K. Hara,* Development and Application of a Sensitive Enzyme Immunoassay for 7-*N*-(p-Hydroxyphenyl)mitomycin C, Cancer Res. *44*, 4172 – 4176 (1984).

[30]   *H. Rosin, P. Naumann, E. Reindjens, M. Kohler,* Comparative Evaluation of Five Aminoglycosides for Treatment, Int. J. Clin. Pharmacol. Biopharm. *13*, 157 – 167 (1976).

1.15   Viomycin and Ampicillin

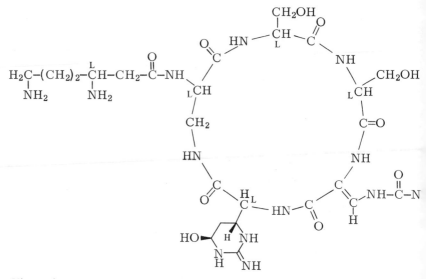

Viomycin

3,6-Diamino-*N*-[6-(aminocarbonylaminomethylene)-3-(2-amino-1,4,5,6-tetrahydro-6-hydroxypyrimidin-4-yl)-9,12-bis(hydroxymethyl)-2,5,8,11,14-pentaoxo-1,4,7,10,13-pentaazacyclohexadec-15-yl]hexanamide

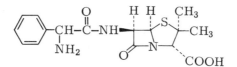

Ampicillin

6-[(Aminophenylacetyl)amino]-3,3-dimethyl-7-oxo-4-thia-1-azabicyclo-[3.2.0]heptane-2-carboxylic acid

Tsunehiro Kitagawa

General

Viomycin was first isolated independently in 1951 by *Finley et al.* and *Bartz et al.* [1, 2]. Viomycin is a strongly basic peptide antibiotic used as a tuberculostatic and differs from most antibiotics in that it is more active against mycobacteria than other groups of bacteria. The mode of action of viomycin is the inhibition of ribosomal protein synthesis. Viomycin shows delayed toxicity, especially towards the kidney, and ototoxicity. Semi-synthetic ampicillin is the first of the broad-spectrum penicillins with activity against many *Gram*-negative as well as *Gram*-positive micro-organisms [3]. Stability of ampicillin in acid media makes possible its oral administration. Absorption from the gastro-intestinal tract is good and about 30% of the antibiotic is excreted in the urine within 6 – 8 h after its oral administration. Ampicillin generally shows little toxicity, except for penicillin-sensitive persons. These useful properties of ampicillin have made it one of the most important drugs for infectious diseases of man and domestic animals. The mode of action of ampicillin is the inhibition of synthesis of cell walls of micro-organisms.

The main reasons for monitoring serum concentrations of viomycin are the large inter-individual differences in pharmacokinetics and the narrow therapeutic range. Monitoring of the serum levels of ampicillin is important for the study of its pharmacokinetics. Ampicillin concentrations have been studied in the body fluids of man and animals, as well as in food of animal origin.

Application of method: in pharmacology, in clinical chemistry and in toxicology for viomycin: in pharmacology, in biochemistry, in bacteriology and environmental toxicology for ampicillin.

Substance properties relevant in analysis: viomycin (M_r = 685.71) is a strong base with high solubility in water. The pK_a of viomycin are 8.2, 10.3 and 12.0. Ampicillin (M_r = 349.42) is sparingly soluble in water; the sodium salt has a high solubility in water.

Methods of determination: analysis of both antibiotics in physiological fluids requires methods with high specificity and sensitivity. Measurements of viomycin concentrations have previously been non-specific spectrophotometric determinations with a detection limit of 10 mg/l [4] and microbiological assays (detection limit 7.5 mg/l) [5]. The latter method was used to determine serum viomycin concentrations. A specific enzyme-immunoassay of viomycin using β-D-galactosidase (EC 3.2.1.23) as the labelling enzyme, is also available.

Various methods for measuring ampicillin concentration are: chromatographic procedures (gas chromatography, thin-layer chromatography, high-pressure liquid chromatography), iodometric assay and microbiological assay. Out of these methods, microbiological assay has been most widely used. High-pressure liquid chromatography [6] appears to be very useful for accurate assay as well as for confirmation analysis in cases of suspected interference. Heterogeneous enzyme-immunoassay in which β-D-galactosidase* is used as marker enzyme is useful in determining serum ampicillin levels.

International reference method and standards: neither standardization at the international level nor the existence of reference standard materials is known so far for viomycin and ampicillin.

---

* β-D-Galactoside galactohydrolase, EC 3.2.1.23.

Assay

Method Design

The method described here operates according to the well known design of a double-antibody enzyme-immunoassay (EIA), based on the competition between viomycin (VM) or ampicillin (ABPC) to be assayed, and a fixed amount of β-D-galactosidase-labelled antibiotic, against a limited number of binding sites on the antibiotic-specific antibodies.

Principle

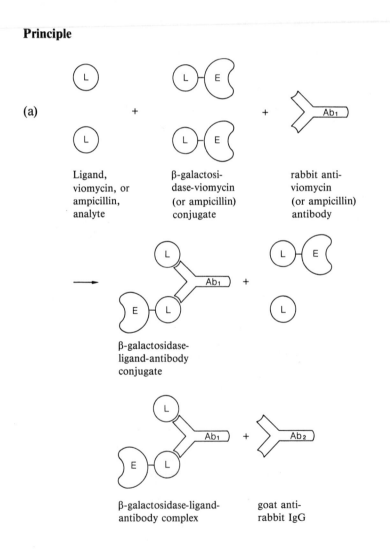

(a)

Ligand,
viomycin, or
ampicillin,
analyte

β-galactosi-
dase-viomycin
(or ampicillin)
conjugate

rabbit anti-
viomycin
(or ampicillin)
antibody

β-galactosidase-
ligand-antibody
conjugate

β-galactosidase-ligand-
antibody complex

goat anti-
rabbit IgG

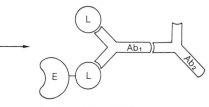

β-galactosidase-labelled
sandwich complex precipitate

(b)   4-Met-um-gal + H$_2$O $\longrightarrow$ 4-met-um + β-D-galactose .

In the competitive EIA, the inhibition of the binding of the enzyme-labelled hapten to antibody is related to the concentration of free hapten present in the mixture. The free and antibody-bound hapten are then separated by a second antibody (goat anti-rabbit IgG). The percentage of the β-D-galactosidase activity present in the bound fraction is determined and compared to the values produced by the assay of a series of standards. Measuring unit is the enzyme activity, i.e., the reaction rate, the increase in fluorescence intensity per unit time, $\Delta F/\Delta t$.

Selection of assay conditions and adaptation to the individual characteristics of the reagents: the assay is designed for sensitive and accurate analysis with the double-antibody method [7]. Antibody is the key reactant and the enzyme-labelled hapten is one of the most important reagents in immunoassay. Antibodies to VM and ABPC were prepared using a new method for preparation of hapten immunogens.

The immunogens to VM and ABPC were prepared by the method of *Kitagawa* [8, 9] with an easy three-step synthesis using a cross-linker of maleimide succinimidyl ester type [10 – 14]. Every reaction for the synthesis was performed in neutral aqueous solutions at 30 °C for 30 min to 2 h (cf. Appendix, pp. 211 – 213). The maleimide group introduced to the amino function(s) of VM or ABPC by a cross-linker was then coupled with free thiol groups of a modified protein [8, 9, 15 – 21]. Three procedures can be used for thiol introduction to prepare the modified protein. Two involve reductive cleavage of disulphide bonds of cystine residues in a protein, by a thiol-exchange reaction [15 – 17] or by the sodium borohydride reduction [18]. A two-step introduction of thiol groups onto amino groups of a protein is also useful [19, 20]. Every process can be performed under very mild conditions and homogeneous hapten immunogen was obtained as judged by disk electrophoresis [17, 18]. Highly specific antibodies to VM and ABPC have been produced by immunizing rabbits with these immunogens in complete *Freund's* adjuvant.

Preparations of the enzyme-labelled haptens were designed to possess sufficient, but not too strong, affinities for their corresponding antigens, since most enzyme-immunoassays (EIA) of antibiotics are of the competitive type. In such immunoassays, the enzyme-labelled hapten should have sufficient affinity for the antibody. However, the affinity for the antibody should not be too strong so that the label can be easily replaced by free antigen to give sensitive and accurate assay results [21].

Enzyme-labellings of VM and ABPC with β-D-galactosidase were performed by a bridge heterologous system [21, 22], with a selective and advanced labelling method [17, 23, 24] which involves a simpler procedure than that for isotopic iodine labellings of antigens and antibodies, using as cross-linker N-(γ-maleimidobutyryloxy)succinimide (GMBS) [24]. This differs from MBS by the substitution of N-butyric acid for benzoic acid which avoids a cross-reaction of anti-VM antibody with MBS, since MBS has been used as the cross-linker for preparation of the VM-immunogen (cf. Appendix, p. 212).

Enzyme activity of 50 µU galactosidase-VM conjugate yielded at 30°C values of $\Delta F/\Delta t$ 100 times greater than that of the reagent blank. This amount of conjugate is incubated with variously diluted solutions (3000 – 300000-fold) of rabbit anti-VM antiserum with or without 50 ng VM. The whole rabbit IgG in the reaction mixture is then precipitated by the double-antibody method [7]. The enzyme activity of the bound conjugate in the precipitate is determined in the presence and absence of VM. The ratio of enzyme activities obtained is maximum at 25000-fold dilution. Under these conditions the bound enzyme activity is about 20 times that of the blank containing all reagents except the antiserum to VM. Those amounts for the enzyme conjugate and antiserum to viomycin are used for the assay of VM. Similarly, 50 µU galactosidase-ABPC conjugate and 10000-fold dilutions of anti-ABPC antiserum are determined for the assay of ABPC.

Binding studies with hapten (VM or ABPC) and conjugate solutions were performed using the optimal quantities and concentrations of the two immunoreagents determined above. The assay range of the antibiotic encompasses doses giving inhibition values between 85% and 15% of the binding of galactosidase-hapten to anti-hapten antibody. The assay conditions such as incubation time and temperature were next determined individually.

The desired assay range is 1 to 1000 µg/l for VM and 3 to 1000 µg/l for ABPC. A serum sample of less than 30 µl for EIA of VM or ABPC gives essentially the same dose-response curve as the standard samples measured in buffer solution without human serum. A serum sample of 5 to 20 µl is adequate for monitoring serum levels of VM and ABPC.

Equipment

Fluorimeter capable of measuring fluorescence intensity with excitation and emission wavelengths of 365 and 448 nm, respectively. Quartz cuvettes suitable for the fluorimeter, spectrophotometer; laboratory centrifuge; water-bath (30°C); stopwatch, pH meter.

Reagents and Solutions

Purity of reagents: cross-linkers N-(3-maleimidobenzoyloxy)succinimide (MBS) and N-(γ-maleimidobutyryloxy)succinimide (GMBS) are from *Sigma Co.,* U.S.A., and

Dojin Chemicals, Kumamoto, Japan, respectively. Antibiotics VM and ABPC are of reagent grade. β-Galactosidase, reagent grade for EIA, is from *Boehringer Mannheim,* W. Germany: from *E. coli,* ca. 600 U/mg protein at 37°C, 4-NPgal as substrate; the lyophilisate contains 1/4 of its weight on protein. Bovine serum albumin (BSA, grade V) is from *Miles Lab.,* U.S.A.

Preparation of solutions (for 100 determinations): all solutions in re-purified water (cf. Vol. II, chapter 2.1.3.2).

1. Phosphate buffer (50 mmol/l, pH 7.0):

   dissolve 2.43 g $NaH_2PO_4 \cdot 2H_2O$ and 8.75 g $Na_2HPO_4 \cdot 12H_2O$ in water and make up to 800 ml.

2. Buffer A (phosphate, 20 mmol/l; NaCl, 0.1 mol/l; $MgCl_2$, 1 mmol/l; BSA, 1 g/l; NaN_3, 1 g/l; pH 7.0):

   dissolve 21.85 g $Na_2HPO_4 \cdot 12H_2O$, 6.08 g $NaH_2PO_4 \cdot 2H_2O$, 29.25 g NaCl, 1.02 g $MgCl_2 \cdot 6H_2O$, 5.0 g BSA, and 5.0 g NaN_3 in water and make up to 1 l; dilute 20 ml of this stock buffer A with water and make up to 100 ml.

3. Buffer B (phosphate, 60 mmol/l; EDTA, 10 mmol/l; BSA, 1 g/l; NaN_3, 1 g/l; pH 7.0):

   dissolve 107.45 g $Na_2HPO_4 \cdot 12H_2O$, 18.6 g $EDTA-Na_2H_2 \cdot 2H_2O$, 5.0 g BSA, and 5.0 g NaN_3 in water and make up to 1 l; dilute 20 ml of this stock buffer B with water and make up to 100 ml.

4. Viomycin standard solutions (VM, 1 to 1000 µg/l; phosphate, 60 mmol/l; EDTA, 10 mmol/l; BSA, 1 g/l; NaN_3, 1 g/l; pH 7.0):

   dissolve 10 mg VM in 10 ml phosphate buffer (1) and serially dilute with buffer B (3) to give concentrations ranging from 1 to 1000 µg/l.

5. Ampicillin standard solutions (ABPC, 1 to 1000 µg/l; phosphate, 60 mmol/l; EDTA, 10 mmol/l; BSA, 1 g/l; NaN_3, 1 g/l; pH 7.0):

   dissolve 10 mg ABPC in 10 ml phosphate buffer (1) and serially dilute with buffer B (3) to give concentrations ranging from 1 to 1000 µg/l.

6. Anti-viomycin antiserum solution (antiserum, 40 µl/l; phosphate, 60 mmol/l; EDTA, 10 mmol/l; BSA, 1 g/l; NaN_3, 1 g/l; pH 7.0):

   dilute anti-VM antiserum prepared according to Appendix, p. 213, 1 : 25000 with buffer B (3).

7. Anti-ampicillin antiserum solution (antiserum, 0.1 ml/l; phosphate, 60 mmol/l; EDTA, 10 mmol/l; BSA, 1 g/l; NaN_3, 1 g/l; pH 7.0):

dilute anti-ABPC antiserum prepared according to Appendix, p. 213, 1 : 10000 with buffer B (3).

8. Galactosidase-VM conjugate (galactosidase, 0.5 U*/l; phosphate, 20 mmol/l; NaCl, 0.1 mol/l; $MgCl_2$, 1 mmol/l; BSA, 1 g/l; NaN_3, 1 g/l; pH 7.0):

   dilute galactosidase-VM conjugate solution prepared according to Appendix, p. 213, with buffer A (2) to give 5 mU in 10 ml.

9. Galactosidase-ampicillin conjugate (galactosidase, 0.5 U/l; phosphate, 20 mmol/l; NaCl, 0.1 mol/l; $MgCl_2$, 1 mmol/l; BSA, 1 g/l; NaN_3, 1 g/l; pH 7.0):

   dilute galactosidase-ABPC conjugate solution prepared according to Appendix, p. 213, with buffer A (2) to give 5 mU in 10 ml.

10. Goat anti-rabbit IgG solution (IgG, ca. 0.6 g/l; phosphate, 60 mmol/l; EDTA, 10 mmol/l; BSA, 1 g/l; NaN_3, 1 g/l; pH 7.0):

    dilute 1 ml anti-RIgG antiserum (IgG ca. 12 g/l) with buffer B (3) and make up to 20 ml.

11. Normal rabbit serum solution (NRS, 3.4 ml/l; phosphate, 60 mmol/l; EDTA, 10 mmol/l; BSA, 1 g/l; NaN_3, 1 g/l; pH 7.0):

    dilute 0.1 ml normal rabbit serum with buffer B (3) and make up to 30 ml.

12. Substrate solution (4-met-um-gal, 0.1 mmol/l; phosphate, 20 mmol/l; NaCl, 0.1 mol/l; $MgCl_2$, 1 mmol/l; BSA, 1 g/l; NaN_3, 1 g/l; pH 7.0):

    dissolve 3.4 mg 4-methylumbelliferyl β-D-galactoside (7-β-D-galactopyranosyl-hydroxy-4-methylcoumarin) in 80 ml water and mix with 20 ml stock buffer A (2).

13. Stopping solution (glycine, 0.2 mol/l, pH 10.4):

    dissolve 4.5 g glycine in 210 ml water, adjust pH at 10.4 with NaOH, 0.1 mol/l, and make up to 300 ml with water.

14. Urea solution (6 mol/l):

    dissolve 36.0 g urea in water and make up to 100 ml.

15. Urea/EDTA solution (urea, 6 mol/l; EDTA, 50 mmol/l):

    dissolve 186 mg EDTA-$Na_2H_2 \cdot 2 H_2O$ in 10 ml urea solution (14).

Stability of reagents and solutions: stock solutions of buffer A and B are stable in stoppered flasks for about 3 months at room temperature providing microbial contamination is avoided. Store buffers A and B in stoppered flasks in a refrigerator at

---

* International units, 30°C.

0°C to 4°C. Galactosidase-labelled VM (8) and ABPC (9) solutions, anti-VM (6), and anti-ABPC (7) serum solutions are stable at 4°C up to 1 month, although VM and ABPC are not stable in aqueous solutions. The enzyme labels must be used within 3 months after their preparation. Prepare the standard solutions for VM (4) and ABPC (5), and the substrate solution (12) every week and keep in dark at 0°C to 4°C.

Procedure

Collection and treatment of specimen: glass tubes may be used for all procedures.

Blood: take blood without stasis and store at −30°C before use. Blood may be used for VM assay after haemolysis with 20 times its volume of water (unpublished data). Add heparin (0.2 mg/ml) to blood immediately after collection, and store in ice-water during the collections. Immediately after collection, haemolyze every 20 µl blood at once with 380 µl water and use immediately for EIA. Haemolysis causes little assay variation.

Tissues: collect tissues from laboratory animals. Homogenize, deproteinize with perchloric acid (cf. Vol. II, chapter 1.2 Cell and Tissue Disintegration). Adjust the supernatant to pH 7.0 – 7.5, assay. Take the dilution and extraction factors into account when calculating.

Urine: take the dilution factor into account when calculating.

Milk: take the dilution factor into account when calculating.

Stability of the analyte in the sample: VM and ABPC are not stable in acidic and alkaline media; therefore keep sample solutions at neutral pH in ice-water before use.

Assay conditions: measure standards and samples in triplicate under identical conditions (within series). Use buffer B (3) as the diluent for samples and immunoreagents except otherwise stated. Initiate immunoreaction of series of samples at 15 s intervals. All solutions must be allowed to come to room temperature before performing the assay.

Immunoreaction: incubation at 25°C overnight and for 3 h, respectively; volume 0.3 ml and 0.5 ml, respectively.

Enzymatic indicator reaction: incubation at 30°C for 30 min. Excitation and emission wavelengths are 365 nm and 448 nm, respectively; final volume 2.7 ml; light path 10 mm; ambient temperature; read against air.

Run a reference with antibody solution for VM (6) or ABPC (7) but without sample, and a blank consisting of assay buffer (3) and galactosidase-labelled VM (8) or ABPC (9) solution, simultaneously. Establish a calibration curve using VM (4) or ABPC (5) standards.

Measurement

Pipette successively into the glass tube:			concentration* in incuba-tion/assay mixture	
sample or standard solution (4) or (5), resp.		0.1 ml	VM or ABPC, resp. phosphate EDTA BSA NaN$_3$	up to 100 µg/l up to 333 µg/l 47 mmol/l 6.7 mmol/l 1 g/l 1 g/l
anti-VM antiserum** galactosidase-VM conjugate solution***	(6) (8)	0.1 ml 0.1 ml	antiserum galactosidase NaCl Mg^{2+}	13.3 ml/l 0.17 U/l 33 mmol/l 0.34 mmol/l
mix well, incubate at 25 °C overnight,				
goat anti-rabbit antiserum (10) normal rabbit serum (11)		0.05 ml 0.05 ml	IgG serum	75 mg/l 0.43 ml/l
mix well, incubate at 25 °C for 3 h; add 1 ml buffer A (2) and centrifuge for 15 min at 2500 rpm; discard supernatants by decantation; repeat this washing once more; use sediment for indicator reaction;				
substrate solution (12)		0.2 ml	4-met-um-gal phosphate NaCl Mg^{2+} BSA NaN$_3$	0.1 mmol/l 20 mmol/l 0.1 mol/l 1 mmol/l 1 g/l 1 g/l
mix, incubate for 30 min at 30°C				
stopping solution (13)		2.5 ml	glycine	185 mmol/l
mix; reaction is stopped; read fluorescence.				

  * The salt and BSA concentrations vary if sample is not diluted with buffer B (3).
 ** For assay of ampicillin use anti-ABPC antiserum (7).
*** For assay of ampicilline use galactosidase-ABPC conjugate solution (9).

If $\Delta F/\Delta t$ of the sample is greater than that of the highest calibrator, dilute the sample with buffer B (3) and re-assay.

Calibration curve: correct readings of standards and reference for readings of blank yielding the increase in fluorescence intensity per unit time of antibody-bound enzyme conjugate for standards, $\Delta F/\Delta t$, and for reference (without standard), $\Delta F_0/\Delta t$. Plot the ratios, $(\Delta F/\Delta t)/(\Delta F_0/\Delta t) = \Delta F/\Delta F_0$ (ordinate) *versus* VM or ABPC mass concentrations in the sample, $\mu g/l$ (abscissa), on a semi-logarithmic graph paper. Fig. 1 shows typical calibration curves for the enzyme-immunoassay of viomycin and ampicillin, respectively.

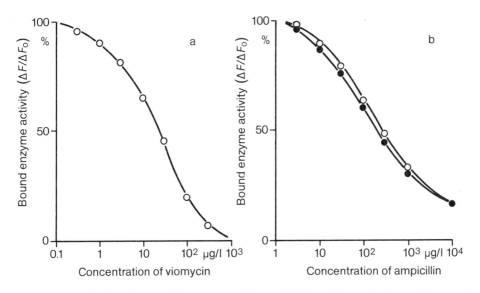

Fig. 1. a) Typical calibration curve for the assay of viomycin in buffer B. b) Typical calibration curve for the assay of ampicillin in milk (open circles) or in buffer B (closed circles).

Various curve-fitting methods are available for the evaluation of the results from enzyme-immunoassays (cf. Vol. I, p. 243).

Calculation: correct readings of sample for that of blank, yielding $\Delta F/\Delta t$ for samples. Calculate $\Delta F/\Delta F_0$ as above and read the corresponding concentration of the antibiotic from the calibration curve. To convert from mass concentration to substance concentration, c ($\mu mol/l$), multiply by 14.6 for VM and by 28.6 for ABPC.

Validation of Method

Precision, accuracy, detection limit and sensitivity: the imprecision of VM assay in buffer B (3) performed within a series (intra-assay) and between days (inter-assay) has been determined to be between 4.4 and 38.2% (relative standard deviations, RSD, $n = 7$) in a range of 0.1 to 30 ng/tube. The intra- and inter-assays imprecision for ABPC in milk has been determined to be RSD = 13.5 to 32.0% ($n = 6$) in a range of 1 to 100 ng/tube. The detection limit of VM and ABPC assays defined as the lowest amount distinguishable from zero at 95% confidence level, was 0.1 ng/tube and 1 ng/tube respectively, and the working range is from 1 µg/l to 1000 µg/l for VM and 3 µg/l to 1000 µg/l for ABPC. The accuracy has been examined by recovery determinations.

Sources of error: no influence of therapeutic measures is known for either assay. The binding of the tracer to the specific antibody must be tested before use, when the tracer is stored for more than one month.

Specificity: each assay is specific for the antigen. Even a minor modification on a functional group of VM or on the substituents of ABPC gives a large change in recognition in the assays, as shown by cross-reactivities of viomycin analogues (Table 1) [17] and ampicillin analogues (Table 2) [16].

Table 1. Cross-reaction values for viomycin analogues and some other antibiotics determined by EIA of viomycin.

Sample	Cross-reactivity %	Sample	Cross-reactivity %
Viomycin	100	Oxoviomycin	0.03
1-Monoacetylviomycin	54.5	Tetrahydroviomycin	0.20
Acetylviomycin	29.0	Tuberactinomycin	1.69
Succinylviomycin	36.7	Capreomycin	0.013
Anhydroviomycin	8.18	Dihydrostreptomycin	< 0.001
Dihydroviomycin	1.38	Kanamycin	< 0.001
Broxoviomycin	0.333	Ampicillin	< 0.001

Calibration curves for viomycin analogues were obtained in a similar way to that for viomycin and each dose giving 50% inhibition was determined from the curve [17]. Cross-reactivity: ratio of the molar concentrations of VM and the test substance effecting 50% displacement of Gal-VM conjugate under routine assay conditions.

Table 2. Cross-reaction values for penicillins and cephalosporins and some other antibiotics determined by EIA of ampicillin.

Sample	Cross-reactivity %	Sample	Cross-reactivity %
Ampicillin	100	Cephaloglycine	0.09
Sulbenicillin	100	Penicilloic acid	0.035
Carbenicillin	13.4	Viomycin	< 0.001
Penicillin G	10.0	Dihydrostreptomycin	< 0.001
Flucloxacillin	1.32	Kanamycin	< 0.001
Cephalexin	0.09		

No cross-reactivity was observed for other antibiotics of different structures in either assay.

Appendix

Two selective modifications of proteins are required to develop enzyme-immuno-assays (EIA) for haptenic antibiotics [8, 9, 25 – 27]. One is enzyme-labelling and the other is preparation of hapten immunogen. The method for preparation of hapten immunogen influences the specificity of the anti-hapten antiserum, and the specificity for the enzyme-labelled hapten is affected by the labelling method [21, 22]. We previously introduced a new method for selective modification of proteins using hetero-bifunctional reagents of a maleimide succinimidyl ester type [7 – 11], and applied the

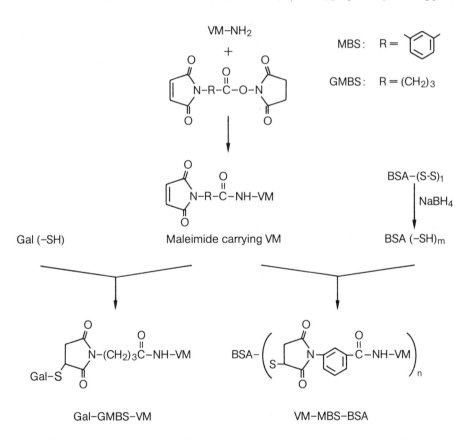

Fig. 2. Scheme for preparation of the VM immunogen and of the galactosidase-labelled VM using hetero-bifunctional cross-linkers of MBS and GMBS, respectively.
Abbreviations: VM, viomycin; MBS, N-(3-maleimidobenzoyloxy)succinimide; GMBS, N-(γ-maleimido-butyryloxy)succinimide; Gal, β-D-galactosidase.

reagents in a new enzyme-labelling method [8 – 13] and also in a new method for preparing hapten immunogens [8, 9, 13 – 21]. The enzyme-labelling method was further improved. This chapter gives surveys for EIA of viomycin (VM) [13, 15, 17, 29] and ampicillin (ABPC) [15, 19, 30] with selective and convenient methods for the improved enzyme-labelling procedure [17, 23, 24] and for preparation of specific antisera to them using a combination of two cross-linkers of a maleimide succinimidyl ester type.

Preparation of the immunogen for viomycin (*Kitagawa* method): the reactions involved in the preparative process are shown in Fig. 2.

Preparation of MBS-acylated VM:

- dissolve 8.5 mg (10 μmol) VM in 1 ml phosphate buffer (1) and keep at 30°C,
- dissolve 3.14 mg (10 μmol) MBS in 0.5 ml tetrahydrofuran and add this solution with vigorous stirring in five portions within 5 min to the VM solution; incubate the mixture at 30°C for 35 min with occasional stirring,
- then remove tetrahydrofuran by flushing with nitrogen, extract free MBS by washing the reaction mixture three times with 5 ml portions of methylene chloride. Use the aqueous layer for the final step.

Preparation of bovine serum albumin (BSA) with reduced disulphide bonds by sodium borohydride reduction:

- dissolve 10.5 mg (0.15 μmol) BSA in 2 ml urea/EDTA solution (15);
- add to this solution with vigorous stirring in five portions of 20 mg $NaBH_4$ with defoaming by occasional addition of total 0.2 ml n-butanol; incubate at 30°C for 30 min;
- decompose excess $NaBH_4$ by adding 1 ml sodium dihydrogen phosphate solution, 0.1 mol/l, and 0.4 ml acetone; use this solution for the next step.

Conjugation of MBS-acylated viomycin to the reduced BSA:

- add to the solution of reduced BSA prepared above, with stirring, the solution of the MBS-acylated VM;
- incubate the mixture at 25°C for 2 h, then chromatograph on a Sephadex G-100 column (25 mm × 570 mm) with urea, 6 mol/l, as an eluent; pool peak protein fractions measured by absorbance at 280 nm and use the solution for immunization of rabbits without further purification.

Preparation of the immunogen for ampicillin

Preparation of the MBS-acylated ABPC:

- dissolve 10.5 mg (30 μmol) ABPC in 2 ml phosphate buffer (1);

- dissolve 9.5 mg (30 μmol) MBS in 1.0 ml tetrahydrofuran and add this solution with vigorous stirring in five portions to the ABPC solution; keep at 30 °C for 30 min with occasional stirring;
- remove tetrahydrofuran by flushing with nitrogen and extract free MBS from the reaction mixture three times with 5 ml portions of ether/methylene chloride solution (2 : 1, v/v); use the aqueous layer for step 3.

Preparation of BSA with reduced disulphide bonds by thiol-exchange reaction with dithiothreitol:

- dissolve 21 mg (0.2 μmol) BSA in 1 ml Tris/HCl buffer, 0.2 mol/l, pH 8.6, containing urea, 8 mol/l;
- add to this solution with stirring a solution of 4.6 mg (30 μmol) dithiothreitol in 1 ml phosphate buffer (1) and incubate for 1 h at room temperature;
- add 2 ml trichloroacetic acid solution, 100 g/l, and centrifuge at 3000 rpm for 10 min, discard the supernatant and wash the precipitate twice with de-ionized water;
- then dissolve the precipitate of the reduced BSA in ca. 6 ml sodium phosphate buffer, 50 mmol/l, pH 7.0, containing urea, 8 mol/l; use this solution for the conjugation step.

Perform conjugation reaction for preparing the immunogen for ABPC by the similar way to that for the immunogen for VM.

Note: the first and the second steps should be designed to be completed at the same time to avoid decomposition of unstable free thiol and maleimide groups [12]. MBS-acylated ABPC is slightly soluble in methylene chloride but insoluble in ether/methylene chloride solution. MBS is very soluble but MBS-acylated VM is insoluble in methylene chloride. Separation of high molecular VM-MBS-BSA conjugate from low molecular VM, MBS or MBS-acylated VM is easily achieved by a usual gel-filtration procedure and the protein fractions identified by absorbance at 280 nm are pooled. The conjugate in the pooled fractions is stable at 4 °C for 3 months.

Immunization: immunize two randomly bred rabbits s.c. and i.m. with 1 mg immunogen emulsified in *Freund's* complete adjuvant and thereafter give booster injections totalling 1.5 mg 3 times at biweekly intervals. 2 weeks after the final booster take blood from ear vein, separate serum by centrifugation and heat at 55 °C for 30 min, store at −30 °C.

Note: content of the immunogen is calculated from absorbance assuming that 1 g immunogen per litre showed $A = 0.7$ at 280 nm.

Preparation of β-galactosidase-labelled viomycin (galactosidase-VM):

- dissolve 0.5 mg (5 μmol) VM in 1 ml phosphate buffer (1);
- add to this solution with stirring 0.01 ml (0.14 mg) GMBS solution in tetrahydrofuran (50 mmol/l) and incubate the mixture at 25 °C for 30 min;

- dissolve 50 μg (1 nmol/l) β-D-galactosidase in 1 ml phosphate buffer (1);
- add 50 μl VM/GMBS mixture into the enzyme solution with vortex mixing; incubate at 30°C for 2 h;
- chromatograph the reaction mixture directly on a Sepharose 6B column (18 mm × 300 mm) with buffer A (2) as an eluent;
- measure the enzymic- and immunoreactive activities by the methods described below using 5 μl of a 1 : 30 solution of each fraction;
- determine the immune specificity of the conjugate by the difference of the bindings of the 5 μl Gal-VM fraction to anti-viomycin antibody with and without 50 ng viomycin using the EIA procedure for VM.

Enzyme-labelling of ampicillin is performed by the previous labelling method [11, 16] which requires skill to master. Re-investigation of the labelling by the improved method using a similar procedure to that for VM gives the enzyme-labelled ABPC more easily than by the former method.

Note: the molar ratio of GMBS to amino groups of VM is limited to one tenth in step 1 so that GMBS reacts exclusively. A part of GMBS-acylated VM formed in the reaction mixture is immediately coupled with one twentieth amount of galactosidase (moles of GMBS-acylated VM can be assumed as 70% moles of GMBS used) (cf. [12]). Five different molar ratios of GMBS-acylated VM to galactosidase (5, 10, 20, 40 or 80 : 1) were reacted at the same time. In most cases the reaction product of the ratio 20 : 1 gives a satisfactory enzyme conjugate. If this is not the case, the best conjugate is separated from the other reaction mixture. The enzyme conjugate fraction of the peak enzyme activity separated by the chromatography is usually diluted with buffer A (2) 200 – 300 times. 50 μl diluted conjugate is used per assay. Keep the conjugate at 4°C and do not freeze.

References

[1] J. Ehrlich, R. M. Smith, M. A. Penner, L. E. Anderson, A. C. Bratton Jr., Antimicrobial Activity of Streptomyces floridae and of Viomycin, Am. Rev. Tuberc. Pulm. Dis. 63, 7 – 12 (1951).
[2] A. C. Finley, G. L. Hobby, F. Hochstein, T. M. Las, T. F. Lenert, J. A. Means, S. Y. P'an, P. P. Regna, J. B. Routien, B. A. Sobin, I. B. Tate, J. H. Kane, Viomycin a New Antibiotic Active against Mycobacteria, Am. Rev. Tuberc. 63, 1 – 6 (1951).
[3] F. P. Doyle, G. R. Fosker, J. H. C. Nayler, H. Smith, Derivatives of 6-Aminopenicillanic Acid. Part 1. alpha-Amino-Benzylpenicillin and Some Related Compounds, J. Chem. Soc. 1440 – 1444 (1962).
[4] A. W. Jackson, Colorimetric Determination of Viomycin, antibiot. Chemother. 4, 1210 – 1215 (1954).
[5] J. Ehrlich, W. P. Iverson, D. Kohberger, Agar Diffusion Method for the Assay of Viomycin, Antibiot. Chemother. 1, 211 – 216 (1951).
[6] K. Tsuji, High-Pressure Liquid Chromatography of Antibiotics, in: S. P. Colowick, N. O. Kaplan (eds.), Methods in Enzymology, Vol. XLIII, Academic Press, New York 1975, pp. 300 – 320.
[7] H. Van Vunakis, L. Levine, Use of the Double Antibody and Nitrocellulose Membrane Filtration Technique to Separate Free Antigen from Antibody Bound Antigen in Radioimmunoassay, in: S. J. Mule, I. Sunshine, M. Braude, R. E. Willete (eds.), Immunological Assay of Drugs of Abuse CRC Press, Cleveland, OH, 1974, pp. 23 – 35.
[8] T. Kitagawa, Competitive Enzyme Immunoassay of Antibiotics, in: E. Ishikawa, T. Kawai, K. Miyai (eds.), Enzyme Immunoassay, Igaku-Shoin, Tokyo 1981, pp. 136 – 145.
[9] T. Kitagawa, Recent Progress in the Methods for Selective Modification of Proteins, J. Synthetic Organic Chem. Japan 42, 283 – 292 (1984) (in Japanese).

[10] *T. Kitagawa,* Enzyme Labeling with *N*-Hydroxysuccinimidyl Ester of Maleimide, in: *E. Ishikawa, T. Kawai, K. Miyai* (eds.), Enzyme Immunoassay, Igaku-shoin, Tokyo 1981, pp. 81 – 89.

[11] *T. Kitagawa, T. Aikawa,* Enzyme Coupled Immunoassay of Insulin Using a Novel Coupling Reagent, J. Biochem. *79,* 233 – 236 (1976).

[12] *T. Kitagawa, T. Shimozono, T. Aikawa, T. Yoshida, H. Nishimura,* Preparation and Characterization of Hetero-bifunctional Cross-linking Reagents for Protein Modifications, Chem. Pharm. Bull. *29,* 1130 – 1135 (1981).

[13] *T. Kitagawa, T. Fujitake, H. Taniyama, T. Aikawa,* Enzyme Immunoassay of Viomycin: New Cross-linking Reagent for Enzyme Labelling and a Preparation Method for Antiserum to Viomycin, J. Biochem. *83,* 1493 – 1501 (1978).

[14] *T. Aikawa, S. Suzuki, M. Murayama, K. Hashiba, T. Kitagawa, E. Ishikawa,* Enzyme Immunoassay of Angiotensin I, Endocrinology *150,* 1 – 6 (1979).

[15] *T. Kitagawa, T. Kanamaru, H. Kato, S. Yano, Y. Asanuma,* Novel Enzyme Immunoassay of Three Antibiotics. New Methods for Preparation of Antisera to the Antibiotics and for Enzyme Labelling Using a Combination of Two Hetero-bifunctional Reagents, in: *S. Pal* (ed.), Enzyme labelled Immunoassay of Hormones and Drugs, Walter de Gruyter and Co., Berlin 1978, pp. 59 – 66.

[16] *T. Kitagawa, T. Kanamaru, H. Wakamatsu, H. Kato, S. Yano, Y. Asanuma,* A New Method for Preparation of an Antiserum to Penicillin and Its Application for Novel Enzyme Immunoassay of Penicillin, J. Biochem. *84,* 491 – 494 (cf. 733 – 735) (1978).

[17] *T. Kitagawa, H. Tanimori, K. Yoshida, H. Asada, T. Miura, K. Fujiwara,* Studies on Viomycin. XV. Comparative Study on the Specificities of Two Anti-viomycin Antisera by Enzyme Immunoassay, Chem. Pharm. Bull. *30,* 2487 – 2491 (1982).

[18] *T. Kitagawa, T. Kawasaki, H. Munechika,* Enzyme Immunoassay of Blasticidin S with a High Sensitivity: A New Convenient Method for Preparation of Immunogenic (Hapten-Protein) Conjugates, J. Biochem. *92,* 585 – 590 (1982).

[19] *K. Fujiwara, M. Yasuno, T. Kitagawa,* Novel Preparation Method of Immunogen for Hydrophobic Hapten, Enzyme Immunoassay of Daunomycin and Adriamycin, J. Immunol. Methods *45,* 195 – 203 (1981).

[20] *K. Fujiwara, H. Saegusa, M. Yasuno, T. Kitagawa,* Enzyme Immunoassay for Quantification of Mitomycin C Using beta-Galactosidase as a Label, Cancer Res. *42,* 1487 – 1491 (1982).

[21] *T. Kitagawa, K. Fujiwara, S. Tomonoh, K. Takahashi, M. Koida,* Enzyme Immunoassays of Kanamycin Group Antibiotics with High Sensitivities Using Anti-kanamycin as a Common Antiserum: Reasoning and Selection of a Heterologous Enzyme Label, J. Biochem. *94,* 1165 – 1172 (1983).

[22] *B. K. VanWeemen, A. H. W. M. Schuurs,* The Influence of Heterologous Combinations of Antiserum and Enzyme Labelled Oestrogen on the Characteristics of Oestrogen Enzyme Immunoassay, Immunochemistry *12,* 667 – 670 (1975).

[23] *H. Tanimori, H. Ishikawa, T. Kitagawa,* A Sandwich Enzyme Immunoassay of Rabbit Immunoglobulin G with an Enzyme Labeling Method and a New Solid Support, J. Immunol. Methods *62,* 123 – 131 (1983).

[24] *H. Tanimori, T. Kitagawa, T. Tsunoda, R. Tsuchia,* Enzyme Immunoassay of Neocarzinostatin Using β-Galactosidase as Label, J. Pharmacobio-Dyn. *4,* 812 – 819 (1981).

[25] *V. P. Butler Jr.,* The Immunological Assay of Drugs, Pharmacol. Rev. *29,* 103 – 184 (1978).

[26] *J. J. Langone, H. V. Vunakis* (eds.), Immunochemical Techniques Part B, in: Methods in Enzymology, Vol. *73,* Academic Press, New York 1981.

[27] *A. J. O'Beirne, H. R. Cooper,* Heterogeneous Enzyme Immunoassay, J. Histochem. Cytochem. *27,* 1148 – 1162 (1979).

[28] *S. B. Pal* (ed.), Enzyme Labelled Immunoassay of Hormones and Drugs, Walter de Gruyter and Co., Berlin 1978.

[29] *T. Kitagawa, H. Tanimori, K. Yoshida, T. Miura, K. Fujiwara,* High Dimensional Structure of the Antigen-binding Site of Anti-viomycin Immunoglobulin Analyzed by Enzyme Immunoassay, J. Biochem. *91,* 1601 – 1605 (1982).

[30] *T. Miura, H. Kouno, T. Kitagawa,* Detection of Residual Penicillin in Milk by Sensitive Enzyme Immunoassay, J. Pharmacobio-Dyn. *4,* 706 – 710 (1981).

1.16   Disopyramide, Lidocaine, Procainamide, *N*-Acetylprocainamide, Quinidine

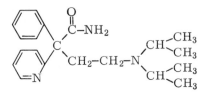

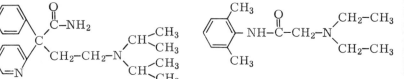

Disopyramide

α-[2-[Bis(1-methylethyl)amino]ethyl]-α-phenyl-2-pyridineacetamide

Lidocaine

2-(Diethylamino)-*N*-(2,6-dimethyl-phenyl)acetamide

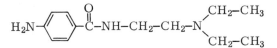

Procainamide

4-Amino-*N*-[2-(diethylamino)ethyl]benzamide

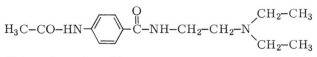

N-Acetylprocainamide

4-Acetylamino-*N*-[2-(diethylamino)ethyl]benzamide

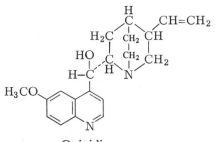

Quinidine

(9*S*)-6-Methoxy-cinhonan-9-ol

Clarence T. Ueda and Mary C. Haven

General

Disopyramide, a type Ia antiarrhythmic, is used to treat ventricular and supraventricular arrhythmias. After oral administration, disopyramide is absorbed rapidly with an average systemic bio-availability of 80% [1]. It is eliminated with a half-life of 6 to 8 h by renal excretion (50%) and metabolism to a mono-N-dealkylated metabolite (25%) which is less active than the parent drug. Disopyramide exhibits dose-dependent pharmacokinetics. The elimination half-life of disopyramide is prolonged in renal and hepatic impairment.

Disopyramide binding to serum proteins is dependent on the drug concentration in the serum. For the therapeutic serum concentration range of disopyramide of 2 to 5 mg/l, the bound drug fraction varies from 55 to 80%. The apparent volume of distribution of disopyramide is about 0.8 l/kg. In patients with cardiac failure and severe renal dysfunction, the distribution volume and clearance of disopyramide are diminished.

Lidocaine, a type Ib antiarrhythmic, is used only in the treatment of ventricular arrhythmias. It is commonly given to patients with acute myocardial infarction to prevent ventricular fibrillation. Lidocaine is also used widely as a local anaesthetic. As an antiarrhythmic, lidocaine is given parenterally because it undergoes extensive first-pass hepatic metabolism. Approximately 70% of the drug is metabolized during a single pass through the liver [2].

Lidocaine is eliminated almost exclusively by hepatic metabolism. Monoethylglycinexylidide (MEGX) and glycinexylidide (GX), which both possess antiarrhythmic and anticonvulsant activities, are the two principal lidocaine metabolites that are found in the blood after drug administration. Renal excretion of unchanged lidocaine is a minor route ($< 5\%$) of elimination. MEGX and GX are further metabolized in the liver to 4-hydroxy-2,6-xylidine which is excreted in the urine [3].

Lidocaine is eliminated with a half-life of about 1.8 h. In liver disease, the elimination half-life is prolonged. The elimination half-lives of MEGX and GX are approximately 2 and 10 h, respectively.

Therapeutic effects with lidocaine are seen when serum drug concentrations are between 1.3 and 5 mg/l. At these concentrations, lidocaine is 60% bound to serum proteins. The apparent volume of distribution of lidocaine averages 1.1 l/kg. The distribution volume is increased in hepatic disease and reduced in cardiac failure. Lidocaine clearance is reduced in both heart failure and liver disease.

Procainamide is a type Ia antiarrhythmic and is given intravenously and orally to treat a variety of ventricular and supraventricular arrhythmias. After oral administration, peak serum concentrations of procainamide are usually reached within 1 to 2 h with a systematic availability of 75 to 90% [4].

Procainamide distributes in an apparent volume of distribution of about 2 l/kg. In patients with congestive heart failure, the distribution volume is smaller. At therapeutic serum concentrations of procainamide of 4 to 8 mg/l, approximately 15% of the drug is bound to serum proteins.

Renal excretion of procainamide accounts for the elimination of 50% of the drug. The major portion of the drug remaining (7 to 34%) is metabolized in the liver to

N-acetylprocainamide (NAPA), a pharmacologically active metabolite, by the polymorphic enzyme *N*-acetyltransferase. The rate of acetylation of procainamide to NAPA is bimodally distributed giving rise to slow and rapid acetylator classes of patients.

The elimination half-life of procainamide is about 3 h. In renal failure, the half-life is prolonged. It may also be altered in liver disease. Procainamide clearance is reduced in patients with renal impairment.

N-Acetylprocainamide is the principal metabolite of procainamide [5]. It is formed in the liver by the polymorphic enzyme *N*-acetyltransferase. Like the parent drug, *N*-acetylprocainamide possesses significant antiarrhythmic activity. The effective serum concentration range of *N*-acetylprocainamide is about 10 to 20 mg/l but higher levels may be required in some subjects.

N-Acetylprocainamide is eliminated almost entirely by the kidneys (85%). A small fraction (2 to 3%) is deacetylated back to procainamide. The elimination half-life of *N*-acetylprocainamide is about 6 h and increases in renal disease. The clearance of *N*-acetylprocainamide is reduced in patients with impaired renal function. *N*-Acetylprocainamide is cleared by haemodialysis.

In the blood, *N*-acetylprocainamide is 10% serum protein bound. Its apparent volume of distribution is 1.4 l/kg.

Quinidine is the prototype class I antiarrhythmic and is used to treat both ventricular and supraventricular arrhythmias. After oral administration, peak serum drug concentrations are reached within 1 to 3 h with an average systemic bio-availability of 70% [6].

Quinidine distributes in an apparent volume of distribution of 3 l/kg. In congestive heart failure, the distribution volume is reduced. It may also be smaller in uraemic patients and larger in patients with cirrhosis. In its therapeutic serum concentration range of 2 to 5 mg/l, about 70% of the drug is serum protein bound.

The principal mode of quinidine elimination is metabolism to the following metabolites: 3-hydroxyquinidine, 2′-oxo-quinidinone, quinidine 10,11-dihydrodiol, quinidine-*N*-oxide and 2-desmethylquinidine. With the exception of quinidine 10,11-dihydrodiol, all of the quinidine metabolites have been found to possess some antiarrhythmic activity. Renal excretion of quinidine accounts for the elimination of 10 to 20% of the drug.

Quinidine is eliminated with a half-life of 6 to 7 h. The elimination half-life of quinidine may be prolonged in patients with cirrhosis. Quinidine clearance is reduced in patients with congestive heart failure.

Quinidine exhibits large inter-patient differences in disposition kinetics. Further, its disposition kinetics may be dose-dependent [7].

Application of method: in clinical chemistry, pharmacology and toxicology.

Substance properties relevant in analysis: disopyramide (molecular weight: 339.5) is a weak base with a pK_a of 10.4 and water solubility of 0.43 g/l. In the form of the phosphate salt, it is freely water soluble.

Lidocaine (molecular weight: 234.3) is a weak base with a pK_a of 7.85. It is insoluble in water as the base but freely soluble as the hydrochloride salt.

Procainamide (molecular weight: 236.3) is a weak base with a pK_a of 9.23. In the form of the hydrochloride salt, procainamide is freely soluble in water.

N-Acetylprocainamide (molecular weight 277.4) has a pK_a of 8.3. It is poorly water-soluble in the form of the free base but readily soluble in acidic solutions.

Quinidine (molecular weight: 324.4) is a weak base with pK_a values of 4.0 and 8.6. Quinidine free base is insoluble in water. The salts are water-soluble.

Methods of determination: disopyramide samples may be assayed by spectrofluorimetry, gas-liquid chromatography, high-pressure liquid chromatography, and enzyme- and fluorescence (substrate-labelled and polarization) immunoassays. The last four methods are specific for disopyramide and are well suited for routine therapeutic drug monitoring. With the gas-liquid and high-pressure liquid chromatographic procedures, it is possible simultaneously to assay the N-dealkylated metabolite of disopyramide which is an advantage of these methods.

Lidocaine samples may be assayed by gas-liquid chromatography, high-pressure liquid chromatography, and enzyme- and fluorescence polarization immunoassays. All of these methods are specific for lidocaine and are well suited for routine therapeutic drug monitoring. With the gas-liquid and high-pressure liquid chromatographic procedures, it is possible simultaneously to assay MEGX and GX which is an advantage that these methods have over the immunoassays.

Procainamide samples may be assayed by spectrophotometry, spectrofluorimetry, gas-liquid chromatography, high-pressure liquid chromatography, and enzyme- and fluorescence (substrate-labelled and polarization) immunoassays. The last four methods are specific for procainamide and are well suited for routine therapeutic drug monitoring. With the gas-liquid and high-pressure liquid chromatographic procedures, it is possible simultaneously to assay NAPA which is an advantage that these methods have over the immunoassays.

N-Acetylprocainamide samples may be assayed by spectrophotometry, spectrofluorimetry, gas-liquid chromatography, high-pressure liquid chromatography, and enzyme- and fluorescence (substrate-labelled and polarization) immunoassays. The last four methods are specific for N-acetylprocainamide and are well suited for routine therapeutic drug monitoring. With the gas-liquid and high-pressure liquid chromatographic procedures, it is possible simultaneously to assay procainamide which is an advantage that these methods have over the immunoassays.

Quinidine samples may be assayed by spectrofluorimetry, gas-liquid chromatography, high-pressure liquid chromatography, thin-layer chromatography and enzyme- and fluorescence (substrate-labelled and polarization) immunoassays. The high-pressure liquid and thin-layer chromatographic methods are the only assays that are specific for quinidine, and of the two, only the high-pressure liquid-chromatographic procedures are suitable for therapeutic drug monitoring. Further, with the high-pressure liquid-chromatographic procedures, it is possible simultaneously to assay many of the quinidine metabolites which is an advantage that this method has over the other assay methods.

International reference method and standards: no reference standards or reference method are presently known.

1.16.1   Determination with Enzyme-multiplied Immunoassay Technique

Assay

Method Design

The method is a modification of the homogeneous enzyme-immunoassay technique first described by *Rubenstein et al.* [8].

Principle: as described in detail on pp. 7, 8.

Glucose-6-phosphate dehydrogenase (G6P-DH) from *Leuconostoc mesenteroides* is used in labelling the drug. When the enzyme-labelled drug is bound to antibody, the enzyme is inactive and cannot react with the added substrate, glucose-6-phosphate. The amount of residual enzyme-labelled drug is determined by measuring the enzyme activity and is related to the concentration of drug in the sample. Enzyme activity is measured by determining the increase in absorbance of NADH at 339 nm per unit time, $\Delta A / \Delta t$.

Selection of assay conditions and adaptation to the individual characteristics of the reagents: all reagents (antibody, enzyme-labelled drug, substrate, coenzyme and calibrators) are prepared as a matched set and should not be interchanged with reagents prepared at a different time. Monoethylglycinexylidide (MEGX), a major metabolite of lidocaine, is added to the lidocaine antibody to reduce cross-reactivity to this metabolite. *N*-Acetylprocainamide may be added to the procainamide antibody and procainamide to the *N*-acetylprocainamide antibody for the same reason. The amounts added depend on the particular antiserum's cross-reactivity to the compound.

The reaction rate is usually determined by measuring the difference between the absorbance at 339 nm at 45 s and the absorbance at 15 s. This time difference could be increased by measuring the absorbance 1 to 2 min after the reaction begins but the initial measurement at 15 s should be maintained. These sampling intervals are optimum when the reaction temperature is 30 °C. The reaction temperature is very important. Decreasing the temperature to 25 °C would necessitate a longer measuring interval; increasing it to 37 °C would necessitate decreasing the measuring interval.

The quantities and concentrations of the reagents must be optimized for each antiserum. A serum sample of 50 µl is used. The desired assay range for disopyramide is 0.5 to 8.0 mg/l, for lidocaine, 1.0 to 12.0 mg/l, for procainamide, 1.0 to 16.0 mg/l, for *N*-acetylprocainamide, 1.0 to 16.0 mg/l and for quinidine, 0.5 to 8.0 mg/l. With these fixed parameters and a goal of achieving a maximum response rate at approximately 2 to 3 mg/l for disopyramide and quinidine, 3 to 4 mg/l for lidocaine and 4 to 5 mg/l for procainamide and N-acetylprocainamide, a specific amount of enzyme-labelled drug is chosen. Standards are then assayed with the fixed enzyme-labelled drug concentration and variable dilutions of antibody. Change in absorbance per unit time, $\Delta A / \Delta t$, for each standard *versus* dilution of antibody is plotted. The dilution of antibody is selected that gives the optimum response where assay sensitivity is most desirable and is functional throughout the assay range. Depending on the specificity of each assay, interfering substances may be added to the antibody reagent to reduce cross-reactivity as previously described.

Equipment

Spectrophotometer capable of measurement at 339 nm or Hg 334 nm with a thermo-statted cuvette chamber; pipettes for measuring 50 µl and 250 µl, or pipetter-diluter for sampling 50 µl and delivering the sample diluted with 250 µl buffer; volumetric pipettes for reconstituting reagents, and logit-log graph paper. Laboratory centrifuge.

Reagents and Solutions

Purity of reagents: chemicals are analytical grade. G6P-DH should be of the best available purity (cf. p. 27).

Preparation of solutions* (for 100 determinations): general reagents are the same for each drug. All solutions in re-purified water (cf. Vol. II, chapter 2.1.3.2).

1. Tris buffer (Tris, 55 mmol/l; Triton X-100, 0.1 ml/l; pH 8.0; pH 6.0 for quinidine):

   dissolve 0.998 g Tris base in about 100 ml water; add 0.015 ml Triton X-100; adjust pH to 8.0 with HCl, 1 mol/l (pH 6.0 for quinidine) and make up to 150 ml with water.

   With the commercial preparations, dilute contents of the buffer vial with water to 150 ml for lidocaine, procainamide and N-acetylprocainamide assays, and to 200 ml for the disopyramide and quinidine assays.

---

* Reagents, calibrators and buffers are commercially available from *Syva Co.,* Palo Alto, CA 94303, U.S.A. The solutions contain sodium azide as a preservative. Separate reagents must be prepared for each antiarrhythmic drug to be assayed.

2. Antibody/substrate solution (sheep antibodies to the drug*, γ-globulin, ca. 50 mg/l; G-6-P, 66 mmol/l; NAD, 40 mmol/l; drug metabolites**, up to about 25 mg/l; Tris, 55 mmol/l, pH 5.0 for N-acetylprocainamide and pH 5.2 for the other antiarrhythmic drugs):

   reconstitute commercial preparation with 6.0 ml water and allow the reagent to equilibrate at room temperature (20°C to 25°C) for 1 h before use, or overnight at refrigerator temperature (2°C to 8°C).

3. Conjugate solution (G6P-DH***, ca. 1 kU/l; Tris, 55 mmol/l; pH 6.2 for quinidine, pH 8.0 for the other antiarrhythmic drugs):

   this reagent must be matched with the antibody/substrate solution (2). Conjugates can be prepared by the N-hydroxysuccinimide ester method [9, 10]. Reconstitute commercial preparation with 6.0 ml water and allow the reagent to equilibrate at room temperature (20°C to 25°C) for one hour before use, or overnight at refrigerator temperature (2°C to 8°C).

4. Drug standard solutions:

   dissolve 40 mg drug in 100 ml Tris buffer (1); dilute 0.2 ml disopyramide standard, 0.3 ml lidocaine standard, 0.4 ml procainamide and N-acetylprocainamide standards and 0.2 ml quinidine standard with drug-free serum containing sodium azide (5 g/l) and Thimerosal (0.5 g/l) to 10 ml. Dilute these standard solutions 1 + 15, 1 + 7, 1 + 3, 1 + 1 with drug-free serum, except for lidocaine which is diluted 1 + 11, 1 + 5, 1 + 3, 1 + 1.4 with drug-free serum to obtain the following concentrations for each drug. Use drug-free serum for the zero calibrator.

Calibrator	Concentration, mg/l					
Disopyramide	0	0.5	1.0	2.0	4.0	8.0
Lidocaine	0	1.0	2.0	3.0	5.0	12.0
Procainamide	0	1.0	2.0	4.0	8.0	16.0
N-Acetylprocainamide	0	1.0	2.0	4.0	8.0	16.0
Quinidine	0	0.5	1.0	2.0	4.0	8.0

Reconstitute the commercial preparations with 1.0 ml water and allow calibrators to equilibrate at room temperature (20°C to 25°C) for one hour before use, or overnight at refrigerator temperature (2°C to 8°C).

---

* Standardized preparation.
** MEGX for lidocaine, N-acetylprocainamide for procainamide and procainamide for N-acetylprocainamide assays. Amount depends on antibody cross-reactivity.
*** Coupled to a derivative of the antiarrhythmic drug to be assayed.

Stability of reagents and solutions: reagents are stable for 12 weeks when stored at $2-8\,°C$. The buffer solutions (1) can be used for 12 weeks when stored at room temperature.

Procedure

Collection and treatment of specimen: each drug to be assayed requires 50 µl serum. Plasma is acceptable if heparin, EDTA or oxalate is used as the anticoagulant.

For therapeutic drug monitoring, serum drug concentrations should be determined after steady-state conditions have been reached. For the determination of trough drug concentrations, a specimen should be collected just before administration of the next dose. Peak drug concentrations may be determined with specimens drawn at the following times after oral administration: disopyramide, $2-3$ h; procainamide and N-acetylprocainamide, $1-2$ h ($4-6$ h after administration of a product with sustained-release properties); quinidine, $1-3$ h ($4-6$ h after administration of products with sustained-release properties). Specimens may also be collected at other times when clinically indicated.

Stability of the analyte in the sample: store sera at refrigerator temperature ($2\,°C$ to $8\,°C$) until analyzed. If the sample is to be transported, keep at refrigerator temperature ($2\,°C$ to $8\,°C$). If stored for longer than 2 weeks, sample may be frozen ($-20\,°C$) for several months without significant loss of drug.

Assay conditions: bring all solutions to room temperature ($20\,°C$ to $25\,°C$) and mix thoroughly before use. Measurements of samples and zero calibrator in duplicate, single determinations of the other standard solutions. A pre-dilution of samples and standard solutions is required: mix 0.05 ml sample or standard (4) with 0.25 ml Tris buffer (1). Establish a calibration curve using the standard solutions (4). Use the zero calibrator as a reagent blank. Separate reagents must be used for each analyte to be measured but the procedure is the same for all of the antiarrhythmic drugs.

Incubation time is 45 s (in the spectrophotometer flow-cell); $30\,°C$; wavelength 339 nm; light path 10 mm; final volume 0.90 ml; measurement for 30 s against water.

Calibration curve: plot the corrected $\Delta A/\Delta t$ readings of the standards *versus* the corresponding drug concentrations (mg/l) on logit-log paper or on the special graph paper* matched with each set of reagents. The calibration curve should be linear (Fig. 1).

Other methods of linearizing the dose-response curve (arcsin-log, probit-log) may be used for preparing the calibration curve as well as methods that approximate the curve, e.g., 4-parameter logistic curve fitting or spline approximation.

---

* Supplied by *Syva*; for using this graph paper, multiply each value of $\Delta A/\Delta t$ by 2.667.

Measurement

Pipette into a suitable container (cuvette or test tube):			concentration in assay mixture
pre-diluted sample or standard solution (4)		0.05 ml	drug up to 148 μg/l
Tris buffer	(1)	0.25 ml	Tris 52 mmol/l
antibody/substrate solution	(2)	0.05 ml	sheep serum variable
			γ-globulin ca. 2.8 mg/l
			G-6-P 3.7 mmol/l
			NAD 2.2 mmol/l
			drug metabolites variable
			up to about 1400 μg/l
Tris buffer	(1)	0.25 ml	
conjugate solution	(3)	0.05 ml	G6P-DH* variable
			ca. 56 U/l
Tris buffer	(1)	0.25 ml	
mix well; transfer immediately to cuvette or flow-cell in spectrophotometer; after a 15 s delay, read first absorbance A_{15}; after further 30 s read second absorbance A_{45}; $A_{45} - A_{15}$ = ΔA per 30 s.			

* Coupled to the antiarrhythmic drug.

If $\Delta A/\Delta t$ of the sample is greater than that of the highest standard, dilute the sample further with zero calibrator and re-assay.

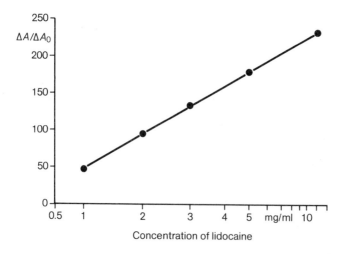

Fig. 1. Typical calibration curve for the assay of lidocaine in serum.

Calculation: read the drug mass concentration, ρ (mg/l), from the calibration curve. To convert from mass concentration to substance concentration, c (µmol/l), multiply the disopyramide concentration in mg/l by 2.9455, lidocaine concentration (mg/l) by 4.2680, procainamide concentration (mg/l) by 4.2319, N-acetylprocainamide concentration (mg/l) by 3.6049 and quinidine concentration (mg/l) by 3.0826.

Validation of Method

Precision, accuracy, detection limit and sensitivity: between the 2nd and 6th calibrators for each calibration curve, the major portion of the curve, day-to-day precision should yield relative standard deviations (RDS) below 8% for all antiarrhythmic drug assays [11 – 14].

Recovery studies on spiked sera revealed between 95 and 105% recovery for the five antiarrhythmic drugs. With the exception of the analysis of quinidine, comparisons with drug assays performed by high-pressure liquid chromatography, gas-liquid chromatography or spectrofluorimetry have yielded correlation coefficients > 0.90 and least-squares regression lines with slopes ranging from 0.90 to 1.10 and intercepts that were insignificant compared to therapeutic serum drug concentrations [11, 12, 14]. For quinidine, *Drayer et al.* [13] found slope and intercept values of 1.22 and 0.27, respectively.

Since data on the detection limit has not been established, the lowest calibrator concentration should be used as the lower limit of the measuring range. Assay sensitivity depends on the shape of the calibration curve. If the dose-response curve is plotted on semilogarithmic graph paper (response versus log drug concentration), a sigmoid curve results. Sensitivity is the highest in the linear portion of the curve, which is usually the middle region of the calibration curve.

Source of error: no clinically significant errors are obtained with sera from haemolyzed blood (< 8 g haemoglobin per litre; < 4 g/l for quinidine), lipaemic (< 11.3 mmol triglycerides per litre) or icteric (< 513 µmol bilirubin per litre) blood.

Specificity: drug specificity depends on the antiserum used in the assay. Of numerous drugs that have been tested [15 – 19], the only significant cross-reactivities or interferences that have been observed for the antiarrhythmic drug assays have been the cross-reactivity of 2-desmethylquinidine and dihydroquinidine in the quinidine assay with the antiserum used.

Therapeutic ranges: usually serum (plasma) concentrations between 2 and 5 mg/l for disopyramide, 1.3 and 5 mg/l for lidocaine, 4 and 8 mg/l for procainamide, and 2 and 5 mg/l for quinidine are found in patients with a satisfactory therapeutic response. No therapeutic range for N-acetylprocainamide has been definitively established. However, a satisfactory therapeutic response has been found in patients whose sum of procainamide and N-acetylprocainamide concentrations in the serum was between 10 and 30 mg/l.

1.16.2   Determination with Substrate-labelled Fluorescent Immunoassay

Assay

Method Design

The method is a modification of the homogeneous substrate-labelled fluorescent immunoassay technique first described by *Burd et al.* [20]. It has been adapted for the determination of disopyramide, procainamide, N-acetylprocainamide and quinidine but not lidocaine.

Principle: as described in detail on pp. 15, 16.

The method involves competition between the drug (L) and umbelliferyl-β-D-galactoside-labelled drug* (S-L), which is non-fluorescent, for limited numbers of binding sites on antibodies to the drug. When the sample containing the drug is mixed with an antibody to the drug (Ab) and drug conjugate*, drug-antibody and substrate-labelled drug-antibody complexes are formed. The substrate-labelled drug bound to antibody, in addition to being non-fluorescent, cannot be hydrolyzed by β-galactosidase**. On the other hand, β-galactosidase can catalyze the breakdown of the excess substrate-drug conjugate to galactose and a fluorescent product, umbelliferyl-drug. The resulting increase in fluorescence intensity per unit time, $\Delta F/\Delta t$, is related to the drug concentration in the sample.

Selection of assay conditions and adaptation to the individual characteristics of the reagents: as described more detailed on pp. 8, 9. All calibrators and samples must be incubated for exactly 20 min, the reaction being initiated by the addition of the substrate-drug conjugate. The reaction is performed at room temperature and the temperature should not be allowed to vary more than $\pm 3\,^{\circ}\mathrm{C}$ during the assay.

---

* Drug-umbelliferyl-β-galactose or 8-[3-(7-β-galactosylcoumarin-3-carboxyamido)propyl]-drug.
** β-D-Galactoside galactohydrolase, EC 3.2.1.23.

Optimal activity of β-galactosidase is obtained when the pH is between 6 and 8. The fluorescence intensity of the hydrolysis product increases up to pH 8 and remains essentially constant to pH 10 [21]. Therefore, the fluorescence intensity and β-galactosidase activity are optimized at a pH of 8.2.

The amount of antibody used in each assay is optimized for antiserum and drug. Sufficient antibody to decrease the fluorescence in the absence of any drug to 10% of that observed in the absence of the antiserum is used [22].

Equipment

Fluorimeter capable of measuring fluorescence intensity in 1.6 ml of total volume at an excitation wavelength of 400 nm and an emission wavelength of 450 nm; cuvettes to hold the 1.6 ml reaction volume to fit the fluorimeter; glass test tubes, 12 mm × 75 mm, for sample and calibrator dilutions; pipettes for measuring 50 µl, 1.5 ml and 2.5 ml; volumetric pipettes for diluting reagents, stopwatch and linear graph paper.

The linearity of the fluorimeter should be checked periodically with 0, 1.0, 5.0, 10.0 and 15.0 nmol/l solutions of 7-hydroxycoumarin-3-[N-(2-hydroxyethyl)] carboxamide in Bicine buffer, 50 mmol/l, pH 8.3.

Reagents and Solutions

Purity of reagents: chemicals are analytical grade. β-Galactosidase from *Escherichia coli* should be of the best available purity.

Preparations of solutions* (for 100 determinations): general reagents are the same for each drug. All solutions in re-purified water (cf. Vol. II, chapter 2.1.3.2).

1. N,N-Bis(2-hydroxyethyl)glycine buffer (Bicine, 1.0 mol/l; NaN_3, 308 mmol/l; pH 8.2):

   dissolve 4.08 g N,N-bis (2-hydroxyethyl)glycine and 0.5 g sodium azide in 20 ml water; adjust pH to 8.2 with NaOH, 12.5 mol/l; make up with water to 25 ml; filter to remove particles. 25 ml of this buffer is provided in each kit.

2. Diluted Bicine buffer (Bicine, 50 mmol/l; NaN_3, 15 mmol/l; pH 8.2):

   mix one part concentrated Bicine buffer (1) with 19 parts water.

---

* Reagents, calibrators and buffers are commercially available from *Ames Company*, Elkhart, IN, U.S.A. 45614. The solutions contain sodium azide as a preservative. Separate reagents must be prepared for each antiarrhythmic drug to be assayed.

3. Antibody/enzyme solution (sheep or rabbit antibody to the drug, γ-globulin*; ga-lactosidase, ca. 3 kU**/l; NaN$_3$, 15 mmol/l; Bicine, 50 mmol/l; pH 8.2):

   5 ml is provided with each kit.

4. Diluted antibody/enzyme solution (sheep or rabbit antibody to the drug, γ-globu-lin*; galactosidase, ca. 0.1 kU/l; NaN$_3$, 15 mmol/l; Bicine, 50 mmol/l; pH 8.2):

   mix one part antibody/enzyme solution (3) with 29 parts of diluted buffer (2).

5. Substrate-drug conjugate solution (β-galactosyl-umbelliferyl-drug, ca. 0.55 μmol/l; formate, 30 mmol/l; pH 3.5):

   5 ml are provided with each kit.

6. Drug standard solutions:

   prepare standards for each drug by dissolving 40 mg drug in 100 ml Bicine buffer (1). Dilute 0.2 ml disopyramide standard, 0.4 ml procainamide standard, 0.75 ml N-acetylprocainamide standard, and 0.2 ml quinidine standard with drug-free serum containing sodium azide (5 g/l) to 10 ml. Further dilute these standard solutions for disopyramide (1 + 3, 1 + 1, 2 + 1); for procainamide (1 + 3, 1 + 1, 3 + 1); for N-acetylprocainamide (1 + 5, 1 + 2, 2 + 1); and for quinidine (1 + 7, 3 + 5, 5 + 3) with drug-free serum containing sodium azide, 5 g/l. Use drug-free serum for the zero calibrator.

Calibrator	Concentrations, mg/l				
Disopyramide	0	2	4	6	8
Procainamide	0	4	8	12	16
N-Acetylprocainamide	0	5	10	20	30
Quinidine	0	1	3	5	8

Standard solutions containing the indicated drug concentrations are provided with each kit.

Stability of solutions: as supplied commercially, the reagents are stable stored at 2°C to 8°C until the expiry date on the labels. When stored at 2°C to 8°C, the diluted buffer (2) is stable for 6 months, the diluted antibody/enzyme solutions (4) are stable for 3 months. The substrate-drug conjugate solution (5) should be protected from light.

---

* Standardized preparation; amount is titre-dependent.

** Enzyme activity defined as in *Wong et al.* [21]. One unit of enzyme activity catalyzes the hydrolysis of 1.0 μmol substrate per min at 25°C in Bicine buffer, 50 mmol/l, pH 8.2, containing, per litre, 15.4 mmol sodium azide and 3 mmol 2-nitrophenyl β-D-galactoside. The molar absorption coefficient of the reaction product under these conditions is 4.27×10^2 l × mol^{-1} × mm^{-1} at 415 nm.

Procedure

Collection and treatment of specimens: each drug to be assayed requires 50 µl serum. Plasma is acceptable if heparin or EDTA is used as the anticoagulant. For further information about specimen collection, cf. chapter 1.16.1, p. 223.

Stability of the substance in the sample: cf. chapter 1.16.1, p. 223.

Assay conditions: bring all solutions to room temperature (20 °C to 25 °C) and mix thoroughly before use. Establish a calibration curve with each assay, using the standard solutions (6). Separate reagents must be used for each analyte to be measured but the procedure is the same for all of the antiarrhythmic drugs. Measure all standards and samples in duplicate.

A pre-dilution of samples and standard solutions (6) is required: mix one part (0.05 ml) sample or standard (6) with 50 parts (2.50 ml) diluted Bicine buffer (2). Incubation time 20 min; ambient temperature. Excitation wavelength 400 nm; emission wavelength 450 nm.

Measurement

Pipette into cuvette:		concentration in assay mixture	
pre-diluted sample or standard (6) diluted antibody/enzyme	0.05 ml	drug	up to 18 µg/l
solution (4)	1.50 ml	β-galactosidase	ca. 91 U/l
		antiserum	variable
		Bicine	48.5 mmol/l
		NaN_3	15 mmol/l
mix well by inversion,			
substrate/drug conjugate			
solution (5)	0.05 ml	β-galactosyl-umbelli- feryl-drug	ca. 17 nmol/l
		formate	938 µmol/l
mix well, incubate exactly for 20 min; measure fluorescence*.			

If $\Delta F/\Delta t$ of the sample is greater than that of the highest standard dilute the pre-diluted sample further with diluted Bicine buffer (2) and re-assay.

---

* Just prior to the end of the 20 min incubation period, pipette 1.5 ml diluted Bicine buffer (2) into a cuvette, read fluorescence and set instrument reading to zero. At about 19 min, read fluorescence of the highest standard and set instrument at 90% of full scale.

Calibration curve: plot mean increase in fluorescence intensity per 20 min, $\Delta F / \Delta t$, of the duplicates *versus* the corresponding drug concentration (mg/l) on linear graph paper. Fig. 2 shows a typical calibration curve.

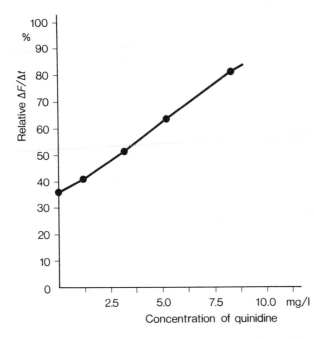

Fig. 2. Typical calibration curve for the assay of antiarrhythmics in serum.

Calculation: read the drug mass concentration, ρ (mg/l) from the calibration curve using the average fluorescence value of each unknown sample. To convert from mass concentration to substance concentration, c (μmol/l), multiply the disopyramide concentration in mg/l by 2.9455, procainamide concentration (mg/l) by 4.2680, N-acetylprocainamide concentration (mg/l) by 3.6049 and quinidine concentration (mg/l) by 3.0826.

Validation of Method

Precision, accuracy, detection limit and sensitivity: between the 2nd and 5th calibrators for each calibration curve, the major portion of the curve, day-to-day precision should yield relative standard derivations (RDS) below 8% for the antiarrhythmic drug assays [23, 24].

Antiarrhythmic drug concentrations have been determined by the substrate-labelled fluorescent immunoassay method and compared to the enzyme-multiplied immunoassay technique and high-performance liquid chromatography (HPLC). Correlation coefficients were greater than 0.96 and the slopes of the least-squares regression lines ranged from 0.86 to 1.17 with intercepts less than 1 mg/l [25, 26, 27].

The lowest detectable concentration of disopyramide is 0.5 mg/l, of procainamide, 1.5 mg/l; of N-acetylprocainamide, 2.0 mg/l; and of quinidine, 1.0 mg/l. Assay sensitivity depends on the shape of the calibration curve with sensitivity being highest in the region of the curve that is the steepest.

Source of error: sera from haemolyzed, icteric or lipaemic blood may interfere with the assays. Triamterene (Dyrenium®, Dyazide®) may cause false elevations of the drug concentrations. For correction for presence of fluorescent compounds, cf. chapter 1.4.1.2, p. 49.

Specificity: the specificity depends on the antiserum used. Of numerous drugs that have been tested [28 – 31], the only significant cross-reactivity observed in the antiarrhythmic drug assays has been the cross-reactivity of procaine and N-acetylprocainamide (serum concentration > 30 mg/l) in the procainamide assay.

Therapeutic ranges: cf. chapter 1.16.1, p. 225.

References

[1] *A. Karim, C. Nissen, D. L. Azarnoff,* Clinical Pharmacokinetics of Disopyramide, J. Pharmacokinet. Biopharm. *10,* 465 – 494 (1982).
[2] *N. L. Benowitz, W. Meister,* Clinical Pharmacokinetics of Lidocaine, Clin. Pharmacokinet. *3,* 177 – 201 (1978).
[3] *J. H. Rodman,* Lidocaine, in: *W. E. Evans, J. J. Schentag, W. J. Jusko* (eds.), Applied Pharmacokinetics, Applied Therapeutics, Inc., San Francisco 1980, pp. 350 – 391.
[4] *E. Karlsson,* Clinical Pharmacokinetics of Procainamide, Clin. Pharmacokinet. *3,* 97 – 107 (1978).
[5] *S. J. Connolly, R. E. Kates,* Clinical Pharmacokinetics of N-Acetylprocainamide, Clin. Pharmacokinet. *7,* 206 – 220 (1982).
[6] *C. T. Ueda,* Quinidine, in: *W. E. Evans, J. J. Schentag, W. J. Jusko* (eds.), Applied Pharmacokinetics, Applied Therapeutics, Inc., San Francisco 1980, pp. 436 – 463.
[7] *P. Bolme, U. Otto,* Dose-Dependence of the Pharmacokinetics of Quinidine, Eur. J. Clin. Pharmacol. *12,* 73 – 76 (1977).
[8] *K. E. Rubenstein, R. S. Schneider, E. F. Ullman,* "Homogeneous" Enzyme Immunoassay. A New Immunochemical Technique, Biochem. Biophys. Res. Commun. *47,* 846 – 851 (1972).
[9] *P. Singh, M. W. Hu, J. B. Gushaw, E. F. Ullman,* Specific Antibodies to Theophylline for Use in a Homogeneous Enzyme Immunoassay, Immunoassay *1,* 309 – 322 (1980).
[10] *J. Chang, S. Gotcher, J. B. Gushaw, R. H. Gadsden, Ch. A. Bradley, Th. C. Stewart,* Homogeneous Enzyme Immunoassay for Theophylline in Serum and Plasma, Clin. Chem. *28,* 361 – 367 (1982).
[11] *L. B. Abbott, W. C. Wenger, J. A. Lott,* Evaluation of EMIT® Procedures for Procainamide, N-Acetylprocainamide and Quinidine on a Flexigem, Clin. Chem. *28,* 1667 (1982).
[12] *B. E. Pape, R. Whiting, K. M. Parker, R. Mitra,* Enzyme Immunoassay and Gas-Liquid Chromatography Compared for Determination of Lidocaine in Serum, Clin. Chem. *24,* 2020 – 2022 (1978).
[13] *D. E. Drayer, B. Lorenzo, M. M. Reidenberg,* Liquid Chromatography and Fluorescence Spectroscopy Compared with a Homogeneous Enzyme Immunoassay Technique for Determining Quinidine in Serum, Clin. Chem. *27,* 308 – 310 (1981).
[14] *W. C. Griffiths, P. Dextroze, M. Hayes, J. Mitchell, I. Diamond,* Assay of Serum Procainamide and N-Acetylprocainamide: A Comparison of EMIT™ and Reverse-Phase High-Performance Liquid Chromatography, Clin. Toxicol. *16,* 51 – 54 (1980).
[15] Emit® Disopyramide Assay, Product Information Insert, *Syva Co.,* 1983.
[16] Emit® Lidocaine Assay, Product Information Insert, *Syva Co.,* 1983.
[17] Emit® Procainamide Assay, Product Information Insert, *Syva Co.,* 1984.
[18] Emit® N-Acetylprocainamide Assay, Product Information Insert, *Syva Co.,* 1983.
[19] Emit® Quinidine Assay, Product Information Insert, *Syva Co.,* 1984.

[20] *J. F. Burd, R. C. Wong, J. E. Feeney, R. J. Carrico, R. C. Boguslaski,* Homogeneous Reactant-Labeled Fluorescent Immunoassay for Therapeutic Drugs Exemplified by Gentamicin Determination in Human Serum, Clin. Chem. *23,* 1402 – 1408 (1977).

[21] *R. C. Wong, J. F. Burd, R. J. Carrico, R. T. Buckler, J. Thoma, R. C. Boguslaski,* Substrate-Labeled Fluorescent Immunoassay for Phenytoin in Human Serum, Clin. Chem. *25,* 686 – 691 (1979).

[22] *T. M. Li, J. L. Benovic, R. T. Buckler, J. F. Burd,* Homogeneous Substrate-Labeled Fluorescent Immunoassay for Theophylline in Serum, Clin. Chem. *27,* 22 – 26 (1981).

[23] *D. Inbar, I. Tabet, B. Fridlender, H. Feinstein, F. E. Ward,* Substrate-Labeled Fluorescent Immunoassay (SLFIA) for Procainamide, Clin. Chem. *28,* 1667 (1982).

[24] *L. E. Csiszar, T. M. Li, J. L. Benovic, R. T. Buckler, J. F. Burd,* Substrate-Labeled Fluorescent Immunoassay for Quinidine, Clin. Chem. *27,* 1087 (1981).

[25] *C. P. Patel, L. C. Cary,* Comparison of EMIT®, Ames TDA®, and HPLC Methods in Determination of Serum Quinidine, Clin. Chem. *28,* 1648 (1982).

[26] *H. Feinstein, H. Hovav, J. Feingers, P. K. Johnson, D. Reed, L. J. Messenger,* SLFIA for Disopyramide in Human Serum: Semi- and Totally Automated Methods, Clin. Chem. *29,* 1197 – 1198 (1983).

[27] *E. E. Pahuski, J. L. Maurer, B. L. Halmo, T. M. Li,* Procainamide and NAPA Determinations by SLFIA on an Automated TDA Analyzer, Clin. Chem. *29,* 1273 (1983).

[28] Ames TDA® Disopyramide Assay, Product Information Insert, *Ames* Division, *Miles Laboratories,* 1984.

[29] Ames TDA® Procainamide Assay, Product Information Insert, *Ames* Division, *Miles Laboratories,* 1984.

[30] Ames TDA® N-Acetylprocainamide Assay, Product Information Insert, *Ames* Division, *Miles Laboratories,* 1984.

[31] Ames TDA® Quinidine Assay, Product Information Insert, *Ames* Division, *Miles Laboratories,* 1984.

1.17   Nortriptyline

3-(10,11-Dihydro-5*H*-dibenzo[*a, d*]cycloheptene-5-ylidene)-*N*-methyl-1-propanamine

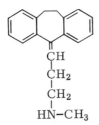

Vincent Marks and Graham P. Mould

General

Nortriptyline belongs to a class of compounds known as the tricyclic antidepressants. These drugs ameliorate various symptoms of depression, particularly of the endogenous type. There are a number of different compounds within this class differing slightly in chemical structure as well as in their clinical effectiveness. Their mode of action is to increase the availability of noradrenaline and 5-hydroxytryptamine as central transmitters by blocking neuronal uptake [1].

Nortriptyline is rapidly absorbed when taken orally, peak concentrations occurring 4 to 8 h after administration [2, 3]. Despite evidence for complete oral absorption [4] the systemic bio-availability of nortriptyline is low. When plasma concentrations of nortriptyline were compared, following similar doses given as an oral dose or as an intravenous infusion in healthy volunteers, the bio-availability varied between 46 and 59% [4]. This reduction in bio-availability following oral administration is assumed to be due to the so-called "first-pass" effect.

Nortriptyline is a highly lipophilic compound which distributes widely in the body and therefore has a high apparent volume of distribution. This value has been calculated to be about 2000 l in healthy volunteers and young or middle-aged depressed patients [4, 5], but is slightly reduced in elderly subjects [6].

Nortriptyline is bound to α_1-acid glycoprotein in plasma; the mean percentage of unbound drug in patients receiving nortriptyline was only 8%, range 5.4 to 10.8% [7].

The major pathway for elimination of nortriptyline from the body is through hepatic metabolism. The first step is hydroxylation at the 10 position on the tricyclic ring followed by conjugation with glucuronic acid in which form it is excreted in the urine [4]. The plasma clearance and therefore the elimination half-life of nortriptyline vary considerably. The variability is due to many factors, such as genetic factors [8], old age [6] and perhaps depression itself, which may affect liver blood flow. The elimination half-life in elderly, mostly female subjects with a mean age of 81 years, was about 44 h with a range of 24 to 79 h [6], whereas in healthy young volunteers, mean age 26 years, it was 26 h with a range of 16 to 38 h [4]. The mean half-life of nortriptyline in plasma of middle-aged depressed patients was 46 h (range 22 to 88 h) in one study [5] and 39 h (range 24 to 86 h) in another [9] but these differences, are probably not significant because of the wide ranges observed. Since steady-state plasma concentrations are dependent on clearance, it can be seen that they will vary considerably in patients receiving similar dosages and this has, indeed, been observed on many occasions [5, 9].

The rationale for monitoring plasma concentrations is that it gives a better indication of therapeutic outcome than does the dose administered and that there is a target range within which most patients experience maximum clinical benefit. Given that a target range exists for nortriptyline, and the evidence is, at best, controversial [5, 10, 11], the indications for routine measurement, i.e. therapeutic drug monitoring (TDM), are the following.

1. Inter-patient variability: there is a wide variability in plasma concentrations between patients receiving the same dose of drug, especially in the elderly. Monitoring of plasma concentration will identify those patients with unusual kinetics or in whom the plasma concentration is affected by other drugs given simultaneously.

2. Reduction in side-effects: the majority of the side-effects of nortriptyline are a direct function of plasma concentration [12] but are sometimes due to unusual sensitivity. In such cases plasma concentrations are low.

3. Patient compliance: it is important to identify patients not taking their prescribed medication since recovery from depression is slow. In a recent study it was shown that up to 35% of patients might not have been complying adequately [13].

4. Improved clinical response: some studies have demonstrated that patients have significantly less chance of recovering from endogenous depression when plasma antidepressant concentrations are above their respective target values [14, 15].

Because of wide variation in plasma concentrations between subjects, various methods have been used to predict dosing regimens in order to achieve targetted steady-state plasma concentration in individual patients. One such method is to use pharmacokinetic data obtained from patient or volunteer studies and, by putting them into a standard pharmacokinetic formula [16], determine steady-state concentration for a given dose. This method has not been used to any great extent for nortriptyline.

Another approach to pharmacokinetic dosing has been reported for nortriptyline by *Cooper & Simpson* [17] and by *Montgomery et al.* [9]. These investigators administered a single oral dose of nortriptyline and measured plasma concentrations of nortriptyline 24, 48 and 72 h after administration. These were compared with steady-state concentrations after a constant maintenance dose of at least 2 weeks. The plasma nortriptyline concentration at 24 h correlated best with the steady-state value in both reports and was subsequently utilized in formulating dosage tables to predict maintenance doses from a single plasma concentration following a test dose. This method has been used by other workers [6, 9].

Application of method: in clinical chemistry, clinical pharmacology and toxicology.

Substance properties relevant in analysis: nortriptyline is a weak base with a pK_a of 8.4. The un-ionized portion is lipid soluble and readily extracts into non-polar solvents, such as hexane or heptane. The molecular weight of the free base is 263.4, of the hydrochloride is 299.9.

Methods of determination: a number of analytical techniques have been applied to the determination of plasma concentrations of nortriptyline. These include isotope-derivative techniques [18], gas-liquid chromatography (GLC) with electron-capture detection [19], or nitrogen detection [20], GLC with mass spectroscopy [21], high-pressure liquid chromatography (HPLC) [22, 23] and various immunoassay techniques [24 – 36].

The first report of an immunoassay was in 1976 [24] and used a radiolabelled ligand. It was later modified to increase the sensitivity [25]. Like subsequent immunoassays, the antiserum used in the early published accounts did not distinguish between amitriptyline and nortriptyline [26 – 28]. In order to measure nortriptyline specifically in the presence of amitriptyline, separation by solvent extraction is necessary at a pH of about 5 [29 – 31]. One manufacturer has developed and marketed a radioimmunoassay (RIA) kit [32] for the measurement of nortriptyline in plasma but its specificity is low.

Only a few non-isotopic immunoassays have been described for tricyclic antidepressants in plasma. *Al-Bassam et al.* [33] used an enzyme-linked label to prepare an enzyme-immunoassay, and an energy-transfer fluoroimmunoassay has also been reported, but not in detail [34]. Various other immunoassays have been described but only in patent applications [35, 36]; none has appeared on the market as of this time.

A recent review has compared methods for the determination of nortriptyline in body fluids [37].

Most of the methods available at present require extraction of the drug from plasma prior to analysis and some also require a derivatization step before assay. They can, therefore, be time-consuming and require sophisticated apparatus for their performance. Nevertheless, specificity is good, as is sensitivity. On the other hand, immunoassays can be performed on material without prior extraction and are consequently simpler. However, specificity is at present poor although sensitivity is as good as with other techniques. As far as TDM is concerned, exquisite sensitivity is rarely required, so that immunoassays are the most potentially useful for this purpose.

Because of doubts about the clinical value of TDM for nortriptyline, immunoassays have not been developed to any great extent. For pharmacokinetic and research purposes, where specificity and selectivity are essential, GLC using a nitrogen detector appears to be the method of choice.

There has been little external assessment of plasma tricyclic antidepressant measurements. An inter-laboratory comparison of nortriptyline measurements by 5 laboratories, each using a different assay technique, was carried out recently [38]. There was up to a 3-fold variation in the results although no statistics were calculated. Several laboratories have separately compared different methods when working up their own methods (e.g. [9, 26, 33]). However, these comparisons are performed as research rather than as routine procedures and throw little light on assay performance under regular working conditions.

International reference method and standards: there are no known international reference methods or standards.

Assay

Method Design

The enzyme-linked immunosorbent assay for nortriptyline described is a competitive enzyme-immunoassay and, as such, is analogous to a classical radioimmunoassay.

Principle

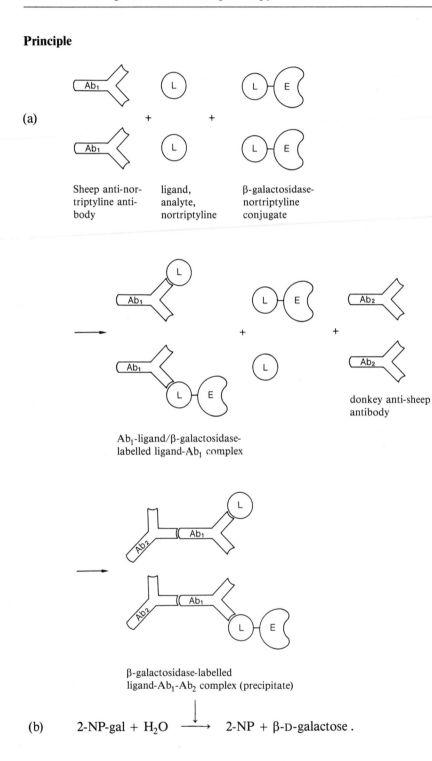

(a)

Sheep anti-nor- ligand, β-galactosidase-
triptyline anti- analyte, nortriptyline
body nortriptyline conjugate

donkey anti-sheep
antibody

Ab₁-ligand/β-galactosidase-
labelled ligand-Ab₁ complex

β-galactosidase-labelled
ligand-Ab₁-Ab₂ complex (precipitate)

(b) 2-NP-gal + H₂O ⟶ 2-NP + β-D-galactose .

Enzyme-labelled antigen competes with unlabelled antigen for binding to a limited quantity of specific antibody sites. Free antigen is separated from antibody-bound antigen and enzyme activity in either fraction is measured as increase in absorbance per unit time, $\Delta A/\Delta t$, and related by comparison with standards treated in the same way to the concentration of unlabelled antigen, the analyte, present in the original sample. Free antigen is separated from antibody-bound antigen by precipitation of the latter by a second antibody directed against the first antibody.

Selection of assay conditions and adaptation to the individual characteristics of the reagents: the assay was based on the radioimmunoassay for nortriptyline [25]. It is expected therefore that the antigen-antibody complex formation is rapid, so that the recommended incubation time of 90 min is generous. Little is known concerning effects of pH and temperature, but conditions are not strict provided pH does not go above 10.

The optimal quantities and concentrations of sample and reagents depend on the characteristics of the antibodies present.

The desired assay range is usually between 1 and 20 µg/l. Samples outside this range, and this will apply to most samples, should be diluted to concentrations within the above range. The volume of the sample or diluted sample and other reagents should be the exact quantities as detailed in the protocol.

The selection of the optimal concentrations of the antiserum for antibody content is made by titrating it with labelled antigen. Increasing dilutions of antiserum are incubated with sufficient label so that the enzyme activity can be measured. After phase separation, the amount of label bound to antibody can be measured and plotted against anti-antiserum dilution. The working dilution of this assay is that at which 50% of the added label is bound to antibody in the absence of added ligand.

Equipment

Spectrophotometer, capable of exact measurements at 420 nm; refrigerated centrifuge; cold tray; accurate air-displacement (*Oxford*) pipettes capable of delivering 100 to 500 µl; plastic tubes, LP3 from *Luckham Ltd.*

Reagents and Solutions

Purity of reagents: use only highly purified β-D-galactosidase, specific activity at least 600 U/mg at 37 °C, 4-nitrophenyl β-D-galactopyranoside* as substrate, pH 7.0. The purity of each preparation is assessed before conjugation. Commercially available preparations (e.g. from *Boehringer Mannheim*) contain at most only very small amounts of impurities and can be used without further purification. The antiserum should be reasonably free from impurities. All other reagents are of A. R. grade.

---

* Note: not 2-nitrophenyl β-D-galactopyranoside as used in this assay.

Preparation of solutions: all solutions in re-purified water (cf. Vol. II, chapter 2.1.3.2).

1. Disodium hydrogen phosphate (100 mmol/l):

   dissolve 35.5 g $Na_2HPO_4 \cdot 12\ H_2O$ in 2.5 l water.

2. Potassium dihydrogen phosphate (100 mmol/l):

   dissolve 34 g KH_2PO_4 in 2.5 l water.

3. Dilution buffer (phosphate, 5 mmol/l; NaCl, 100 mmol/l; $MgCl_2$, 10 mmol/l; 2-mercaptoethanol, 10 mmol/l; BSA, 1 g/l; pH 7.4):

   mix 40.2 ml solution (1), 9.9 ml solution (2) and approximately 890 ml water; add 9 g NaCl, 2.03 g $MgCl_2 \cdot 6\ H_2O$ and 0.78 g 2-mercaptoethanol; adjust pH to 7.4 with NaOH, 1 mol/l; make up to 1 litre with water; dissolve 1 g bovine serum albumin in this solution at 37 °C.

4. Phosphate buffer (100 mmol/l, pH 7.4):

   mix 803 ml solution (1) and 197 ml solution (2).

5. Tris buffer (Tris, 50 mmol/l; $MgCl_2$, 10 mmol/l; NaCl, 100 mmol/l; 2-mercaptoethanol, 10 mmol/l; pH 7.5):

   dissolve 6.05 g Tris base, 5.8 g NaCl, 0.95 g $MgCl_2 \cdot 6\ H_2O$ and 0.78 g 2-mercaptoethanol in 1 l water; adjust with glacial acetic acid to pH 7.5.

6. Carbonate buffer (100 mmol/l; $MgCl_2$, 10 mmol/l; 2-mercaptoethanol, 10 mmol/l; pH 9.9):

   dissolve 5.3 g Na_2CO_3, 4.2 g $NaHCO_3$, 0.95 g $MgCl_2 \cdot 6\ H_2O$ and 0.78 g 2-mercaptoethanol in 1 litre water.

7. Anti-nortriptyline sheep antiserum (γ-globulin, ca. 30 µg/l; phosphate, 5 mmol/l; NaCl, 100 mmol/l; 2-mercaptoethanol, 10 mmol/l; BSA, 1 g/l; pH 7.4):

   add 500 µl neat antiserum (bleed S26, 10/6/76, purchased from *Guildhay Antisera,* Guildford, Surrey, U.K., or prepare according to Appendix, p. 243) to 150 ml dilution buffer (3).

8. Conjugate solution, β-galactosidase-labelled desmethylnortriptyline (galactosidase, ca. 600 U/l; phosphate, 100 mmol/l; pH 7.4):

   dilute the solution obtained from the column (cf. Appendix, p. 244) with phosphate buffer (4) appropriately.

9. Nortriptyline standard solutions (1 to 40 µg/l):

   prepare a stock solution of nortriptyline by dissolving 11.4 mg nortriptyline hydrochloride (corresponding to 10 mg nortriptyline) in 25 ml water; make two successive dilutions 1 : 100 to a final concentration of 40 µg/l. Dilute this with dilution buffer (3) to give a range of concentrations of 1 to 20 µg/l.

10. Second antibody (γ-globulin, ca. 10 µg/l; phosphate, 5 mmol/l; NaCl, 100 mmol/l; $MgCl_2$, 10 mmol/l; 2-mercaptoethanol, 10 mmol/l; BSA, 1 g/l; pH 7.4):

    dilute 400 µl antibody solution (donkey anti-sheep antibody, purchased from *Guildhay Antisera,* Guildford, Surrey, U.K.) to 16 ml with dilution buffer (3) to give a working dilution of 1 : 40.

11. Substrate solution (2-NPgal, 2.3 mmol/l; phosphate, 100 mmol/l; $MgCl_2$, 10 mmol/l; 2-mercaptoethanol, 10 mmol/l; pH 7.4):

    dissolve 692.3 mg 2-nitrophenyl β-D-galactopyranoside, 95.0 mg $MgCl_2 \cdot 6\ H_2O$ and 780 mg 2-mercaptoethanol in 1 l phosphate buffer (4). Ensure that pH remains at 7.4.

Stability of solutions: store all solutions at 4 °C in a refrigerator. Solutions (1) and (2) are stable indefinitely. Buffers (3), (4), (5) and (6) are stable for 6 months. Nortriptyline standards (9) are made up freshly from the stock solution for each assay. The stock solution itself is stable for at least 3 months. The activity of the undiluted conjugate (8) is maintained within 90% of maximum when stored at 4 °C for 6 months. It must be diluted immediately prior to use in the assay on each occasion, as are the anti-nortriptyline antiserum (7) and second antibody (10). Both the anti-nortriptyline antiserum and second (donkey anti-sheep) antibody are stable indefinitely at 4 °C (since they contain 1% sodium azide as a preservative).

Procedure

Collection and Treatment of Specimens

Plasma or serum: blood specimens are usually taken in the "steady-state" immediately before administration of the next dose. Take blood from a vein and obtain plasma or serum. The addition of heparin (0.2 g/l plasma) is without effect on results of the assay.

Saliva: free-flowing saliva is collected by spitting into a pot, and cooling to − 20 °C as soon as possible. After thawing it is centrifuged and the clear supernatant is used in the assay.

Stability of the substance in the sample: nortriptyline is stable in plasma (or serum) for 7 days at 4°C, and for several months at −20°C.

Assay conditions: all measurements in duplicate.

Immunoreaction: incubate for 90 min at 20°C and for 16 h at 4°C.

Enzymatic indicator reaction: incubate for 2 h at 20°C; wavelength 420 nm; light path 10 mm; measurements against dilution buffer (3).

With each series of measurement run

- two tubes containing undiluted conjugate ["total tubes"],
- two tubes containing 0.1 ml conjugate diluted with 0.5 ml buffer (3) ["non-specific binding tube"],
- two tubes containing all constituents except sample ["zero tubes"].

Establish a calibration curve with standard solution (9) instead of sample under the same conditions (within the series).

Calibration curve: divide readings of standards (samples) and "zero tubes" by readings of "total tubes", yielding $\Delta A/\Delta t$ and $\Delta A_0/\Delta t$, respectively. Calculate the ratio of $(\Delta A/\Delta t)/(\Delta A_0/\Delta t) = \Delta A/\Delta A_0$; plot $100 \times \Delta A/\Delta A_0$ *versus* nortriptyline concentration, µg/l, on a linear graph paper.

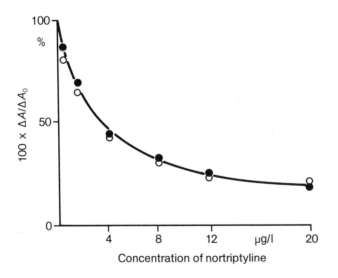

Fig. 1. Typical calibration curve for the assay of nortriptyline in plasma, serum or saliva. Percentage binding of zero ($100 \times \Delta A/\Delta A_0$) against concentration of nortriptyline (µg/l). With (○) and without (●) the addition of 5 µl drug-free plasma.

Measurement

Pipette successively into plastic tubes*:			concentration in incubation/ assay mixture	
conjugate solution	(8)	0.1 ml	β-galactosidase	ca. 100 U/l
dilution buffer	(3)	0.3 ml	phosphate	21 mmol/l
			NaCl	83 mmol/l
			Mg^{2+}	8.3 mmol/l
			2-mercapto-ethanol	8.3 mmol/l
			BSA	0.8 g/l
diluted antiserum	(7)	0.1 ml	γ-globulin	ca. 5 µg/l
sample or standard (9)		0.1 ml	nortriptyline	up to 6.7 µg/l
vortex for 5 s and incubate at room temperature (20 °C) for 90 min;				
second antibody**	(10)	0.1 ml	γ-globulin	ca. 1.4 µg/l
incubate overnight (16 h) in a refrigerator at 4 °C; centrifuge for 10 min at 2500 rpm and aspirate supernatant; wash with approximately 1 ml diluent phosphate buffer (3); vortex for 5 s; recentrifuge for 10 min at 2500 rpm; aspirate supernatant;				
substrate solution	(11)	1.0 ml	2-NPgal	2.3 mmol/l
			phosphate	100 mmol/l
			2-mercapto-ethanol	10 mmol/l
			Mg^{2+}	10 mmol/l
vortex for 10 min; incubate at 20 °C for 2 h;				
carbonate buffer	(6)	1.5 ml	carbonate	60 mmol/l
			Mg^{2+}	6 mmol/l
			2-mercapto-ethanol	6 mmol/l
reaction is stopped; centrifuge and read absorbance of the supernatant.				

* LP3 (*Luckham, Ltd.*).
** To all tubes except "total tubes".

If $\Delta A/\Delta t$ of the sample is greater than that of the highest standard, dilute the sample with buffer (3) and re-assay.

The non-specific binding tubes give an indication of the assay performance, so a high value ($>10\%$) indicates deteriorating assay performance especially if it occurs gradually between assays. It is not necessary to use the non-specific binding value in the calculation, provided it is low and constant from assay to assay.

Fig. 1 represents a typical calibration curve and illustrates that 1 µg nortriptyline per litre inhibits binding of enzyme label by bout 15%.

Calculation: calculate $\Delta A/\Delta A_0$ and read the nortriptyline mass concentration, ρ (µg/l), from the calibration curve. For conversion from mass concentration to substance concentration, c (nmol/l), multiply by 3.7965.

Validation of Method

Precision, accuracy, detection limit and sensitivity: imprecision within-series was determined by measuring the same sample several times at three different concentrations of nortriptyline (30, 125 and 250 µg/l, respectively). The relative standard deviations (RSD) for within-series assays were less than 5% in all cases. When the same samples were assayed on 4 different days, RSD was less than 8%.

Accuracy was estimated from the recovery of nortriptyline added to drug-free plasma which averaged 97% over the range 40 – 250 µg/l. A series of 21 plasma samples taken from patients receiving nortriptyline and assayed by both this technique and gas-liquid chromatography with electron capture detection [19] showed good correlation with a regression line of y = 0.90x + 11.6 µg/l. The correlation coefficient was 0.97.

The detectability of the assay as measured by the amount of drug producing a 10% inhibition in binding was 0.5 µg/l. The lower limit of the measuring range is 1.0 µg/l.

Sources of error: no interference by therapeutic agents – apart from other tricyclics – is known. Provided the plasma sample is diluted at least 4 fold, there is no interference by sera from icteric, haemolyzed or lipaemic blood.

Specificity: the cross-reactivity in the assay of a number of other drugs is listed in Table 1.

Table 1. Cross-reaction of various tricyclic antidepressant drugs in enzyme-immunoassay.

Compound	Percentage cross reaction*	Compound	Percentage cross reaction*
Nortriptyline	100	Desipramine	24
Amitriptyline	153	Protriptyline	42
Desmethylnortriptyline	45	Trimipramine	26
Imipramine	89		

* Cross-reaction: ratio of the molar concentrations of nortriptyline and the test substance effecting 50% displacement of nortriptyline-β-galactosidase conjugate under assay conditions.

The lack of specificity with regard to other tricyclic antidepressants is in part explained by the method of preparation of the immunogen. Since nortriptyline and amitriptyline differ only in their side chain, the fact that their cross-reactivity is similar is not surprising. When the remaining methyl group is removed, as in desmethylnortriptyline the cross-reactivity is reduced to 45%. Other tricyclic compounds such as imipramine and protriptyline, with a ring structure with only minor changes from that of nortriptyline, showed considerable cross-reactivity in the assay and it could indeed be used for measuring them providing appropriate standards were used and no other interfering compounds were present. Substitution in the ring, as in clomipramine or 10-hydroxyamitriptyline, on the other hand does produce a low cross-reactivity, as would be expected [25].

Therapeutic range: the accepted steady-state target concentration (therapeutic range) for plasma nortriptyline concentration in patients with endogenous depression receiving long-term nortriptyline therapy is 50 to 150 µg/l (190 to 570 nmol/l).

Appendix

Preparation of immunogen: reflux nortriptyline hydrochloride together with *N*-4-(bromobutyl)phthalimide for 20 min in 15 ml dry absolute ethanol. The reaction mixture contains a small quantity of [^{14}C]*N*-methylnortriptyline hydrochloride in order to monitor the reaction. Add 50 ml 85% (w/v) aqueous solution of hydrazine hydrate and reflux the mixture for a further 2 h. Acidify with 0.5 ml HCl, 2 mol/l, and centrifuge, remove the precipitate by filtration. Mix the supernatant gently with twice its volume of chloroform to remove unreacted compounds, discard the chloroform layer. Repeat before adjusting the pH of the aqueous phase to 6.9 with NaOH. Then add 5 ml aqueous BSA solution, 288 µg/l, and 2 ml aqueous solution containing 160 mg ethyl (dimethylaminopropyl) carbodiimide per litre. Stir the mixture at room temperature (20 °C) for 16 h. Place the aqueous phase into dialysis tubing and dialyze against water for 24 h. Freeze-dry the dialyzate under vacuum at 0 °C. Measurement of the radioactivity in the final mixture indicates that 2 to 3 nortriptyline residues are coupled to each BSA molecule.

Immunization procedure: inject sheep with an emulsion of 5 mg nortriptyline conjugate and 3 mg BCG (bacillus *Calmette-Guérin*) vaccine in 1 ml sterile saline with 2 ml Marcol 52 adjuvant at numerous sites in the leg muscles. Inject intradermally at the same time, 0.5 ml pertussis vaccine. After 4 weeks give a boosting dose of 2 mg immunogen. Repeat this at 4 week intervals for the next 2 months and thereafter at intervals of approximately 6 months. Take samples of serum from the sheep at different times after the various boosting doses. With each bleed, investigate the potency of the antisera. One sheep produces an antiserum such that when used at an initial dilution of 1 : 300 it produces a calibration curve sufficiently sensitive for clinical use. This antiserum and later more potent ones are available commercially from *Guildhay Antisera Ltd.,* Guildford GU2 5XH, Surrey, England.

Preparation of enzyme conjugate: the preparation of enzyme labelled antigens and antibodies involves the use of cross-linking reactions which should, ideally, produce a good yield of conjugate with minimal damage to both enzyme and ligand. In the assay described here a derivative of nortriptyline, desmethylnortriptyline, was conjugated to β-D-galactosidase by use of dimethyladipimidate for use as a label.

Incubate a solution containing 50 μg desmethylnortriptyline, 488 μg dimethyladipimidate and 18 mg N-ethylmorphine in 0.4 ml dry methanol at 20°C. After 30 min, add a 100 μl aliquot to 100 μg β-D-galactosidase in 1 ml carbonate buffer (6). Maintain the solution at 20°C for 90 min then terminate the reaction by adding 1 ml Tris buffer (5). Purify the conjugate by adding the solution to a Sephadex G-25 column and eluting with Tris buffer (5). Collect the fractions containing the enzyme activity, pool and dilute with Tris buffer (5).

References

[1] *W. C. Bowman, M. J. Rand,* Textbook of Pharmacology, Blackwell Publications Oxford 1980.

[2] *G. Alvan, O. Borga, M. Lind, L. Palmer, B. Siwers,* First-pass Hydroxylation of Nortriptyline: Concentrations of Parent Drug and Major Metabolites in Plasma, Eur. J. Clin. Pharmacol. *11*, 219 – 224 (1977).

[3] *J. E. Burch, R. P. Hullin,* Amitriptyline Pharmakokinetics, A Cross-over Study with Single Dose of Amitriptyline and Nortriptyline, Psychopharmacol. *74*, 35 – 42 (1981).

[4] *L. F. Gram, K. F. Overo,* First-pass Metabolism of Nortriptyline in Man, Clin. Pharmacol. Ther. *18*, 305 – 314 (1974).

[5] *R. A. Braithwaite, S. Montgomery, S. Dawling,* Nortriptyline in Depressed Patients with High Plasma Levels, II, Clin. Pharmacol. Ther. *23*, 303 – 308 (1978).

[6] *S. Dawling, P. Crome, R. A. Braithwaite, R. R. Lewis,* Nortriptyline Therapy in Elderly Patients: Dosage Predictions after Single Dose Pharmacokinetic Study, Eur. J. Clin. Pharmacol. *18*, 147 – 150 (1978).

[7] *E. Pike, E. Skuterrud,* Plasma Binding Variations of Amitriptyline and Nortriptyline, Clin. Pharmacol. Ther. *32*, 228 – 234 (1982).

[8] *B. Alexanderson,* Prediction of Steady-state Levels of Nortriptyline from Single Oral Dose Kinetics. A Study in Twins, Eur. J. Clin. Pharmacol. *6*, 44 – 53 (1973).

[9] *S. A. Montgomery, R. McAuley, D. B. Montgomery, R. A. Braithwaite, S. Dawling,* Dosage Adjustment from Single Nortriptyline Spot Level Predictor Tests in Depressed Patients, Clin. Pharmacokinet. *4*, 129 – 136 (1979).

[10] *M. Asberg, B. Cronholm, F. Sjoqvist, D. Tuck,* Relationship between Plasma Level and Therapeutic Effect of Nortriptyline, Br. Med. J. *3*, 331 – 334 (1971).

[11] *G. D. Burrows, B. Davies, B. A. Scoggins,* Plasma Concentration of Nortriptyline and Clinical Response in Depressive Illness, Lancet *II*, 619 – 623 (1972).

[12] *V. E. Ziegler, J. R. Taylor, R. D. Wetzel, J. T. Briggs,* Nortriptyline Plasma Levels and Subjective Side Effects, Br. J. Psychiatry *132*, 55 – 60 (1978).

[13] *J. C. Voris, C. Morin, J. S. Keil,* Monitoring Outpatients Plasma Antidepressant Drug Concentrations as a Measure of Compliance, Am. J. Hosp. Pharm. *40*, 119 – 121 (1983).

[14] *P. Kragh-Sorensen, C. Eggert-Hansen, P. C. Baastrup, E. F. Hvidberg,* Self-inhibiting Action of Nortriptyline's Antidepressive Effect at High Plasma Levels, Psychopharmacologia *45*, 305 – 321 (1976).

[15] *S. Montgomery, R. A. Braithwaite, S. Dawling,* High Plasma Nortriptyline Levels in the Treatment of Depression. 1., Clin. Pharmacol. Ther. *23*, 309 – 314 (1978).

[16] *M. Gibaldi,* Biopharmaceutics and Clinical Pharmacokinetics, Lea & Febiger, Philadelphia, PA, U.S.A. 1984.

[17] *T. B. Cooper, G. M. Simpson,* Prediction of Individual Dosage of Nortriptyline, Am. J. Psychiatry *135,* 333 – 335 (1978).

[18] *W. H. Hammer, B. B. Brodie,* Application of Isotope Derivative Technique to the Assay of Secondary Amines. Estimation of Desipramine by Acetylation with 3H-Acetic Anhydride, J. Pharmacol. Exp. Ther. *157,* 503 – 508 (1967).

[19] *R. A. Braithwaite, B. Widdop,* A Specific Gas-Chromatographic Method for the Measurement of "Steady-State" Plasma Levels of Amitriptyline and Nortriptyline in Patients, Clin. Chim. Acta *35,* 461 – 472 (1971).

[20] *S. Dawling, R. A. Braithwaite,* Simplified Method for Monitoring Tricyclic Antidepressant Therapy Using Gas-Liquid Chromatography with Nitrogen-Detection, J. Chromatogr. *146,* 449 – 456 (1979).

[21] *O. Borga, M. Garle,* A Gas Chromatographic Method for the Quantitative Determination of Nortriptyline and Some of Its Metabolites in Human Plasma and Urine, J. Chromatogr. *68,* 77 – 88 (1972).

[22] *R. R. Brodie, L. F. Chasseaud, D. R. Hawkins,* Separation and Measurement of Tricyclic Antidepressant Drugs in Plasma by High-Performance Liquid Chromatography, J. Chromatogr. *143,* 535 – 539 (1977).

[23] *D. M. Boss, N. B. Patton, W. W. Noll,* Rapid Method for Tricyclic Antidepressant Drugs in Serum Using Reverse Phase Liquid Chromatography, Clin. Chem. *28,* 1645 (1982).

[24] *G. W. Aherne, E. Piall, V. Marks,* The Radioimmunoassay of Tricyclic Antidepressants, Br. J. Clin. Pharmacol. *3,* 561 – 565 (1976).

[25] *G. P. Mould, G. Stout, G. W. Aherne, V. Marks,* Radioimmunoassay of Amitriptyline and Nortriptyline in Body Fluids, Ann. Clin. Biochem. *15,* 221 – 225 (1978).

[26] *K. P. Maguire, G. D. Burrows, T. R. Norman, B. A. Scoggins,* A Radioimmunoassay for Nortriptyline (and Other Tricyclic Antidepressants) in Plasma, Clin. Chem. *24,* 549 – 554 (1978).

[27] *R. S. Kamel, J. Landon, D. S. Smith,* Novel 125-I-labelled Nortriptyline Derivatives and Their Use in Liquid-phase or Magnetisable Solid-phase Second Antibody Radioimmunoassays, Clin. Chem. *25,* 1997 – 2002 (1979).

[28] *K. K. Midha, J. C. Loo, C. Charette, M. L. Rowe, J. W. Hubbard, I. J. McGilveray,* Monitoring of Therapeutic Concentrations of Psychotropic Drugs in Plasma Radioimmunoassays, J. Anal. Toxicol. *2,* 185 – 192 (1978).

[29] *D. J. Brunswick, B. Needelman, J. Mendels,* Specific Radioimmunoassay of Amitriptyline and Nortriptyline, Br. J. Clin. Pharmacol. *7,* 343 – 348 (1979).

[30] *R. Lucek, R. Dixon,* Specific Radioimmunoassay for Amitriptyline and Nortriptyline in Plasma, Res. Commun. Chem. Pathol. Pharmacol. *18,* 125 – 136 (1977).

[31] *R. Virtanen,* Radioimmunoassay for Tricyclic Antidepressants, Scand. J. Clin. Lab. Invest. *40,* 191 – 197 (1980).

[32] *K. P. Maguire, G. D. Burrows, T. R. Norman, B. A. Scoggins,* Evaluation of a Kit for Measuring Tricyclic Antidepressants, Clin. Chem. *26,* 529 (1980).

[33] *M. N. Al-Bassam, M. J. O. O'Sullivan, E. Gnemmi, J. W. Bridges, V. Marks,* Double-antibody Enzyme Immunoassay for Nortriptyline, Clin. Chem. *24,* 1590 – 1594 (1978).

[34] *J. N. Miller, C. S. Lim, J. W. Bridges,* Fluorescamine and Fluorescein as Labels in Energy-transfer Immunoassay, Analyst *105,* 91 – 92 (1980).

[35] *C. H. J. Wang, S. D. Stroupe, M. E. Jolley,* Fluorescein Derivatives and Fluorescence Polarisation Immunoassay Methods, U.S. Applic. *329,* 975 11 Dec. 1981, 53 pp.

[36] *D. S. Smith,* Eur. Pat. Appl. 34050 19 Aug. 1981.

[37] *B. A. Scoggins, K. P. Maguire, T. R. Norman, G. D. Burrows,* Measurement of Tricyclic Antidepressants. Part 1. A Review of the Methodology, Clin. Chem. *26,* 5 – 17 (1980).

[38] *P. Baumann, U. Breyer-Pfaff, B. Muller-Oerlinghausen, M. Sandoz,* Quality Control of Amitriptyline and Nortriptyline Plasma Level Assessments: A Multicenter Study, Pharmacopsychiatry *15,* 156 – 160 (1982).

1.18   Aprotinin

Pancreatic basic trypsin inhibitor

1.18.1   Enzyme-linked Immunosorbent Assay

Werner Müller-Esterl

General

The bovine pancreatic trypsin inhibitor (BPTI), commonly known by its generic name aprotinin (Trasylol®), is present in a variety of bovine organs including lung [1], pancreas and parotid gland (for recent review, cf. [2]). Aprotinin was first characterized by *Kunitz & Northrop* in 1936 [3]. Since then, aprotinin has been the object of extensive experimental studies. Analysis of the amino acid sequence of the inhibitor revealed that the protein backbone is built up of 58 amino acids with a resulting molecular mass of 6512 [4]. X-ray analysis of the three-dimensional structure indicated that aprotinin is a pyriform molecule with a compact tertiary structure [5]. Cloning and sequence analysis of the genomic DNA coding for aprotinin suggested that aprotinin is biosynthesized as a larger precursor molecule which is proteolytically processed to the mature inhibitor [6].

Aprotinin is an extremely basic protein ($pI = 10.5$) with an outstanding stability against heat, acid or alkaline treatment, organic solvents and proteolytic breakdown [2, 7]. The primary target enzymes of the inhibitor are serine proteinases such as trypsin, kallikrein, plasmin and chymotrypsin [2]. This inhibition spectrum classifies aprotinin as a broad-specificity inhibitor. Despite the elucidation of many of its physical characteristics, the functional role of aprotinin in bovine tissues has remained obscure.

As early as 1953, aprotinin was introduced into clinical practice. It is now used as a therapeutic agent in diseases related to imbalance of proteolytic systems, e.g. hyperfibrinolysis, septicaemia, and acute pancreatitis.

Application of method: in clinical and experimental biochemistry. Aprotinin is usually measured in plasma, serum and biological fluids such as urine, liquor and ascites following therapeutic application in man or in experimental animals. In the latter case, aprotinin may also be measured in tissue extracts.

Substance properties relevant in analysis: aprotinin can be quantitated by enzyme-inhibition assays using trypsin [8] (cf. Vol. V, pp. 123, 126), kallikrein [9] (cf. Vol. V, pp. 135, 141) or related serine proteinases (cf. chapter 1.18.2, p. 257 – 263). The specific inhibitor activity of aprotinin is 7.14×10^6 KIU*/g. Aprotinin concentrations are often given in KIU; multiplication by a factor of 0.14 transforms these figures into the corresponding microgram quantities of aprotinin.

Methods of determination: a variety of methods for aprotinin quantitation have been published, including enzyme-inhibition assays [8, 9] and immunological methods such as radioimmunoassay [10, 11] or enzyme-immunoassay [12, 13]. No commercial tests are available yet.

International reference method or standards: no reference method has been established so far.

Assay

Method Design

A competitive enzyme-immunoassay of the ELISA type is described here, cf. Vol. I, chapter 2.7, p. 233.

Principle

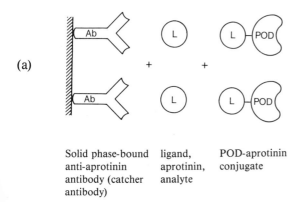

(a)

Solid phase-bound anti-aprotinin antibody (catcher antibody)	ligand, aprotinin, analyte	POD-aprotinin conjugate

---

* 1 Kallikrein inhibitor unit (KIU) is defined as that amount of aprotinin which inhibits one enzyme unit, U; cf. Vol. V, pp. 135, 136.

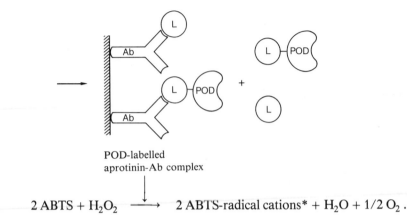

POD-labelled
aprotinin-Ab complex

(b) $2\,ABTS + H_2O_2 \longrightarrow 2\,ABTS\text{-radical cations}^* + H_2O + 1/2\,O_2$.

Microtitre plates are coated with affinity-purified antibodies directed against aprotinin. Samples containing aprotinin are mixed with an equal volume of peroxidase**-
aprotinin conjugate and placed into the wells of the microtitre plate. The solid phase-
bound antibodies (catcher antibodies) adsorb aprotinin and the POD-aprotinin conjugate in amounts reflecting their relative concentrations in the sample. All unreacted
material and conjugate is washed out. The quantity of POD bound to the solid phase
via aprotinin and its antibody is measured as increase in absorbance per unit time,
$\Delta A/\Delta t$, by incubation with ABTS®***/H_2O_2. The concentration of aprotinin present
in the sample is inversely related to the enzyme activity present on the solid phase.

**Selection of assay conditions and adaptation to the individual characteristics of the
reagents:** incubation time for the immunoreaction is 3 h, and for the enzymatic
reaction 30 min. All procedures are carried out at 37°C. The pH optimum of the
enzymatic reaction is 4.5.

The desired working range for the assay is from 50 µg/l to 5 mg/l. Depending on
the aprotinin concentration in the sample, a starting dilution is made (usually 1:20 for
plasma and 1:2 for urine and other secretory fluids). Using an aprotinin concentration in the upper working range (i.e. 5 mg/l) and a sample size of 200 µl, the optimal
concentrations for the coating antibody and the POD-aprotinin conjugate are determined in the following way:

(a) Serial dilutions (2^n) of the coating antibody (range 0.3 µg/l to 20 mg/l) are
delivered into the wells of a microtitre plate (200 µl each) and incubated overnight
at 4°C. Then the competitive ELISA is performed (see below) using a fixed concentration of the conjugate (100 µg/l). The resulting values of $\Delta A/\Delta t$ are plotted
against the concentration of the coating IgG (Fig. 1a). The minimum IgG concentration resulting in a maximum response ($\Delta A/\Delta t = 1.5$) is chosen as the optimal
coating concentration.

---

  * Cf. chapter 1.10, p. 116.
 ** POD, donor: hydrogen-peroxide oxidoreductase, EC 1.11.1.7.
*** ABTS®, diammonium 2,2'-azinobis-[3-ethyl-2,3-dihydrobenzothiazole-6-sulphonate].

(b) Serial dilutions (2^n) of the POD-aprotinin conjugate are prepared (range $1 : 10^2$ to $1 : 6 \times 10^6$) and pipetted into *Eppendorf* tubes (110 µl each). An equal volume of buffer (3) (without aprotinin) is added, the resulting mixture is thoroughly mixed, 200 µl is removed from each tube and placed in a corresponding well of a micro-titre plate pre-coated with the optimum concentration of the anti-aprotinin IgG. Following incubation for 3 h, the indicator reaction is performed (see below). The velocities $\Delta A/\Delta t$ are plotted against the conjugate concentrations (Fig. 1b). The minimum concentration (i.e. the maximum dilution) of the conjugate giving rise to a maximum response ($\Delta A/\Delta t = 1.5$) is selected as the optimal conjugate con-centration.

Example: in our assay system for aprotinin, we use a concentration of 0.6 mg/l for the coating (catcher) antibody and a conjugate concentration of 20 µg/l. This corres-ponds to a conjugate dilution of $1 : 100000$ from a stock solution (cf. Appendix, p. 255). The highest concentration of the antigen standard is 10 mg/l.

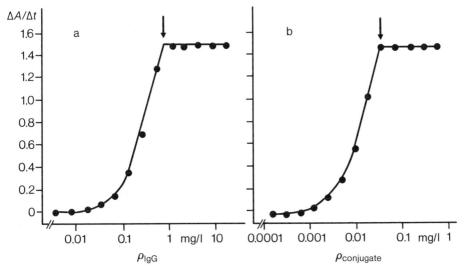

Fig. 1. Optimization a) of the concentration of the coating IgG and b) of the POD-aprotinin conjugate. Arrows point to the concentrations selected for the optimized assay system.

Equipment

Photometer for microplates (Microplate Reader MR 600, *Dynatech*); microplate washer (Titertek® Microplate Washer 120, *Flow*); automatic dispenser (Autodrop, *Flow*); incubator (37 ± 0.1 °C, *Binder*); digital diluter-dispenser (Microlab 1000, *Hamilton*); multi-channel pipettes with variable volumes from 50 to 200 µl (Titer-tek® Multichannel Pipettes, *Flow*); Socorex® microdispenser, *Dynatech*; microtitre plates (Immunolon F, cobalt sterilized, *Dynatech*).

Reagents and Solutions

Purity of reagents: aprotinin isolated from bovine lung tissue is a research product from *Bayer*. Homogeneity of the inhibitor preparations is routinely monitored by gel electrophoresis in the presence of sodium dodecylsulphate. POD from horseradish is of best quality available (grade I, RZ > 3.0, *Boehringer Mannheim*). ABTS® (diammonium salt) is chromatographically pure (*Boehringer Mannheim*). Bovine serum albumin is electrophoretically homogeneous (>98%) (*Boehringer Mannheim*). CNBr-activated Sepharose 4B is a product of *Pharmacia*. All other reagents are at least of analytical grade.

Double distilled water is used throughout (cf. Vol. II, chapter 2.1.3.2, p. 113).

Preparation of solutions (for 10 microtitre plates corresponding to 960 wells):

1. Coating buffer (carbonate, 50 mmol/l, pH 9.6):

   dissolve 0.32 g Na_2CO_3 and 0.59 g $NaHCO_3$ in 200 ml water.

2. Washing buffer (phosphate, 0.1 mol/l; NaCl, 0.15 mol/l; Tween 20, 0.5 ml/l; pH 7.4):

   dissolve 178 g $Na_2HPO_4 \cdot 2\ H_2O$, 138 g $NaH_2PO_4 \cdot H_2O$, 87.6 g NaCl and 5 g Tween 20 in 10 l water.

3. Diluting buffer (phosphate, 100 mmol/l; NaCl, 150 mmol/l; Tween 20, 0.5 ml/l; BSA, 20 g/l; pH 7.4):

   dissolve 20 g bovine serum albumin in 1 l washing buffer (2).

4. Substrate buffer (citrate, 0.1 mol/l; phosphate, 0.1 mol/l; pH 4.5):

   dissolve 4.2 g citric acid, monohydrate, 3.56 g $Na_2HPO_4 \cdot 2\ H_2O$ and 2.76 g $NaH_2PO_4 \cdot H_2O$ in 200 ml water.

5. ABTS® solution (36 mmol/l):

   dissolve 200 mg ABTS® in 10 ml water.

6. Hydrogen peroxide solution (0.3 mol/l):

   mix 1 ml 30% (w/v) H_2O_2 with 50 ml water.

7. Substrate solution (ABTS®, 1.8 mmol/l; H_2O_2, 6 mmol/l; citrate, 0.1 mol/l; phosphate, 0.1 mol/l; pH 4.5):

   mix 10 ml ABTS® solution (5) and 4 ml hydrogen peroxide solution (6) with 190 ml substrate buffer (4).

8. IgG solution (IgG, 0.6 mg/l; carbonate, 50 mmol/l; pH 9.6):

for coating mix 120 µl IgG stock solution (1 g/l; for preparation, cf. Appendix, p. 255) with 200 ml coating buffer (1).

9. Aprotinin solution (aprotinin, 10 mg/l; phosphate, 0.1 mol/l; NaCl, 0.15 mol/l; Tween 20, 0.5 ml/l; BSA, 20 g/l; pH 7.4):

for preparation of the standard mix 0.1 ml aprotinin stock solution (1 g/l) with 9.9 ml diluting buffer (3) resulting in a final concentration of 10 mg/l.
For precision controls spike citrated plasma with the appropriate amounts of aprotinin (from the stock solution) to yield final concentrations of 0.5 to 100 mg/l.

10. Conjugate solution (conjugate, 20 µg/l; POD, 7 U*/l; phosphate, 0.1 mol/l; NaCl, 0.15 mol/l; Tween 20, 0.5 ml/l; BSA, 20 g/l; pH 7.4):

for working dilution mix 10 µl conjugate stock solution (2 g/l, cf. Appendix, p. 255) with 10 ml diluting buffer (3); final concentration 2 mg/l. Mix 2 ml of the diluted aprotinin solution (9) with 200 ml diluting buffer (3); final concentration 20 µg/l (7 U/l).

11. Coupling buffer (NaHCO$_3$, 0.1 mol/l; NaCl, 0.5 mol/l; pH 8.3):

dissolve 4.2 g NaHCO$_3$ and 14.6 g NaCl in 500 ml water.

12. HCl solution (1 mmol/l):

mix 100 µl HCl, 2 mol/l with 200 ml water.

13. Termination buffer (ethanolamine, 1 mol/l, pH 8.0):

mix 2.4 ml ethanolamine with 40 ml water.

14. Acetate buffer (sodium acetate, 0.1 mol/l; NaCl, 1 mol/l; pH 4.0):

dissolve 8.2 g sodium acetate and 58.4 g NaCl in 1 l water.

15. Borate buffer (Na$_2$B$_4$O$_7$, 0.1 mol/l; NaCl, 1 mol/l; pH 8.0):

dissolve 20.1 g Na$_2$B$_4$O$_7$ and 58.4 g NaCl in 1 l water.

Stability of reagents and solutions: the IgG stock solution (8), the conjugate stock solution (10) and the aprotinin stock solution (9) are stable for at least 1 year at −20°C. The ABTS® solution (5) and the hydrogen peroxide solution (6) are freshly prepared every week and kept at 4°C in the dark. The diluted conjugate (cf. solution (10)), the substrate buffer (4), and the diluted aprotinin standards (cf. solution (9)) are

---

* Guaiacol and H$_2$O$_2$ as substrates, 25°C.

freshly prepared every day. Buffers (11), (13) are freshly prepared for each coupling procedure. Buffers (1), (14), (15) are stable for at least one month if anti-microbial agents such as NaN$_3$ (3 mmol/l) or Merthiolate® (0.1 g/l) are present. However, do not add NaN$_3$ to solutions (2) – (7) that come in contact with POD as NaN$_3$ is a strong inhibitor of POD.

Procedure

Collection and treatment of specimen: anticoagulated plasma is used for aprotinin determinations. Application of citrate, heparin or EDTA is appropriate and does not interfere with the assay. Also, plasma from icteric, slightly haemolyzed or lipaemic blood does not interfere with the assay.

Stability of the analyte in the sample: plasma samples can be stored at $-70\,°C$ for at least one year without appreciable loss of measurable antigen. Storage at $4\,°C$ should be strictly avoided to prevent cold-activation and subsequent complex formation between aprotinin and activated proteinases. Other specimens such as urine, pancreatic juice, seminal plasma and tissue extracts can be stored at $-70\,°C$ for at least 3 months without detectable loss of measurable antigen.

Assay conditions: all measurements in duplicate at three different dilutions.

Immunoreaction: incubation at $37\,°C$ for 2 h each.

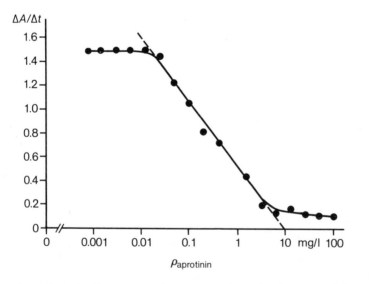

Fig. 2. Typical calibration curve for the assay of aprotinin in plasma. Only $\Delta A/\Delta t$ values between 0.2 and 1.3 are used for the calculation of analyte concentrations from a regression line ($--$).

Enzymatic indicator reaction: incubation at 37 °C for 30 min in the dark. Wavelength 405 nm.

Establish a calibration curve under exactly the same conditions.

Measurement

Pipette into each well of the microplate:		concentration in the incubation/assay mixture	
sample or standard solution (9) in diluting buffer (3)	200 µl	aprotinin phosphate NaCl Tween 20 BSA	up to 10 mg/l 0.1 mol/l 0.15 mol/l 0.5 ml/l 20 g/l
cover the microplate, incubate for 2 h at 37 °C; wash with approximately 100 ml washing buffer (2); remove water as completely as possible,			
diluted conjugate (10)	200 µl	conjugate POD phosphate NaCl Tween 20 BSA	20 µg/l 7 U/l 0.1 mol/l 0.15 mol/l 0.5 ml/l 20 g/l
incubate and wash as detailed above,			
substrate solution (7)	200 µl	ABTS® H_2O_2 citrate phosphate	1.8 mmol/l 6 mmol/l 0.1 mol/l 0.1 mol/l
incubate in the dark for exactly 30 min at 37 °C, read absorbance; measurement must be completed in 1 min; otherwise stop reaction by acidification with 100 µl H_2SO_4, 1 mol/l.			

Calibration curve: plot the increases in absorbance over 30 min, $\Delta A/\Delta t$, *versus* aprotinin concentrations in a semi-logarithmic scale. The curve is a straight line over a broad range ($\Delta A/\Delta t = 0.2 - 1.3$) with a reasonable correlation coefficient (Fig. 2).

Calculation: read aprotinin concentration from the calibration curve. On-line computer programmes have been developed in our laboratory which process the data from the microplate reader.

Validation of Method

Precision, accuracy, detection limit and sensitivity: for concentrations within the therapeutic range (ca. 1 µmol/l corresponding to 6.5 mg/l or 42.9 KIU/l), the relative standard deviations, RSD, are 5.8 to 7.6% (within-run imprecision) and 10.6 to 10.9% (day-to-day imprecision); $n = 20$.

Accuracy has been studied by recovery experiments using citrated plasma supplemented with aprotinin. The mean recovery was 98% (range 86 to 124%) for a concentration range of 0.5 mg/l to 100 mg/l.

The detection limit, defined as the lowest concentration of aprotinin giving rise to a value of $\Delta A/\Delta t$ significantly different from that of the zero standard, was calculated to be 20 µg/l (i.e. 4 ng/well) corresponding to 132 KIU/l.

Sensitivity largely depends on the shape of the calibration curves; it is lowest in steep calibration curves and highest in flat calibration curves. The steepness of the calibration curves can be manipulated by varying the concentration of the POD-aprotinin conjugate.

Inter-method comparison of the enzyme-immunoassay with the enzyme-inhibition assay for aprotinin (cf. chapter 1.18.2, p. 257) revealed an adequate correlation for the two assay systems ($r_{xy} = 0.779$, $n = 178$).

Sources of error: no interfering condition due to the samples has been found yet. In particular, we have not observed any interference with the assay upon complex formation of aprotinin with proteinases. However, a detailed analysis of complex formation on the accuracy of the assay awaits further experiments.

Specificity: the assay is specific for aprotinin. Control experiments with aprotinin-free plasma did not give any cross-reactivity with unrelated plasma proteins. Similarly, application of other inhibitors of non-human origin used in clinical or experimental medicine, such as eglin and hirudin (from leeches) and trypsin inhibitor (from soybeans) failed to interfere with the assay. The fact that strictly parallel curves are obtained with complex biological media containing appreciable amounts of aprotinin (e.g. urine, ascites) lends further support to the conclusion that the assay is specific for aprotinin.

Therapeutic ranges: following continuous intravenous administration of aprotinin (1515 mg in 36 hours), mean plasma concentrations of 19.8 ± 1.4 mg/l resulted; $n = 7$ [14].

Appendix

Preparation of the immunoaffinity gel: for affinity purification of anti-aprotinin IgG, immobilization of the antigen on insoluble supports such as Sepharose 4 B is required [15]. Usually 10 to 50 mg antigen can be coupled per 1 g Sepharose. The procedure given is rapid and ensures high yields of coupled protein.

- Weigh out 1 g (dry weight) of CNBr-activated Sepharose 4 B and add 200 ml HCl solution (12); pour the gel on a sintered-glass filter (G 3), wash it with 200 ml HCl solution (12), transfer it to an appropriate reaction vessel and re-suspend it in 5 ml coupling buffer (11);
- dissolve 15 mg aprotinin in 7.5 ml coupling buffer (11) and immediately add it to the gel suspension. Incubate for 2 h at room temperature or at 4 °C overnight under constant mixing, e.g. by an end-over-end mixer. Do not use a magnetic stirrer because this will damage the gel;
- wash the reacted gel on a sintered-glass filter with 100 ml coupling buffer (11) to wash out the unreacted protein. Transfer the gel into a clean reaction vessel, add 20 ml termination buffer (13) and incubate it for 2 h at room temperature under constant mixing. This blocks the remaining active groups on the gel;
- wash the gel with 100 ml acetate buffer (14), followed by 100 ml borate buffer (15). This washing procedure is repeated 5 times. Suspend the gel in PBS (including NaN$_3$, 3 mmol/l, to prevent microbial contamination). Store the gel at 4 °C until used for immuno-selection of anti-aprotinin IgG. Under these conditions, the affinity gel is stable for at least 2 years without considerable loss of binding capacity.

Preparation of IgG: antibodies against aprotinin are affinity-purified from rabbit antisera following the procedure given elsewhere (Vol. IX, chapter 4.10, p. 313).

Briefly, treat 30 ml antiserum with diisopropyl fluorophosphate and subject to aprotinin-Sepharose (*vide supra*). Wash out the unbound proteins, and elute the specific antibodies attached to the affinity gel, at pH 2.2. Dialyze the resulting IgG fraction against PBS, 0.1 mol/l, overnight at 4 °C, concentrate by vacuum dialysis at 4 °C to a final concentration of 1 g/l, shock-freeze in liquid nitrogen and store at − 20 °C.

Preparation of POD-aprotinin conjugate: covalent coupling of aprotinin and POD is carried out using the hetero-bifunctional cross-linker SPDP®* [16] essentially as described elsewhere (Vol. IX, chapter 4.10, p. 314).

Briefly, let 5 mg aprotinin react with SPDP and chromatograph on a Sephadex G 25 column in the presence of DTT; in a second step, separate free DTT from the aprotinin-SPDP conjugate by another gel filtration. Similarly, prepare a conjugate of 5 ng POD and SPDP except that DTT is omitted. Finally, mix the SPDP-reacted POD and aprotinin and incubate overnight. Centrifuge the resulting POD-aprotinin conjugate (2 g/l; 700 kU/l), aliquot the supernatant and transfer into sterile tubes, shock-freeze

---

* SPDP®, *N*-succinimidyl-3-(2-pyridyldithio)-propionate.

in liquid nitrogen and store at $-20\,^\circ$C for at least 6 months without detectable loss of activity.

As judged from radioactive tracer experiments, approximately 3 to 4 molecules of aprotinin are covalently attached to a single POD molecule. Note that the conjugate linker is susceptible to reductive cleavage, e.g. by DTT [16].

Coating of the solid phase: the procedure for IgG coating exactly follows the scheme presented for anti-kininogen IgG (Vol. IX, chapter 4.10, p. 315).

Deliver the IgG working solution (8) to the wells of the microtitre plates and incubate overnight. Following extensive rinsing with washing buffer (2), use the plates directly or store at $-20\,^\circ$C.

References

[1] *H. Fritz, J. Kruck, I. Rüsse, H. G. Liebich,* Immunofluorescence Studies Indicate that the Basic Trypsin-Kallikrein Inhibitor of Bovine Organs (Trasylol®) Originates from Mast Cells, Hoppe-Seyler's Z. physiol. Chem. *360*, 437 – 444 (1979).

[2] *H. Fritz, G. Wunderer,* Biochemistry and Applications of Aprotinin, the Kallikrein Inhibitor from Bovine Organs, Drug Res. *33*, 479 – 494 (1983).

[3] *M. Kunitz, J. H. Northrop,* Isolation from Beef Pancreas of Crystalline Trypsinogen, Trypsin, Trypsin Inhibitor, and an Inhibitor Trypsin Compound, J. Gen. Physiol. *19*, 991 – 1007 (1936).

[4] *B. Kassell, M. Laskowski, Sr.,* The Basic Trypsin Inhibitor of Bovine Pancreas. IV. The Linear Sequence of the 58 Amino Acids, Biochem. Biophys. Res. Commun. *18*, 255 – 258 (1965).

[5] *R. Huber, D. Kukla, O. Epp, H. Formanek,* The Basic Trypsin Inhibitor of Bovine Pancreas. I. Structure Analysis and Conformation of the Polypeptide Chain, Naturwissenschaften *57*, 389 – 392 (1970).

[6] *S. Anderson, B. Kingston,* Isolation of a Genomic Clone for Bovine Pancreatic Trypsin Inhibitor by Using a Unique-Sequence Synthetic DNA Probe, Proc. Natl. Acad. Sci. U.S.A. *80*, 6838 – 6842 (1983).

[7] *K. Wüthrich, G. Wagner,* Nuclear Magnetic Resonance of Labile Protons in the Basic Pancreatic Trypsin Inhibitor, J. Mol. Biol. *130*, 1 – 18 (1979).

[8] *H. Fritz, I. Trautschold, E. Werle,* Protease Inhibitors, in: *H. U. Bergmeyer* (ed.), Methods of Enzymatic Analysis, Vol. 2, Verlag Chemie, Weinheim, and Academic Press, New York 1974, pp. 1064 – 1080.

[9] *M. Jochum, V. Janokova, H. Dittmer, H. Fritz,* An Enzymatic Assay Convenient for the Control of Aprotinin Levels during Proteinase Inhibitor Therapy, Fresenius Z. anal. Chem. *317*, 719 – 720 (1984).

[10] *E. Fink, L. J. Greene,* Measurement of the Bovine Pancreatic Trypsin Inhibitor by Radioimmuno-assay, in: *H. Fritz, H. Tschesche, L. J. Greene, E. Truscheit* (eds.), Proteinase Inhibitors, Springer-Verlag, Berlin 1974, pp. 243 – 249.

[11] *I. Trautschold,* Radioimmunoassay for Kallikrein Inhibitors, Life Sci. *16*, 830 – 831 (1975).

[12] *T. Shikimi,* Sandwich Enzyme-Immunoassay of Aprotinin, J. Pharmacobio-dyn. *5*, 708 – 715 (1982).

[13] *W. Müller-Esterl, A. Oettl, E. Truscheit, H. Fritz,* Monitoring of Aprotinin Plasma Levels by an Enzyme-Linked Immunosorbent Assay (ELISA), Fresenius Z. anal. Chem. *317*, 718 – 719 (1984).

[14] *H. Maier, D. Adler, T. Lenarz, W. Müller-Esterl,* New Concepts in the Treatment of Chronic Recurrent Parotitis, Arch. Otorhinolaryngol. *242*, 321 – 328 (1985).

[15] *R. Axén, J. Porath, S. Ernback,* Chemical Coupling of Peptides and Proteins to Polysaccharides by Means of Cyanogen Halides, Nature *214*, 1302 – 1304 (1967).

[16] *J. Carlsson, H. Drevin, R. Axén,* Protein Thiolation and Reversible Protein-Protein Conjugation. *N*-Succinimidyl-3-(2-pyridyldithio)-propionate, a New Heterobifunctional Agent, Biochem. J. *173*, 723 – 737 (1978).

1.18.2   Inhibition Assay

Marianne Jochum

General

The biomedical use of the low molecular-weight (M_r = 6500) proteinase inhibitor aprotinin (Trasylol®), from bovine tissue cells is based on its broad inhibitory specificity. Major target enzymes are trypsin, chymotrypsin, plasmin, tissue kallikrein and plasma kallikrein. Hence, administration of aprotinin to patients is recommended as part of the therapeutic regimen of various diseases, in particular in shock syndromes, hyperfibrinolysis or acute pancreatitis (for review cf. [1]). Serum or plasma levels of aprotinin obtained after intravenous injection decline rather rapidly due to distribution of the inhibitor in the extracellular fluid and subsequent accumulation, primarily in the kidney [2, 3]. Therefore, one major prerequisite for the optimization of the proteinase inhibitor therapy is the quantitative assessment of the inhibitor levels in patients' plasma or other body fluids.

Application of method: in biochemistry, clinical chemistry and pharmacology.

Substance properties relevant in analysis: aprotinin can be exposed to solutions with extreme pH values [1]. Hence, acid treatment of aprotinin-containing plasma samples allows a total recovery and thereby a specific measurement of aprotinin as the only tissue kallikrein inhibitor present in these specimens.

Methods of determination: so far, aprotinin has been measured in biological samples by the trypsin inhibition assay [1, 4] or by immunological methods [5, 6]. However, these assays are inappropriate for monitoring inhibitor levels in routine clinical diagnosis because of the limited specificity of the enzyme inhibition assay and the prolonged incubation (> 40 h) necessary in the immunoassays. The enzymatic determination described here circumvents these difficulties and therefore can conveniently be used as a specific and rapid bedside control of high-dosage proteinase-inhibitor therapy [7].

An enzyme-linked immunosorbent assay was developed only recently ([8], cf. chapter 1.18.1) which provides a versatile means of measuring very precisely and relatively rapidly (4 h) large numbers of aprotinin-containing samples from plasma and other body fluids. This assay is suitable for special purposes, e.g. for monitoring lower dose inhibitor therapy in retrospective studies.

International reference method and standards: not yet available.

Assay

Method Design

Principle

(a)   Aprotinin + tissue kallikrein (excess) $\xrightarrow{\text{5 min}}$

aprotinin-tissue kallikrein + tissue kallikrein (remaining)

(b)   D-Val-Leu-Arg-4-NA + H_2O $\xrightarrow{\text{tissue kallikrein (remaining)}}$

D-Val-Leu-ArgOH + 4-nitroaniline .

After incubation of acid-treated plasma with tissue kallikrein (EC 3.4.21.35) in excess, the remaining amount of the enzyme is determined by its amidolytic activity on the substrate D-Val-Leu-Arg-4-NA (S-2266). The initial rate at which 4-nitroaniline is released is measured photometrically at 405 nm.

The reaction rate decreases linearly with increasing concentration of aprotinin in the range of 20000 – 80000 KIU* per litre plasma. The concentration of aprotinin is calculated from a calibration curve prepared by diluting normal plasma to which aprotinin (2×10^5 KIU/l) has been added before acid treatment.

Optimized conditions for measurement: an enzyme activity of 1 KU** is used in the assay mixture; taking K_m into account, the substrate concentration of 0.15 mmol/l ensures minimal influence of substrate depletion. Pre-incubation of acid-treated aprotinin-containing samples with the enzyme for 5 min at pH 8.2 is sufficient to achieve complete complex formation. Since the kallikrein activity towards the chromogenic peptide substrate is increased by the addition of plasma, all dilutions of aprotinin-containing plasma samples have to be performed with aprotinin-free plasma instead of buffer: in this way, the same amount of plasma (e.g. 25 µl) is always added to the assay system.

Temperature conversion factors: the assay should be performed at 37 °C. Conversion factors for other temperatures have not yet been determined.

Equipment

Spectrophotometer or spectral-line photometer capable of exact measurement at 405 nm, provided with a thermostatted cuvette holder; water-bath, centrifuge (*Eppendorf* 5412); recorder or stopwatch; semi-microcuvettes.

---

* KIU biological kallikrein inhibitor unit (cf. [1]).
** KU biological kallikrein unit (cf. [1]).

Reagents and Solutions

Purity of reagents: all chemicals should be of the highest analytical grade commercially available.

Preparation of solutions (for about 280 determinations): all solutions in re-purified water (cf. Vol. II, chapter 2.1.3.2).

1. Buffer (Tris, 0.2 mol/l, pH 8.2):

   dissolve 24.2 g tris(hydroxymethyl)aminomethane in 800 ml water, adjust to pH 8.2 with HCl, 1 mol/l (approximately 100 ml), and dilute to 1000 ml with water (25 °C).

2. Substrate solution (D-Val-Leu-Arg-4-NA, 1.5 mmol/l):

   dissolve 25 mg S-2266 (from *AB Kabi Diagnostica,* Stockholm, Sweden) in 28.8 ml water.

3. Porcine pancreatic kallikrein solution* (10^6 KU/l):

   dissolve the lyophilized enzyme with sterile physiological saline (adjusted to pH 7.5 with triethanolamine) to a stock solution of 10^6 KU/ml. Prepare further dilutions for the assay (5×10^4 KU/l $\triangleq$ 1 KU/20 µl) freshly each day.

4. Normal human plasma:

   take blood specimens from veins of at least ten healthy blood donors (9 vol blood and 1 vol sodium citrate solution, 0.1 mol/l). Prepare plasma by centrifugation at 2000 g for 20 min at 4 °C. Mix equal amounts of plasma from each donor and dispense in small volumes. Treat with perchloric acid (6) as described for samples under Collection and treatment of specimens.

5. Aprotinin standard solutions:

   use ampoules with 2×10^5 KIU/10 ml, pH 5, from *Bayer AG,* Leverkusen; add 10 µl (200 KIU) to 1 ml normal plasma (4) for preparation of the calibration curve. Dilute this stock solution (10^5 KIU/l) with aprotinin-free normal plasma (4) according to the following scheme:

---

* Research product from *Bayer AG,* Wuppertal-Elberfeld.

Stock solution $(10^5$ KIU/l) μl	Normal plasma μl	KIU per assay	KIU ($\times$ 10^3) per l untreated plasma
10	190	0.125	10
10	90	0.25	20
20	80	0.50	40
30	70	0.75	60
40	60	1.00	80
50	50	1.25	100

6. Perchloric acid solution (PCA, 30 g/l):

dilute 8.57 ml 70% (w/v) PCA with 191.43 ml water.

7. Potassium carbonate solution (K_2CO_3, 5 mol/l):

dissolve 6.91 g K_2CO_3 in 8 ml water and dilute to 10 ml with water.

Stability of solutions: store all solutions at 2 °C to 8 °C. Solutions (1) and (7) are stable for two months if not contaminated with micro-organisms. Solution (2) is stable for six months if prepared with sterile water and kept in the dark. Solution (3) is stable for at least two weeks. If not contaminated, solutions (5) and (6) are stable indefinitely. Normal plasma (4) is stable for at least 3 – 4 months at − 20 °C or more than 6 months at − 70 °C. Commercially available lyophilized normal plasma may also be used.

Procedure

Collection and treatment of specimens: take blood specimens from the vein without stasis (9 vol blood + 1 vol sodium citrate solution, 0.1 mol/l). Prepare plasma by centrifugation at 2000 g for 20 min at 4 °C.

Acid treatment of plasma samples: incubate plasma (e.g. 0.5 ml) for 10 min at 25 °C with an equal volume of PCA solution (6). Centrifuge for 10 min at 10 000 g, neutralize ca. 0.980 ml supernatant with ca. 20 μl K_2CO_3 solution (7), keep at 4 °C to 8 °C for further 30 min and thereafter centrifuge again for 10 min. Freeze the resulting supernatant at − 20 °C or below.

Stability of aprotinin in the sample: aprotinin in plasma is stable at room temperature at least for 1 day or at 4 °C for 2 days. Aprotinin solutions of pH < 4 should be used within a few hours. Neutralized aprotinin-containing plasma samples can be stored for 1 month at − 20 °C or for 6 months at − 70 °C. Repeated freezing and thawing (n = 10) has no influence on the inhibitory activity of aprotinin in plasma.

Details for measurements in other samples: the same procedure may be used for measurements in serum and other specimens.

Assay conditions: wavelength Hg 405 nm; light path 10 mm; final volume 1 ml; temperature 37 °C (thermostatted cuvette holder). Measure against air. Before starting the assay, adjust temperature of solutions to 37 °C.

Establish a calibration curve using aprotinin solution (5) instead of sample.

Measurement

Pipette successively into the cuvette:		blank	sample	concentration in assay mixture	
Tris buffer	(1)	0.855 ml	0.855 ml	Tris	171 mmol/l
normal plasma	(4)	0.025 ml	–		
sample or standard	(5)	–	0.025 ml	aprotinin up to 1250 KIU/l	
kallikrein solution	(3)	0.020 ml	0.020 ml	kallikrein	1000 KU/l
mix thoroughly with a plastic spatula, wait for 5 min,					
substrate solution	(2)	0.100 ml	0.100 ml	D-Val-Leu-Arg-4-NA	0.15 mmol/l
mix and read absorbance each min or monitor the reaction on a recorder over a period of 5 min.					

If the amount of aprotinin in the sample exceeds 1.0 KIU per assay, plasma samples must be diluted with aprotinin-free, acid-treated normal plasma (4).

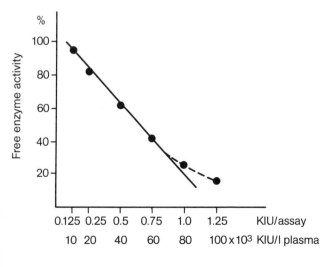

Fig. 1. Typical calibration curve for the assay of aprotinin in plasma.

Calibration curve: calculate $(\Delta A/\Delta t)_{sample}$ in percent of $(\Delta A/\Delta t)_{blank}$. Plot the percent of free enzyme activities of the standards against their activity concentrations of aprotinin (Fig. 1).

Calculation: the inhibitory activity of aprotinin in the sample (patients' plasma) is taken from the calibration curve.

Validation of Method

Precision, accuracy, detection limit and sensitivity: the within-run imprecision for aprotinin, 4×10^4 KIU/l was $1.8 - 10.8\%$ in 8 series; the between-series imprecision ($n = 8$) was 4.4%. The recovery for 2×10^4 to 8×10^4 KIU/l (working range) was $90 - 125\%$. Data on accuracy are not available since standard reference material is not established yet.

The detection limit is 0.125 KIU per assay, i.e. 10^4 KIU/l of untreated plasma. Sensitivity is found to be $\Delta A/\Delta t = 0.002/min$ at 405 nm (*Eppendorf* photometer). the mean 1.23-fold increase caused by pooled plasma. With an individual increase in

Sources of error: since kallikrein activity is increased 1.23-fold ($n = 90$; RSD = 7.2\%) by addition of acid-treated normal plasma, all dilutions of aprotinin-containing plasma samples must be made with aprotinin-free, acid-treated plasma. Plasma samples without acid treatment cannot be used because of a rather high and irregular stimulation of the kallikrein activity in the assay system. The same holds true for serum. Dilution of aprotinin-containing plasma samples with buffer or isotonic saline solution instead of plasma yields too low inhibitor concentrations.

Occassionally, aprotinin-free plasma samples from a single individual may stimulate kallikrein activity to a lesser (up to 1.1-fold) or greater (up to 1.5-fold) extent than kallikrein activity lower than 1.15-fold or higher than 1.3-fold, the calibration curve and dilutions of aprotinin-containing plasma samples should be performed with the individual patient's plasma (pre-aprotinin infusion sample) instead of pooled plasma.

Addition of NaN_3 to aprotinin-containing plasma as a preservative before acid treatment partly destroys the inhibitor activity during the subsequent procedure.

Specificity: under the given conditions the substrate S-2266 is split specifically by the tissue kallikrein used in the assay.

Therapeutic ranges: continuous i.v. infusion of 250000 KIU/h in polytraumatized patients resulted in a mean plasma concentration of 4.5×10^4 KIU/l [7].

References

[1] *H. Fritz, G. Wunderer,* Biochemistry and Applications of Aprotinin, the Kallikrein Inhibitor from Bovine Organs, Drug Res. 33 (I), *4*, 479 – 494 (1983).

[2] *H. Fritz, K.-H. Oppitz, D. Meckl, B. Kemkes, H. Haendle, H. Schult, E. Werle,* Verteilung und Ausscheidung von natürlich vorkommenden und chemisch modifizierten Proteinaseinhibitoren nach intravenöser Injektion bei Ratte, Hund (und Mensch), Hoppe-Seyler's Z. physiol. Chem. *350*, 1541 – 1550 (1969).

[3] *H. Kaller, K. Patzschke, L. A. Wegner, F. A. Horster,* Pharmacokinetic Observations Following Intravenous Administration of Radioactive Labelled Aprotinin in Volunteers, Eur. J. Drug Metab. Pharmacokinet. *2*, 79 – 85 (1978).

[4] *H. Fritz, I. Trautschold, E. Werle,* Protease-Inhibitors, in: *H. U. Bergmeyer* (ed.), Methods of Enzymatic Analysis, Vol. *II*, Verlag Chemie, Weinheim, and Academic Press, New York 1974, pp. 1064 – 1080.

[5] *T. Shikimi,* Sandwich Enzyme Immunoassay of Aprotinin, J. Pharmcobio-Dyn. *5*, 708 – 715 (1982).

[6] *A. Eddeland, K. Ohlsson,* A Radioimmunoassay for Measurement of Human Pancreatic Secretory Trypsin Inhibitor in Different Body Fluids, Hoppe-Seyler's Z. physiol. Chem. *359*, 671 – 675 (1978).

[7] *M. Jochum, V. Jonáková, H. Dittmer, H. Fritz,* An Enzymatic Assay Convenient for the Control of Aprotinin Levels during Proteinase Inhibitor Therapy, Fresenius Z. anal. Chem. *317*, 718 – 719 (1984).

[8] *W. Müller-Esterl, A. Oettl, E. Truscheit, H. Fritz,* Monitoring of Aprotinin Plasma Levels by an Enzyme-Linked Immunosorbent Assay (ELISA), Fresenius Z. Anal. Chem. *317*, 717 – 718 (1984).

2 Drugs of Abuse and of Toxicological Relevance

2.1   Introduction

Christopher P. Price, Peter M. Hammond, R. Stewart Campbell
and Tony Atkinson

In its broadest sense toxicology refers to the investigation of the adverse effects of environmental and therapeutic agents on living organisms. Toxicological investigations associated with therapeutic agents can be considered in two distinct categories: 1. assessment of the potential toxicity of a chemical prior to widespread availability, and 2. assessment of the effects consequent upon excessive usage once it has become available. The latter category is included within the sphere of clinical toxicology. The clinical toxicologist is most commonly concerned with the acute symptoms of a potentially intoxicated patient. This may require immediate recognition and quantitation of the toxic agent. In many cases, however, the patient's immediate medical condition will relegate the nature of the toxic agent to a matter of secondary importance. When the acute emergency has passed, the clinical toxicologist is faced with other issues such as the origin of the toxic agent and the reasons for exposure. In these matters the clinical toxicologist will be a part of a larger investigative team.

Clinical Toxicology and Therapeutic Agents

Clinical toxicology is concerned with the measurement of drugs in body fluids, either as an aid to the detection of non-compliance and individualization of dosage or as a method of detection of the use of illegal drugs and finally as a help in the management of the intoxicated patient [1].

Therapeutic drug monitoring is particularly concerned with non-compliance and dosage individualization by measurement of serum drug levels; principles are dealt with in chapter 1.1.1, p. 2. The following chapters deal with the detection of the abuse of prescribed drugs and recognition of intoxication with therapeutic agents.

Drug Assays in Occupational Health Schemes

The increase of drug measurements in occupational health schemes in the last decade is a consequence of the rising abuse of certain drugs. This practice has risen to prominence in situations where impairment of human faculties due to the abuse of drugs may endanger the working environment or the population in general. The drugs that have been monitored include hypnotic agents, sedatives, tranquilizers, hallucinogens and stimulants. Analytical techniques in this field are directed towards the rapid recognition of the presence of a drug or its metabolites rather than accurate quantitation.

Drug Assays in Emergency Hospital Admissions

Drug measurements in emergency hospital admissions are made 1. to exclude the possible involvement of drugs in a "coma of unknown origin", and 2. to determine whether a patient suspected of taking a drug overdose requires active treatment. The latter is important in cases of paracetamol and salicylate intoxication where the treatment regime is dependent on the plasma drug concentration.

Treatment of Drug Overdose

In cases of drug overdose a common approach is to provide supportive therapy with particular attention given to the management of secondary complications. However, there are certain situations where more active treatment is required. Forced diuresis may be instituted if increased elimination of the drug can be achieved as in the case of forced alkaline diuresis for the treatment of severe salicylate intoxication [2, 3]. Dialysis or haemoperfusion, although much less common, are alternative means of effecting elimination of the drug from the body [4].

In certain cases there are antidotes available that will either 1. enhance the biotransformation of the drug, or 2. reverse the main pharmacological effects of the drug. Thus, N-acetylcysteine will reduce the accumulation of the toxic intermediary metabolite of paracetamol by acting as an alternative nucleophile when the hepatic glutathione levels are depleted. However, there are cases of anaphylactic shock reported following administration of this antidote, and thus initial recognition and quantitation of the paracetamol in the blood is important prior to treatment. Similarly, naloxone is an opiate antagonist and has become an important antidote in the treatment of opiate toxicity.

Factors Affecting Drug Handling

There are many factors that will affect the level of a drug and its metabolites in the blood or urine; this may limit the usefulness of a single blood or urine measurement. Clearly, the level of a drug in the circulation will be affected by the amount of drug administered; it will also be affected by the route by which the drug is administered and the time at which the blood sample is collected, as in the case of marijuana. Smoking of cannabis produces peak plasma concentrations of the active constituent, Δ^9-tetrahydrocannabinol (Δ^9-THC), in 10 – 30 minutes, whereas after oral administration peak levels are not reached until 2 – 3 hours after ingestion [5].

Despite the availability of detailed "normal" pharmacokinetic data there are several metabolic and pathological effects which may conspire to alter the handling of a drug by any one patient. Thus, the presence of liver dysfunction and reduced renal function will have a significant bearing on the elimination of a drug [6, 7].

In contrast, the hepatic handling of a range of drugs may be enhanced in patients who have also been exposed to substances such as phenobarbital and ethanol which stimulate the hepatic microsomal enzyme system. This may contribute to an alteration in any relationship that may exist between dose, blood concentration and pharmacological effect. Additionally, it may increase the risk of tissue damage from drugs whose toxicity is mediated through metabolites. Thus, the hepatic damage due to paracetamol poisoning will be enhanced in the patient who takes the drug with alcohol.

Analytical Methods

Ideally, the techniques available for the recognition of drugs or metabolites should be rapid. It might be suggested that they should also be capable of use in the doctor's office or in the hospital emergency room.

Chromatographic techniques have played an important part in the development of clinical toxicology. Thin-layer techniques have been devised that can give a rapid screen for many drugs encountered in the patient suspected of taking an overdose. Quantitation of drugs and metabolites has been achieved by the use of gas-liquid chromatography and more recently of high-performance liquid chromatography. The disadvantage of these methods is the need for investment in capital equipment and expertise; this has been overcome in the United Kingdom with the establishment of Regional Poisons Units where the necessary investment can be made.

There are certain enzymatic end-point methods that offer a reasonable degree of specificity for particular drugs; the best examples of these techniques are those for salicylate (cf. p. 374) and paracetamol (cf. p. 364).

A recent interesting development is the recognition that micro-organisms in the environment can degrade drugs. This observation has led to the development of specific assays for paracetamol and salicylate using enzymes obtained from micro-organisms as reagents, with simple colorimetric reactions to detect the products.

Immunoassay techniques have also been used for the quantitation of certain drugs (for example, cannabis, digoxin, amphetamine and paraquat). However, the most significant advance in drug detection has been achieved with the development of homogeneous enzyme-immunoassays. The technology has been refined to the extent that it is reasonable to consider the use of such analytical systems in the doctor's office (with appropriate training). However, confirmation of positive results by alternative, preferably non-immunological procedures is necessary. The majority of the techniques described here are semi-quantitative and are designed to detect drugs or metabolites in urine.

Methods of Enzymatic Analysis

Errata

Volume VII

page 78, formula:

read

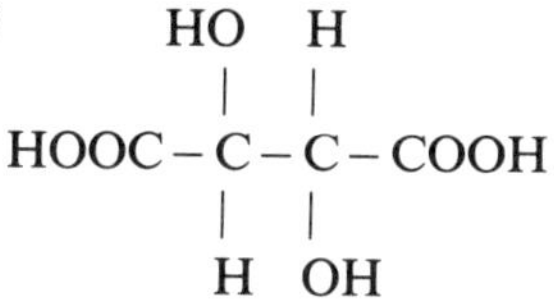

$$\underset{\underset{H\ \ \ OH}{|\ \ \ |}}{\overset{\overset{HO\ \ \ H}{|\ \ \ |}}{HOOC - C - C - COOH}}$$

page 85, formula:

read

$$\underset{\underset{HO\ \ \ H}{|\ \ \ |}}{\overset{\overset{H\ \ \ OH}{|\ \ \ |}}{HOOC - C - C - COOH}}$$

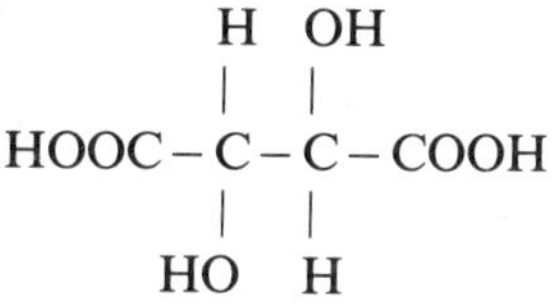

page 82, line 4 from top:

read L-TDH, 40 kU/l

page 83, right column:

read L-TDH, 345 U/l

page 205, line 8 from top:

read mmol/l

page 291, line 15 from top:

read Flavine-adenine Dinucleotide

Volume VIII

page 500, lines 5 and 6 from top:

read $\Delta A_{365} = 0.006$ which corresponds to 12 µmol/l or 1.4 mg/l. The sensitivity $\Delta A_{365} = 0.002$ corresponds to 4 µmol/l or 0.5 mg/l

Volume IX

Title page:

read Editorial Consultant

page 41, line 10:

read ... WHO do so on ...

page 57, equation (a), left-hand legend:

read Solid phase-bound rabbit anti-human albumin IgG

page 57, equation (a), right-hand legend:

read AP-rabbit anti-human albumin IgG conjugate

page 59, line 8:

read ... purified rabbit anti-human albumin IgG

page 65, equation, left-hand legend:

read Solid phase-bound rabbit anti-human RBP IgG

page 65, equation, right-hand legend:

read AP-rabbit anti-human RBP IgG conjugate

page 74, Appendix, line 1:

read Preparation of AP-rabbit anti-human RBP IgG conjugate:

page 242, line 3:

read ... 1.6 g NaH_2PO_4 ...

page 250, equations (a) and (b):

replace by equations (a) and (b) from page 257

page 257, equations (a) and (b):

replace by equations (a) and (b) from page 250

page 352, Fig. 2, abscissa:

read pg/0.1 ml

page 361, line 3:

read ... (cf. p. 356).

page 372, footnote:

read POD, donor: hydrogen-peroxide oxidoreductase

page 375, equation, bottom right-hand legend:

delete Resin

page 388, Reagents and Solutions, line 5:

read ... 10 mmol/l; KCl,

page 394, Table 4, line "mid-cycle peak", right-hand column:

read 61.0 (46 – 76)

pages 407 to 417, running heads:

read TSH

page 423, equation (c), right of arrow:

read penicilloic acid-I_2

page 425, line 13:

read dissolve 3.56 g $Na_2HPO_4 \cdot 2\ H_2O$ in 100 ml water.

page 471, equation:

read
$$\Delta F/\Delta F_0 = \frac{\Delta F - \Delta F_{blank}}{\Delta F_0 - \Delta F_{blank}} \times 100 \quad \%$$

page 471, footnote:

read
$$\frac{\Delta F/\Delta t}{\Delta F_0/\Delta t} = \Delta F/\Delta F_0$$

page 488, Fig. 1, abscissa:

read Antiserum dilution

page 518, line 8:

read Sources of error: samples from haemolyzed blood cannot ...

page 518, line 9:

read ... and lipaemic ...

page 525, lines 29/30:

read ... to give an increase in absorbance per unit time of $\Delta A/\Delta t = 0.075$ ...

page 531, line 6 from bottom:

read twice ... $\Delta A/\Delta t$ ...

page 566, right-hand column, line 18 from bottom:
read . . . 447

page 568, left-hand column, line 10 from bottom:
read . . . 292

page 568, left-hand column, line 9 from bottom:
read . . . 291

page 568, right-hand column, line 11:
read . . . 256

page 570, right-hand column, line 13:
add 538.

VCH
Verlagsgesellschaft

References

[1] *J. Koch-Weser,* Serum Drug Concentrations in Clinical Perspective, in: *A. Richens, V. Marks* (eds.), Therapeutic Drug Monitoring, Churchill Livingstone, Edinburgh 1981, pp. 1 – 22.
[2] *T. J. Meredith, J. A. Vale,* Salicylate Poisoning, in: *J. A. Vale, T. J. Meredith* (eds.), Poisoning: Diagnosis and Treatment, Update Books, London 1981, pp. 97 – 103.
[3] *S. S. Brown,* Clinical Chemistry of Salicylate Poisoning, Ann. Clin. Biochem. *6,* 13 – 17 (1969).
[4] *J. A. Vale, T. J. Meredith,* Forced Diuresis, Dialysis and Haemoperfusion, in: *J. A. Vale, T. J. Meredith* (eds.), Poisoning: Diagnosis and Treatment, Update Books, London 1981, pp. 59 – 68.
[5] *J. H. Jaffe,* Drug Addiction and Drug Abuse: Cannabinoids (Marihuana), in: *A. Goodman Gilman, L. S. Goodman, A. Gilman* (eds.), The Pharmacological Basis of Therapeutics, 6th edit., MacMillan Publishing, New York 1980, pp. 560 – 563.
[6] *C. F. George, P. J. Watt,* The Liver and Response to Drugs, in: *R. Wright, K. G. M. M. Alberti, S. Karran, G. H. Millward-Sadler* (eds.), Liver and Biliary Disease: Pathophysiology, Diagnosis Management, W. B. Saunders Co. Ltd., London 1979, pp. 344 – 377.
[7] *A. F. Lant,* Renal Excretion and Toxicity of Drugs, in: *D. Black, N. F. Jones* (eds.), Renal Disease, 4th edit., Blackwell Scientific Publications, Oxford 1979, pp. 617 – 639.

2.2   Amphetamine (and Methamphetamine)

(±)-α-Methylbenzeneethanamine

and

(S)-N-α-dimethylbenzeneethanamine

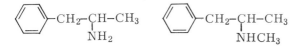

Amphetamine   Methamphetamine

Christopher P. Price and R. Stewart Campbell

General

The amphetamines are a small class of compounds in the group of sympathomimetic amines whose effects are more long-lasting than those of catecholamines; the most common are amphetamine and methamphetamine. They exhibit marked central stimulatory effects in doses that are not associated with peripheral side-effects [1].

Amphetamines can be given orally and are readily absorbed. Metabolism occurs in the liver, including hydroxylation, *N*-demethylation, deamination and conjugation; however a significant proportion of the drug can be excreted unchanged in the urine. Excretion of unchanged drug in the urine is increased in patients when the urine is acid [2].

Amphetamines have been used clinically in the treatment of narcolepsy and in hyperkinetic states in children. In the past amphetamines have been used as appetite depressants.

Amphetamines have been abused as a result of their known stimulatory properties. Symptoms of amphetamine poisoning can include restlessness, irritability, nausea and other abdominal symptoms, hyperpyrexia, cardiac arrhythmias, hallucination and convulsions. Treatment of amphetamine poisoning is generally supportive, following gastric lavage. The use of forced acid diuresis may be helpful in the severely poisoned patient [3].

Application of method: in clinical chemistry and toxicology.

Substance properties relevant in analysis: amphetamine, molecular weight 135.2, pK_a 9.8; amphetamine sulphate, molecular weight 368.5, soluble in water to 110 g/l, in ethanol to 1.95 g/l. Methamphetamine, molecular weight 149.2, pK_a 10.1, soluble in water to 20 g/l.

Methods of determination: the drug can be detected by thin-layer chromatography [4]. Semi-quantitative determinations in urine can be made by homogeneous enzyme-immunoassay [5]. Quantitative determinations can be made using gas-liquid [6] or high-performance liquid chromatographic techniques [7].

International reference method and standards: no reference methods or standards are known to date.

Assay

Method Design

The method is based on the enzyme-multiplied immunoassay technique (EMIT®) first described by *Rubenstein et al.* [8]. The assay is designed for the semi-quantitative assay of amphetamines in urine.

The first homogeneous enzyme-immunoassay for amphetamines described used a lysozyme label; the assay described here employs a glucose-6-phosphate dehydrogenase label.

Principle

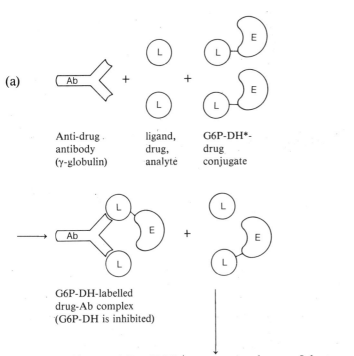

| (a) | Anti-drug
antibody
(γ-globulin) | ligand,
drug,
analyte | G6P-DH*-
drug
conjugate |

G6P-DH-labelled
drug-Ab complex
(G6P-DH is inhibited)

(b)   D-Glucose-6-P + NAD$^+$ $\longrightarrow$ D-glucono-δ-lactone-6-P + NADH + H$^+$.

The activity of the marker enzyme, glucose-6-phosphate dehydrogenase, G6P-DH, is reduced if the enzyme-amphetamine conjugate (E-L) is coupled to the antibody (Ab). Amphetamine (L) in a sample competes with G6P-DH-labelled amphetamine (E-L) for a limited concentration of antibody (Ab). The activity of G6P-DH measured as increase in absorbance per unit time, $\Delta A/\Delta t$, is related to the amphetamine concentration of the sample (cf. Vol. I, chapter 2.7, p. 244). The method is intended for use as a semi-quantitative assay.

Selection of assay conditions and adaptation to the individual characteristics of the reagents: the assay is designed for rapid analysis. Antibodies have been produced by immunizing sheep with amphetamine chemically coupled to a macromolecular carrier. Enzyme-drug conjugate is obtained by coupling an amphetamine derivative to G6P-DH from *Leuconostoc mesenteroides*. Interference from endogenous G6P-DH is avoided by use of the coenzyme NAD which is converted only by the bacterial enzyme.

The method involves pre-dilution of the sample in the reaction vessel.

---

* D-Glucose-6-phosphate: NADP$^+$ 1-oxidoreductase, EC 1.1.1.49.

The optimal quantities and concentrations of sample and reagents are dependent on the characteristics of the antibodies present in the antiserum. The desired assay range is from 0.3 to 2.0 mg/l. The sample volume is 50 μl.

To obtain the desired assay range, the optimum ratio of antibody to enzyme-drug conjugate in the presence of different concentrations of amphetamine must be determined. The first step is to determine the approximate range of antibody concentrations needed to modulate the activity of a given concentration of enzyme-drug conjugate. The next step is to determine the response of the system in the presence of analyte standards of different concentrations in the desired range, for several ratios of antibody to enzyme-drug conjugate. The degree of enzyme inhibition achieved on binding of enzyme-drug conjugate to antibody and the inhibition achieved at different analyte concentrations is judged on a fairly empirical basis, taking into account the requirements of the assay [5]. Finally, the ratios of antibody to enzyme-drug conjugate are compared at various conjugate concentrations to select the optimum quantity of both reagents.

Equipment

Spectrophotometer or spectral-line photometer capable of measurement at 334 or 339 nm. It should be equipped with a thermostatted flow-cell which consistently maintains $30.0 \pm 0.1\,°C$ throughout the working day. Work rack and disposable 2 ml beakers with conical bottoms. Pipetter-diluter capable of delivering 50 ± 1 μl sample and 250 ± 5 μl buffer with a relative standard deviation of less than 0.25%. Delivery must be of sufficient force to ensure adequate mixing of the components.

Reagents and Solutions

Purity of reagents*: G6P-DH from *Leuconostoc mesenteroides* should be of the best available purity, e.g. 600 U/mg at 25 °C, G-6-P as substrate, NAD as coenzyme. Chemicals are of analytical grade.

Preparation of solutions* (for 100 determinations): all solutions in re-purified water (cf. Vol. II, Chapter 2.1.3.2).

1. Tris buffer (Tris, 55 mmol/l; NaN$_3$, 0.5 g/l; Triton X-100, 0.1 ml/l; pH 8.0):

   dissolve 1.996 g Tris base in 150 ml water; add 0.03 ml surfactant Triton X-100 and 0.15 g sodium azide; adjust to pH 8.0 with HCl, 1.0 mol/l; make up with water to 300 ml.

   Alternatively, dilute buffer concentrate provided with each kit to 200 ml with water.

---

* Reagents, calibrators and buffer are commercially available from *Syva Co.,* Palo Alto, U.S.A. The solutions contain preservatives.

2. Antibody/substrate solution (γ-globulin*; G-6-P, 66 mol/l; NAD, 40 mmol/l; Tris, 55 mmol/l; NaN$_3$, 0.5 g/l; pH 5.2):

   reconstitute available lyophilized preparation with 6.0 ml water.

3. Conjugate solution (G6P-DH**; Tris, 55 mmol/l; NaN$_3$, 0.5 g/l; pH 8.0):

   reconstitute available lyophilized preparation with 6.0 ml water. This reagent is standardized to match the antibody/substrate reagent (cf. p. 272).

4. Amphetamine standard solutions (0.3 and 2.0 mg/l):

   dissolve 22.9 mg D,L-amphetamine sulphate in 100 ml water (84 mg amphetamine per litre). Use this stock solution to prepare two working standards of 0.3 and 2.0 mg amphetamine per litre by appropriate dilution with pooled drug-free human urine containing 0.5 g NaN$_3$ per litre. The urine diluent is used as a zero calibrator.

   Alternatively, reconstitute each of the available lyophilized calibrators with 3.0 ml water; re-stopper the bottles and swirl gently to assist reconstitution.

Stability of solutions: store all reagents, stoppered in a refrigerator at 2 °C to 8 °C. All reagents should be allowed to equilibrate to room temperature before use. The buffer (1), antibody and enzyme-drug conjugate reagents (2) and (3) and calibrators (4) are all stable for 12 weeks after reconstitution.

Procedure

Collection and treatment of specimen: urine may be collected in plastic or glass containers. The effect of urine preservatives has not been established and therefore their use is not recommended. Turbid samples should be centrifuged before analysis.

Stability of the substance in the sample: samples may be stored at 4 °C for up to 3 days; storage for longer periods may result in false negative results in the case of samples with levels near to the low calibrator value [9].
   Samples should be within the pH range of 5 – 8. Samples with pH outside this range should be adjusted to fall within this range by addition of hydrochloric acid, 1.0 mol/l, or sodium hydroxide, 1.0 mol/l, prior to analysis.

Assay conditions: measurement of samples and standards in duplicate; incubation time 30 s (in a reaction cuvette); room temperature; 0.6 ml; wavelength Hg 334 or 339 nm; light path 10 mm; final volume 0.9 ml; measurements against water; 30 s at 30 °C.

---

  * Standardized preparation from immunized sheep, concentration is titre-dependent.
 ** Enzyme activity concentration is dependent on final choice of conjugate.

Run the standards with each series of analyses. The zero calibrator assay serves as a reagent blank; correct all measurements for this value.

Measurement

Pipette* successively into the cuvette:			concentration in assay mixture	
sample or standard (4)		0.05 ml	drug	up to 111 µg/l
buffer	(1)	0.25 ml	Tris	52 mmol/l
			NaN$_3$	0.47 g/l
antibody/substrate solution	(2)	0.05 ml	γ-globulin	0.10 to 0.25 mg/l
			G-6-P	3.7 mol/l
			NAD	2.2 mmol/l
buffer	(1)	0.25 ml		
incubate at room temperature for 30 s,				
conjugate solution	(3)	0.05 ml	G6P-DH	variable
buffer	(1)	0.25 ml		
place cuvette in spectrophotometer immediately or aspirate contents into flow-cell, read absorbance after 15 s and 45 s.				

* Adequate mixing should be achieved by the addition of each reagent and no further agitation is necessary.

If $\Delta A/\Delta t$ of the sample is greater than that of the highest calibrator, dilute the sample further with Tris buffer (1) and re-assay.

Calibration curve: plot the corrected values of increase in absorbance per 30 s, $\Delta A/\Delta t$ of the calibrators *versus* the corresponding amphetamine concentrations (mg/l) using semi-logarithmic graph paper. A typical curve is shown in Fig. 1.

Calculation: read the amphetamine mass concentration of the sample from the calibration curve. To convert from mass concentration to substance concentration, c (µmol/l), multiply by 7.396 (molecular weight of amphetamine is 135.2).

It is important to confirm a positive result by an alternative, preferably non-immunological, procedure.

Interpretation: amphetamine can be detected in urine samples up to 24 hours after administration of a 10 mg therapeutic dose. Amphetamine appears in the urine shortly after oral administration; if a large dose has been taken amphetamine will appear in the urine for as long as seven days [10]. The maximum urine concentration following a single 10 or 15 mg dose is between 3 and 5 mg/l. The urine of subjects abusing amphetamine may demonstrate concentrations ten times these levels.

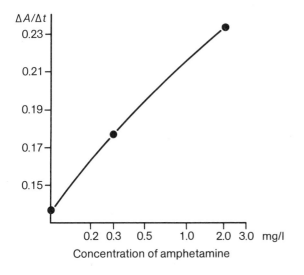

Fig. 1. Typical calibration curve for the assay of amphetamine in urine.

The blood concentration of amphetamine following a therapeutic dose will be in the range of 0.04 to 0.06 mg/l; blood from persons suspected of abusing amphetamine may demonstrate concentrations of twice this level depending on the amount of drug ingested [10, 11].

The assay is designed as a semi-quantitative procedure. The assay is designed in the knowledge that several factors will influence the precision of an assay. The formulation of the assay and success in its use centre on the recognition of a minimum $\Delta A/\Delta t$, indicating the presence of the analyte in the sample. Samples that give a signal above this cut-off will be considered as positive. If this cut-off is set too low then the probability of identifying a positive sample is increased but at the risk of producing false positive results. The low calibrator is generally set at this cut-off point. This approach has been demonstrated for several analytes [5, 12]. The original EMIT® procedure employed a lysozyme label; it is considered that the assays using a G6P-DH label show enhanced sensitivity [5] and this is borne out in the recommended concentration range in which the assay can be used. In the original assay the cut-off level for amphetamine was 2.0 mg/l; at this concentration 95 per cent of samples containing the drug would be detected as positive. The strategy recommended for interpretation of results has altered slightly and it is now considered that a sample giving a response equal to or greater than the low calibrator (0.3 mg/l) will be considered as positive [9]. The reagent manufacturer has reported [9] that, for each lot of reagent, between 25 – 50 amphetamine-free urine samples and 25 – 50 urine samples containing amphetamine (0.5 mg/l) are tested and 99% of samples are correctly identified.

Validation of Method

Precision, accuracy, detection limit and sensitivity: the reagent manufacturer has reported [9] that in a clinical trial involving 173 urine samples (positive = 84, negative = 89) there was agreement in 99% of cases when compared with results by TLC, GLC and/or GC/MS procedures. The reagents for this assay have only recently become available and there are no independent published performance data.

The relative standard deviation of the original "lysozyme label" assay using a partially automated system was found to be 32.9% (within batch) and 44.7% (between batch) at 2.94 mg/l [13]. The detection limit, defined as the signal plus three times the standard deviation of the signal derived from 20 determinations on a drug-free urine, gave a value of 0.4 mg/l [13].

A comparison of the earlier method with a TLC procedure on a total of over 700 urine samples gave 12.5% false positives with less than 1% false negatives [14]. The authors concluded that the high false positive rate may be due to the poorer sensitivity of the TLC procedure. *Oellerich et al.* [13] in a study of urine samples from 300 patients found only one false positive with 12 false negatives when compared with TLC; the latter was thought to be due to interference in the TLC method.

Broughton & Ross [15] using a 2.5 mg/l cut-off in the original EMIT® "lysozyme" method found less than 4% of results to be discrepant when compared with a TLC procedure.

Sources of error: false negative results may be obtained if the sample is not stored correctly [9]. The urine pH should be within the range 5 – 8.

It has been recognized with the earlier EMIT® employing the lysozyme label that urines with high ionic strengths can produce falsely low absorbance changes; this can be achieved by adding sodium chloride to urine. It is expected that this effect will be seen with the dehydrogenase-linked assays [16] although this has not been confirmed.

Specificity: the assay is designed to detect amphetamine and methamphetamine. The assay has been shown to detect certain other phenylethylamines. The following data are provided by the reagent manufacturer [9] as giving readings equivalent to > 0.3 mg/l DL-amphetamine: phentermine (> 0.5 mg/l), mephentermine (> 0.5 mg/l), DL-ephedrine (> 1.0 mg/l), nylidrin (> 2.0 mg/l), phenylpropanolamine (> 1.0 mg/l), isoxsuprine (> 6.0 mg/l) and phenmetrazine (> 1.0 mg/l).

The reagent manufacturer has also reported [9] that the following compounds at concentrations of 200 mg/l or greater gave a negative response: benzoylecgonine, dextromethorphan, diethylpropion, methadone, morphine, oxazepam, phencyclidine, propoxyphene, secobarbital.

Oellerich et al. [13] studying the earlier "lysozyme label" assay showed that the system detected methamphetamine with seven tenths the sensitivity of amphetamine. In addition they found that the following compounds gave a signal equivalent to that of 1 mg amphetamine per litre: fenethylline hydrochloride (0.9 mg/l), fenfluramine hydrochloride (7.0 mg/l), propylhexedrine hydrochloride (10.5 mg/l), ephedrine hydrochloride (13.5 mg/l), D-phenylethylamine (35.0 mg/l), L-phenylethylamine (68.0 mg/l). *Mule et al.* [14], with the lysozyme system, found that phenmetrazine at 0.95 mg/l and mephentermine (1.6 mg/l) gave signals equivalent to 1 mg amphetamine per litre. *Allen & Stiles* [17] confirmed the cross-reactivities found by other authors. In a study of over 60 different amphetamine related compounds *Budd* [18] found that addition of a single alkyl group to the nitrogen of amphetamine increased reactivity, whereas any changes to other positions of the amphetamine molecule will decrease reactivity.

References

[1] *N. Weiner,* Norepinephrine, Epinephrine and the Sympathomimetic Amines, in: *A. Goodman Gilman, L. S. Goodman, A. Gilman* (eds.), The Pharmacological Basis of Therapeutics, 6th edit., MacMillan Publishers, New York 1980, pp. 138 – 175.

[2] *A. H. Beckett, M. Rowland,* Urinary Excretion Kinetics of Amphetamine in Man, J. Pharm. Pharmacol. *17,* 628 – 639 (1965).

[3] *J. A. Vale, T. J. Meredith,* Poisoning due to Non-Catecholamine Sympathomimetic Drugs, in: *J. A. Vale, T. J. Meredith* (eds.), Poisoning Diagnosis and Treatment, Update Books, London 1981, pp. 125 – 127.

[4] *I. Sunshine,* Amphetamine and Methamphetamine Type A Procedure, in: *I. Sunshine* (ed.), Methodology for Analytical Toxicology, CRC Press, Palm Beach 1975, pp. 22 – 23.

[5] *D. S. Kabakoff, H. M. Greenwood,* Recent Advances in Homogeneous Enzyme Immunoassay, in: *K. G. M. M. Alberti, C. P. Price* (eds.), Recent Advances in Clinical Biochemistry 2, Churchill Livingstone, Edinburgh 1981, pp. 1 – 30.

[6] *W. Harrington,* Amphetamines Type C Procedure, in: *I. Sunshine* (ed.), Methodology for Analytical Toxicology, CRC Press, Palm Beach 1975, pp. 27 – 29.

[7] *C. C. Cuppett, G. Krisko,* HPLC Assay of Amphetamine and 7 Other Stimulant Drugs, Clin. Chem. *28,* 1671 (1982).

[8] *K. E. Rubenstein, R. S. Schneider, E. F. Ullman,* "Homogeneous" Enzyme Immunoassay. A New Immunochemical Technique, Biochem. Biophys. Res. Commun. *47,* 846 – 851 (1972).

[9] Product literature, EMIT®-d.a.u. Amphetamine Assays, *Syva Co., Palo Alto, U.S.A.,* 1984.

[10] *G. A. Alles, B. B. Wisegarver,* Amphetamine Excretion Studies in Man, Toxicol. Appl. Pharmacol. *3,* 678 – 683 (1961).

[11] *N. Jain, T. Sneath, R. Budd,* Amphetamine and Metamphetamine Type B Procedure, in: *I. Sunshine* (ed.), Methodology for Analytical Toxicology, CRC Press, Palm Beach 1978, pp. 24 – 26.

[12] *R. J. Bastiani, R. C. Phillips, R. S. Schneider, E. F. Ullman,* Homogeneous Immunochemical Drug Assays, Am. J. Med. Technol. *29,* 211 – 216 (1973).

[13] *M. Oellerich, W. R. Kulpmann, R. Haeckel,* Drug Screening by Enzyme Immunoassay (EMIT) and Thin-layer Chromatography (Drug Skreen), J. Clin. Chem. Clin. Biochem. *15,* 275 – 283 (1977).

[14] *S. J. Mule, M. L. Bastos, D. Jukofsky,* Evaluation of Immunoassay Methods for Detection in Urine, of Drugs Subject to Abuse, Clin. Chem. *20,* 243 – 248 (1974).

[15] *A. Broughton, D. L. Ross,* Drug Screening by Enzyme Immunoassay with a Centrifugal Analyser, Clin. Chem. *21,* 186 – 189 (1975).

[16] *W. Godolphin,* Enzyme Multiplied Immunoassay Technique (EMIT®), in: *I. Sunshine, P. I. Jatlow* (eds.), Methodology for Analytical Toxicology, Vol. *II,* CRC Press, Palm Beach 1982, pp. 189 – 203.

[17] *L. V. Allen, M. L. Stiles,* Specificity of the EMIT Drug Abuse Urine Assay Methods, Clin. Toxicol. *18,* 1043 – 1065 (1981).

[18] *R. D. Budd,* Amphetamine EMIT – Structure Versus Reactivity, Clin. Toxicol. *18,* 91 – 110 (1981).

2.3   Barbiturates

Derivatives of 2,4,6-trioxohexahydropyrimidine

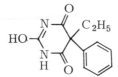

Phenobarbital

5-Ethyl-5-phenyl-2,4,6(1H, 3H, 5H)-pyrimidinetrione

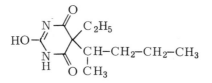

Pentobarbital

5-Ethyl-5-(1-methylbutyl)-2,4,6(1H,3H,5H)-pyrimidinetrione

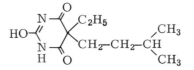

Amobarbital

5-Ethyl-5-(3-methylbutyl)-2,4,6(1H,3H,5H)-pyrimidinetrione

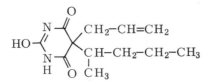

Secobarbital

5-(1-Methylbutyl)-5-(2-propenyl)-2,4,6(1H,3H,5H)-pyrimidinetrione

Christopher P. Price and R. Stewart Campbell

General

The barbiturates are a group of drugs derived from barbituric acid (pheno-, penta-, amo- and secobarbitone are examples). The major effects of the barbiturates are on the central nervous system ranging from mild sedation to general anaesthesia. Certain of the barbiturates have anticonvulsant properties. There has been a convention which divides the barbiturates into groups, depending on whether the pharmacological effects are short, medium or long lasting; this is now rarely used. Some of the barbiturates are also known to induce the synthesis of enzymes in the hepatic microsomal system; the property has been used to stimulate the synthesis of bilirubin metabolizing enzymes in the neonate [1].

Barbiturates are readily absorbed and there are several routes of administration. In the plasma a variable proportion of the drug may be bound to albumin. The more water-soluble derivatives may be excreted in the urine unchanged, whilst the remainder are metabolized in the liver. The major metabolic route is through oxidation to alcohols, ketones, phenols or carboxylic acids; these metabolites are then conjugated to glucuronic acid [2, 3].

The main symptoms of barbiturate overdose are associated with impairment of consciousness, depression of respiration, hypotension and hypothermia. If the drug is taken with alcohol the effects are greater. Some of the complications of barbiturate overdosage have resulted in renal failure and death in the past [4].

The treatment of barbiturate poisoning is mainly supportive, following gastric lavage. Forced alkaline diuresis may be helpful in the case of overdosage with certain barbiturates [4].

Application of methods: clinical chemistry, toxicology.

Substance properties relevant in analysis: amobarbital, molecular weight 226.3, soluble in water to 0.67 g/l and in ethanol to 200 g/l; barbituric acid, molecular weight 128.1, barely soluble in cold water; phenobarbital, molecular weight 232.2, soluble in water to 1 g/l and in ethanol to 100 g/l; pentobarbital, molecular weight 226.3, soluble in water to 0.5 g/l and at least 500 g/l in ethanol; secobarbital, molecular weight 238.3, soluble in water to at least 500 g/l and in ethanol to 300 g/l.

Methods of determination: the barbiturates can be detected in blood and urine using thin-layer chromatography [5] and UV-spectrophotometry after solvent extraction [5]. Semi-quantitative determinations can be made on blood and urine by homogeneous enzyme-immunoassay [6].

Quantitative measurements can be made by high-performance [7] or gas-liquid [7] chromatography and radioimmunoassay [8].

International reference methods and standards: no reference methods or standards are known so far.

2.3.1   Determination in Serum with Enzyme-multiplied Immunoassay Technique

Assay

Method Design

The method is based on the enzyme-multiplied immunoassay technique (EMIT®) first described by *Rubenstein et al.* [9]. The method is intended for use as a semi-quantitative assay.

Principle: as described in detail on pp. 7, 8, 271.

The activity of the marker enzyme, glucose-6-phosphate dehydrogenase, G6P-DH, is reduced if the enzyme-secobarbital conjugate (E-L) is coupled to the antibody (Ab). Barbiturate (L) in a sample competes with G6P-DH-labelled secobarbital (E-L) for a limited concentration of antibody (Ab). The activity of G6P-DH measured as increase in absorbance per unit time, $\Delta A/\Delta t$, is related to the barbiturate concentration.

Selection of assay conditions and adaptation to the individual characteristics of the reagents: as described in general on pp. 8, 9. Enzyme-drug conjugate is obtained by coupling secobarbital to G6P-DH from *Leuconostoc mesenteroides*. The method involves pre-dilution of the sample in the reaction vessel. The desired assay range is from 3.0 to 6.0 mg/l. The sample volume is 50 µl.

Equipment

The same equipment is used as described on p. 272.

Reagents and Solutions

Purity of reagents: as described on pp. 27, 272.

Preparation of solutions* (for 100 determinations): all solutions in re-purified water (cf. Vol. II, chapter 2.1.3.2).

---

* Reagents, calibrators and buffer are commercially available from *Syva Co.,* Palo Alto, U.S.A. The solutions contain perservatives.

1. Tris buffer (Tris, 55 mmol/l; NaN_3, 0.5 g/l; Triton X-100, 0.1 ml/l; pH 8.0):

   dissolve 1.996 g Tris base in 150 ml water; add 0.03 ml Triton X-100 and 0.15 g sodium azide; adjust to pH 8.0 with HCl, 1.0 mol/l; make up with water to 300 ml.

   Alternatively, dilute buffer concentrate provided with each kit to 200 ml with water.

2. Antibody/substrate solution (γ-globulin*; G-6-P, 66 mmol/l; NAD, 40 mmol/l; Tris, 55 mmol/l, pH 5.2):

   reconstitute available lyophilized preparation with 3.0 ml water.

3. Conjugate solution (G6P-DH**; Tris, 55 mmol/l; NaN_3, 0.5 g/l; pH 8.0):

   this reagent is standardized to match the antibody/substrate reagent (cf. p. 272). Reconstitute available lyophilized preparation with 3.0 ml water.

4. Secobarbital standard solutions (3 mg/l; 6 mg/l):

   dissolve 65.5 mg monosodium secobarbital (corresponding to 60.0 mg secobarbital) in 100 ml water. Use this stock solution to prepare two working standards of 3.0 mg/l and 6.0 mg/l by appropriate dilution with pooled normal human serum containing 0.5 g NaN_3/l. The serum diluent is used as a zero calibrator.

   Alternatively, reconstitute each of the available lyophilized calibrators with 3.0 ml water; re-stopper the bottles and swirl gently to assist reconstitution.

Stability of solutions: store all reagents, stoppered, in a refrigerator at 2°C to 8°C. All reagents should be allowed to equilibrate to room temperature before use. The buffer (1), antibody, and enzyme-drug conjugate solutions (2) and (3), and the standard solutions (4) are all stable for 12 weeks after reconstitution.

Procedure

Collection and treatment of specimen: collect blood and treat in the usual manner to obtain serum or plasma.

Stability of the substance in the sample: there is evidence to suggest that barbiturates may be stable in serum or plasma at room temperature for several weeks; *Wilensky* [10] showed that phenobarbitone was stable when stored at room temperature (in the dark) for six months.

---

 * Standardized preparation from immunized sheep, concentration is titre-dependent.
** Activity concentration of the enzyme depends on final choice of conjugate.

Assay conditions: measurement of samples and standards in duplicate.

A pre-dilution of samples and standard solutions is required: mix one part (0.05 ml) sample or standard solution(s) with 5 parts (0.25 ml) Tris buffer (1).

Incubation time 30 s (in a reaction cuvette); room temperature; 0.6 ml; wavelength Hg 334 or 339 nm; light path 10 mm; final volume 0.9 ml; measurements against water; 30 s at 30°C.

Run the standards with each series of analyses. The zero calibrator serves as a reagent blank; correct all measurements for this value.

Measurement

Pipette* successively into the cuvette:			concentration in assay mixture	
pre-diluted sample or standard solution (4)		0.05 ml	secobarbital	up to 55 µg/l
buffer	(1)	0.25 ml	Tris	52 mmol/l
			NaN$_3$	0.47 g/l
antibody/substrate solution	(2)	0.05 ml	γ-globulin	variable
			G-6-P	3.7 mmol/l
			NAD	2.2 mmol/l
buffer	(1)	0.25 ml		
incubate at room temperature for 30 s;				
conjugate solution	(3)	0.05 ml	G6P-DH	variable
buffer	(1)	0.25 ml		
place cuvette in spectrophotometer immediately or aspirate contents into flow-cell, read absorbance after 15 s and 45 s.				

* Adequate mixing should be achieved by the addition of each reagent and no further agitation is necessary.

If ΔA per 30 s of the sample is greater than that of the highest calibrator, dilute the sample further with Tris buffer (1) and re-assay.

Calibration curve: plot the corrected values of the increase in absorbance per 30 s, $\Delta A/\Delta t$, of the calibrators *versus* the corresponding secobarbital concentration (mg/l) using semi-logarithmic graph paper. A typical curve is shown in Fig. 1.

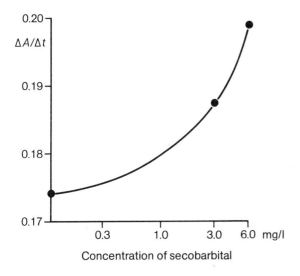

Fig. 1. Typical calibration curve for the assay of barbiturate in serum.

Calculation: read the mass concentration of barbiturate in the sample as secobarbital equivalents from the calibration curve. It is important that a positive result is confirmed by an alternative, non-immunochemical procedure. In view of the wide specificity, conversion to units of substance concentration is not recommended.

Interpretation: it must be remembered that this assay demonstrates a variable response at a given concentration depending on the barbiturate present. The half-lives of barbiturates in the circulation vary for each individual and between the various barbiturates [11]; in addition, half-lives are affected by various factors including repeated use. Toxicity is associated with variable levels of barbiturates in the blood, depending on the barbiturate as well as its rate of elimination [11].

The assay is designed as a semi-quantitative assay and it is not recommended for the determination of the amount of drug present. The optimum sensitivity of the assay is between secobarbital concentrations of 3.0 to 6.0 mg/l [12]. If $\Delta A/\Delta t$ for a sample is less than that for the low calibrator, the sample is considered to contain no barbiturate [12].

Validation of Method

Precision, accuracy, detection limit and sensitivity: the detection limit of this system has been reported for several barbiturates and varies from 3.0 mg/l for secobarbital to 62.0 mg/l for barbital [13]. The precision of the assay reported by a group involved in the development of the assay indicated relative standard deviations (within batch) of less than 10% at levels of 3.0 and 6.0 mg/l [14].

An alternative configuration of this method has been developed whereby a single vial of reagent is reconstituted, the sample added and the assay performed immediately. *Sutheimer et al.* [15] found a within-day relative standard deviation of 14% at a serum secobarbital concentration of 6.0 mg/l. The day-to-day precision gave relative standard deviations of 29 and 22% at concentrations of 4.7 and 5.7 mg/l, respectively. These authors considered that the lowest concentration of barbiturate that could be distinguished from the low calibrator (3.0 mg/l) was 4.0 mg/l. These authors compared results on 100 samples assayed by EMIT® and GLC procedures and found no discrepancies.

Specificity: it is stated by the manufacturer [12] that the following barbiturates (at the concentrations stated) give readings equivalent to 3.0 mg/l secobarbital: amobarbital ($\leqslant 15$ mg/l), butabarbital ($\leqslant 7.0$ mg/l), butalbital ($\leqslant 8$ mg/l), pentobarbital ($\leqslant 6$ mg/l), phenobarbital (12 – 25 mg/l), and talbutal ($\leqslant 5$ mg/l). In addition, the following compounds were tested against the same criterion: acetyl salicylate ($\geqslant 1000$ mg/l), amitriptyline ($\geqslant 100$ mg/l), diazepam ($\geqslant 100$ mg/l), glutethimide ($\geqslant 100$ mg/l), methaqualone ($\geqslant 100$ mg/l), morphine ($\geqslant 100$ mg/l), phenytoin ($\geqslant 100$ mg/l), primidone ($\geqslant 100$ mg/l), propoxyphene ($\geqslant 100$ mg/l). *Sutheimer et al.* [15], in addition, investigated the effects of paracetamol, carbamazepine, codeine, ethanol, ethosuximide, *N*-acetyl procainamide, quinine and theophylline and found no significant interference.

Sources of error: serum samples from severely haemolyzed, icteric or lipaemic blood may give erroneous results [12].

2.3.2   Determination in Urine with Enzyme-multiplied Immunoassay Technique

Assay

Method Design

Principle: as described on pp. 7, 8.

Selection of assay conditions and adaptation to the individual characteristics of the reagents: as described on pp. 8, 9. The desired assay range is from 0.3 to 1.0 mg/l. The sample volume is 50 µl. To obtain the desired assay range, follow p. 272.

Equipment

The same equipment is used as described on p. 272 for the determination in serum.

Reagents and Solutions

Purity of reagents: cf. pp. 27, 272.

Preparation of solutions* (for 100 determinations): all solutions in re-purified water (cf. Vol. II, chapter 2.1.3.2).

1. Tris buffer (Tris, 55 mmol/l; NaN_3, 0.5 g/l; Triton X-100, 0.1 ml/l; pH 8.0):

   dissolve 1.996 g Tris base in 150 ml water; add 0.03 ml Triton X-100 and 0.15 g sodium azide; adjust to pH 8.0 with HCl, 1.0 mol/l; make up with water to 300 ml.

   Alternatively, dilute buffer concentrate provided with each kit to 200 ml with water.

2. Antibody/substrate solution (γ-globulin**; G-6-P, 66 mmol/l; NAD, 40 mmol/l; NaN_3, 0.5 g/l; Tris, 55 mmol/l; pH 5.2):

   reconstitute available lyophilized preparation with 6.0 ml water.

3. Conjugate solution (G6P-DH***; NaN_3, 0.5 g/l; Tris, 55 mmol/l; pH 8.0):

   this reagent is standardized to match the antibody/substrate reagent (cf. p. 272). Reconstitute available lyophilized preparation with 6.0 ml water.

4. Secobarbital standard solutions (0.3 mg/l; 1.0 mg/l):

   dissolve 10.92 mg monosodium secobarbital (corresponding to 10.0 mg secobarbital) in 100 ml water. Use this stock solution to prepare two working standards of 0.3 mg/l and 1.0 mg/l by appropriate dilution with pooled normal human urine containing 0.5 g NaN_3/l. The urine diluent is used as a zero calibrator.

   Alternatively, reconstitute each of the available lyophilized calibrators with 3.0 ml water; re-stopper the bottles and swirl gently to assist reconstitution.

---

* Reagents, calibrators and buffer are commercially available from *Syva Co.*, Palo Alto, U.S.A. The solutions contain preservatives.
** Standardized preparation from immunized sheep, concentraton is titre dependent.
*** Activity concentration of the enzyme depends on final choice of the conjugate.

Stability of solutions: store all reagents, stoppered in a refrigerator at 2 °C to 8 °C. All reagents should be allowed to equilibrate to room temperature before use. The buffer (1), antibody and enzyme-drug conjugate solutions (2) and (3), and the standard solution (4) are all stable for 12 weeks after reconstitution.

Procedure

Collection and treatment of specimen

Urine: collect urine samples in plastic or glass containers. The effect of urine preservatives has not been established and therefore their use is not recommended. Centrifuge turbid samples before analysis. If pH is outside the range 5 – 8 bring it back into range with hydrochloric acid, 1.0 mol/l, or sodium hydroxide, 1.0 mol/l.

Tissue: homogenize 5 g tissue with 20 ml phosphate buffer, 0.1 mol/l, pH 7.0. Then extract the homogenate with 50 ml chloroform. Wash the chloroform layer with 10 ml sodium hydrogen carbonate, 0.476 mol/l; re-extract with 5 ml sodium hydroxide, 0.13 mol/l. Separate the aqueous phase and adjust pH to 7.0; re-extract with 50 ml chloroform. Separate the chloroform phase and evaporate to dryness; reconstitute the residue in 1 ml water.

Stability of the substance in the sample: samples may be stored for up to 3 days at 4 °C. Storage for greater periods may lead to false negative results with samples in which the barbiturate concentration is near that of the low calibrator.

Assay conditions: measurement of samples and standards in duplicate; incubation time 30 s (in a reaction cuvette); room temperature; 0.6 ml; wavelength Hg 334 or 339 nm; light path 10 mm; final volume 0.9 ml; measurements against water; 30 s at 30 °C.

Run the standards with each series of analyses. The zero calibrator serves as a reagent blank; correct all measurements for this value.

Calibration curve: as described on p. 274. A typical curve is shown in Fig. 2.

Measurement

Pipette* successively into the cuvette:			concentration in assay mixture	
sample or standard solution (4) buffer	(1)	0.05 ml 0.25 ml	secobarbital Tris NaN$_3$	17 to 56 μg/l 52 mmol/l 0.47 g/l
antibody/substrate solution	(2)	0.05 ml	γ-globulin G-6-P NAD	variable 3.7 mmol/l 2.2 mmol/l
buffer	(1)	0.25 ml		
incubate at room temperature for 30 s;				
conjugate solution buffer	(3) (1)	0.05 ml 0.25 ml	G6P-DH	variable
place cuvette in spectrophotometer immediately or aspirate contents into flow-cell, read absorbance after 15 s and 45 s.				

* Adequate mixing should be achieved by the addition of each reagent and no further agitation is necessary.

If ΔA per 30 s of the sample is greater than that of the highest calibrator, dilute the specimen with Tris buffer (1) and re-assay.

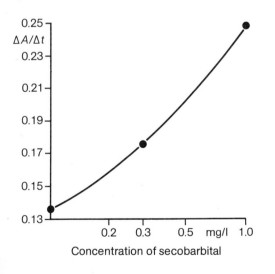

Concentration of secobarbital

Fig. 2. Typical calibration curve for the assay of barbiturate in urine.

Calculation: as described on p. 274.

Interpretation: it is reported that barbiturates are present in the urine within 3.5 hours after ingestion. Data suggest that a blood concentration of 3 mg/l is accompanied by a urine barbiturate level of greater than 1 mg/l [16].

The assay is designed as a semi-quantitative procedure. The assay is designed in the knowledge that several factors will influence the precision of an assay. The formulation of the assay and success in its use centre on the recognition of a minimum $\Delta A/\Delta t$ indicating the presence of the analyte in the sample. Samples that give a signal above this cut-off will be considered positive. If this cut-off is set too low, the probability of identifying a positive sample is increased but at the risk of producing false positive results. The low calibrator is generally set at this cut-off point. This approach has been demonstrated for several analytes [6, 17].

The original EMIT® procedure employed a lysozyme label; it is considered that the assays using a G6P-DH label show enhanced sensitivity and this is borne out in the recommended concentration range in which the cut-off level for barbiturate was 2.0 mg/l [6]; at this concentration 95% of samples would be detected as positive. The strategy recommended for the interpretation of results has altered slightly and it is now considered that a sample giving a response equal to or greater than that of the low calibrator (0.3 mg/l) will be considered as positive [18]. The reagent manufacturer has reported [18] that, for each lot of reagents, 25 to 50 barbiturate-free urines and 25 to 50 urines containing secobarbital (0.5 mg/l) are tested and 99% of samples are correctly identified.

Validation of Method

Precision, accuracy, detection limit and sensitivity: the reagent manufacturer has reported [18] that in a clinical trial involving 175 urine samples (positive 62, negative 113), there was agreement in 99% of cases when results were compared between EMIT® and TLC or GC/MS methods. The reagents for this assay have only recently become available and there are no independent published performance data.

In a study of the original EMIT® procedure using lysozyme *Oellerich et al.* [19] found relative standard deviations of 31.4% (within-series) and 39.1% (day-to-day) at 1.67 mg/l secobarbital in spiked urine samples. The detection limit, defined as $\Delta A/\Delta t$ equivalent to that of the zero calibrator plus three standard deviations of the variation in the zero calibrator, was found to be 0.5 mg/l. These authors examined 300 urine samples, comparing the original EMIT® procedure with a TLC procedure, and found that 10% of results did not agree. On further investigation using a GLC procedure the EMIT® assay agreed with the GLC in 64.5% of cases, whereas the TLC procedure only agreed in 35.5% of cases. The GLC procedure was known to have a lower detection limit and this contributed to the lack of agreement in determining whether a result was positive in some cases when compared with the EMIT® method.

Mule et al. [20], in a study of over 600 urine samples, found agreement with a TLC procedure in 94.7% of cases with a total of 34 false positives; the lack of complete agreement was thought to be due to the poorer sensitivity of the TLC procedure.

In a study of the lysozyme procedure *Walberg* [21] found agreement in 190 out of 203 samples studied by the EMIT® method comparing results with a serum spectrophotometric procedure. *Broughton & Ross* [22], investigating the same method, found agreement with a TLC procedure in 92% of cases.

Slightom [23] found the detection limit of the earlier EMIT® procedure to be 2.6 mg/l for pentobarbital and 1.9 mg/l for phenobarbital; improved sensitivity could be achieved by employing extraction and concentration of the sample.

Sources of error: samples that are not stored correctly may give false negative results.

It has been recognized with the earlier EMIT® assays employing lysozyme that urines with high ionic strengths can produce falsely low absorbance changes. This can be achieved by adding sodium chloride to urine. It is expected that this will be seen with the dehydrogenase-linked assays [24] although this has not been confirmed.

Specificity: the assay is designed to detect the major barbiturates and their metabolites in human urine. The reagent manufacturer claims [18] that the following barbiturates will give a response (at the concentrations stated) equal to that of 0.3 mg/l secobarbital: butabarbital (1.0 mg/l), pentobarbital (1 mg/l) amobarbital (2.0 mg/l), talbutal (2.0 mg/l), allylbarbital (3.0 mg/l), phenobarbital (3.0 mg/l).

In a study of the original EMIT® procedure, *Oellerich et al.* [19] found that most barbiturates cross-react with the method at concentrations in the range 1 to 10 mg/l. Glutethimide was the only non-barbiturate compound that exhibited any cross-reactivity, a concentration of 40 mg/l giving a result equivalent to that of 1.0 mg/l secobarbital.

In the present assay a glutethimide concentration of 25 mg/l was found to produce a response less than the 0.3 mg/l secobarbital calibrator.

Allen & Stiles [25], again studying the original EMIT® procedure, found several compounds that cross-reacted in the assay, including anticonvulsant and anti-inflammatory agents; the cross-reactivity was only demonstrated at concentrations of 1000 mg/l and was not considered to be significant.

References

[1] *L. Stern, N. N. Khanna, G. Levy, S. J. Yaffe,* Effect of Phenobarbital on Hyperbilirubinaemia and Glucuronide Formation in Newborns, Am. J. Dis. Child. *120*, 26 – 31 (1970).

[2] *R. I. Freudenthal, F. I. Carroll,* Metabolism of Certain Commonly Used Barbiturates, Drug Metab. Rev. *2*, 265 – 278 (1973).

[3] *B. K. Tang, T. Inaba, W. Kalow,* N-Hydroxylation of Pentobarbital in Man, Drug Metab. Dispos. *5*, 71 – 74 (1977).

[4] *T. J. Meredith, J. A. Vale,* Poisoning due to Hypnotics, Sedatives, Tranquillizers and Anticonvulsants, in: *J. A. Vale, T. J. Meredith* (eds.), Poisoning: Diagnosis and Treatment, Update Books, London 1981, pp. 84 – 89.

[5] *R. Bath, G. Kananen,* Barbiturates Type B Procedure, in: *I. Sunshine* (ed.), Methodology for Analytical Toxicology, CRC Press, Palm Beach 1975, pp. 36 – 40.

[6] *D. S. Kabakoff, H. M. Greenwood,* Recent Advances in Homogeneous Enzyme Immunoassay, in: *K. G. M. M. Alberti, C. P. Price* (eds.), Recent Advances in Clinical Biochemistry 2, Churchill Livingstone, Edinburgh 1981, pp. 1 – 30.

[7] *E. Szabo, I. Sunshine,* Analysis of Neutral and Acidic Drugs by Gas Chromatography and Reversed-Phase High Pressure Liquid Chromatography, in: *I. Sunshine, P. I. Jatlow* (eds.), Methodology for Analytical Toxicology, Vol. *II*, CRC Press, Palm Beach 1982, pp. 11 – 17.

[8] *E. J. Flynn, S. Spector,* Determination of Barbiturate Derivatives by Radioimmunoassay, J. Pharmacol. Exp. Ther. *181*, 547 – 554 (1974).

[9] *K. E. Rubenstein, R. S. Schneider, E. F. Ullman,* "Homogeneous" Enzyme Immunoassay. A New Immunochemical Technique, Biochem. Biophys. Res. Commun. *47*, 846 – 851 (1972).

[10] *A. J. Wilensky,* Stability of Some Anti-epileptic Drugs in Plasma, Clin. Chem. *24*, 722 – 723 (1978).

[11] *S. C. Harvey,* Hypnotics and Sedatives, in: *A. Goodman Gilman, L. S. Goodman, A. Gilman* (eds.), The Pharmacological Basis of Therapeutics, 6th edit., MacMillan Publishing, New York 1980, pp. 339 – 375.

[12] Product literature EMIT®-tox. Serum Barbiturate Assay, *Syva Co.,* Palo Alto, U.S.A., 1982.

[13] *C. P. Crowl, I. Gibbons, R. S. Schneider,* Recent Advances in Homogeneous Enzyme Immuno-assays for Haptens and Proteins, in: *R. M. Nakamura, W. R. Dito, E. S. Tucker* (eds.), Immuno-assays: Clinical Laboratory Techniques for the 1980's, Alan Liss, New York 1980, pp. 89 – 126.

[14] *H. Tom, R. S. Schneider, R. Ernst, W. Khan, P. Singh, D. Kabakoff,* Homogeneous Enzyme Immunoassays for Barbiturates and Benzodiazepines in Serum, Clin. Chem. *25*, 1094 (1979), Abstract.

[15] *C. Sutheimer, B. R. Hepler, I. Sunshine,* Clinical Application and Evaluation of the EMIT®-st Drug Detection System, Am. J. Clin. Pathol. *77*, 731 – 735 (1982).

[16] *K. D. Parker, M. Crim, H. W. Elliott, J. A. Wright, N. Nomof, C. H. Hine,* Blood and Urine Concentrations of Subjects Receiving Barbiturates, Meprobamate, Glutethimide and Diphenyl Hydantoin, Clin. Toxicol. *3*, 131 – 136 (1970).

[17] *R. J. Bastiani, R. C. Phillips, R. S. Schneider, E. F. Ullman,* Homogeneous Immunochemical Drug Assays, Am. J. Med. Technol. *29*, 211 – 216 (1973).

[18] Product literature EMIT®-d.u.a. Barbiturate Assay, *Syva Co.,* Palo Alto, U.S.A., 1984.

[19] *M. Oellerich, W. R. Külpmann, R. Haeckel,* Drug Screening by Enzyme Immunoassay (EMIT) and Thin-Layer Chromatography (Drug Skreen), J. Clin. Chem. Clin. Biochem. *15*, 275 – 283 (1977).

[20] *S. J. Mule, M. L. Bastos, D. Jukofsky,* Evaluation of Immunoassay Methods for Detection, in Urine, of Drugs Subject to Abuse, Clin. Chem. *20*, 243 – 248 (1974).

[21] *C. B. Walberg,* Correlation of the "EMIT" Urine Barbiturate Assay with a Spectrophotometric Serum Barbiturate Assay in Suspected Overdose, Clin. Chem. *20*, 305 – 306 (1974).

[22] *A. Broughton, D. L. Ross,* Drug Screening by Enzymatic Immunoassay with the Centrifugal Analyser, Clin. Chem. *21*, 186 – 189 (1975).

[23] *E. L. Slightom,* The Analysis of Drugs in Blood, Bile, and Tissue with an Indirect Homogeneous Enzyme Immunoassay, J. Forensic Sci. *23*, 292 – 303 (1978).

[24] *W. Godolphin,* Enzyme Multiplied Immunoassay Technique (EMIT®), in: *I. Sunshine, P. I. Jatlow* (eds.), Methodology for Analytical Toxicology, Vol. *II*, CRC Press, Palm Beach 1982, pp. 189 – 203.

[25] *L. V. Allen, M. L. Stiles,* Specificity of the EMIT® Drug Abuse Urine Assay Methods, Clin. Toxicol. *18*, 1043 – 1065 (1981).

2.4 Benzodiazepines

7-Chloro-1,3-dihydro-1-methyl-5-phenyl-2*H*-1,4-benzodiazepin-2-one
7-Chloro-1,3-dihydro-3-hydroxy-5-phenyl-2*H*-1,4-benzodiazepin-2-one

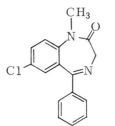
Diazepam

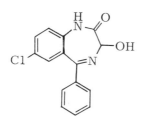
Oxazepam

Christopher P. Price and R. Stewart Campbell

General

The benzodiazepines are a group of drugs (of which diazepam and oxazepam are examples) that are used in clinical practice as mild tranquilizers and sedatives. The group also includes bromazepam, clonazepam, desmethyldiazepam, flunitrazepam, flurazepam, lorazepam, medazepam, nitrazepam, prazepam and temazepam [1, 2].

The major effects of the benzodiazepines are on the central nervous system. The most prominent effects are sedation, hypnosis, decreased anxiety, muscle relaxation and anticonvulsant activity.

The majority of benzodiazepines are administered orally and are readily absorbed. The peak plasma levels are reached in about two hours; the pharmacokinetics will, however, vary from type to type. All of the drugs are bound to serum albumin, the degree varying according to the drug (e.g. diazepam approximately 90%, flurazepam approximately 3%).

Biotransformation of the benzodiazepines takes place in the liver with the major change being to the 1,4-diazepine moiety. The drug is conjugated to give glucuronides with a trace of sulphate. The wide variety of drugs within the group yields a large number of metabolites, of which several are active (e.g. oxazepam) [3].

Symptoms of benzodiazepine toxicity are mild and include dizziness and slurred speech. Respiratory depression and coma may develop and are more likely to be pronounced if the drug is taken with alcohol [4]. The treatment of benzodiazepine overdose is generally supportive following gastric lavage.

Application of method: clinical chemistry, toxicology.

Substance properties relevant in analysis: diazepam, molecular weight 282.8; pK_a 3.3; slightly soluble in water, soluble in ethanol to at least 500 g/l. Oxazepam, molecular weight 286.7, pK_a 1.7, 11.6, practically insoluble in water, soluble in ethanol to at least 50 g/l and in chloroform to at least 50 g/l.

Methods of determination: many of the benzodiazepines can be detected by thin-layer chromatography [5]. A semi-quantitative homogeneous enzyme-immunoassay is described for the detection of benzodiazepines in serum and urine [6].

Quantitative measurements of the benzodiazepines are available using radio-immunoassay [7], high-performance [8] and gas-liquid [9] chromatography.

International reference methods and standards: no reference methods or standards are known so far.

2.4.1   Determination of Diazepam in Serum with Enzyme-multiplied Immunoassay Technique

Assay

Method Design

The method is based on the enzyme-multiplied immunoassay technique (EMIT®) first described by *Rubenstein et al.* [10].

Principle: as described in detail on pp. 7, 8, 271.

The activity of the marker enzyme, glucose-6-phosphate dehydrogenase, G6P-DH, is reduced if the enzyme-drug conjugate (E-L) is coupled to the antibody (Ab). Benzodiazepine (L) in a sample competes with G6P-DH labelled drug (E-L) for a limited concentration of antibody (Ab). The activity of G6P-DH measured as increase in absorbance per unit time, $\Delta A/\Delta t$, is related to the benzodiazepine concentration. The method is intended for use as a semi-quantitative assay; the assay is not designed to determine the level of intoxication.

Selection of assay conditions and adaptation to the individual characteristics of the reagents: as described in general on pp. 8, 9. Enzyme-drug conjugate is obtained by coupling a benzodiazepine derivative to G6P-DH from *Leuconostoc mesenteroides* (cf. p. 272). The method involves pre-dilution of the sample in the reaction vessel. The desired assay range is from 0.3 to 2.0 mg/l. The sample volume is 50 μl.

Equipment

The same equipment is used as described on p. 272.

Reagents and Solutions

Purity of reagents: as described on pp. 27, 272.

Preparation of solutions* (for 100 determinations): all solutions in re-purified water (cf. Vol. II, chapter 2.1.3.2).

1. Tris buffer (Tris, 55 mmol/l; NaN_3, 0.5 g/l; Triton X-100, 0.1 ml/l; pH 8.0):

   dissolve 1.996 g Tris base in 150 ml water; add 0.03 ml Triton X-100 and 0.15 g sodium azide; adjust to pH 8.0 with HCl, 1.0 mol/l; make up with water to 300 ml.

   Alternatively, dilute buffer concentrate provided with each kit to 200 ml with water.

2. Antibody/substrate solution (γ-globulin**; G-6-P, 66 mmol/l; NAD, 40 mmol/l; Tris, 55 mmol/l; NaN_3, 0.5 g/l; pH 5.2):

   reconstitute available lyophilized preparation with 6.0 ml water, or the volume stated on the bottle if different.

3. Conjugate solution (G6P-DH***; Tris, 55 mmol/l; NaN_3, 0.5 g/l; pH 8.0):

   this reagent is standardized to match the antibody/substrate reagent (cf. p. 272). Reconstitute available lyophilized preparation with 6.0 ml water or the volume stated on the bottle if different.

4. Diazepam standard solutions (0.3 mg/l; 2.0 mg/l):

   dissolve 10 mg diazepam in 2 ml hydrochloric acid, 2 mol/l, make up to 50 ml with water. Use this stock solution to prepare two working standards of 0.3 mg/l and 2.0 mg/l by appropriate dilution with pooled normal human serum containing 0.5 g NaN_3 per litre. The serum diluent is used as a zero calibrator.

   Alternatively, reconstitute each of the available lyophilized calibrators with 3.0 ml water; re-stopper the bottles and swirl gently to assist reconstitution.

---

* Reagents, calibrators and buffer are commercially available from *Syva Co.,* Palo Alto, U.S.A. The solutions contain preservatives.
** Standardized preparation from immunized sheep, concentration is titre-dependent.
*** Activity concentration of the enzyme depends on final choice of the conjugate.

Stability of solutions: store all reagents stoppered in a refrigerator at 2 °C to 8 °C. All reagents should be allowed to equilibrate to room temperature before use. The buffer (1), reagent solutions (2) and (3), and standard solutions (4) are all stable for 12 weeks after reconstitution.

Procedure

Collection and treatment of specimen: collect blood and process in the usual manner to obtain plasma or serum. Heparin, fluoride/oxalate or EDTA may be used as anti-coagulants.

Stability of the substance in the sample: the stability of benzodiazepines in serum or plasma is not documented. It is recommended that samples are stored at 2 °C to 8 °C until analysis.

Assay conditions: measurement of samples and standards in duplicate.

A pre-dilution of samples and standard solutions is required: mix one part (0.05 ml) sample or standard solution(s) with 5 parts (0.25 ml) Tris buffer (1).

Incubation time 30 s (in a reaction cuvette); room temperature; 0.6 ml; wavelength Hg 334 or 339 nm; light path 10 mm; final volume 0.9 ml; measurements against water; 30 s at 30 °C.

Run the standards with each series of analyses. The zero calibrator assay serves as a reagent blank; correct all measurements for this value.

Calibration curve: plot the corrected values of the increase in absorbance per 30 s, $\Delta A/\Delta t$, of the calibrators *versus* the corresponding diazepam concentration* (mg/l) using semi-logarithmic graph paper**. A typical curve is shown in Fig. 1.

Calculation: read the mass concentration of diazepam equivalents of the sample from the calibration curve. In view of the wide specificity of the assay, conversion from one system of units to another is not recommended.

---

* For determination in urine: oxazepam concentration.
** For using this paper multiply each value of $\Delta A/\Delta t$ by 2.667.

Measurement

Pipette* successively into the reaction cuvette:			concentration in assay mixture	
pre-diluted sample or standard solution (4) buffer	 (1)	0.05 ml 0.25 ml	diazepam Tris NaN$_3$	up to 19 µg/l 52 mmol/l 470 mg/l
antibody/substrate solution	(2)	0.05 ml	γ-globulin G-6-P NAD	variable 3.7 mmol/l 2.2 mmol/l
buffer	(1)	0.25 ml		
incubate for 30 s at room temperature,				
conjugate solution buffer	(3) (1)	0.05 ml 0.25 ml	G6P-DH**	variable
place cuvette in spectrophotometer immediately or aspirate contents into flow-cell, read absorbance after 15 s and 45 s.				

* Adequate mixing should be achieved by the addition of each reagent and no further agitation is necessary.
** Coupled to a benzodiazepine derivative.

If ΔA per 30 s of the sample is greater than that of the highest calibrator, dilute the sample further with Tris buffer (1) and re-assay.

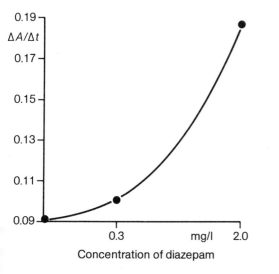

Fig. 1. Typical calibration curve for the assay of diazepam in serum.

Interpretation: the serum concentration of diazepam required to achieve a demonstrable effect is 0.4 mg/l [11]. However, there is a large inter-individual variation in the way in which benzodiazepines are metabolized; this may be up to 30-fold [2]. Serum concentrations up to 2.0 mg/l may be expected following a dosage of 40 mg of diazepam per day; peak concentrations are obtained within one hour of administration. Serum concentrations of 10 mg/l are associated with coma [2].

The assay is designed as a semi-quantitative assay. If the $\Delta A/\Delta t$ for a sample is less than that for the 0.3 mg/l calibrator, the sample is considered to contain no benzodiazepine [12]. This strategy is slightly different from that used for the original drug assays in urine [6, 13].

Validation of Method

Precision, accuracy, detection limit and sensitivity: the relative standard deviation of the assay has been reported as less than 10% (within-day) across the range 0.3 to 2.0 mg diazepam per litre [14]. A second study demonstrated relative standard deviations of 5% (within-assay) and 8% (between-assay) at a benzodiazepine concentration of 0.1 mg/l [15]. This study also reported a comparison with a liquid-chromatographic procedure with a correlation coefficient of 0.961 and a slope of 1.07 [15]. A comparison with a spectrophotometric procedure on 56 positive specimens gave no discrepancies in the interpretation of semi-quantitative results [14].

The detection limits of this assay for various benzodiazepines have been reported [15]; the assay showed similar sensitivity for diazepam, oxazepam, N-desmethyl diazepam and demoxepam.

There is a configuration of this method in which all reagents for the assay are lyophilized in one vial and are reconstituted prior to use (EMIT®-st Drug Detection System, *Syva Co.,* Palo Alto, U.S.A.). An evaluation of this assay [17] gave a relative standard deviation of 30% at 0.5 mg/l for within-day precision and 33 and 16% at 0.5 and 1.0 mg/l, respectively, for between-day precision.

Sources of error: serum samples from grossly haemolyzed, lipaemic or icteric blood may produce false negative results.

Specificity: it is claimed by the reagent manufacturer [12] that this assay will give a signal equivalent to 0.3 mg diazepam per litre for the following diazepines at the concentrations stated:

Clonazepam	2 mg/l	Medazepam	2 mg/l
Chlordiazepoxide	5 mg/l	Demoxepam	4 mg/l
Temazepam	2 mg/l	Nitrazepam	2 mg/l
Oxazepam	2 mg/l	Prazepam	2 mg/l
N-Desmethyldiazepam	2 mg/l	Des-alkylflurazepam	2 mg/l
Flurazepam	2 mg/l	Lorazepam	3 mg/l

In addition, the following compounds have been tested and found to give a signal equivalent to that of 0.3 mg diazepam per litre, at the concentrations stated:

Methaqualone	100 mg/l	Imipramine	100 mg/l
Aspirin	1000 mg/l	Secobarbital	100 mg/l
Phencyclidine	1000 mg/l	Ethchlorvynol	300 mg/l
Phenytoin	100 mg/l	Amphetamine	100 mg/l
Amitriptyline	100 mg/l	Acetaminophen	1500 mg/l
Propoxyphene	100 mg/l	Glutethimide	200 mg/l
Morphine	25 mg/l		

The specificity of the assay may vary from lot to lot of reagents.

2.4.2 Determination of Oxazepam in Urine with Enzyme-multiplied Immunoassay Technique

Assay

Method Design

Principle: as described in detail on pp. 7, 8, 271.

Selection of assay conditions and adaptation to the individual characteristics of the reagents: as described on pp. 8, 9. The desired assay range is from 0.3 – 1.0 mg/l. The sample volume is 50 μl. To obtain the desired assay range, follow p. 272.

Equipment

The same equipment is used as described on p. 272 for the determination in serum.

Reagents and Solutions

Purity of reagents: cf. pp. 27, 272.

Preparation of solutions* (for 100 determinations): all solutions in re-purified water (cf. Vol. II, chapter 2.1.3.2).

1. Tris buffer (Tris, 55 mmol/l; NaN$_3$, 0.5 g/l; Triton X-100, 0.1 ml/l; pH 8.0):

   dissolve 1.996 g Tris base in 150 ml water, add 0.093 ml Triton X-100 and 0.15 g sodium azide, adjust to pH 8.0 with HCl, 1.0 mol/l; make up with water to 300 ml.

   Alternatively, dilute buffer concentrate provided with each kit to 200 ml with water.

2. Antibody/substrate solution (γ-globulin**; G-6-P, 66 mmol/l; NAD, 40 mmol/l; NaN$_3$, 0.5 g/l; Tris, 55 mmol/l; pH 5.2):

   reconstitute available lyophilized preparation with 6.0 ml water.

3. Conjugate solution (G6P-DH***; NaN$_3$, 0.5 g/l; Tris, 55 mmol/l; pH 8.0):

   this reagent is standardized to match the antibody/substrate reagent (cf. p. 272). Reconstitute available lyophilized preparation with 6.0 ml water.

4. Oxazepam standard solutions (0.3 mg/l; 1.0 mg/l):

   dissolve 10 mg oxazepam in 2.0 ml hydrochloric acid, 1 mol/l, and make up to 100 ml with water. Use this stock solution to prepare two working standards of 0.3 mg/l and 1.0 mg/l by appropriate dilution with pooled drug-free human urine, containing 0.5 g NaN$_3$ per litre. The urine diluent is used as a zero calibrator.

   Alternatively, reconstitute each of the available lyophilized calibrators with 3.0 ml water; re-stopper the bottles and swirl gently to assist reconstitution.

Stability of solutions: store all reagents stoppered in a refrigerator at 2 °C to 8 °C. All reagents should be allowed to equilibrate to room temperature before use. The buffer (1), antibody and enzyme-drug conjugate solutions (2) and (3), and the standard solutions (4) are all stable for 12 weeks after reconstitution.

Procedure

Collection and treatment of specimen

Urine: collect urine samples in plastic or glass containers. The effect of urine preservatives has not been established and therefore their use is not recommended. Centrifuge

---

* Reagents, calibrators and buffer are commercially available from *Syva Co.,* Palo Alto, U.S.A. The solutions contain preservatives.

** Standardized preparation from immunized sheep, concentration is titre-dependent.

*** Enzyme activity concentration is dependent on final choice of conjugate.

turbid samples before analysis. If pH is outside the range 5 to 8, adjust to within the range using HCl, 1 mol/l, or NaOH, 1.0 mol/l, before analysis.

Tissue: homogenize 5 g tissue with 20 ml phosphate buffer, 0.1 mol/l, pH 7.5 and extract with 50 ml chloroform. Wash the chloroform phase with 10 ml hydrochloric acid, 0.1 mol/l, followed by 10 ml sodium hydroxide, 0.13 mol/l, and then evaporate to dryness; reconstitute the residue with 0.5 ml water.

Stability of the substance in the sample: urine samples may be stored at 4 °C for up to 3 days. Storage for longer periods may result in false negative results being obtained for samples with a benzodiazepine concentration close to that of the low calibrator.

Assay conditions: measurement of samples and standards in duplicate; incubation time 30 s (in a reaction cuvette); room temperature; 0.6 ml; wavelength Hg 334 or 339 nm; light path 10 mm; final volume 0.9 ml; measurements against water; 30 s at 30 °C.

Run the standards with each series of analyses. The zero calibrator serves as a reagent blank; correct all measurements for this value.

Measurement

Pipette* successively into the cuvette:			concentration in assay mixture	
sample or standard solution (4)		0.05 ml	oxazepam	up to 56 µg/l
buffer	(1)	0.25 ml	Tris	52 mmol/l
			NaN$_3$	470 mg/l
antibody/substrate solution	(2)	0.05 ml	γ-globulin	variable
			G-6-P	3.7 mmol/l
			NAD	2.2 mmol/l
buffer	(1)	0.25 ml		
incubate for 30 s at room temperature,				
conjugate solution	(3)	0.05 ml	G6P-DH**	variable
buffer	(1)	0.25 ml		
place cuvette in spectrophotometer immediately or aspirate contents into flow-cell, read absorbance after 15 s and 45 s.				

*   Adequate mixing should be achieved by the addition of each reagent and no further agitation is necessary.

** Coupled to a benzodiazepine derivative.

If ΔA per 30 s of the sample is greater than that of the highest calibrator, dilute the specimen further with Tris buffer (1) and re-assay.

Calibration curve: as described on p. 274. A typical curve is shown in Fig. 2.

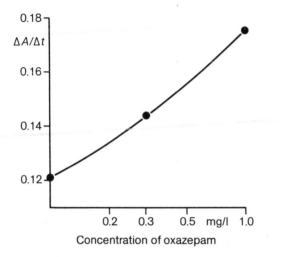

Fig. 2. Typical calibration curve for the assay of oxazepam in urine.

Calculation: read the mass concentration of benzodiazepine in the sample, as equivalents of oxazepam from the calibration curve. It is important to confirm a positive result by an alternative, non-immunochemical procedure. It is recommended that in view of the wide specificity of the assay no conversion to units of substance concentration is made.

Interpretation: the level of benzodiazepine metabolites in the urine is very variable depending clearly on type and dose of drug. *Haden et al.* [18] found that an EMIT® assay response equivalent to >0.5 mg oxazepam per litre was achieved in urine samples collected one hour after ingestion of 5 mg of Valium®. The study also showed there was a variation between individuals in the period over which benzodiazepines were detected in the urine, varying from 1 to 3 days.

The assay is designed as a semi-quantitative procedure. The assay is designed in the knowledge that several factors will influence the precision of an assay. The formulation of the assay and success in its use centre on the recognition of the analyte in the sample. Samples that give a signal above this cut-off will be considered positive. If this cut-off is set too low, the probability of identifying a positive sample is increased but at the risk of producing false positive results. The low calibrator is generally set at the cut-off point. This approach has been demonstrated for several analytes [6, 13]. The original EMIT® procedure employed a lysozyme label; it is considered that assays using a G6P-DH label show enhanced sensitivity [6] and this is borne out in the recommended concentration range in which the assay can be used. In the original assay the cut-off level for benzodiazepine was 0.7 mg/l; at this concentration 95% of samples would be detected as positive. The strategy recommended for interpretation of results has altered slightly and it is now considered that a sample giving a response equal to or greater than that of the low calibrator (0.3 mg/l) will be considered positive [19]. The reagent manufacturer has reported [19] that, for each lot of reagents, between 25 to 50 benzodiazepine-free urine samples and 25 to 50 urine samples containing 0.5 mg oxazepam per litre were tested and 99% of samples were correctly identified.

Validation of Method

Precision, accuracy, detection limit and sensitivity: the reagent manufacturer has reported [19] that in clinical trials involving 175 urine samples (positive 80, negative 95) there was agreement in instances when results with the EMIT® assay were compared with TLC and/or GC/MS procedures. The detection limit of the original lysozyme-based assay was found to be 0.8 mg/l for both diazepam and nordiazepam; by using an extraction procedure the detection limit could be reduced to 0.02 mg/l [20].

Haden et al. [18], using the original EMIT® urine procedure, found a detection limit of 0.7 mg/l employing the criteria already described [13]. These authors also found a relative standard deviation (within series) of 4% at a concentration of 1 mg/l.

The reagents recommended for the determination of serum benzodiazepine have also been used for the measurement of benzodiazepines in urine [15]. Comparison of quantitative results with a liquid-chromatographic procedure yielded a correlation coefficient of 0.85 with a slope of 0.62. The apparent discrepancy was considered to be due to the fact that the assay designed for serum is less sensitive to oxazepam; these authors claimed a 50% cross-reactivity for oxazepam.

Sources of error: erroneous results may be obtained if the samples are not stored correctly or if the urine pH is outside the range of 5 to 8.

It has been recognized with the earlier EMIT® assays employing lysozyme that urines with high ionic strengths can produce falsely low absorbance changes; this can be achieved by adding sodium chloride to urine. It is expected that this effect will be seen with the dehydrogenase-linked assays [21] although this has not been confirmed.

Specificity: the assay shows greatest sensitivity for oxazepam but is capable of detecting several benzodiazepines with similar sensitivity. The reagent manufacturer [19] reports that the following benzodiazepines give responses equivalent to 0.3 mg oxazepam per litre at the stated concentrations: clonazepam (≤ 2.0 mg/l), demoxepam (≤ 2.0 mg/l), desalkylflurazepam (≤ 2.0 mg/l), N-desmethyldiazepam (≤ 2.0 mg/l), diazepam (≤ 2.0 mg/l), flunitrazepam (≤ 2.0 mg/l), flurazepam (≤ 2.0 mg/l), nitrazepam (≤ 2.0 mg/l), chlordiazepoxide (≤ 2.0 mg/l) and lorazepam (≤ 2.0 mg/l).

In a study of the original EMIT® procedure using a lysozyme label *Allen & Stiles* [22] found a total of 26 drugs that cross-reacted in the assay; the majority only showed an effect at the 1000 mg/l level and none was considered to represent a significant problem. *Budd* [23] studied a total of 52 benzodiazepine-like molecules to determine factors affecting the reactivity of the antibody used in the original EMIT® assay; the results confirmed the wide specificity toward benzodiazepine, with 16 compounds showing greater than 50% cross-reactivity compared with oxazepam.

The reagent manufacturer has also reported that several compounds tested to levels of 1000 mg/l showed no response in the assay described here [19]; these compounds included amphetamine, benzoylecgonine, dextromethorphan, methadone, morphine, phencyclidine, propoxyphene and secobarbital.

References

[1] *L. H. Sternbach,* Chemistry of 1,5-Benzodiazepines and Some Aspects of the Structure-Activity Relationship, in: *S. Garattini, E. Mussini, L. O. Randall* (eds.), The Benzodiazepines, Raven Press, New York 1973, pp. 1 – 26.

[2] *D. J. Greenblatt, R. I. Shader, D. R. Abernethy,* Current Status of Benzodiazepines, N. Eng. J. Med. *309,* 354 – 358 (1983).

[3] *L. E. Hollister,* Clinical Pharmacology of Psychotherapeutic Agents, Churchill Livingstone, New York 1978.

[4] *T. J. Meredith, J. A. Vale,* Poisoning Due to Hypnotics, Sedatives, Tranquillizers and Anticonvulsants, in: *J. A. Vale, T. J. Meredith* (eds.), Poisoning: Diagnosis and Treatment, Update Books, London 1981, pp. 84 – 89.

[5] *P. R. Sedgwick,* Benzodiazepines Type B Procedure, in: *I. Sunshine* (ed.), Methodology for Analytical Toxicology, CRC Press, Palm Beach 1975, pp. 45 – 47.

[6] *D. S. Kabakoff, H. M. Greenwood,* Recent Advances in Homogeneous Enzyme Immunoassay, in: *K. G. M. M. Alberti, C. P. Price* (eds.), Recent Advances in Clinical Biochemistry 2, Churchill Livingstone, Edinburgh 1981, pp. 1 – 30.

[7] *R. Sellman, A. Pekkarinen, L. Kangas, E. Jaijola,* Reduced Concentrations of Plasma Diazepam in Chronic Alcoholic Patients Following an Oral Administration of Diazepam, Acta Pharmacol. Toxicol. *36,* 25 – 32 (1975).

[8] *C. Sutheimer, I. Sunshine,* Benzodiazepines by HPLC, in: *I. Sunshine, P. I. Jatlow* (eds.), Methodology for Analytical Toxicology, Vol. *II,* CRC Press, Palm Beach 1982, pp. 25 – 30.

[9] *D. J. Greenblatt,* Analysis of Benzodiazepines by Electron-Capture Gas Liquid Chromatography, in: *I. Sunshine, P. I. Jatlow* (eds.), Methodology for Analytical Toxicology, Vol. II, CRC Press, Palm Beach 1982, pp. 19 – 24.

[10] *K. E. Rubenstein, R. S. Schneider, E. F. Ullman,* "Homogeneous" Enzyme Immunoassay. A New Immunochemical Technique, Biochem. Biophys. Res. Commun. *47,* 846 – 851 (1972).

[11] *M. Mandelli, G. Tognoli, S. Garatini,* Clinical Pharmacokinetics of Diazepam, Clin. Pharmacokinet. *3,* 72 – 91 (1978).

[12] Product literature. EMIT®-tox Serum Benzodiazepine Assay, *Syva Co.,* Palo Alto, U.S.A., 1981.

[13] *R. J. Bastiani, R. C. Phillips, R. S. Schneider, E. F. Ullman,* Homogeneous Immunochemical Drug Assays, Am. J. Med. Technol. *29,* 211 – 216 (1973).

[14] *H. Tom, R. S. Schneider, R. Ernst, W. Khan, P. Singh, D. Kabakoff,* Homogeneous Enzyme Immunoassays for Barbiturates and Benzodiazepines in Serum, Clin. Chem. *25,* 1094 (1979), Abstract.

[15] *J. E. Wallace, S. C. Harris, E. L. Shimetz,* Evaluation of an Enzyme Immunoassay for Determining Diazepam and Nordiazepam in Serum and Urine, Clin. Chem. *26,* 1905 – 1907 (1980).

[16] *C. P. Crowl, I. Gibbons, R. S. Schneider,* Recent Advances in Homogeneous Enzyme Immunoassays for Haptens and Proteins, in: *R. M. Nakamura, W. R. Dito, E. S. Tucker* (eds.), Immunoassays, Clinical Laboratory Techniques for the 1980s, Alan R. Liss Inc., New York 1980, pp. 89 – 126.

[17] *C. Sutheimer, B. R. Hepler, I. Sunshine,* Clinical Application and Evaluation of the EMIT®-st Drug Detection System, Am. J. Clin. Pathol. *77,* 731 – 735 (1982).

[18] *B. H. Haden, K. G. McNeil, N. A. Huber, W. A. Khan, P. Singh, R. S. Schneider,* An EMIT® Assay for Benzodiazepines in Urine, Clin. Chem. *22,* 1200 (1976), Abstract.

[19] Product Literature. EMIT®-d.a.u. Benzodiazepine Assay, *Syva Co.,* Palo Alto, U.S.A., 1984.

[20] *E. L. Slightom,* The Analysis of Drugs in Blood, Bile and Tissue with an Indirect Homogeneous Enzyme Immunoassay, J. Forensic. Sci. *23,* 292 – 303 (1978).

[21] *W. Godolphin,* Enzyme Multiplied Immunoassay Technique (EMIT®), in: *I. Sunshine, P. I. Jatlow* (eds.), Methodology for Analytical Toxicology *II,* CRC Press, Palm Beach 1982, pp. 189 – 203.

[22] *L. V. Allen, M. L. Stiles,* Specificity of the EMIT Drug Abuse Urine Assay Methods, Clin. Toxicol. *18,* 1043 – 1065 (1981).

[23] *R. D. Budd,* Benzodiazepine Structure *versus* Reactivity with EMIT Oxazepam Antibody, Clin. Toxicol. *18,* 643 – 655 (1981).

2.5 Methadone

(±)-6-(Dimethylamino)-4,4-diphenyl-3-heptanone

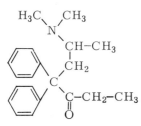

Christopher P. Price and R. Stewart Campbell

General

Methadone is a synthetic narcotic agent. It is a potent analgesic although its main clinical use is in the maintenance therapy of heroin addicts. The pharmacological effects of single doses of methadone are very similar to those of morphine [1]. It has also been employed in linctuses as a cough suppressant.

Methadone is easily absorbed from the gastro-intestinal tract and peak plasma levels are reached at about 4 hours; in the bloodstream about 85% of the drug is bound to plasma proteins. The drug undergoes extensive biotransformation in the liver and the major metabolites are excreted in the urine and the bile [2].

The side-effects of methadone are similar to those of morphine, in cases of over-dosage. Thus the most common feature of methadone overdosage is respiratory depression accompanied by cyanosis, coma and diminished pupil aperture [3].

Application of method: clinical chemistry, toxicology.

Substance properties relevant in analysis: molecular weight 309.5, as hydrochloride 345.9; pK_a 8.3, soluble in water to 120 g/l, in ethanol to 80 g/l and in isopropanol to 24 g/l.

Methods of determination: the drug can be detected in urine using thin-layer chromatography [4] and UV-spectrophotometry after solvent extraction [5]. A semi-quantitative result can be obtained by homogeneous enzyme-immunoassay [6]. Quantitation of methadone in urine can be achieved using gas-liquid chromatography [7] and high-performance liquid chromatography [8].

International reference methods and standards: no reference methods or standards are known so far.

Assay

Method Design

The method is based on the enzyme-multiplied immunoassay technique (EMIT®) first described by *Rubenstein et al.* [9].

Principle: as described in detail on pp. 7, 8, 271.

The activity of the marker enzyme, glucose-6-phosphate dehydrogenase, G6P-DH, is reduced if the enzyme-methadone conjugate (E-L) is coupled to the antibody (Ab). Methadone (L) in a sample competes with G6P-DH-labelled methadone (E-L) for a limited concentration of antibody (Ab). The activity of G6P-DH measured as increase in absorbance per unit time, $\Delta A/\Delta t$, is related to the methadone concentration (cf. Vol. I, chapter 2.7, p. 244). The method is intended for use as a semi-quantitative assay for urine samples; the assay is not designed to determine the level of intoxication.

Selection of assay conditions and adaptation to the individual characteristics of the reagents: as described in general on pp. 8, 9. Enzyme-drug conjugate is obtained by coupling methadone to G6P-DH from *Leuconostoc mesenteroides* (cf. p. 272). The method involves pre-dilution of the sample in the reaction vessel. The desired assay range is from 0.3 to 1.0 mg/l. The sample volume is 50 μl.

Equipment

The same equipment is used as described on p. 272.

Reagents and Solutions

Purity of reagents: as described on pp. 27, 272.

Preparation of solutions* (for 100 determinations): all solutions in re-purified water (cf. Vol. II, chapter 2.1.3.2).

1. Tris buffer (Tris, 55 mmol/l; NaN$_3$, 0.5 g/l; Triton X-100, 0.1 ml/l; pH 8.0):

   dissolve 1.996 g Tris base in 150 ml water, add 0.03 ml Triton X-100 and 0.15 g sodium azide; adjust to pH 8.0 with HCl, 1.0 mol/l; make up with water to 300 ml.

   Alternatively, dilute buffer concentrate provided with each kit to 200 ml with water.

2. Antibody/substrate solution (γ-globulin**; G-6-P, 66 mmol/l; NAD, 40 mmol/l; NaN$_3$, 0.5 g/l; Tris, 55 mmol/l; pH 5.2):

   reconstitute available lyophilized preparation with 6.0 ml water.

3. Conjugate solution (G6P-DH***; NaN$_3$, 0.5 g/l; Tris, 55 mmol/l; pH 8.0):

   this reagent is standardized to match the antibody/substrate reagent (cf. p. 272). Reconstitute available lyophilized preparation with 6.0 ml water.

4. Methadone standard solutions (0.3 mg/l; 1.0 mg/l):

   dissolve 111.8 mg methadone hydrochloride (corresponding to 100.0 mg methadone) in 100 ml water. Use this stock solution to prepare two working standards of 0.3 mg/l and 1.0 mg/l by appropriate dilution with pooled normal human urine. The urine diluent is used as a zero calibrator.

   Alternatively, reconstitute each of the available lyophilized calibrators with 3.0 ml water; re-stopper the bottles and swirl gently to assist reconstitution.

Stability of solutions: store all reagents stoppered in a refrigerator at 2 °C to 8 °C. All reagents should be allowed to equilibrate to room temperature before use. The buffer (1), antibody and enzyme-drug conjugate solutions (2) and (3), and the standard solutions (4) are all stable for 12 weeks after reconstitution.

---

* Reagents, calibrators and buffer are commercially available from *Syva Co.,* Palo Alto, U.S.A. The solutions contain preservatives.
** Standardized preparation from immunized sheep, concentration is titre-dependent.
*** Enzyme activity concentration is dependent on final choice of conjugate.

Procedure

Collection and treatment of specimen: collect urine samples in plastic or glass containers. The effect of urine preservatives has not been established and therefore their use is not recommended. Centrifuge turbid samples before analysis.

Samples should be within the pH range of 5 to 8; if not, adjust to fall within the range by addition of HCl, 1.0 mmol/l, or NaOH, 1.0 mol/l, before analysis.

Stability of the substance in the sample: samples may be stored at 4°C for up to 3 days. Storage for longer periods may result in false negative results in the case of samples with a methadone level close to that of the low calibrator [10].

Assay conditions: measurement of samples and standards in duplicate; incubation time 30 s (in a reaction cuvette); room temperature; 0.6 ml; wavelength Hg 334 or 339 nm; light path 10 mm; final volume 0.9 ml; measurements against water; 30 s at 30°C.

Run the standards with each series of analyses. The zero calibrator assay serves as a reagent blank; correct all measurements for this value.

Measurement

Pipette* successively into the cuvette:			concentration in assay mixture	
sample or standard solution (4)		0.05 ml	methadone	up to 56 µg/l
buffer	(1)	0.25 ml	Tris	52 mmol/l
			NaN₃	470 mg/l
antibody/substrate solution	(2)	0.05 ml	γ-globulin	variable
			G-6-P	3.7 mmol/l
			NAD	2.2 mmol/l
buffer	(1)	0.25 ml		
incubate for 30 s at room temperature;				
conjugate solution	(3)	0.05 ml	G6P-DH	variable
buffer	(1)	0.25 ml		
place cuvette in spectrophotometer immediately or aspirate contents into flow-cell, read absorbance after 15 s and 45 s.				

* Adequate mixing should be achieved by the addition of each reagent and no further agitation is necessary.

If ΔA per 30 s of the sample is greater than that of the highest calibrator, dilute the specimen further with Tris buffer (1) and re-assay.

Calibration curve: plot the corrected values of the increase in absorbance per 30 s, $\Delta A/\Delta t$, of the calibrators *versus* the corresponding methadone concentration (mg/l) using semi-logarithmic graph paper. A typical curve is shown in Figure 1.

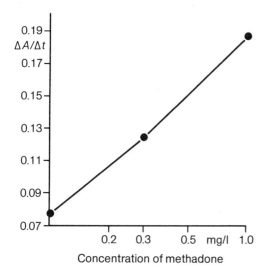

Fig. 1. Typical calibration curve for the assay of methadone in urine.

Calculation: read the mass concentration of methadone in the sample from the calibration curve. It is important to confirm a positive result by an alternative, non-immunochemical procedure. To convert from mass concentration to substance concentration, c (μmol/l), multiply by 2.891 (molecular weight of methadone is 345.9).

Interpretation: the concentration range of methadone in urine from addicts on maintenance doses of methadone is between 4.0 and 40 mg/l. In a study by *Reynolds* [11] 98% of urines from addicts gave a concentration of less than 50 mg/l.

The assay is designed as a semi-quantitative procedure. The assay is designed in the knowledge that several factors will influence the precision of an assay. The formulation of the assay and success in its use centre on the recognition of the analyte in the sample. Samples that give a signal above this cut-off will be considered positive. If this cut-off is set too low, the probability of identifying a positive sample is increased but at the risk of producing false positive results. The low calibrator is generally set at the cut-off point. This approach has been demonstrated for several analytes [6, 12]. The original EMIT® assay employed a lysozyme label; it is considered that the assays using a G6P-DH label show enhanced sensitivity [6] and this is borne out in the recommended concentration range in which the assay can be used. In the original assay the cut-off level for methadone was 0.5 mg/l; at this concentration 95% of samples would be detected as positive. The strategy recommended for interpretation has altered slightly in that it is now considered that a sample giving a response equal to or greater than that of the low calibrator (0.3 mg/l) will be considered positive [13]. The reagent manufacturer has reported [12] that, for each lot of reagent, between 25 – 50 methadone-free urine samples and 25 to 50 urine samples containing methadone (0.5 mg/l) are tested and 99% of samples are correctly identified.

Validation of Method

Precision, accuracy, detection limit and sensitivity: the reagent manufacturer has reported [12] that in a clinical trial involving 166 urine samples (positive 80, negative 86) there was agreement in all instances when results were compared between EMIT® and TLC procedures. The reagents for this assay have only recently become available and there are no independent published performance data.

The original procedure was demonstrated to give relative standard deviations of 13.3% within series and 34.4% between days at a mean methadone level of 1.95 mg/l [14]. The detection limit, expressed as three times the standard deviation of the reagent baseline signal, was found to be 0.2 mg/l.

Comparison of the original "lysozyme label" assay with a thin-layer chromatographic procedure gave divergent results in less than 5% of the cases studied [14, 10]. Discrepancies between EMIT® and TLC assays are likely to be due to the poor sensitivity of the TLC assays [14].

In an automated version of the original EMIT® procedure, *Broughton et al.* [15] using a cut-off of 1.0 mg/l found less than 1% of cases discrepant when compared with a TLC procedure.

Sources of error: false negative results may be obtained if the samples are not stored correctly. The urine pH should be within the range of 5 to 8.

It has been recognized with the earlier EMIT® assays employing the lysozyme label that urines with high ionic strengths can produce falsely low absorbance changes; this can be achieved by adding sodium chloride to urine. It is expected that this effect will be seen with the dehydrogenase-linked assays [16] although this has not been confirmed.

Specificity: it is claimed by the reagent manufacturer [13] that the following compounds give a signal less than the low calibrator at the concentrations stated:

Acetaminophen	1000 mg/l	Benzoylecgonine	400 mg/l
α-Acetyl-N,N-dinormethadol	25 mg/l	Codeine	500 mg/l
Amitriptyline	50 mg/l	Dextromethorphan	300 mg/l
Amphetamine	500 mg/l	Diphenhydramine	100 mg/l
Aspirin	1000 mg/l	Doxylamine	200 mg/l

Meperidine	200 mg/l	Oxazepam	250 mg/l
Methaqualone	100 mg/l	Phencyclidine	500 mg/l
Morphine	200 mg/l	Promethazine	75 mg/l
Naloxone	500 mg/l	Propoxyphene	300 mg/l
Noracetylmethadol	5 mg/l	Secobarbital	100 mg/l

The assay does not detect metabolites of the long acting form of methadone, 1α-acetyl-methadol, in concentrations that would be found in urine from patients on this treatment.

In the original "lysozyme-based" assay, *Oellerich et al.* [14] found the following compounds gave a signal equivalent to 0.5 mg methadone per litre: normethadone (4.0 mg/l), promethazine hydrochloride (130 mg/l), chlorpromazine hydrochloride (150 mg/l) and morphine hydrochloride (190 mg/l).

Budd [17] investigated fifteen compounds structurally related to methadone in order to determine the relationship between immunoreactivity and structure; the original "lysozyme-based" assay was used. The assay was found to be very specific for methadone, with no methadone-related compounds (either structural or metabolic) giving any significant reactivity. This opinion was also shared by *Mule et al.* [14] in their comparison of results on over 8000 urine samples from drug abuse centres using EMIT® and TLC procedures.

Allen & Stiles [18] found several compounds that would give a positive result in the original EMIT® assay; however, a concentration of 1000 mg/l was needed to produce a response. These levels are not likely to be experienced in any biological samples.

Kelner [19] has found that diphenhydramine at a concentration of 10 mg/l in urine will give a result equal to or greater than 0.5 mg methadone per litre. It has been noted that this varied from lot to lot of reagent [19, 20]. However, *Ajel* [20] has indicated that the assay has been re-formulated so that concentrations of diphenhydramine in excess of 100 mg/l are required to achieve this response.

References

[1] *J. H. Jaffe, W. R. Martin,* Opioid Analgesics and Antagonists, in: *A. Goodman Gilman, L. S. Goodman, A. Gilman* (eds.), The Pharmacological Basis of Therapeutics, 6th edit., MacMillan Publishing, New York 1980, pp. 518 – 520.

[2] *K. Vereby, J. Volavka, S. Mule, R. Resnick,* Methadone in Man: Pharmakokinetic and Excretion Studies in Acute and Chronic Treatment, Clin. Pharmacol. Ther. *18*, 180 – 190 (1975).

[3] *D. W. Fraser,* Methadone Overdose, J. Am. Med. Assoc. *217*, 1387 – 1389 (1972).

[4] *N. H. Choulis,* Identification Procedures of Drugs of Abuse, European Press, Ghent 1977, p. 245.

[5] *J. E. Wallace, H. E. Hamilton, J. T. Payte, K. Blum,* Sensitive Spectrophotometric Method for Determining Methadone in Biological Specimens, J. Pharm. Sci. *61*, 1397 – 1400 (1972).

[6] *D. S. Kabakoff, H. M. Greenwood,* Recent Advances in Homogeneous Enzyme Immunoassay, in: *K. G. M. M. Alberti, C. P. Price* (eds.), Recent Advances in Clinical Biochemistry 2, Churchill Livingstone, Edinburgh 1981, pp. 1 – 30.

[7] *C. E. Inturissi, K. Verebely,* A Gas Chromatographic Method for the Quantitative Determination of Methadone in Human Plasma and Urine, J. Chromatogr. *65*, 361 – 369 (1972).

[8] *J. W. Hsich, J. K. H. Ma, J. P. O'Donnell, N. H. Choulis,* High-performance Liquid-chromatographic Analysis of Methadone in Sustained Release Formulations, J. Chromatogr. *161*, 366 – 370 (1978).

[9] *K. E. Rubenstein, R. S. Schneider, E. F. Ullman,* "Homogeneous" Enzyme Immunoassay. A New Immunochemical Technique, Biochem. Biophys. Res. Commun. *47*, 846 – 851 (1972).

[10] *S. J. Mule, M. L. Bastos, D. Jukofsky,* Evaluation of Immunoassay Methods for Detection, in Urine, of Drugs Subject to Abuse, Clin. Chem. *20*, 243 – 248 (1974).

[11] *P. C. Reynolds,* Methadone Type C Procedure, in: *I. Sunshine* (ed.), Methodology for Analytical Toxicology, Vol. *I*, CRC Press, Palm Beach 1978, pp. 233 – 235.

[12] *R. J. Bastiani, R. C. Phillips, R. S. Schneider, E. F. Ullman,* Homogeneous Immunochemical Drug Assays, Am. J. Med. Technol. *29*, 211 – 216 (1973).

[13] *Syva* Product Literature, EMIT®-d.a.u. Methadone Assays, *Syva Co.,* Palo Alto, California, U.S.A., 1984.

[14] *Oellerich, W. R. Kulpmann, R. Haeckel,* Drug Screening by Enzyme Immunoassay (EMIT®) and Thin-layer Chromatography (Drug Skreen), J. Clin. Chem. Clin. Biochem. *15*, 216 – 283 (1977).

[15] *A. Broughton, D. L. Ross,* Drug Screening by Enzyme Immunoassay with a Centrifugal Analyser, Clin. Chem. *21*, 186 – 189 (1975).

[16] *W. Godolphin,* Enzyme Multiplied Immunoassay Technique (EMIT®), in: *I. Sunshine, P. I. Jatlow* (eds.), Methodology for Analytical Toxicology, Vol. *II*, CRC Press, Palm Beach 1982, pp. 189 – 203.

[17] *R. D. Budd,* Methadone EMIT®-Structure versus Reactivity, Clin. Toxicol. *18*, 783 – 792 (1981).

[18] *L. V. Allen, M. L. Stiles,* Specificity of the EMIT Drug Abuse Urine Assay Methods, Clin. Toxicol. *18*, 1043 – 1065 (1981).

[19] *M. J. Kelner,* Positive Diphenhydramine Interference in the EMIT®-d.a.u. Assay, Clin. Chem. *30*, 1430 (1984), letter.

[20] *L. A. Ajel,* Positive Diphenhydramine Interference in the EMIT®-d.a.u. Assay, Clin. Chem. *31*, 340 – 341 (1985), letter.

2.6   Methaqualone

2-Methyl-3-(2-methylphenyl)-4(3*H*)-quinazolinone

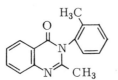

Christopher P. Price and R. Stewart Campbell

General

Methaqualone was first developed as a tranquilizer and initially it was considered to be non-addictive. In addition to hypnotic properties it is also considered to have anti-convulsant, anaesthetic and weak anti-histamine properties.

The drug does not have significant analgesic properties, although it will enhance the analgesia of codeine. In anaesthetic doses methaqualone also depresses myocardial action, resulting in hypotension. Methaqualone will induce drowsiness in less than one hour and the overall effect is heightened by the concurrent ingestion of alcohol [1].

Methaqualone is absorbed quite rapidly; up to 90% of the drug is bound to serum protein in the circulation. The majority of the drug is metabolized in the liver microsomal system with the major metabolites being 4-hydroxymethaqualone and the N-oxide of methaqualone; there are at least eight other metabolites. The 4-hydroxy metabolite is excreted in the bile, while other metabolites are excreted in the urine [2].

Mild overdosage with methaqualone produces depressed consciousness, slurred speech and difficulty with movement of the limbs. Severe poisoning produces muscle tension, increased limb reflexes and exaggerated muscle contractions. Methaqualone also inhibits blood clotting and therefore toxic doses can cause haemorrhage [3].

Treatment of patients with methaqualone poisoning is mainly supportive. In patients who are severely poisoned, haemoperfusion may be effective in aiding removal of the drug.

Application of method: clinical chemistry, toxicology.

Substance properties relevant in analysis: molecular weight 250.3, as hydrochloride 286.8; pK_a 2.5. Substance is virtually insoluble in water but is soluble in ethanol to 83 g/l, in ether to 20 g/l and in chloroform to 1000 g/l.

Methods of determination: methaqualone can be detected using thin-layer chromatography [4]. Homogeneous enzyme-immunoassay provides a rapid semi-quantitative screening technique for urine [5]. Quantitative results can be obtained using radio-immunoassay [6], UV-spectrophotometry [7], gas-liquid [8] and high-performance liquid chromatography [9].

International reference method and standards: no reference methods or standards are known so far.

Assay

Method Design

The method is based on the enzyme-multiplied immunoassay technique (EMIT®) described by *Rubenstein et al.* [10].

The assay is designed as a semi-quantitative assay for methaqualone in urine samples; the assay is not designed to determine the level of intoxication.

Principle: as described in detail on pp. 7, 8, 271.

The activity of the enzyme label glucose-6-phosphate dehydrogenase, G6P-DH, is reduced if the enzyme-methaqualone conjugate (E-L) is coupled to the antibody (Ab). Methaqualone in a sample competes with G6P-DH-labelled methaqualone (E-L) for a limited quantity of antibody (Ab). The activity of G6P-DH measured as increase in absorbance per unit time, $\Delta A/\Delta t$, is related to the methaqualone concentration (cf. Vol. I, chapter 2.7, p. 244).

Selection of assay conditions and adaptation to the individual characteristics of the reagents: as described in general on pp. 8, 9. Enzyme-drug conjugate is obtained by coupling methaqualone to G6P-DH from *Leuconostoc mesenteroides*. The desired assay range is 0.75 to 1.50 mg/l. The sample volume is 50 µl.

Equipment

The same equipment is used as described on p. 272.

Reagents and Solutions

Purity of reagents: as described on pp. 27, 272.

Preparation of solutions* (for 100 determinations): all solutions in re-purified water (cf. Vol. II, chapter 2.1.3.2).

1. Tris buffer (Tris, 55 mmol/l; NaN$_3$, 0.5 g/l; Triton X-100, 0.1 ml/l; pH 8.0):

   dissolve 1.996 g Tris base in 150 ml water, add 0.03 ml Triton X-100 and 0.15 g sodium azide; adjust to pH 8.0 with HCl, 1.0 mol/l; make up with water to 300 ml.

   Alternatively, dilute buffer concentrate provided with each kit to 200 ml with water.

2. Antibody/substrate solution (γ-globulin**; G-6-P, 66 mmol/l; NAD, 40 mmol/l; Tris, 55 mmol/l; NaN$_3$, 0.5 g/l; pH 5.3):

   reconstitute available lyophilized preparation with 6.0 ml water.

3. Conjugate solution (G6P-DH***; Tris, 55 mmol/l; NaN$_3$, 0.5 g/l; pH 8.0):

   this reagent is standardized to match the antibody/substrate reagent (cf. p. 272). Reconstitute available lyophilized preparation with 6.0 ml water.

---

* Reagents, standards and buffer are commercially available from *Syva Co.,* Palo Alto, U.S.A. The solutions contain preservatives.

** Standardized preparation from immunized sheep, concentration is titre-dependent.

*** The activity concentration of the enzyme is dependent on final choice of conjugate dilution.

4. Methaqualone standard solutions (0.3 mg/l; 1.5 mg/l):

dissolve 17.25 mg methaqualone hydrochloride (corresponding to 15.0 mg metha-qualone) in 10 ml methanol. Use this stock solution to prepare two working stand-ards of 0.3 and 1.5 mg/l by appropriate dilution with pooled drug-free human urine, containing 0.5 g sodium azide per litre. The urine diluent is used as a zero calibrator.

Alternatively, reconstitute each of the available lyophilized calibrators with 3.0 ml water; re-stopper the bottles and swirl gently to assist reconstitution.

Stability of solutions: store all reagents stoppered in a refrigerator at 2°C to 8°C. All reagents and calibrators should be allowed to equilibrate to room temperature before use. The buffer (1), antibody/substrate and conjugate solutions (2) and (3), and standard solutions (4) are all stable for 12 weeks after reconstitution.

Procedure

Collection and treatment of specimen

Urine: collect urine specimens in plastic or glass containers. The effect of urine preser-vatives has not been established and therefore their use is not recommended. Centri-fuge turbid specimens before analysis.

Blood: collect blood in a lithium heparin or a plain tube; centrifuge; place 1 ml serum or plasma in a glass stoppered centrifuge tube containing 3.5 ml sodium hydroxide, 0.5 mol/l, and 10.0 ml hexane. Stopper, shake for 5 min and allow the layers to separate. Remove and discard the aqueous layer and then add 1.0 ml hydrochloric acid, 1.0 mol/l, and shake for 5 min. Use the acid extract for analysis.

Stability of the substance in the sample urine should be stored frozen if analysis is not possible within two days of collection. Methaqualone is stable in urine for at least six months when stored at −20°C. Methaqualone in serum is stable for at least twelve months when stored at −20°C.

Assay conditions: measurements of samples and standards in duplicate; incubation time 30 s (in a reaction cuvette); room temperature; 0.6 ml; wavelength Hg 334 or 339 nm; light path 10 mm; final volume 0.9 ml; measurements against water; 30 s at 30°C.

Run the standards with each series of analyses. The zero calibrator assay serves as a reagent blank; correct all measurements for this value.

Measurement

Pipette* successively into the cuvette:			concentration in assay mixture	
sample or calibrator		0.05 ml	methaqualone	up to 83 µg/l
buffer	(1)	0.25 ml	Tris	52 mmol/l
			NaN₃	470 mg/l
antibody/substrate solution	(2)	0.05 ml	γ-globulin	variable
			G-6-P	3.7 mmol/l
			NAD	2.2 mmol/l
buffer	(1)	0.25 ml		
incubate for 30 s at room temperature,				
conjugate solution	(3)	0.05 ml	G6P-DH**	variable
buffer	(1)	0.25 ml		
place cuvette in spectrophotometer immediately or aspirate contents into flow-cell, read absorbance after 15 s and 45 s.				

* Adequate mixing should be achieved by the addition of each reagent and no further agitation necessary.
** Coupled to methaqualone.

If ΔA per 30 s of sample exceeds that of the highest calibrator dilute the sample further in Tris buffer (1) and re-assay.

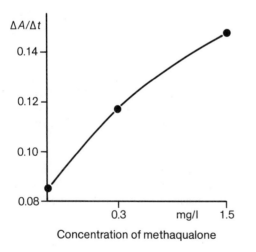

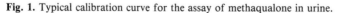

Concentration of methaqualone

Fig. 1. Typical calibration curve for the assay of methaqualone in urine.

Calibration curve: plot the corrected values of the increase in absorbance per 30 s, $\Delta A/\Delta t$, of the calibrators *versus* the corresponding methaqualone concentration (mg/l) using semi-logarithmic graph paper. A typical curve is shown in Fig. 1.

Calculation: read the mass concentration of methaqualone in the sample from the calibration curve. As the assay system also detects mecloqualone and certain methaqualone metabolites the result will be in equivalents of methaqualone. For conversion from mass concentration to substance concentration, c (μmol/l), multiply by 3.995 (molecular weight 250.3).

Interpretation: the concentration of unchanged methaqualone in the serum after the use of a therapeutic dose (150 mg) will be in the range 2.0 to 3.0 mg/l. Serum concentrations of greater than 8.0 mg/l are associated with unconsciousness; values up to 22 mg/l have been found [11].

Approximately 30% of a dose is excreted in the urine in the first 24 h after ingestion, 60% being excreted in the first 3 days [12]. Levels of methaqualone in urine at concentrations of 0.5 to 1.0 mg/l may be detected for up to 2 weeks after a single 300 mg dose [13].

The assay is designed as a semi-quantitative procedure. The assay is designed in the knowledge that several factors will influence the precision of an assay. The formulation of the assay and success in its use centre on the recognition of a minimum absorbance change indicating the presence of the analyte in the sample. Samples that give a signal above this cut-off will be considered positive. If this cut-off is set too low, the probability of identifying a positive sample is increased, but at the risk of producing false positive results. The low calibrator is generally set at the cut-off point. This approach has been demonstrated for several analytes [5, 14]. In the case of methaqualone the cut-off level is 0.3 mg/l.

It is important to confirm a positive result by an alternative, preferably non-immunological, procedure.

Validation of Method

Precision, accuracy, detection limit and sensitivity: it has been found that in 100 urine samples spiked with 0.3 mg methaqualone per litre the assay gave a range of results with a mean value of 0.3 mg/l and a standard deviation of 0.2 mg/l. With this cut-off, less than 5% false positive will be detected and the assay will detect more than 95% of samples containing 0.75 mg methaqualone per litre [15].

Sources of error: turbid urine samples may produce an erroneous result and should therefore be centrifuged prior to sampling.

It has been recognized with the earlier EMIT® assays employing the lysozyme label that urines with high ionic strengths can produce falsely low absorbance changes; this can be shown by adding sodium chloride to urine. It is expected that this effect will be seen with the dehydrogenase-linked assays [16] although this has not been confirmed.

Specificity: the assay is designed to detect methaqualone. The assay will also detect methaqualone metabolites and a pharmaceutically similar drug, mecloqualone [15]. The potential cross-reactivity with several structurally related and unrelated compounds has been assessed by the reagent manufacturer [15]. The concentrations of these substances that will give a response equivalent to 0.3 mg methaqualone per litre are as follows: mecloqualone ⩽ 1.0 mg/l, 3-hydroxymethaqualone ⩽ 1.0 mg/l, 4-hydroxymethaqualone ⩽ 1.0 mg/l, 2-hydroxymethyl-methaqualone ⩽ 5.0 mg/l. The following compounds have been found not to cross-react (against the same criterion):

Oxazepam (Benzodiazepine metabolite)	⩾ 250 mg/l	Meperidine	⩾ 500 mg/l
Amitriptyline	⩾ 500 mg/l	Methadone	⩾ 500 mg/l
Amphetamine	⩾ 500 mg/l	Morphine	⩾ 500 mg/l
Benzoylecgonine		Promethazine	⩾ 500 mg/l
(Cocaine metabolite)	⩾ 500 mg/l	Propoxyphene	⩾ 500 mg/l
Codeine	⩾ 500 mg/l	Secobarbital	⩾ 500 mg/l
Dextromethorphan	⩾ 500 mg/l	Phencyclidine (PCP)	⩾ 750 mg/l
Glutethimide	⩾ 500 mg/l	Acetaminophen	⩾ 1000 mg/l

References

[1] S. Roden, M. E. Williams, M. Mitchard, The Influence of Ethanol on the Absorption and Disposition of Methaqualone, Proc. Eur. Soc. Toxicol. 18, 145 – 147 (1977).
[2] O. Ericsson, B. Danielsson, Urinary Excretion Patterns of Methaqualone Metabolites in Man, Drug Metab. Dispos. 5, 497 – 502 (1977).
[3] D. G. Mills, Effects of Methaqualone on Blood Platelet Function, Clin. Pharmacol. Ther. 23, 685 – 691 (1978).
[4] H. K. Sleeman, J. A. Cella, J. L. Harvey, D. J. Beach, Thin-layer Chromatographic Detection and Identification of Methaqualone Metabolites in Urine, Clin. Chem. 27, 76 – 80 (1981).
[5] D. S. Kabakoff, H. M. Greenwood, Recent Developments in Homogeneous Enzyme Immunoassay, in: K. G. M. M. Alberti, C. P. Price (eds.), Recent Advances in Clinical Biochemistry 2, Churchill Livingstone, Edinburgh 1981, pp. 1 – 30.
[6] A. R. Berman, J. P. McGrath, R. C. Permisohn, J. A. Cella, Radioimmunoassay of Methaqualone and Its Monohydroxy Metabolites in Urine, Clin. Chem. 21, 1878 – 1881 (1975).
[7] D. N. Bailey, P. I. Jatlow, Methaqualone Type B Procedure, in: I. Sunshine (ed.), Methodology for Analytical Toxicology, CRC Press, Palm Beach 1975, pp. 241 – 243.
[8] D. J. Berry, Methaqualone Type C Procedure, in: I. Sunshine (ed.), Methodology for Analytical Toxicology, CRC Press, Palm Beach 1975, pp. 243 – 244.
[9] C. C. Cuppett, C. Fisher, K. J. Custer, HPLC Assay of Chlordiazepoxide, Diazepam and Methaqualone at Drug Abuse Levels in Serum, Clin. Chem. 27, 1103 (1981).
[10] K. E. Rubenstein, R. S. Schneider, E. F. Ullman, "Homogeneous" Enzyme Immunoassay. A New Immunochemical Technique, Biochem. Biophys. Res. Commun. 47, 846 – 851 (1972).
[11] D. Bailey, P. Jatlow, Methaqualone Overdose. Analytical Methodology and the Significance of Serum Drug Concentrations, Clin. Chem. 19, 615 – 620 (1973).
[12] R. D. Smyth, P. B. Chemburkar, P. P. Mathur, A. F. DeLang, A. Polk, P. B. Shah, R. S. Joslin, N. H. Reavey-Cantwell, Bioavailability and Biological Deposition of Methaqualone, J. Int. Med. Res. 2, 85 – 99 (1974).

[13] *R. D. Budd, F. C. Yang, W. J. Leung,* Mass Screening and Confirmation of Methaqualone and Its Metabolites in Urine by Radioimmunoassay-Thin Layer Chromatography, J. Chromat. *190,* 129 – 132 (1980).

[14] *R. J. Bastiani, R. C. Phillips, R. S. Schneider, E. F. Ullman,* Homogeneous Immunochemical Drug Assays, Am. J. Med. Technol. *29,* 211 – 216 (1973).

[15] Product literature. EMIT®-d.a.u., Methaqualone Assay, *Syva Co.,* Palo Alto, U.S.A., 1982.

[16] *W. Godolphin,* Enzyme Multiplied Immunoassay Technique (EMIT®), in: *I. Sunshine, P. I. Jatlow* (eds.), Methodology for Analytical Toxicology *II,* CRC Press, Palm Beach 1982, pp. 189 – 203.

2.7 Morphine and Codeine

(5α,6α)-7,8-Didehydro-4,5-epoxy-17-methylmorphinan-3,6-diol
and
(5α,6α)-7,8-Didehydro-4,5-epoxy-3-methoxy-17-methylmorphinan-6-ol

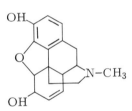

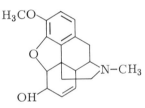

Morphine   Codeine

Christopher P. Price and R. Stewart Campbell

General

Morphine is a narcotic analgesic agent; codeine, in addition to being employed as an analgesic, it is also a cough suppressant. A proportion of codeine (about 10%) is metabolized to morphine. Both are members of the opiate group of drugs [1].

Morphine and codeine are both readily absorbed from the gastro-intestinal tract (as well as the nasal mucosa and lungs), and after subcutaneous or intramuscular injection. After oral administration, peak blood levels are reached after about one hour; about one third of the drug is bound to plasma proteins [1].

Morphine is conjugated in the liver with glucuronic acid. The majority of the drug or its glucuronide is excreted in the urine; a small proportion is excreted in the bile [1].

The most common features of an opiate overdose are respiratory depression accompanied by cyanosis, coma and diminished pupil aperture. The respiratory depression in morphine overdosage may be associated with pulmonary oedema; although less common, it has also been reported in cases where codeine has been administered intravenously. Codeine overdosage may also be associated with convulsions, particularly in children. Other less common features are hypothermia and hypoglycaemia [1].

The central feature of the treatment of opiate poisoning is the use of naloxone, a pure narcotic antagonist. Naloxone is given immediately to counteract severe respiratory depression and coma. The drug is given intravenously or intramuscularly and the effects will be apparent in two or three minutes with increased respiratory rate. The duration and effect of naloxone is less than that of the opiates and therefore it is important to monitor the patient's respiratory rate and level of consciousness with a view to further doses if necessary [2].

Aspiration of gastric contents is not considered as a major priority because of the risks in the light of the pulmonary oedema that may be present.

The recovery of the patient is also dependent on intense supportive treatment with particular respect to the respiratory system.

Application of method: clinical chemistry, toxicology.

Substance properties relevant in analysis: morphine, molecular weight 285.3, pK_a 8.0, 9.9, as monohydrate slightly soluble to 0.91 g/l in boiling water, to 4.76 g/l in ethanol, to 0.82 g/l in chloroform. Codeine, molecular weight 299.4, pK_a 8.2, soluble to 8.3 g/l in water, to 500 g/l in ethanol and several organic solvents.

Methods of determination: morphine can be detected using fluorimetry following solvent extraction [3]. Morphine and codeine can both be detected using thin-layer chromatography [4].

A semi-quantitative determination using homogeneous enzyme-immunoassay will allow detection of morphine and codeine in urine [5, 6]. An immunoassay based on haemagglutination inhibition has also been described for the detection of morphine [7].

Gas- [8], high-performance [9] liquid chromatography and radioimmunoassay [10] are the only techniques that will allow accurate quantitation of the drugs.

International reference method and standards: no reference methods or standards are known so far.

Assay

Method Design

The method is based on the enzyme-multiplied immunoassay technique (EMIT®) first described by *Schneider et al.* [5]. The assay is designed to detect a class of drug rather than a specific drug and will detect abuse with morphine and codeine.

The assay is designed as semi-quantitative. The first homogeneous enzyme-immunoassay for opiates described used a lysozyme label; this assay employs a glucose-6-phosphate dehydrogenase label.

Principle: as described in detail on pp. 7, 8, 271.

The activity of the marker enzyme, glucose-6-phosphate dehydrogenase, G6P-DH, is reduced if the enzyme-morphine conjugate (E-L) is coupled to the antibody (Ab). Morphine or codeine (L) in a sample competes with G6P-DH-labelled morphine (E-L) for a limited concentration of antibody (Ab). The activity of G6P-DH measured as increase in absorbance per unit time, $\Delta A/\Delta t$, is related to the morphine (or codeine) concentration (cf. Vol. I, chapter 2.7, p. 244). The method is intended for use as a semi-quantitative assay for urine samples; the assay is not designed to determine the level of intoxication.

Selection of assay conditions and adaptation to the individual characteristics of the reagents: as described in general on pp. 8, 9. Enzyme-drug conjugate is obtained by coupling morphine to G6P-DH from *Leuconostoc mesenteroides*. The method involves pre-dilution of the sample in the reaction vessel. The desired assay range is from 0.3 to 1.0 mg/l. The sample volume is 50 µl.

Equipment

The same equipment is used as described on p. 272.

Reagents and Solutions

Purity of reagents: as described on pp. 27, 272.

Preparation of solutions* (for 100 determinations): all solutions in re-purified water (cf. Vol. II, chapter 2.1.3.2).

1. Tris buffer (Tris, 55 mmol/l; NaN$_3$, 0.5 g/l; Triton X-100, 0.1 ml/l; pH 8.0):

   dissolve 1.996 g Tris base in 150 ml water, add 0.03 ml Triton X-100 and 0.15 g sodium azide; adjust to pH 8.0 with HCl, 1.0 mol/l; make up with water to 300 ml.

   Alternatively, dilute buffer concentrate provided with each kit to 200 ml with water.

---

* Reagents, calibrators and buffer are commercially available from *Syva Co.,* Palo Alto, U.S.A. The solutions contain preservatives.

2. Antibody/substrate solution (γ-globulin*; G-6-P, 66 mmol/l; NAD, 40 mmol/l; Tris, 55 mmol/l; NaN$_3$, 0.5 g/l; pH 5.2):

reconstitute available lyophilized preparation with 6.0 ml water.

3. Conjugate solution (G6P-DH**; Tris, 55 mmol/l; NaN$_3$, 0.5 g/l; pH 8.0):

this reagent is standardized to match the antibody/substrate reagent (cf. p. 272). Reconstitute available lyophilized preparation with 6.0 ml water.

4. Morphine standard solutions (0.3 mg/l; 1.0 mg/l):

dissolve 10 mg morphine in 100 ml water. Use this stock solution to prepare two working standards of 0.3 mg/l and 1.0 mg/l by appropriate dilution with pooled normal human urine, containing 0.5 g sodium azide per litre. The urine diluent is used as a zero calibrator.

Alternatively, reconstitute each of the available lyophilized calibrators with 3.0 ml water; re-stopper the bottles and swirl gently to assist reconstitution.

Stability of solutions: store all reagents, stoppered, in a refrigerator at 2°C to 8°C. All reagents should be allowed to equilibrate to room temperature before use. The buffer (1), antibody and enzyme-drug conjugate solutions (2) and (3), standard solutions (4) are all stable for 12 weeks after reconstitution.

Procedure

Collection and treatment of specimen

Urine: collect urine samples into plastic or glass containers. The effect of urine preservatives is not known and therefore their use is not recommended. Centrifuge turbid samples before analysis.

Urine samples should be within the pH range of 5 – 8; if outside adjust to fall within the range by addition of HCl, 1.0 mol/l, NaOH, 1.0 mol/l, before analysis.

Blood: collect blood into siliconized tubes. Adjust a 10 ml aliquot of blood to pH 8.5 with sodium hydroxide, 0.1 mol/l, and extract with 70 ml chloroform in the case of codeine; the solvent for morphine is 70 ml chloroform: ethanol in a ratio of 9:1. In either case wash the solvent phase with 10 ml disodium hydrogen phosphate, 67 mmol/l, and then extract with 10 ml sulphuric acid, 1 mol/l. Adjust the sulphuric acid phase to pH 8.5 with sodium hydroxide, 1.0 mol/l, and then extract with the organic solvent used in the initial extraction. After separation, evaporate the organic solvent to dryness and reconstitute with 0.5 ml water.

---

* Standardized preparation from immunized sheep, concentration is titre-dependent.
** The activity concentration of the enzyme is dependent on final choice of conjugate dilution.

Tissue: homogenize 5 g tissue with 20 ml phosphate buffer, 0.1 mol/l, pH 8.5. Then extract the homogenate in exactly the same way as blood, described above.

Stability of the substance in the sample: store urine samples at 4 °C for up to 3 days. Storage beyond this period may result in samples which contain morphine levels near the concentration of the low calibrator giving a negative result.

Blood and tissue should be analyzed as soon as possible; the stability of morphine in blood and tissue is not known.

Assay conditions: measurement of samples and standards in duplicate; incubation time 30 s (in a reaction cuvette); room temperature; 0.6 ml; wavelength Hg 334 or 339 nm; light path 10 mm; final volume 0.9 ml; measurements against water; 30 s at 30 °C.

Run the standards with each series of analyses. The zero calibrator assay serves as a reagent blank; correct all measurements for this value.

Measurement

Pipette* successively into the cuvette:			concentration in assay mixture	
sample or standard solution (4)		0.05 ml	morphine	up to 56 µg/l
buffer	(1)	0.25 ml	Tris	52 mmol/l
			NaN$_3$	470 mg/l
antibody/substrate solution	(2)	0.05 ml	γ-globulin	variable
			G-6-P	3.7 mmol/l
			NAD	2.2 mmol/l
buffer	(1)	0.25 ml		
incubate for 30 s at room temperature,				
conjugate solution	(3)	0.05 ml	G6P-DH	variable
buffer	(1)	0.25 ml		
place cuvette in spectrophotometer immediately or aspirate contents into flow-cell, read absorbance after 15 s and 45 s.				

* Adequate mixing should be achieved by the addition of each reagent and no further agitation is necessary.

If ΔA per 30 s of the sample is greater than that of the highest calibrator, dilute the specimen further with Tris buffer (1) and re-assay.

Calibration curve: plot the corrected values of the increase in absorbance per 30 s, $\Delta A/\Delta t$, of the calibrators *versus* the corresponding morphine concentration (mg/l) using semi-logarithmic graph paper. A typical curve is shown in Fig. 1.

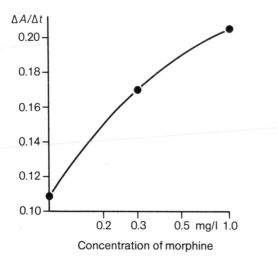

Fig. 1. Typical calibration curve for the assay of morphine derivatives in urine.

Calculation: read the mass concentration of morphine in the sample as morphine equivalents from the calibration curve.

It is important to confirm a positive result by an alternative, non-immunochemical procedure.

Interpretation: levels of morphine in the blood of up to 140 µg/l are found in persons taking heroin. In acute heroin fatalities the concentration in the blood ranges from 100 to 930 µg/l [11].

The concentration of morphine and morphine-like substances, as measured by RIA, in urines from people known to be taking heroin gave a range from 25 to 5000 µg/l. The higher levels were seen in the samples collected in the first 36 h after administration of the drug.

The assay is designed as a semi-quantitative procedure. The assay is designed in the knowledge that several factors will influence the precision of an assay. The formulation of the assay and success in its use centre on the recognition of a minimum $\Delta A/\Delta t$ indicating the presence of the analyte in the sample. Samples that give a signal above this cut-off will be considered positive. If this cut-off is set too low, the probability of identifying a positive sample is increased but at the risk of producing false positive results. The low calibrator is generally set at the cut-off point. This approach has been demonstrated for several analytes [5, 6, 12].

The original EMIT® employed a lysozyme label; it is considered that the assays using a G-6-PDH label show enhanced sensitivity [6] and this is borne out in the recommended concentration range in which the assay can be used. In the original assay the cut-off level for morphine was 0.5 mg/l; at this concentration 95% of samples would be detected as positive. The strategy recommended for interpretation has altered slightly in that it is now considered that a sample giving a response equal to or greater than the low calibrator (0.3 mg/l) will be considered positive [13]. The reagent manufacturer has reported [13] that for each lot of reagent, between 25 to 30 opiate-free urine samples and 25 to 50 urine samples containing 0.5 mg morphine per litre are tested and 99% of samples are correctly identified.

Validation of Method

Precision, accuracy, detection limit and sensitivity: the reagent manufacturer has reported [13] that in a clinical trial involving 98 urine samples (positive 58, negative 40) there was complete agreement between the EMIT® results when compared with TLC, RIA and GC/MS results. The reagents for this assay have only recently become available and there are no independent published performance data.

The relative standard deviation of the original assay using a partially automated analytical system has been found to be 13.7% within-batch and 23.8% between days at 3.0 mg/l [14]. Relative standard deviations of 5.0% and 9.4% (within-batch) at concentrations of 0.49 and 2.98 mg/l, respectively, have also been reported [5]. A third study reported a relative standard deviation of 12.7% on recovery-experiment replicates at 0.45 mg/l [15]. The detection limit has been determined by different groups as 0.3 mg/l [14], 0.5 mg/l [5] and 0.4 mg/l [7]. Employing an extraction and concentration technique, *Slightom* [16] was able to obtain a detection limit of 0.025 mg/l. In a single vial dry reagent configuration of the assay, relative standard deviations of 29.0% and 10.0% were obtained at 0.5 and 1.8 mg/l (between-batch) respectively [17] with a detection limit of 0.3 mg/l.

In a comparison of the original assay with thin-layer chromatography divergent results have been obtained in less than 5% of cases [14, 18]. In a study of 100 samples compared with a GC procedure no discrepant results were obtained [17]. Greater than 95% agreement with RIA was found in another study [5].

Broughton & Ross [19], using a 1.0 mg/l cut-off in the original EMIT® lysozyme assay, found 6% of results to be discrepant when compared with a TLC procedure.

Sources of error: false negative results may be obtained if the sample is not stored correctly [13]. The urine pH should be within the range of 5 to 8.

It has been recognized with the earlier EMIT® assays employing the lysozyme label that urines with high ionic strengths can produce falsely low absorbance changes; this can be achieved by adding sodium chloride to urine. It is expected that this effect will be seen with the dehydrogenase-linked assays [20] although this has not been confirmed.

Specificity: antiserum has been chosen in this assay to enable recognition of morphine, morphine glucuronide and codeine. The reagent manufacturer claims [13] that the assay will detect codeine and hydrocodone with one third, and hydromorphone, levorphanol and morphine glucuronide with one tenth, of the sensitivity for morphine. The following were also tested and found to give a signal less than the low calibrator at the concentrations stated:

Amphetamine	1000 mg/l	Methadone	500 mg/l
Benzoylecgonine	1000 mg/l	Nalorphine	20 mg/l
Chlorpromazine	12 mg/l	Naloxone	150 mg/l
Chloroquine	1000 mg/l	Oxazepam	250 mg/l
Dextromethorphan	175 mg/l	Phencyclidine	1000 mg/l
Doxylamine	1000 mg/l	Propoxyphene	1000 mg/l
Meperidine	20 mg/l	Secobarbital	1000 mg/l

In the original assay *Oellerich et al.* [14] showed that the following compounds gave a signal equivalent to 0.5 mg morphine per litre: codeine phosphate (0.4 mg/l), levorphanol tartrate (2.5 mg/l), ketobemidone (50.0 mg/l), pethidine hydrochloride (70.0 mg/l), levallorphan tartrate (100 mg/l), promethazine hydrochloride (260.0 mg/l) and chlorpromazine hydrochloride (310.0 mg/l).

Schneider et al. [6] showed that the original assay gave a signal 40% higher with codeine than with morphine, but in the case of morphine glucuronide it only gave 70% of the signal for a comparable concentration of morphine. Different sensitivities toward other compounds were also found.

Allen & Stiles [21] list a further 19 compounds that they found to cross-react in the original assay. Clearly the specificity of the assay will depend on the antiserum used and may vary from batch to batch of manufactured reagent.

References

[1] *J. H. Jaffe, W. R. Martin,* Opioid Analgesics and Antagonists, in: *A. Goodman Gilman, L. S. Goodman, A. Gilman* (eds.), The Pharmacological Basis of Therapeutics, 6th edit., MacMillan Publishers, New York 1980, pp. 494 – 513.
[2] *T. J. Meredith, J. A. Vale,* Poisoning due to Opiates and Morphine Derivatives, in: *T. J. Meredith, J. A. Vale* (eds.), Poisoning Diagnosis and Treatment, Update Books, London 1981, pp. 113 – 116.
[3] *J. Monforte, R. G. Turk,* Morphine, in: *I. Sunshine* (ed.), Methodology for Analytical Toxicology, CRC Press, Palm Beach 1975, pp. 267 – 271.
[4] *R. J. Kokoski, M. Jain,* Comparison of Results for Morphine Urinalyses by Radioimmunoassay and Thin Layer Chromatography in a Narcotic Clinic Setting, Clin. Chem. *21*, 417 – 419 (1975).
[5] *R. S. Schneider, P. Lindquist, E. T. Wong, K. E. Rubenstein, E. F. Ullman,* Homogeneous Enzyme Immunoassay for Opiates in Urine, Clin. Chem. *19*, 821 – 825 (1973).
[6] *D. S. Kabakoff, H. M. Greenwood,* Recent Advances in Homogeneous Enzyme Immunoassay, in: *K. G. M. M. Alberti, C. P. Price* (eds.), Recent Advances in Clinical Biochemistry 2, Churchill Livingstone, Edinburgh 1981, pp. 1 – 30.
[7] *F. L. Adler, C. T. Lin, D. H. Catlin,* Immunological Studies of Heroin Addition. I. Methodology and Application of a Haemagglutination Inhibition Test for Detection of Morphine, Clin. Immunol. Immunopathol. *1*, 53 – 58 (1972).
[8] *N. C. Jain, R. D. Budd, T. C. Sneath,* Morphine Type C Procedure, in: *I. Sunshine* (ed.), Methodology for Analytical Toxicology, CRC Press, Palm Beach 1975, pp. 271 – 274.

[9] *J. E. Wallace, S. C. Harris,* Morphine by Liquid Chromatography with Electrochemical Detection, in: *I. Sunshine* (ed.), Methodology for Analytical Toxicology, CRC Press, Palm Beach 1975, pp. 139 – 142.

[10] *D. Catlin, R. Cleeland, E. Grunberg,* A Sensitive, Rapid Radioimmunoassay for Morphine and Immunologically Related Substances in Urine and Serum, Clin. Chem. *19*, 216 – 220 (1973).

[11] *J. C. Garriott, W. Q. Sturner,* Morphine Concentrations and Survival Periods in Acute Heroin Fatalities, N. Eng. J. Med. *289*, 1276 – 1281 (1973).

[12] *R. J. Bastiani, R. C. Phillips, R. S. Schneider, E. F. Ullman,* Homogeneous Immunochemical Drug Assays, Am. J. Med. Technol. *29*, 211 – 216 (1973).

[13] Product literature. EMIT®-d.a.u. Opiate Assays, *Syva Co.,* Palo Alto, U.S.A. 1984.

[14] *M. Oellerich, W. R. Külpmann, R. Haeckel,* Drug Screening by Enzyme Immunoassay (EMIT) and Thin-Layer Chromatography (Drug Skreen), J. Clin. Chem. Clin. Biochem. *15*, 275 – 283 (1977).

[15] *V. R. Spiehler, D. Reed, R. H. Cravey, W. P. Wilcox, R. F. Shaw, S. Holland,* Comparison of Results for Quantitative Determination of Morphine by Radioimmunoassay, Enzyme Immunoassay and Spectrofluorimetry, J. Forensic Sci. *20*, 647 – 655 (1975).

[16] *E. L. Slightom,* The Analysis of Drugs in Blood, Bile and Tissue with an Indirect Homogeneous Enzyme Immunoassay, J. Forensic Sci. *23*, 292 – 303 (1978).

[17] *C. Sutheimer, B. R. Hepler, I. Sunshine,* Clinical Application and Evaluation of the EMIT®-s.t. Drug Detection System, Am. J. Clin. Pathol. *77*, 731 – 735 (1982).

[18] *S. J. Mule, M. L. Bastos, D. Jukofsky,* Evaluation of Immunoassay Methods for Detection, in Urine, or Drugs Subject to Abuse, Clin. Chem. *20*, 243 – 248 (1974).

[19] *A. Broughton, D. L. Ross,* Drug Screening by Enzyme Immunoassay with a Centrifugal Analyser, Clin. Chem. *21*, 1043 – 1065 (1981).

[20] *W. Godolphin,* Enzyme Multiplied Immunoassay Technique (EMIT®), in: *I. Sunshine, P. I. Jatlow* (eds.), Methodology for Analytical Toxicology *II*, CRC Press, Palm Beach 1982, pp. 189 – 203.

[21] *L. V. Allen, M. L. Stiles,* Specificity of the EMIT Drug Abuse Urine Assay Methods, Clin. Toxicol. *18*, 1043 – 1065 (1981).

2.8   Benzoylecgonine

3-(Benzoyloxy)-8-methyl-8-azabicyclo[3.2.1]octane-2-carboxylic acid

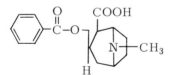

Christopher P. Price and R. Stewart Campbell

General

Benzoylecgonine is the major metabolite of cocaine; cocaine is benzoylmethyl-ecgonine. The major action of cocaine is its ability to block the initiation or conduction of the nerve impulse following local application; one of its major uses has been as a local anaesthetic. Its major systemic effect is the stimulation of the central nervous system; thus, there is a feeling of general well-being and euphoria. Small amounts of cocaine may slow the heart rate but as the amount of drug increases there is a marked increase in heart rate. Cocaine is also known to produce pyrexia [1].

Cocaine is absorbed from all sites of application including mucous membranes. After absorption cocaine is degraded by plasma and hepatic enzymes. A small amount of unchanged cocaine may be excreted in the urine [2, 3].

The known stimulatory effects of cocaine have led to the abuse of its use. The symptoms of toxicity due to cocaine are virtually indistinguishable from those associated with amphetamine intoxication. Thus, patients may demonstrate restlessness, irritability, nausea, hyperpyrexia, cardiac arrhythmias, hallucination and convulsions [4]. Treatment of cocaine toxicity is generally supportive.

Application of method: clinical chemistry, toxicology.

Substance properties relevant in analysis: molecular weight 289.3, as hydrochloride 325.8; pK_a 8.6. Substance is soluble in hot water to 10 g/l and is neutral to litmus.

Methods of determination: the metabolites of cocaine can be detected in urine using thin-layer chromatography [5]. A semi-quantitative method based on homogeneous enzyme-immunoassay can be used for determination of benzoylecgonine in urine [6, 7]. Quantitative determinations can be made using gas-liquid chromatography [8] and radioimmunoassay [5].

International reference method and standards: no reference methods or standards are known so far.

Assay

Method Design

The method is based on the enzyme-multiplied immunoassay technique (EMIT®) first described by *Rubenstein et al.* [9].

The assay is designed as a semi-quantitative assay to detect benzoylecgonine in urine samples. The assay is not designed to determine the level of intoxication.

The first homogeneous enzyme-immunoassay described for benzoylecgonine used a lysozyme label. The current assay employs a glucose-6-phosphate dehydrogenase label.

Principle: as described on pp. 7, 8, 271.

The activity of the marker enzyme, glucose-6-phosphate dehydrogenase, G6P-DH, is reduced if the enzyme-benzoylecgonine conjugate (E-L) is coupled to the antibody (Ab). Benzoylecgonine (L) in a sample competes with G6P-DH-labelled benzoyl-ecgonine (E-L) for a limited concentration of antibody (Ab). The activity of G6P-DH measured as increase in absorbance per unit time, $\Delta A/\Delta t$, is related to the benzoyl-ecgonine concentration (cf. Vol. I, chapter 2.7, p. 244).

Selection of assay conditions and adaptation to the individual characteristics of the reagents: as described in general on pp. 8, 9. Enzyme-drug conjugate is obtained by coupling benzoylecgonine to G6P-DH from *Leuconostoc mesenteroides*. The method involves pre-dilution of the sample in the reaction vessel. The desired assay range is 1.6 to 3.0 mg/l. The sample volume is 50 µl.

Equipment

The same equipment is used as described on p. 272.

Reagents and Solutions

Purity of reagents: as described on pp. 27, 272.

Preparation of solutions* (for 100 determinations): all solutions in re-purified water (cf. Vol. II, chapter 2.1.3.2).

1. Tris buffer (Tris, 55 mmol/l; NaN$_3$, 0.5 g/l; Triton X-100, 0.1 ml/l; pH 8.0):

    dissolve 1.996 g Tris base in 150 ml water, add 0.03 ml Triton X-100 and 0.15 g sodium azide, adjust to pH 8.0 with HCl, 1.0 mol/l; make up with water to 300 ml.

    Alternatively, dilute buffer concentrate provided with each kit to 200 ml with water.

2. Antibody/substrate solution (γ-globulin**; G-6-P, 66 mmol/l; NAD, 40 mmol/l; Tris, 55 mmol/l; NaN$_3$, 0.5 g/l; pH 5.2):

    reconstitute available lyophilized preparation with 6.0 ml water.

3. Conjugate solution (G6P-DH***; Tris, 55 mmol/l; NaN$_3$, 0.5 g/l; pH 8.0):

    this reagent is standardized to match the antibody/substrate solution (cf. p. 272). Reconstitute available lyophilized preparation with 6.0 ml water.

---

 * Reagents, calibrators and buffer are commercially available from *Syva Co.,* Palo Alto, U.S.A. The solutions contain preservatives.
 ** Standardized preparation from immunized sheep, concentration is titre-dependent.
*** The activity concentration of the enzyme is dependent on final choice of conjugate dilution.

4. Benzoylecgonine standard solutions (0.3 mg/l; 3.0 mg/l):

dissolve 67.2 mg benzoylecgonine hydrochloride (corresponding to 60.0 mg ben-zoylecgonine) in 1000 ml water. Use this stock solution to prepare two working standards of 0.3 mg/l and 3.0 mg/l by appropriate dilution with pooled drug-free human urine, containing 0.5 g sodium azide per litre. The urine diluent is used as a zero calibrator.

Alternatively, reconstitute each of the available lyophilized calibrators with 3.0 ml water; re-stopper the bottles and swirl gently to assist reconstitution.

Stability of solutions: store all reagents, stoppered, in a refrigerator at 2°C to 8°C. All reagents should be allowed to equilibrate to room temperature before use. The buffer (1), antibody and enzyme-drug conjugate solutions (2) and (3), and standard solutions (4) are all stable for 12 weeks after reconstitution.

Procedure

Collection and treatment of specimen: collect urine specimens in plastic or glass containers. The effect of urine preservatives has not been established and therefore their use is not recommended. Centrifuge turbid specimens before analysis.

Stability of the substance in the sample: store samples at 4°C for up to 3 days; storage for longer periods may result in false negative results in the case of samples with concentrations of benzoylecgonine near the low calibrator value [10].
Samples should be within the pH range of 5 to 8, if outside, adjust to fall within this range by addition of HCl, 1.0 mol/l, or NaOH, 1.0 mol/l, before analysis.

Assay conditions: measurement of samples and standards in duplicate; incubation time 30 s (in a reaction cuvette); room temperature; 0.6 ml; wavelength Hg 334 or 339 nm; light path 10 mm; final volume 0.9 ml; measurements against water; 30 s at 30°C.

Run the standards with each series of analyses. The zero calibrator assay serves as a reagent blank; correct all measurements for this value.

Calibration curve: plot the corrected values of the increase in absorbance per 30 s, $\Delta A/\Delta t$, of the calibrators *versus* the corresponding benzoylecgonine concentration (mg/l) using semi-logarithmic graph paper. A typical curve is shown in Fig. 1.

Measurement

Pipette* successively into the cuvette:			concentration in assay mixture	
sample or standard solution (4)		0.05 ml	benzoyl-ecgonine	up to 167 µg/l
buffer	(1)	0.25 ml	Tris	52 mmol/l
			NaN₃	470 mg/l
antibody/substrate solution	(2)	0.05 ml	γ-globulin	variable
			G-6-P	3.7 mmol/l
			NAD	2.2 mmol/l
buffer	(1)	0.25 ml		
incubate for 30 s at room temperature,				
conjugate solution	(3)	0.05 ml	G6P-DH**	variable
buffer	(1)	0.25 ml		
place cuvette in spectrophotometer immediately or aspirate contents into flow-cell, read absorbance after 15 s and 45 s.				

* Adequate mixing should be achieved by the addition of each reagent and no further agitation is necessary.
** Coupled to benzoylecgonine.

If ΔA per 30 s of the sample is greater than that of the highest calibrator, dilute the specimen further with Tris buffer (1) and re-assay.

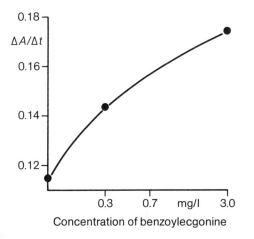

Fig. 1. Typical calibration curve for the assay of benzoylecgonine in urine.

Calculation: read the mass concentration of benzoylecgonine in the sample from the calibration curve. To convert from mass concentration to substance concentration, c (μmol/l), multiply by 3.457 (molecular weight is 289.3).

Interpretation: the peak of benzoylecgonine excretion following administration of cocaine will depend on the route of administration. In one study following intranasal application levels peaked at about 10 hours and remained positive for over 24 h. Typically when administered at a level of 1.5 mg/kg body weight the peak excretion was 10 mg/l; the peak level of excretion is dose dependent [11]. This data suggested that cocaine abuse can only be detected if a urine sample is obtained within 24 hours of administration.

The assay is designed as a semi-quantitative procedure. The assay is designed in the knowledge that several factors will influence the precision of an assay. The formulation of the assay and success in its use centre on the recognition of the analyte in the sample. Samples that give a signal above this cut-off will be considered as positive. If this cut-off is set too low, the probability of identifying a positive sample is increased but at the risk of producing false positive results. The low calibrator is generally set at the cut-off point. This approach has been demonstrated for several analytes [7, 12]. The original EMIT® procedure employed a lysozyme label; it is considered that assays using a G6P-DH label show enhanced sensitivity [7] and this is borne out in the recommended concentration range in which the assay can be used. In the original assay the cut-off level for benzoylecgonine was 1.6 mg/l; at this concentration 95% of samples would be detected as positive. The strategy recommended for interpretation of results has altered slightly and it is now considered that a sample giving a response equal to or greater than the low calibrator (0.3 mg/l) will be considered positive [10].

The reagent manufacturer has reported [10] that for each lot of reagents, between 25 to 50 benzoylecgonine-free urine samples and 25 to 50 urine samples containing 0.75 mg/l are tested and 99% of samples are correctly identified.

It is important to confirm a positive result by an alternative, preferably non-immunological, procedure.

Validation of Method

Precision, accuracy, detection limit and sensitivity: the reagent manufacturer has reported [10] that in clinical trials involving 175 urine samples (positive 92, negative 83) there was agreement in all but one instance when results with the EMIT® assay were compared with TLC and/or RIA. The reagents for this assay have only recently become available and there are no independent published performance data.

The original method has been shown to give a relative standard deviation (within series) of less than 7.0% at a benzoylecgonine concentration of 2.0 mg/l [6]. No false positives were found in samples containing barbiturates, amphetamines, methadone or morphine.

In over 100 samples from subjects known to have taken cocaine no false negatives were found. *Mule et al.* [13] found a false positive rate of 10% compared with TLC, although this was considered to be due to the poor sensitivity of the latter method. In a study of 200 urine samples *Mule et al.* [5] found that the EMIT® procedure agreed with RIA in 84.5% of cases and with GLC in 86.0% of cases.

Sources of error: false negative results may be obtained if the sample is not stored correctly [10]. The urine pH should be within the range of 5 to 8.

It has been recognized with the earlier EMIT® assays employing the lysozyme label that urines with high ionic strengths can produce falsely low absorbance changes; this can be achieved by adding sodium chloride to urine. It is expected that this effect will be seen with the dehydrogenase-linked assays [14] although this has not been confirmed.

Specificity: the assay is designed to detect the major metabolite of cocaine in urine. It has been claimed by the reagent manufacturers that the assay will also detect ecgonine at about one quarter of the sensitivity for benzoylecgonine. In addition, several compounds have been tested for possible cross-reactivity; concentrations that gave a response equivalent to less than 0.3 mg benzoylecgonine per litre were as follows [10]:

Acetaminophen	1000 mg/l	MEGX*	1000 mg/l
Acetylsalicylic acid	1000 mg/l	Methadone	500 mg/l
4-Aminobenzoic acid	1000 mg/l	Methaqualone	100 mg/l
Amitriptyline	100 mg/l	Morphine	200 mg/l
Amphetamine	500 mg/l	Oxazepam	250 mg/l
Cocaine	25 mg/l	Phencyclidine	750 mg/l
Codeine	500 mg/l	Procainamide	1000 mg/l
Dextromethorphan	175 mg/l	Propoxyphene	500 mg/l
Ecgonine	5 mg/l	Secobarbital	1000 mg/l

Mule et al. [13] found that the earlier lysozyme label assay detected ecgonine with only one tenth the sensitivity of benzoylecgonine and confirmed claims made by the reagent manufacturer for other compounds. The lack of any significant cross-reactivity was confirmed by *Allen & Stiles* [15].

References

[1] *J. M. Ritchie, N. M. Greene,* Local Anesthetics, in: *A. Goodman Gilman, L. S. Goodman, A. Gilman* (eds.), The Pharmacological Basis of Therapeutics, 6th edit., MacMillan Publishing, New York 1980, pp. 300 – 320.
[2] *C. Van Dyke, P. G. Barash, P. I. Jatlow, R. Byck,* Cocaine: Plasma Concentrations after Intranasal Application in Man, Science *191*, 859 – 861 (1976).
[3] *R. J. de Jong,* Local Anesthetics, Charles C. Thomas, Springfield 1977.

---

* Monoethylglycine xylidide.

[4] *J. A. Vale, T. J. Meredith,* Poisoning due to Non-Catecholamine Sympathomimetic Drugs, in: *J. A. Vale, T. J. Meredith* (eds.), Poisoning Diagnosis and Treatment, Update Books, London 1981, pp. 125–127.
[5] *S. J. Mule, D. Jukofsky, M. Kogan, A. DePace, K. Vereby,* Evaluation of the Radioimmunoassay for Benzoylecgonine (a Cocaine Metabolite in Human Urine), Clin. Chem. *23*, 796–801 (1977).
[6] *R. J. Bastiani, R. S. Schneider, K. R. James, M. J. Soffer,* Homogeneous Enzyme Immunoassay for Cocaine Metabolite in Urine, Clin. Chem. *19*, 663 (1973) Abstract.
[7] *D. S. Kabakoff, H. M. Greenwood,* Recent Advances in Homogeneous Enzyme Immunoassay, in: *K. G. M. M. Alberti, C. P. Price* (eds.), Recent Advances in Clinical Biochemistry 2, Churchill Livingstone, Edinburgh 1981, pp. 1–30.
[8] *P. I. Jatlow, D. N. Bailey,* Gas Chromatographic Analysis for Cocaine in Human Plasma, with Use of a Nitrogen Detector, Clin. Chem. *21*, 1918–1921 (1975).
[9] *K. E. Rubenstein, R. S. Schneider, E. F. Ullman,* "Homogeneous" Enzyme Immunoassay. A New Immunochemical Technique, Biochem. Biophys. Res. Commun. *47*, 846–851 (1972).
[10] Product literature. EMIT®-d.a.u. Cocaine Metabolite Assays, *Syva Co.,* Palo Alto, U.S.A., 1984.
[11] *C. Van Dyke, R. Byck, P. G. Barash, P. I. Jatlow,* Urinary Excretion of Immunologically Reactive Metabolite(s) after Intranasal Administration of Cocaine, as Followed by Enzyme Immunoassay, Clin. Chem. *23*, 241–244 (1977).
[12] *R. J. Bastiani, R. C. Phillips, R. S. Schneider, E. F. Ullman,* Homogeneous Immunochemical Drug Assays, Am. J. Med. Technol. *29*, 211–216 (1973).
[13] *S. J. Mule, M. L. Bastos, D. Jukofsky,* Evaluation of Immunoassay Methods for Detection in Urine of Drugs Subject to Abuse, Clin. Chem. *20*, 243–248 (1974).
[14] *W. Godolphin,* Enzyme Multiplied Immunoassay Technique (EMIT®), in: *I. Sunshine, P. I. Jatlow* (eds.), Methodology for Analytical Toxicology *II*, CRC Press, Palm Beach 1982, pp. 189–203.
[15] *L. V. Allen, M. L. Stiles,* Specificity of the EMIT® Drug Abuse Urine Assay Methods, Clin. Toxicol. *18*, 1043–1065 (1981).

2.9   Phencyclidine

1-(1-Phenylcyclohexyl)piperidine

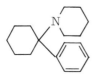

Christopher P. Price and R. Stewart Campbell

General

Phencyclidine is an arylcyclohexylamine which acts on the central nervous system; it can act either as a stimulant or a depressant and its action is considered to be dose-dependent. The drug was initially formulated and tested as a potent analgesic and anaesthetic [1]. However, it was found to produce serious and prolonged post-anaesthetic confusion and delirium and its clinical use was abandoned [2].

The use of phencyclidine has subsequently been abused for its potential psycho-mimetic effects; however, serious side-effects have limited its popularity.

Phencyclidine is easily absorbed following all routes of administration. The parent drug is hydroxylated and the metabolites are conjugated in the liver with glucuronic acid; only a small proportion of the drug is excreted unchanged [3].

Symptoms of phencyclidine abuse range from agitation, irritability, difficulty with speech, and horizontal and vertical nystagmus with low-grade abuse. Higher-dose symptoms may also include coma, stupor, muscle rigidity and psychosis. In certain cases clinical problems have been reported with phencyclidine intoxication, including *status epilepticus*, cerebral haemorrhage, acute renal failure, cardiac arrhythmias and apnoea.

Treatment of phencyclidine intoxication varies to a certain extent with the degree of abuse. After gastric lavage the management of the patient is mainly supportive with regular monitoring of the vital signs until normal mental state is achieved. Forced acid diuresis may promote the elimination of phencyclidine from the body [4].

Application of method: clinical chemistry, toxicology.

Substance properties relevant in analysis: molecular weight 243.4; phencyclidine hydrochloride (M_r = 279.86) is soluble in water to 166 g/l, in alcohol to 143 g/l and in chloroform to 500 g/l.

Methods of determination: phencyclidine may be detected in urine using thin-layer chromatography [5]. Homogeneous enzyme-immunoassay has been used to provide a rapid semi-quantitative method [6, 7].

Gas-liquid chromatography [8] and radioimmunoassay [9] have provided accurate quantitative results, but are time-consuming.

International reference method and standards: no reference methods or standards are known to date.

Assay

Method Design

The method is based on the enzyme-multiplied immunoassay technique (EMIT®) first described by *Rubenstein et al.* [10].

Principle: as described in detail on pp. 7, 8, 271.

The activity of the marker enzyme, glucose-6-phosphate dehydrogenase, G6P-DH, is reduced if the enzyme-phencyclidine conjugate (E-L) is coupled to the antibody (Ab). Phencyclidine (L) in a sample competes with G6P-DH-labelled phencyclidine (E-L) for a limited concentration of antibody (Ab). The activity of G6P-DH measured as increase in absorbance per unit time, $\Delta A/\Delta t$, is related to the phencyclidine concentration. The method is intended for use as a semi-quantitative assay for urine samples; the assay is not designed to determine the level of intoxication.

Selection of assay conditions and adaptation to the individual characteristics of the reagents: as described in general on pp. 8, 9. Enzyme-drug conjugate is obtained by coupling phencyclidine to G6P-DH from *Leuconostoc mesenteroides*. The method involves pre-dilution of the sample in the reaction vessel.

The desired assay range is 75 to 400 µg/l. The sample volume is 50 µl.

Equipment

The same equipment is used as described on p. 272.

Reagents and Solutions

Purity of reagents: as described on pp. 27, 272.

Preparation of solutions* (for 100 determinations): all solutions in re-purified water (cf. Vol. II, chapter 2.1.3.2).

1. Tris buffer (Tris, 55 mmol/l; NaN$_3$, 0.5 g/l; Triton X-100, 0.1 ml/l; pH 8.0):

   dissolve 1.996 g Tris base in 150 ml water, add 0.03 ml Triton X-100; adjust to pH 8.0 with HCl, 1.0 mol/l; make up with water to 300 ml.

   Alternatively, dilute buffer concentrate provided with each kit to 200 ml with water.

2. Antibody/substrate solution (γ-globulin**; G-6-P, 66 mmol/l; NAD, 40 mmol/l; Tris, 55 mmol/l; NaN$_3$ 0.5 g/l; pH 5.2):

   reconstitute available lyophilized preparation with 6.0 ml water.

---

  * Reagents, calibrators and buffer are commercially available from *Syva Co.,* Palo Alto, U.S.A. The solutions contain preservatives.

** Standardized preparation from immunized sheep, concentration is titre-dependent.

3. Conjugate solution (G6P-DH*; Tris, 55 mmol/l; NaN$_3$, 0.5 g/l; pH 8.0):

   this reagent is standardized to match the antibody/substrate solution (cf. p. 272). Reconstitute available lyophilized preparation with 6.0 ml water.

4. Phencyclidine standard solutions (75 µg/l; 400 µg/l):

   dissolve 46 mg phencyclidine hydrochloride (corresponding to 40.0 mg phencyclidine) in one litre water. Use this stock solution to prepare two working standards of 75 µg/l and 400 µg/l by appropriate dilution with pooled drug-free human urine, containing 0.5 g sodium azide per litre. The urine diluent is used as a zero calibrator.

   Alternatively, reconstitute each of the available lyophilized calibrators with 3.0 ml water; re-stopper the bottles and swirl gently to assist reconstitution.

Stability of solutions: store all reagents, stoppered, in a refrigerator at 2 °C to 8 °C. All reagents should be allowed to equilibrate to room temperature before use. The buffer (1), antibody and enzyme-drug conjugate solutions (2) and (3) are all stable for 12 weeks after reconstitution. The standard solutions (4) are stable for 14 days after reconstitution.

Procedure

Collection and treatment of specimen

Urine: collect urine specimens in plastic or glass containers. The effect of urine preservatives has not been established and therefore their use is not recommended. Centrifuge turbid specimens before analysis.

Blood: adjust 10 ml blood to pH 8.5 with sodium hydroxide, 0.1 mol/l, and extract with 70 ml diethyl ether. Wash the organic phase with 10 ml disodium hydrogen phosphate, 67 mmol/l, and then extract with 10 ml sulphuric acid, 1.0 mol/l. Adjust the sulphuric acid phase to pH 8.5 and then extract with 20 ml diethyl ether. After separation evaporate the organic solvent to dryness and reconstitute with 10 ml water.

Tissue: homogenize 5 g tissue with 20 ml phosphate buffer, 0.1 mol/l, pH 8.5. Extract the homogenate in the same way as blood, described above.

Stability of the substance in the sample: the stability of phencyclidine in biological materials is not known.

Assay conditions: measurement of samples and standards in duplicate; incubation time 30 s (in a reaction cuvette); room temperature; 0.6 ml; wavelength Hg 334 or

---

* The activity concentration of the enzyme is dependent on final choice of conjugate dilution.

339 nm; light path 10 mm; final volume 0.9 ml; measurements against water; 30 s at 30°C.

Run the standards with each series of analyses. The zero calibrator assay serves as a reagent blank; correct all measurements for this value.

Measurement

Pipette* successively into the cuvette:			concentration in assay mixture	
sample or standard solution (4)		0.05 ml	phencyclidine	up to 22 µg/l
buffer	(1)	0.25 ml	Tris	52 mmol/l
			NaN₃	470 mg/l
antibody/substrate solution	(2)	0.05 ml	γ-globulin	variable
			G-6-P	3.7 mmol/l
			NAD	2.2 mmol/l
buffer	(1)	0.25 ml		
incubate for 30 s at room temperature,				
conjugate solution	(3)	0.05 ml	G6P-DH**	variable
buffer	(1)	0.25 ml		
place cuvette in spectrophotometer immediately or aspirate contents into flow-cell, read absorbance after 15 s and 45 s.				

* Adequate mixing should be achieved by the addition of each reagent and no further agitation is necessary.
** Coupled to phencyclidine.

If ΔA per 30 s of the sample is greater than that of the highest calibrator, dilute the specimen further with Tris buffer (1) and re-assay.

Calibration curve: plot the corrected values of the increase in absorbance per 30 s, $\Delta A/\Delta t$, of the calibrators *versus* the corresponding phencyclidine concentration (mg/l) using semi-logarithmic graph paper. A typical curve is shown in Fig. 1.

Calculation: read the mass concentration of phencyclidine in the sample from the calibration curve. To convert from mass concentration to substance concentration, c (µmol/l), multiply by 4.108 (molecular weight is 243.4).

Interpretation: in fatal cases where no other drugs were detected concentrations of phencyclidine in the range 0.2 to 6.0 mg/l have been found [11]. Serum concentrations of 25 µg/l are associated with a feeling of disorientation and concentrations of greater than 100 µg/l are observed with stages of coma.

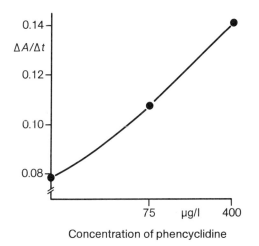

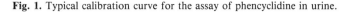

Concentration of phencyclidine

Fig. 1. Typical calibration curve for the assay of phencyclidine in urine.

Urine phencyclidine concentrations of 0.5 to 5.0 mg/l have been found in persons taking phencyclidine [12] with concentrations up to 330 mg/l in fatal overdosage.

The assay is designed as a semi-quantitative assay. The assay is designed in the knowledge that several factors will influence its precision. The formulation of the assay and success in its use centre on the recognition of a minimum $\Delta A/\Delta t$ indicating the presence of the analyte in the sample. Samples that give a signal above this cut-off will be considered positive. If this cut-off is set too low, the probability of identifying a positive sample is increased but at the risk of producing false positive results. The low calibrator is generally set at the cut-off point. This approach has been demonstrated for several analytes [7, 13]. In the case of phencyclidine the cut-off level is 75 µg/l.

It is important to confirm a positive result by an alternative, non-immunochemical procedure.

Validation of Method

Precision, accuracy, detection limit and sensitivity: it has been found that, in 100 urine samples spiked with 75 µg phencyclidine per litre, the assay gives a range of results with a mean value of 75 µg/l with a standard deviation of 37 µg/l. With this cut-off less than 5% false positives will be detected and the assay will detect more than 95% of samples containing 150 µg phencyclidine per litre [14].

The method has been shown to give a within-series relative standard deviation of 3.9% at 200 µg/l [15]. *Tom et al.* [6] found a relative standard deviation (within series) of the enzymatic rate of 0.52% and 0.70%, at the equivalent of 75 and 400 µg/l. Both groups used a manual procedure.

Walberg & Gupta [15] found results by the enzyme-immunoassay to be 20% higher than the result by gas chromatography; in a comparison of 95 samples with phencyclidine levels between 100 µg/l and 19.3 mg/l all results were positive, while in 61 samples found not to contain phencyclidine by GC the immunoassay produced 5 results greater than 75 µg/l. In a comparison between TLC, GC, GC/MS and the method described on 167 urine samples *Tom et al.* [6] found that there was agreement in 98% of cases between EMIT® and the other methods.

Sources of error: it has been recognized with the earlier EMIT® assays employing the lysozyme label that urines with high ionic strengths can produce falsely low $\Delta A/\Delta t$; this can be achieved by adding sodium chloride to urine [14]. It is expected that this effect will be seen with the dehydrogenase-linked assays [16] although this has not been confirmed.

Specificity: the assay is designed to detect phencyclidine. At high concentrations certain phencyclidine metabolites and analogues may produce a positive result. The cross-reactivity of a variety of substances has been assessed by the reagent manufacturer [14] and the minimum concentration to give a response equal to 75 µg/l phencyclidine was found to be as follows:

1-[1-(2-Thienyl)-cyclohexyl]morpholine (TCM)	⩽ 5.0 mg/l
N,N-diethyl-1-phenylcyclohexylamine	⩽ 3.0 mg/l
1-(4-Hydroxypiperidino)phenylcyclohexane	⩽ 3.0 mg/l
4-Phenyl-4-piperidinocyclohexanol	⩽ 2.0 mg/l
1-[1-(2-Thienyl)-cyclohexyl]pyrrolidine (TCPy)	⩽ 1.0 mg/l
1-(1-Phenylcyclohexyl) morpholine (PCM)	⩽ 1.0 mg/l
1-(1-Phenylcyclohexyl)pyrrolidine (PCPy)	⩽ 1.0 mg/l
1-Piperidinocyclohexane carbonitrile (PCC)	> 50 mg/l
1-Phenylcyclohexylamine (PCA)	> 50 mg/l
Dextromethorphan	⩾ 60 mg/l
Merperidine	⩾ 100 mg/l
Ketamine	⩾ 100 mg/l
Promethazine	⩾ 125 mg/l
Morphine	⩾ 200 mg/l

References

[1] *F. E. Greifenstein, M. De Vault, V. Yoshitaki, J. E. Gajewski,* A Study of 1-Arylcyclohexylamine for Anaesthesia, Anaesth. Analg. *37*, 283 – 294 (1958).

[2] *G. Chem, C. R. Ensor, B. Hohner,* The Neuropharmacology of 2-(*O*-Chlorophenyl)-2-methyl-amino-cyclohexanone hydrochloride, J. Pharmacol. Exp. Ther. *52*, 332 – 342 (1966).

[3] *J. H. Jaffe,* Drug Addiction and Drug Abuse, in: *A. Goodman Gilman, L. S. Goodman, A. Gilman* (eds.), The Pharmacological Basis of Therapeutics, 6th edit., MacMillan Publishing, New York 1980, pp. 567 – 568.

[4] *J. A. Vale, T. J. Meredith,* Forced Diuresis, Dialysis and Haemoperfusion, in: *J. A. Vale, T. J. Meredith* (eds.), Poisoning Diagnosis and Treatment, Update Books, London 1981, pp. 59 – 68.

[5] *H. I. Finkle,* Phencyclidine Identification by Thin Layer Chromatography. A Rapid Screening Procedure for Emergency Toxicology, Am. J. Clin. Pathol. *70,* 287 – 290 (1978).

[6] *H. Tom, D. S. Kabakoff, C. I. Lin. P. Singh, M. White, P. Westkamper,* Homogeneous Enzyme Immunoassay for Phencyclidine in Urine, Clin. Chem. *25,* 1144 (1979), Abstract.

[7] *D. S. Kabakoff, H. M. Greenwood,* Recent Advances in Homogeneous Enzyme Immunoassay, in: *K. G. M. M. Alberti, C. P. Price* (eds.), Recent Advances in Clinical Biochemistry, Churchill Livingstone, Edinburgh 1981, pp. 1 – 30.

[8] *D. N. Bailey, J. J. Guba,* Gas Chromatographic Analysis for Phencyclidine in Plasma, with Use of a Nitrogen Detector, Clin. Chem. *26,* 437 – 440 (1980).

[9] *S. M. Owens, J. Woodworth, M. Mayersohn,* Radioimmunoassay for Phencyclidine (PCP) in Serum, Clin. Chem. *28,* 1509 – 1513 (1982).

[10] *K. E. Rubenstein, R. S. Schneider, E. G. Ullman,* "Homogeneous" Enzyme Immunoassay. A New Immunochemical Technique, Biochem. Biophys. Res. Commun. *47,* 846 – 851 (1972).

[11] *G. F. Kessler, L. M. Demers, C. Berlin,* Phencyclidine and Fatal Status Epilepticus, N. Eng. J. Med. *291,* 979 – 983 (1974).

[12] *P. C. Reynolds,* Clinical and Forensic Experiences with Phencyclidine, Clin. Toxicol. *9*(4), 547 – 552 (1976).

[13] *R. J. Bastiani, R. C. Phillips, R. S. Schneider, E. F. Ullman,* Homogeneous Immunochemical Drug Assays, Am. J. Med. Technol. *29,* 211 – 216 (1973).

[14] Product Literature, EMIT®-d.a.u. Phencyclidine Urine Assay, *Syva Co.,* Palo Alto, U.S.A., 1982.

[15] *C. Walberg, R. Gupta,* Qualitative Estimation of Phencyclidine in Urine by Homogeneous Enzyme Immunoassay (EMIT), Clin. Chem. *25,* 1144 (1979), Abstract.

[16] *W. Godolphin,* Enzyme Multiplied Immunoassay Technique (EMIT®), in: *I. Sunshine, P. I. Jatlow* (eds.), Methodology for Analytical Toxicology *II,* CRC Press, Palm Beach 1982, pp. 189 – 203.

2.10   Propoxyphene

[S-(R*,S*)]-α-[2-(Dimethylamino)-1-methylethyl]-α-phenylbenzeneethanol propanoate

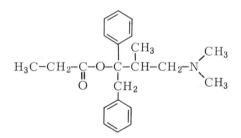

Christopher P. Price and R. Stewart Campbell

General

Propoxyphene is a drug that binds to opioid receptors and has analgesic and other central nervous system effects similar to those seen with codeine and other opiates. It has no anti-inflammatory or antipyretic effects [1]. Propoxyphene and propoxyphene napsylate may be prescribed as mildly effective narcotic analgesics.

Propoxyphene is absorbed after oral or parenteral administration; peak concentrations in serum are reached about two hours after administration. The major route of metabolism is *N*-demethylation to yield norpropoxyphene which is then excreted in the urine [1].

The most common preparation of propoxyphene is in combination with paracetamol. Overdosage with propoxyphene is most commonly found in combination with paracetamol.

The symptoms of propoxyphene overdosage are similar to those associated with opiate poisoning and include depressed respiration, diminished pupil size and loss of consciousness. In cases of more severe toxicity the picture may be complicated by convulsions. Toxic doses have also been associated with delirium, hallucination, and cardiac arrhythmias. When the drug is ingested with alcohol the respiratory depression is more pronounced [2].

Respiratory depression may be reversed by the use of naloxone. In view of the possibility of respiratory complications, gastric aspiration and lavage are not usually performed in the unconscious patient. Further treatment is in the form of intense supportive therapy with careful monitoring of the respiratory system [2].

Application of method: clinical chemistry, toxicology.

Substance properties relevant in analysis: molecular weight 339.5, as hydrochloride 376.0; pK_a 6.3. It is readily soluble in water as the hydrochloride to 3000 g/l, in ethanol to 666 g/l and in chloroform to 1667 g/l.

Methods of determination: propoxyphene can be detected by ultraviolet spectrophotometry following a complicated extraction procedure [3]. A semi-quantitative technique has been described using homogeneous enzyme-immunoassay [4]. Quantitative procedures are generally based on gas-liquid [5] or high-performance liquid [6] chromatography.

International reference method and standards: no reference methods or standards are known to date.

Assay

Method Design

The method is based on the enzyme-multiplied immunoassay technique (EMIT®) described by *Rubenstein et al.* [7].

The assay is designed to detect the presence of propoxyphene, propoxyphene salts such as propoxyphene napsylate, and the major metabolite norpropoxyphene. The assay is designed as a semi-quantitative technique for urine samples.

Principle: as described in detail on pp. 7, 8, 271.

The activity of the marker enzyme, glucose-6-phosphate dehydrogenase, G6P-DH, is reduced if the enzyme-propoxyphene conjugate (E-L) is coupled to the antibody (Ab). Propoxyphene in a sample competes with G6P-DH-labelled propoxyphene (E-L) for a limited quantity of antibody (Ab). The activity of G6P-DH measured as increase in absorbance per unit time, $\Delta A/\Delta t$, is related to the propoxyphene concentration (cf. Vol. I, chapter 2.7, p. 244).

Selection of assay conditions and adaptation to the individual characteristics of the reagents: as described in general on pp. 8, 9. Antibodies to propoxyphene have been produced by immunizing sheep with propoxyphene chemically coupled to a macromolecular carrier.

Enzyme-drug conjugate is obtained by coupling a derivative of propoxyphene to G6P-DH from *Leuconostoc mesenteroides* (cf. p. 272). Interference from endogenous G6P-DH is avoided by use of the coenzyme NAD which is converted only by the bacterial enzyme.

The optimal quantities and concentrations of sample and reagents are dependent on the characteristics of the antibodies present in the antiserum. The desired assay range is 0.3 to 1.0 mg/l. The sample is 50 µl.

Equipment

The same equipment is used as described on p. 272.

Reagents and Solutions

Purity of reagents: as described on pp. 27, 272.

Preparation of solutions* (for 100 determinations): all solutions in re-purified water (cf. Vol. II, chapter 2.1.3.2).

---

* Reagents, calibrators and buffer are commercially available from *Syva Co.,* Palo Alto, U.S.A. The solution contain preservatives.

1. Tris buffer (55 mmol/l; NaN$_3$, 0.5 g/l; Triton X-100, 0.1 ml/l; pH 8.0):

   dissolve 1.996 g Tris base in 150 ml water; add 0.03 ml Triton X-100 and 0.15 g sodium azide; adjust to pH 8.0 with HCl, 1.0 mol/l; make up with water to 300 ml.

   Alternatively, dilute buffer concentrate provided with each kit to 200 ml with water.

2. Antibody/substrate solution (γ-globulin*; G-6-P, 66 mmol/l; NAD, 40 mmol/l; Tris, 55 mmol/l; NaN$_3$, 0.5 g/l; pH 5.2):

   reconstitute available lyophilized preparation with 6.0 ml water.

3. Conjugate solution (G6P-DH**; Tris, 55 mmol/l; NaN$_3$, 0.5 g/l; pH 8.0):

   this reagent is standardized to match the antibody/substrate solution (cf. p. 272). Reconstitute available lyophilized preparation with 6.0 ml water.

4. Propoxyphene standard solution (0.3 mg/l; 1.0 mg/l):

   dissolve 11.1 mg propoxyphene hydrochloride (corresponding to 10.0 mg propoxyphene) in 100 ml water. This stock solution is then used to prepare two working standards of 0.3 and 1.0 mg/l by appropriate dilution with pooled normal human urine containing 0.5 g sodium azide/l. The urine diluent is used as a zero calibrator.

   Alternatively, reconstitute each of the available lyophilized calibrators with 3.0 ml water. After addition of water the bottles should be re-stoppered and gently swirled to assist reconstitution.

Stability of solutions: store all reagents, stoppered, in a refrigerator at 2°C to 8°C. All reagents and calibrators should be allowed to equilibrate to room temperature before use. The buffer (1), antibody/substrate and conjugate reagents (2) and (3), and standard solutions (4) are all stable for 12 weeks after reconstitution.

Procedure

Collection and treatment of specimen

Urine: collect urine into plastic or glass containers. The effect of urine preservatives is not known and therefore their use is not recommended. Centrifuge turbid specimens before analysis.

---

 * Standard preparation from immunized sheep, concentration is titre-dependent.
** Enzyme activity concentration is dependent on final choice of conjugate.

Blood: adjust 10 ml blood to pH 8.5 with potassium carbonate solution, 50 mmol/l, and extract with 75 ml chloroform. Wash the organic phase with 10 ml disodium hydrogen phosphate, 65 mmol/l, and extract with 10 ml sulphuric acid, 25 mmol/l. Adjust to pH 8.5 with sodium hydroxide, 0.5 mol/l, and extract with chloroform. Remove the organic solvent phase and evaporate to dryness; reconstitute the residue subsequently in 0.5 ml water.

Assay conditions: measurements of samples and standards in duplicate; incubation time 30 s (in a reaction cuvette); room temperature; 0.6 ml; wavelength Hg 334 or 339 nm; light path 10 mm; final volume 0.9 ml; measurements against water; 30 s at 30°C.

Run the standards with each series of analyses. The zero calibrator assay serves as a reagent blank; all measurements are corrected for this value.

Measurement

Pipette* successively into the reaction cuvette:			concentration in assay mixture	
sample or calibrator (4)		0.05 ml	drug	up to 56 µg/l
buffer	(1)	0.25 ml	Tris	52 mmol/l
			NaN$_3$	470 mg/l
antibody/substrate solution	(2)	0.05 ml	γ-globulin	variable
buffer	(1)	0.25 ml	G-6-P	3.7 mmol/l
			NAD	2.2 mmol/l
incubate at room temperature for 30 s;				
conjugate solution	(3)	0.05 ml	G6P-DH**	variable
buffer	(1)	0.25 ml		
place cuvette in spectrophotometer immediately or aspirate contents into flow-cell, read absorbance after 15 s and 45 s.				

* Adequate mixing should be achieved by the addition of each reagent and no further agitation is necessary.
** Coupled to a derivative of propoxyphene.

If ΔA per 30 s of sample exceeds that of the highest calibrator, dilute the sample further with Tris buffer (1) and re-assay.

Calibration curve: plot the corrected values of increase in absorbance per 30 s, $\Delta A/\Delta t$, of the calibrators *versus* the corresponding propoxyphene concentration (mg/l) using semi-logarithmic graph paper. A typical curve is shown in Fig. 1.

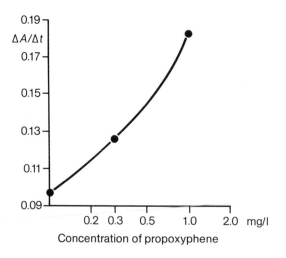

Fig. 1. Typical calibration curve for the assay of propoxyphene in urine.

Calculation: read the propoxyphene mass concentration of the sample from the calibration curve. As the assay system also detects propoxyphene salts, result will be in equivalents of propoxyphene and conversion to substance concentration is not appropriate.

Interpretation: the serum concentration of propoxyphene can reach 0.2 mg/l in one hour following an oral dose of 195 mg propoxyphene; a level of 0.3 mg/l is achieved in 30 min after a 50 mg intravenous dose [8]. In fatal poisonings blood concentrations of at least 2.0 mg/l are attained with levels up to 20 mg/l possible; levels in the liver may reach 30 mg/100 g tissue.

The assay is designed as a semi-quantitative procedure. The assay is designed in the knowledge that several factors will influence its precision. The formulation of the assay and success in its use centre on the recognition of a minimum $\Delta A/\Delta t$ indicating the presence of the analyte in the sample. Samples that give a signal above this cut-off will be considered positive. If this cut-off is set too low, the probability of identifying a positive sample is increased but at the risk of producing false positive results. The low calibrator is generally set at the cut-off point. This approach has been demonstrated for several assays [4, 9]. In the original EMIT procedure the cut-off level was 2.0 mg/l; with this cut-off less than 5% false positives will be detected and the assay will detect more than 95% of samples containing 2.0 mg propoxyphene per litre [9].

It is considered that the assays using a G6P-DH label show enhanced sensitivity [4] and this is borne out in the recommended concentration range in which the assay can be used. The strategy recommended for interpretation of results has altered slightly and it is now considered that a sample giving a response equal to or greater than the low calibrator (0.3 mg/l) will be considered as positive [10]. The reagent manufacturer has reported [10] that, for each lot of reagent, between 25 to 50 urine samples containing 0.5 mg propoxyphene per litre are tested and 99% of samples are correctly identified.

It is important that a positive result is confirmed by an alternative, preferably non-immunological, procedure.

Validation of Method

Precision, accuracy, detection limit and sensitivity: the reagent manufacturer has reported [10] that in a clinical trial involving 172 urine samples (positive, 84; negative, 88) there was agreement in 97% of cases when compared with a TLC procedure; the TLC procedure was found to have an inferior limit of detection (1.0 mg/l). The discrepant samples were analyzed further and in each case norpropoxyphene was shown to be present by GLC. One sample had given a positive result on the EMIT whilst the other five were negative; in the latter cases $\Delta A/\Delta t$ was greater than the zero but less than the low (0.3 mg/l) calibrator. The reagents for this assay have only recently become available and there are no independent published performance data.

Slightom [11] found with the earlier EMIT, employing the lysozyme label, that a detection limit of 2.0 mg/l was attainable and by use of an extraction and concentration procedure this could be reduced to 50 µg/l.

Sources of error: the optimum pH (5.5 to 8.0) of urine samples to be assayed must be maintained. Turbid samples may produce an erroneous result and should therefore be centrifuged prior to assay.

It has been recognized with the earlier EMIT employing the lysozyme label that urines with high ionic strengths can produce falsely low absorbance changes; this situation can be achieved by adding sodium chloride to urine. It is expected that this effect will be seen with the dehydrogenase-linked assay [12] although this has not been confirmed.

Specificity: the assay has been designed to detect propoxyphene, propoxyphene salts and the major urinary metabolite norpropoxyphene. Cross-reactivity has been tested by the manufacturer [10] with a number of non-propoxyphene substances and each has been shown to give a response less than the 0.3 mg calibrator per litre (at the concentration stated).

Compound	Concentration mg/l	Compound	Concentration mg/l
Amphetamine	1000	Methaqualone	1000
Benzoylecgonine	1000	Morphine	1000
Caffeine	1000	Oxazepam	300
Codeine	500	Phencyclidine	1000
Dextromethorphan	100	Secobarbital	1000
Methadone	100		

It has been pointed out that methadone may interfere with the assay at a toxic level of greater than 100 mg/l. Urine concentration of methadone and its metabolite 2-ethyli-dene-1,5-dimethyl-3,3-diphenylpyrrolidine from 1 to 50 mg/l are commonly found in patients on maintenance doses of methadone [10].

Employing the original EMIT with the lysozyme label, *Allen & Stiles* [13] found that the following compounds gave a positive result (i.e. greater than 2.0 mg/l at a concentration of 1000 mg/l): brompheniramine maleate, cyproheptadine hydrochloride, desipramine hydrochloride, imipramine hydrochloride, promethazine hydrochloride, triethyperazine maleate, tripellenamine HCl. It is unlikely that these concentrations would ever be encountered in urine samples.

References

[1] *A. J. McBay,* Propoxyphene and Norpropoxyphene Concentrations in Blood and Tissues in Cases of Fatal Overdose, Clin. Chem. *22,* 1319 – 1321 (1976).

[2] *J. H. Jaffe, W. R. Martin,* Opioid Analgesics and Antagonists, in: *A. Goodman Gilman, L. S. Goodman, A. Gilman* (eds.), The Pharmacological Basis of Therapeutics, 6th edit., MacMillan Publishers, New York 1980, pp. 494 – 534.

[3] *A. J. McBay, R. F. Turk, B. W. Corbett, P. Hudson,* Determination of Propoxyphene in Biological Materials, J. Forensic Sci. *19,* 81 – 85 (1974).

[4] *D. S. Kabakoff, H. M. Greenwood,* Recent Advances in Homogeneous Enzyme Immunoassay, in: *K. G. M. M. Alberti, C. P. Price* (eds.), Recent Advances in Clinical Biochemistry 2, Churchill Livingstone, Edinburgh 1981, pp. 1 – 30.

[5] *B. S. Finkle,* Propoxyphene Type C Procedure, in: *I. Sunshine* (ed.), Methodology for Analytical Toxicology, CRC Press, Palm Beach 1975, pp. 324 – 327.

[6] *G. Krisko, J. Depra, C. C. Cuppett,* Optimization of an HPLC Assay for Propoxyphene and Norpropoxyphene, Clin. Chem. *29,* 1203 – 1204 (1983).

[7] *K. E. Rubenstein, R. S. Schneider, E. F. Ullman,* "Homogeneous" Enzyme Immunoassay. A New Immunochemical Technique. Biochem. Biophys. Res. Commun. *47,* 846 – 851 (1972).

[8] *R. L. Wolen, C. M. Gruber, G. F. Kiplinger, N. E. Scholz,* Concentration of Propoxyphene in Human Plasma Following Oral, Intramuscular and Intravenous Administration, Toxicol. Appl. Pharmacol. *19,* 493 – 498 (1971).

[9] *R. J. Bastiani, R. C. Phillips, R. S. Schneider, E. F. Ullman,* Homogeneous Immunochemical Drug Assays, Am. J. Med. Technol. *29,* 211 – 216 (1973).

[10] Product Literature. EMIT®-d.a.u. Propoxyphene Assay, *Syva Co.,* Palo Alto, U.S.A. 1985.

[11] *E. L. Slightom,* The Analysis of Drugs in Blood, Bile and Tissue with an Indirect Homogeneous Enzyme Immunoassay, J. Forensic Sci. *23,* 292 – 303 (1978).

[12] *W. Godolphin,* Enzyme Multiplied Immunoassay Technique (EMIT®), in: *I. Sunshine, P. I. Jatlow* (eds.), Methodology for Analytical Toxicology, Vol. *II,* CRC Press, Palm Beach 1982, pp. 189 – 203.

[13] *L. V. Allen, M. L. Stiles,* Specificity of the EMIT Drug Abuse Urine Assay Methods, Clin. Toxicol. *18,* 1043 – 1065 (1981).

2.11 Δ^9-Tetrahydrocannabinol

Tetrahydro-6,6,9-trimethyl-3-pentyl-6*H*,dibenzo[*b,d*]pyran-1-ol

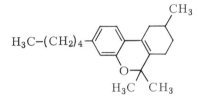

Christopher P. Price and R. Stewart Campbell

General

Δ^9-Tetrahydrocannabinol (Δ^9-THC) is generally considered to be the major psycho-active agent in marijuana; other constituents may also be important in modifying its effects. The effects of Δ^9-THC are mainly on the central nervous and cardiovascular systems; the effects vary quite significantly according to dose and mode of administration: smoking is the most common method of receiving Δ^9-THC. It is not used for any clinical purposes, although some analogues have been considered for their antiemetic and bronchodilator properties [1].

Despite the fact that the most effective route of administration of Δ^9-THC is inhalation, less than 50% is actually absorbed; peak concentrations in the blood are achieved in less than 30 minutes.

Δ^9-THC is rapidly metabolized to an active metabolite, 11-hydroxy-Δ^9-THC; this is subsequently converted to an inactive metabolite, 8,11-dihydroxy-THC, which is excreted in the urine and the faeces. Metabolites excreted in the bile may be re-absorbed. Traces of Δ^9-THC metabolites remain in the body for several days after administration and can be detected in urine [2].

There is no relationship between urinary levels of metabolites and symptoms, hence any assay is only of value in detecting abuse [3].

Symptoms of Δ^9-THC abuse are very varied but will include effects on mood, memory, motor co-ordination, time- and self-perception. Balance is affected at low doses of drug as are sensory faculties. Higher levels of drug may induce hallucinations and thinking becomes disorganized. A panic state may be observed and there is invariably an increase in heart rate and blood pressure [4].

Treatment of patients with evidence of drug abuse will vary considerably depending on the symptoms demonstrated. Intensive supportive therapy will clearly be important with careful attention to the cardiovascular complications [4].

Application of method: clinical chemistry, toxicology.

Substance properties relevant in analysis: molecular weight of tetrahydrocannabinol is 314.5; pK_a 10.6; insoluble in water, soluble in ethanol to at least 5 g/l. Molecular weight of 11-nor-Δ^8-tetrahydrocannabinol carboxylic acid is 361; this derivative is soluble in water to at least 0.2 g/l.

Methods of determination: tetrahydrocannabinol metabolites can be measured in blood and urine using radioimmunoassay [5]. Gas chromatography with mass spectrometry has also been employed in the detection of cannabinoids in urine [6] although the sample requirement is high.

High-pressure liquid chromatography has also been used to identify cannabinoids in blood [7]. Several thin-layer chromatographic procedures have also been described [8, 9], fluorescent derivatives being used to obtain the required sensitivity.

A semi-quantitative homogeneous enzyme-immunoassay has been developed for the rapid detection of cannabinoid metabolites in urine [3].

Assay

Method Design

The method is based on the enzyme-multiplied immunoassay technique (EMIT®) described by *Rubenstein et al.* [10].

The assay has been designed to detect a group of cannabinoid compounds in urine samples. The assay will detect abuse of marijuana. The assay is designed as a semi-quantitative assay.

Principle: as described in detail on pp. 7, 8, 271. However, malate dehydrogenase*, MDH, is used as label. The conjugated enzyme catalyzes the following reaction:

$$\text{Malate} + \text{NAD}^+ \xrightarrow{\text{MDH*}} \text{oxaloacetate} + \text{NADH} + \text{H}^+ .$$

The activity of the marker enzyme is reduced if the enzyme-Δ^9-THC conjugate (E-L) is coupled to the antibody (Ab). Δ^9-THC (L) in a sample competes with MDH-labelled Δ^9-THC (E-L) for a limited concentration of antibody (Ab). The activity of the MDH measured as increase in absorbance per unit time, $\Delta A/\Delta t$, is related to the Δ^9-THC concentration (cf. Vol. I, chapter 2.7, p. 244).

Selection of assay conditions and adaptation to the individual characteristics of the reagents: as described in general on pp. 8, 9, 271.

---

* L-malate: NAD⁺ oxidoreductase, EC 1.1.1.37.

Enzyme-drug conjugate is prepared by conjugation of pig-heart mitochondrial malate dehydrogenase to a derivative of Δ^9-THC [3]. The desired assay range is from 20 to 75 µg/l. The sample volume is 50 µl.

Equipment

The same equipment is used as described on p. 272.

Reagents and Solutions

Purity of reagents: MDH should be of the best available purity (e.g. 1200 U/mg at 25 °C, oxaloacetate as substrate). Chemicals are of analytical grade.

Preparation of solutions* (for 100 determinations): all solutions in re-purified water (cf. Vol. II, chapter 2.1.3.2).

1. Tris buffer/substrate solution (Tris, 150 mmol/l; malate, 143 mmol/l; NaN_3, 0.5 g/l; Triton X-100, 0.1 ml/l; pH 8.8):

   dissolve 5.444 g Tris base and 5.749 g malic acid in 150 ml water; add 0.03 ml Triton X-100 and 0.15 g sodium azide; adjust to pH 8.8 with NaOH, 1.0 mol/l; make up with water to 300 ml.

   Alternatively, dilute buffer concentrate provided with each kit to 200 ml with water.

2. Antibody/coenzyme solution (γ-globulin**; NAD, 100 mmol/l; glycine, 150 mmol/l; pH 5.0):

   reconstitute available lyophilized preparation with 8.0 ml water or to the volume stated on the bottle if different.

3. Conjugate solution (MDH***; Tris, 10 mmol/l; pH 7.4):

   this reagent must be standardized to match the antibody/coenzyme solution (cf. p. 272). Or reconstitute available lyophilized preparation with 6.0 ml water or to the volume stated on the bottle if different.

---

* Reagents, calibrators and buffer are commercially available from *Syva Co.,* Palo Alto, U.S.A. The solutions contain preservatives.
** Standardized preparation from immunized sheep, concentration is titre-dependent.
*** The activity concentration of the enzyme is dependent on final choice of conjugate dilution.

4. Cannabinoid standard solutions (20 µg/l; 75 µg/l):

the more soluble 11-nor-Δ^8-tetrahydrocannabinol 9-carboxylic acid is used as a standard in preference to Δ^9-THC.

Dissolve 7.5 mg 11-nor-Δ^8-THC 9-carboxylic acid in 1 litre water. Use this stock solution to prepare two working standards of 20 µg/l and 75 µg/l by appropriate dilution with pooled drug-free human urine, containing 0.5 g sodium azide per litre. The urine diluent is used as zero calibrator.

Alternatively, reconstitute each commercial calibrator with 3.0 ml water. Re-stopper the bottles and swirl gently to assist reconstitution.

Stability of solutions: store all reagents stoppered in a refrigerator at 2°C to 8°C. All reagents should be allowed to equilibrate to room temperature before use. The buffer (1), antibody and enzyme-drug conjugate solutions (2) and (3) are all stable for 12 weeks after reconstitution. The standard solutions (4) are stable for 14 days after reconstitution.

Procedure

Collection and treatment of specimen: collect urine in plastic or glass containers. Bring fresh specimens with pH outside the range 5 to 8 within range by the use of hydrochloric acid, 1.0 mol/l, or sodium hydroxide, 1.0 mol/l. Urine should be analyzed within 24 h of collection or stored at −20°C until analysis. Centrifuge turbid specimens before analysis.

The enzyme malate dehydrogenase is known to be inactivated by the preservative Thiomersal; the effects of other preservatives on this assay have not been established and their use is not recommended. This method has not been applied to other body fluids or tissue; consequently there is no validated extraction procedure available.

Stability of substance in the sample: samples known to contain cannabinoids may exhibit reduced recovery of analyte when stored in direct sunlight or at elevated temperatures.

Assay conditions: measurement of samples and standards in duplicate; incubation time 30 s (in a reaction cuvette); room temperature; 0.6 ml; wavelength Hg 334 or 339 nm; light path 10 mm; final volume 0.9 ml; measurements against water; 30 s at 30°C.

Run the standards with each series of assays. The zero calibrator assay serves as a reagent blank; correct all measurements for this value.

Measurement

Pipette* successively into the cuvette:			concentration in assay mixture	
sample or calibrator (4) buffer/substrate solution (1)		0.05 ml 0.25 ml	cannabinoids up to ca. 4 µg/l Tris 125 mmol/l NaN$_3$ 470 mg/l malate 119 mmol/l	
antibody/coenzyme solution (2)		0.05 ml	γ-globulin variable NAD 5.6 mmol/l glycine 8.3 mmol/l	
buffer/substrate solution (1)		0.25 ml		
incubate for 30 s,				
conjugate solution (3) buffer/substrate solution (1)		0.05 ml 0.25 ml	MDH** variable	
place cuvette in spectrophotometer immediately or aspirate contents into flow-cell, read absorbance after 15 s and 45 s.				

* Adequate mixing should be achieved by the addition of each reagent and no further agitation is necessary.
** Coupled to a derivative of Δ^9-THC.

If ΔA per 30 s of the sample is greater than that of the highest calibrator, dilute the specimen further with Tris buffer (1) and re-assay.

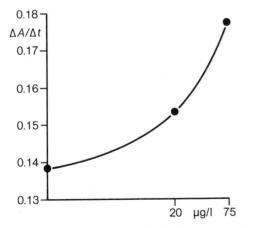

Concentration of 11-nor-Δ^8-THC 9-carboxylic acid

Fig. 1. Typical calibration curve for the assay of Δ^9-THC metabolites in urine.

Calibration curve: plot the corrected values of increase in absorbance per 30 s, $\Delta A/\Delta t$, of the calibrators *versus* the corresponding Δ^9-THC concentration (µg/l) using semi-logarithmic graph paper. A typical curve is shown in Fig. 1.

Calculation: read the mass concentration of 11-nor-Δ^8-THC 9-carboxylic acid equivalents in the sample from the calibration curve. It is not appropriate to convert mass concentration to substance concentration (assay detects various metabolites).

Interpretation: Δ^9-THC is rapidly absorbed and is almost completely metabolized, less than 1% of the unchanged compound being recovered. It has been shown that approximately 30% of metabolites are excreted in the faeces over the first 72 h following administration, up to 15% are excreted in the urine [11].

Smoking marijuana is the most effective route of administration; serum levels of Δ^9-THC peak immediately after smoking and decline rapidly: peak levels ranged from 7 to 70 µg/l in four subjects after smoking one cigarette impregnated with 5 mg tetrahydrocannabinol [12]. Urine excretion was greatest over the first 24 h period with levels ranging from 15 to 55 µg/l. Levels can be detected in the urine up to 10 days after exposure.

The assay is designed as a semi-quantitative procedure. The assay is designed in the knowledge that several factors will influence its precision. The formulation of the assay and success in its use centre on the recognition of a minimum $\Delta A/\Delta t$ indicating the presence of the analyte in the sample. Samples that give a signal above this cut-off will be considered as positive. If this cut-off is set too low, the probability of identifying a positive result is increased but at the risk of producing false positive results. The low calibrator is generally set at the cut-off point. This approach has been demonstrated for several analytes [3, 13, 14]. In the case of this assay the cut-off level is 20 µg/l.

It is important to confirm a positive result by an alternative, preferably non-immunological, procedure.

Validation of Method

Precision, accuracy, detection limit and sensitivity: it has been shown that, in 100 urine samples spiked with 20 µg Δ^9-THC per litre, the assay gives a range of results with a mean of 20 µg/l with a standard deviation of 15 µg/l. Thus, with this cut-off, less than 5% false positives will be detected and the assay will detect more than 95% of samples containing 50 µg Δ^9-THC per litre [15].

Rodgers et al. [3] found a within-run relative standard deviation of 4.0% at 15 µg/l and 2.0% at 75 µg/l. Using spiked samples the relative standard deviations were 17.0% at 25 µg/l and 12.0% at 45 µg/l. These authors found that a cut-off value of 15 µg/l was sufficient to obtain the differentiation as explained above.

Sources of error: the addition of sodium chloride in concentrations greater than 20 g/l to urine will inhibit any reaction creating the possibility of a false negative result [15].

The enzyme, malate dehydrogenase, is affected by the preservative Thiomersal and therefore care must be taken to flush out the spectrophotometer flow-cell with at least 10 ml of water before running this assay.

Specificity: the assay has been designed to detect cannabinoids in urine. Various cannabinoid compounds and metabolites have been tested by the reagent manufacturer [15] and shown to give a response equivalent to 20 µg 11-nor-Δ^8-THC 9-carboxylic acid per litre:

11-Nor-Δ^9-THC 9-carboxylic acid	< 25 µg/l
11-Hydroxy-Δ^9-THC	< 50 µg/l
11-Hydroxy-Δ^8-THC	< 50 µg/l
8-β-Hydroxy-Δ^9-THC	< 50 µg/l
8-β-11-Dihydroxy-Δ^9-THC	< 50 µg/l
Δ^9-THC	< 75 µg/l
Cannabinol	> 100 µg/l
Cannabidiol	> 500 µg/l

In addition, one of the clinical trials [14] observed cross-reactivities with additional cannabinoid metabolites using the same criterion, as follows:

Δ^9-THC	66 µg/l
Δ^8-THC	71 µg/l
2'-Hydroxy-Δ^9-THC	302 µg/l
3'-Hydroxy-Δ^9-THC	271 µg/l
4'-Hydroxy-Δ^9-THC	112 µg/l
1'2'3'4'5'-Pentanor-Δ^9-THC-3-COOH	> 10 mg/l
1'2'3'4'5'-Pentanor-Cannabinol-3-COOH	> 10 mg/l
11-nor-Cannabinol-9-COOH	39 µg/l
1'-Hydroxycannabinol	> 10 mg/l

The following unrelated compounds were tested and minimum concentrations required to give an insignificant response, i.e. less than that due to 20 µg 11-nor-Δ^9-THC 9-carboxylic acid per litre, were found to be as follows:

Acetyl salicylate	> 1000 mg/l	Methaqualone	> 500 mg/l
Amphetamine	> 100 mg/l	Morphine	> 200 mg/l
Amitriptyline	> 1000 mg/l	Phencyclidine	> 1000 mg/l
Benzoylecgonine	> 400 mg/l	Propoxyphene	> 100 mg/l
Diazepam	> 1000 mg/l	Secobarbital	> 1000 mg/l
Meperidine	> 1000 mg/l		

References

[1] *S. E. Sallan, C. Cronin, M. Zelen, N. E. Zinberg,* Antiemetics in Patients Receiving Chemotherapy for Cancer, N. Eng. J. Med. *302,* 135 – 138 (1980).

[2] *L. Lemberger, R. E. Crabtree, H. M. Rowe,* 11-Hydroxy-Δ^9-tetrahydrocannabinol: Pharmacology, Distribution, and Metabolism of a Major Metabolite of Marijuana in Man, Science *177,* 62 – 64 (1972).

[3] *R. Rodgers, C. P. Crowe, W. M. Eimstad, M. W. Hu, J. K. Kam, R. C. Ronald, G. L. Rowley, E. F. Ullman,* Homogeneous Enzyme Immunoassay for Cannabinoids in Urine, Clin. Chem. *24,* 95 – 100 (1978).

[4] *J. H. Jaffe,* Drug Addiction and Drug Abuse, in: *A. Goodman Gilman, L. S. Goodman, A. Gilman* (eds.), The Pharmacological Basis of Therapeutics, 6th edit., MacMillan, New York 1980, pp. 535 – 584.

[5] *J. D. Teale, E. J. Forman, L. J. King, V. Marks,* Production of Antibodies to Tetrahydrocannabinol as the Basis for Its Radioimmunoassay, Nature *249,* 154 – 155 (1974).

[6] *D. Rosenthal, T. M. Harvey, J. T. Bursey,* Comparison of Gas-Chromatography Mass Spectrometry Methods for the Determination of Δ^9-Tetrahydrocannabinol in Plasma, Biomed. Mass Spectrom. *5,* 312 – 316 (1978).

[7] *E. R. Garrett, C. A. Hunt,* Separation and Sensitive Analysis of Tetrahydrocannabinol in Biological Fluids by HPLC and GLC, in: *R. E. Willette* (ed.), Cannabinoid Assays in Humans, NIDA Research Monograph 7, U.S. Department of Health, Education and Welfare, Washington D.C., 1976, pp. 33 – 39.

[8] *L. Lombrozo, S. L. Kanter, L. E. Hollister,* Marijuana Metabolites in Urine of Man. VI Separation of Cannabinoids by Sequential Thin Layer Chromatography, Res. Commun. Chem. Pathol. Pharmacol. *15,* 697 – 705 (1976).

[9] *S. L. Kanter, L. E. Hollister,* Marijuana Metabolites in Urine of Man. VII Excretion Patterns and Acid Metabolites Detected by Sequential Thin Layer Chromatography, Res. Commun. Chem. Pathol. Pharmacol. *17,* 421 – 428 (1977).

[10] *K. E. Rubenstein, R. S. Schneider, E. F. Ullman,* "Homogeneous" Enzyme Immunoassay. A New Immunochemical Technique, Biochem. Biophys. Res. Commun. *47,* 846 – 851 (1972).

[11] *M. E. Wall, D. R. Brine, M. Perez-Reyes,* Metabolism of Cannabinoids in Man, in: *M. C. Braude, S. Szara* (eds.), The Pharmacology of Marijuana, Raven Press, New York 1976, pp. 93 – 116.

[12] *J. D. Teale, L. J. King, E. J. Forman, V. Marks,* Radioimmunoassay of Cannabinoids in Blood and Urine, Lancet *II,* 553 – 555 (1974).

[13] *R. J. Bastiani, R. C. Phillips, R. S. Schneider, E. F. Ullman,* Homogeneous Immunochemical Drug Assays, Am. J. Med. Technol. *29,* 211 – 216 (1973).

[14] *D. S. Kabakoff, H. M. Greenwood,* Recent Advances in Homogeneous Enzyme Immunoassay, in: *K. G. M. M. Alberti, C. P. Price* (eds.), Recent Advances in Clinical Biochemistry 2, Churchill Livingstone, Edinburgh 1981, pp. 1 – 30.

[15] Product Literature EMIT®-d.a.u. Cannabinoid Urine Assay, *Syva Co.,* Palo Alto, U.S.A., 1982.

2.12 Nortriptyline (and Amitriptyline)

3-(10,11-Dihydro-5*H***-dibenzo[*a, d*]-cyclohepten-5-ylidene)-*N*-methyl-1-propanamine**

and

3-(10,11-Dihydro-5*H***-dibenzo[*a, d*]-cyclohepten-5-ylidene) *N, N*-dimethyl-1-propanamine**

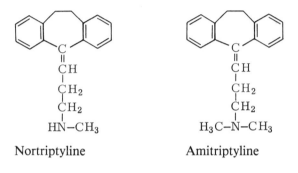

Nortriptyline   Amitriptyline

Christopher P. Price and R. Stewart Campbell

General

The tricyclic antidepressants are a group of drugs of which the most common are amitriptyline, nortriptyline, imipramine, desipramine, trimipramine, protriptyline, amoxapine and doxepin.

The major effects of the tricyclic antidepressants are sedation, anticholinergic effects and the ability to inhibit the uptake of noradrenaline and/or serotonin into presynaptic neurones [1].

The major clinical use of the tricyclic antidepressants is the reduction of the symptoms of depression [2].

The tricyclic antidepressants are absorbed rapidly from the gastro-intestinal tract. Peak levels generally occur in the serum within five hours of administration; the majority of the drug is bound to serum protein, varying from 70 to 97%. The drug is metabolized in the liver with only a small proportion of the drug appearing unchanged in the urine. The major metabolic routes are through demethylation of the amine side chain and hydroxylation of the ring nucleus. Metabolism of some of the parent drugs leads to active metabolites. Several of the metabolites are conjugated with glucuronic acid and excreted in the urine [1, 3].

The widespread use of tricyclic antidepressants has been associated with an increased incidence of overdosage in recent years. The symptoms of mild overdosage include blurred vision, dilated pupils, warm dry skin and abnormal deep tendon reflexes. More severe toxicity is associated with coma, and possibly with metabolic acidosis, respiratory depression, agitation or delirium and various cardiac disturbances [4].

The treatment of overdosage with tricyclic antidepressant is mainly supportive with particular monitoring of respiratory and cardiac problems.

Application of method: clinical chemistry, toxicology.

Substance properties relevant in analysis: amitriptyline hydrochloride, molecular weight 313.9, pK_a 9.4, soluble in water to 1000 g/l, in ethanol to 660 g/l and in chloroform to 830 g/l. Nortriptyline hydrochloride, molecular weight 299.8, pK_a 10.0, soluble in water to 20 g/l, in ethanol to 100 g/l and in chloroform to 1000 g/l.

Methods of determination: methods for the detection and quantitation of the tricyclic antidepressants and metabolites have recently been reviewed [5]. The most commonly used methods are gas-liquid chromatography [6], high-pressure liquid chromatography [7] and radioimmunoassay [8]. A highly specific gas-liquid chromatography/ mass-spectrometry method [9] and a qualitative homogeneous enzyme-immunoassay has also been described [10].

International reference method and standards: no reference method or standards are known to date.

Assay

Method Design

The method is based on the enzyme-multiplied immunoassay technique (EMIT®) first described by *Rubenstein et al.* [11].

Principle: as described in detail on pp. 7, 8, 271.

The activity of the marker enzyme, glucose-6-phosphate dehydrogenase, G6P-DH, is reduced if the enzyme-desipramine conjugate (E-L) is coupled to the antibody (Ab). Nortriptyline or other tricyclic antidepressants (L) in a sample compete with G6P-DH-labelled drug (E-L) for a limited concentration of antibody (Ab). The activity of G6P-DH measured as increase in absorbance per unit time, $\Delta A/\Delta t$, is related to the nortriptyline concentration. The method is intended for use as a semi-quantitative assay for serum samples; the assay is not designed to determine the level of intoxication.

Selection of assay conditions and adaptation to the individual characteristics of the reagents: as described in general on pp. 8, 9, 271.

Enzyme-drug conjugate is obtained by coupling a derivative of desipramine to G6P-DH from *Leuconostoc mesenteroides*. The method involves pre-dilution of the sample in the reaction vessel. The desired assay range is 0.3 to 1.0 mg/l. The sample volume is 50 µl.

Equipment

The same equipment is used as described on p. 272.

Reagents and Solutions

Purity of reagents: as described on pp. 27, 272.

Preparation of solutions* (for 100 determinations): all solutions in re-purified water (cf. Vol. II, chapter 2.1.3.2).

1. Tris buffer (Tris, 55 mmol/l; NaN$_3$, 0.5 g/l; Triton X-100, 0.1 ml/l; pH 8.0):

   dissolve 1.996 g Tris base in 150 ml water, add 0.03 ml Triton X-100, adjust to pH 8.0 with HCl, 1.0 mol/l; make up with water to 300 ml.

   Alternatively, dilute buffer concentrate provided with each kit to 200 ml with water.

2. Antibody/substrate solution (γ-globulin**, 2 to 5 mg/l; G-6-P, 66 mmol/l; NAD, 40 mmol/l; Tris, 55 mmol/l; pH 5.5):

   reconstitute available lyophilized preparation with 3.0 ml water.

3. Conjugate solution (G6P-DH***; Tris, 55 mmol/l; NaN$_3$, 0.5 g/l; pH 6.7):

   this reagent is standardized to match the antibody/substrate solution (cf. p. 272). Reconstitute available lyophilized preparation with 3.0 ml water.

---

 * Reagents, calibrators and buffer are commercially available from *Syva Co.,* Palo Alto, U.S.A. The solutions contain preservatives.
 ** Standardized preparation from immunized sheep, concentration is titre-dependent.
*** The activity concentration of the enzyme is dependent on final choice of conjugate dilution.

4. Nortriptyline standard solutions (0.3 mg/l; 1.0 mg/l):

dissolve 50.0 mg nortriptyline (as free base) in 500 ml methanol. Use this stock solution to prepare two working standards of 0.3 mg/l and 1.0 mg/l by appropriate dilution with pooled drug-free human serum containing 0.5 g sodium azide per litre. The serum diluent is used as a zero calibrator.

Alternatively, reconstitute each of the available lyophilized calibrators with 1.0 ml water; re-stopper the bottles and swirl gently to assist reconstitution.

Stability of solutions: store all reagents and calibrators in a refrigerator at $2-8\,^\circ$C. All reagents should be allowed to equilibrate to room temperature before use. When used in this fashion all reagents (1), (2) and (3), and standard solutions (4) are stable for 12 weeks after reconstitution.

Procedure

Collection and treatment of specimen

Blood: collect blood and treat to yield serum or plasma, in the usual manner.

Urine: collect urine in plastic or glass containers. The effect of urine preservatives has not been established and therefore their use is not recommended.

Tissue: homogenize 5 g tissue with 20 ml water, adjust the homogenate to pH 11 with NaOH, 1 mol/l. Shake a 2 ml aliquot of the homogenate with 50 ml hexane in a separating funnel. Transfer 10 ml of the organic phase to a glass-stoppered centrifuge tube and shake with 0.5 ml hydrochloric acid, 50 mmol/l, for 5 min; centrifuge; remove the aqueous layer, adjust to pH 7. Use this extract for analysis.

Stability of the substance in the sample: the stability of tricyclic antidepressants in biological materials is not known.

Assay conditions: measurement of samples and standards in duplicate.
    A pre-dilution of samples and standard solutions is required for serum samples: mix one part (0.05 ml) sample or standard solution(s) with 5 parts (0.25 ml) Tris buffer.
    Incubation time 30 s (in a reaction cuvette); room temperature; 0.6 ml; wavelength Hg 334 or 339 nm; light path 10 mm; final volume 0.9 ml; measurements against water; 30 s at 30°C.

Run the standards with each series of analyses. The zero calibrator assay serves as a reagent blank; correct all measurements for this value.

Measurement

Pipette* successively into the cuvette:			concentration in assay mixture	
pre-diluted sample or standard solution (4)		0.05 ml	nortriptyline	up to 9 µg/l
buffer	(1)	0.25 ml	Tris	52 mmol/l
			NaN₃	470 mg/l
antibody solution	(2)	0.05 ml	γ-globulin	variable
			G-6-P	3.7 mmol/l
			NAD	2.2 mmol/l
buffer	(1)	0.25 ml		
incubate for 30 s at 30 °C,				
conjugate solution	(3)	0.05 ml	G6P-DH**	variable
buffer	(1)	0.25 ml		
place cuvette in spectrophotometer immediately or aspirate contents into flow-cell, read absorbance after 15 s and 45 s.				

* Adequate mixing should be achieved by the addition of each reagent and no further agitation is necessary.
** Coupled to a derivative of desipramine.

If ΔA per 30 s of the sample is greater than that of the highest calibrator, dilute the sample further with Tris buffer (1) and re-assay.

Calibration curve: plot the corrected values of the increase in absorbance per 30 s, $\Delta A/\Delta t$, of the calibrators *versus* the corresponding nortriptyline concentration (mg/l) using semi-logarithmic graph paper. A typical curve is shown in Fig. 1.

Calculation: read the mass concentration of nortriptyline equivalents in the sample from the calibration curve. In view of the wide specificity of the method it is not appropriate to convert from one form of units to another; the result is presented in equivalents of nortriptyline.

It is important to confirm a positive result by an alternative, non-immunochemical procedure.

Interpretation: the wide range of tricyclic antidepressants used, together with the many dosage regimes employed, means that levels of drug found in the blood are quite varied. The peak concentrations are found in the period 2 to 8 h after taking the drug, although this may be delayed for several hours. Therapeutic effects are attained with levels of 100 to 300 µg/l with toxic effects being associated with serum levels of

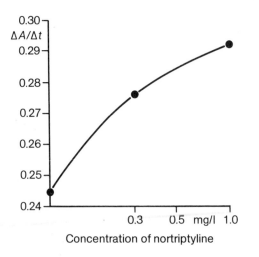

Concentration of nortriptyline

Fig. 1. Typical calibration curve for the assay of tricyclic antidepressants in serum.

1 mg/l. The half-lives of the tricyclic antidepressants vary from 10 to 80 h, depending on the drug and inter-individual variability in pharmacokinetics.

The assay is designed as a qualitative assay and it is not recommended for the determination of the amount of drug present. The assay is designed such that the calibrator concentration of 0.3 mg/l represents the cut-off value of the system. Thus, if $\Delta A/\Delta t$ of a sample is less than that of this calibrator the sample is considered to contain either no tricyclic antidepressant, or only a very low concentration [12].

Validation of Method

Precision, accuracy, detection limit and sensitivity: as part of a clinical trial conducted by the reagent manufacturer, the assay was tested in three laboratories [12]. In the trial, patients' samples were analyzed by the immunoassay method and a quantitative method (one GC, one GC/MS and one HPLC). In the concentration range 0 to ≤ 99 µg/l (42 samples) the EMIT® procedure gave 7 positive results; in the range 100 to 199 µg/l (65 samples) the EMIT® procedure gave 59 positive results; in the range 200 to 299 µg/l (41 samples) the EMIT® procedure gave 39 positive results and, finally, in the range ≥ 300 µg/l (159 samples) the EMIT® procedure gave 158 positive results. It should be stressed that the specificities of each of the methods used was not identical; the EMIT® result is cumulative for any tricyclic antidepressants present with some metabolites also being detected.

Sources of error: serum samples from grossly lipaemic, icteric or haemolyzed blood may yield results with poor reproducibility.

It has been found that high therapeutic and toxic levels of chlorpromazine (for example 200 to 300 µg/l) may give a positive result [12]. It is known that tricyclic antidepressant drugs bind to glass; plastic ware should therefore be used where possible. Alternatively, siliconized glassware may be used.

It is recommended by the reagent manufacturer that only EMIT® Serum Tricyclic Antidepressants Calibrator should be used to standardize the assay.

Specificity: the reagent manufacturer has stated [12] that the following approximate concentrations of tricyclic antidepressants and metabolites give responses equivalent to that of nortriptyline at 0.3 mg/l.

Parent Compounds
Amitriptyline	0.3 mg/l
Desipramine	0.25 mg/l
Imipramine	0.25 mg/l

Metabolites
10-Hydroxyamitriptyline	< 1 mg/l
2-Hydroxydesipramine	< 1.25 mg/l
2-Hydroxyimipramine	< 0.75 mg/l
10-Hydroxynotriptyline	< 1.25 mg/l

Other Tricyclic Antidepressants
Amoxapine	< 0.5 mg/l
Clomipramine	< 0.5 mg/l
Doxepin	< 0.5 mg/l
Protriptyline	< 0.5 mg/l
Trimipramine	< 0.6 mg/l

The following structurally related compounds give responses equivalent to that of nortriptyline (0.3 g/l): carbamazepine (> 100 mg/l), chlorpromazine (> 200 µg/l), cyclobenzaprine (> 200 µg/l), perphenazine (> 350 µg/l) and promethazine (> 500 µg/l).

Drugs of abuse and of unrelated structure which demonstrate no cross-reactivity; that is, no response at the level tested, are as follows: amphetamine (500 mg/l), dextromethorphan (1000 mg/l), diazepam (500 mg/l), ethchlorvynol (500 mg/l), methaqualone (100 mg/l), morphine (100 mg/l), paracetamol (1500 mg/l), phencyclidine (1000 mg/l), phenytoin (100 mg/l), propoxyphene (500 mg/l) and secobarbital (500 mg/l).

References

[1] *L. DeVane,* Tricyclic Antidepressants, in: *W. E. Evans, J. J. Schentag, W. J. Jusko* (eds.), Applied Pharmacokinetics: Principles of Therapeutic Drug Monitoring, Applied Therapeutics Inc., San Francisco 1981, pp. 549–586.
[2] *H. E. Hollister,* Current Antidepressant Drugs: Their Clinical Use, Drugs *22*, 129–152 (1981).

[3] *R. J. Baldessarini*, Drugs and the Treatment of Psychiatric Disorders, in: *A. Goodman Gilman, L. S. Goodman, A. Gilman* (eds.), The Pharmacological Basis of Therapeutics, 6th edit., MacMillan Publishers, New York 1980, pp. 391 – 447.

[4] *D. G. Spiker, A. N. Weiss, S. S. Chang, J. F. Ruwitch, Jr., J. T. Biggs,* Tricyclic Antidepressant Overdose: Clinical Presentation and Plasma Levels, Clin. Pharmacol. Ther. *18*, 539 – 546 (1976).

[5] *B. A. Scoggins, K. P. Maguire, T. R. Norman, G. D. Burrows,* Measurement of Tricyclic Antidepressants, Part 1: A Review of Methodology, Clin. Chem. *26*, 5 – 17 (1980).

[6] *S. Dawling, R. A. Braithwaite,* Simplified Method for Monitoring Tricyclic Antidepressant Therapy Using Gas-Liquid Chromatography with Nitrogen Detection, J. Chromatogr. *146*, 449 – 456 (1978).

[7] *T. A. Sutfin, R. D'Ambrosio, W. J. Jusko,* Liquid Chromatographic Determination of Eight Tri- and Tetracyclic Antidepressants and Their Major Active Metabolites, Clin. Chem. *30*, 471 – 474 (1984).

[8] *K. K. Midha, J. C. K. Loo, C. Charette, M. L. Rowe, J. W. Hubbard, I. J. McGilveray,* Monitoring of Therapeutic Concentrations of Psychotropic Drugs in Plasma by Radioimmuno-assays, J. Anal. Toxicol. *2*, 185 – 192 (1978).

[9] *D. M. Chinn, T. A. Jennison, D. J. Crouch, M. A. Peat, G. W. Thatcher,* Quantitative Analysis for Tricyclic Antidepressant Drugs in Plasma or Serum by Gas Chromatography-Chemical Ionization Mass Spectrometry, Clin. Chem. *26*, 1201 – 1204 (1980).

[10] *D. S. Kabakoff, H. M. Greenwood,* Recent Advances in Homogeneous Enzyme Immunoassay, in: *K. G. M. M. Alberti, C. P. Price* (eds.), Recent Advances in Clinical Biochemistry 2, Churchill Livingstone, Edinburgh 1981, pp. 1 – 30.

[11] *K. E. Rubenstein, R. S. Schneider, E. F. Ullman,* "Homogeneous" Enzyme Immunoassay. A New Immunochemical Technique, Biochem. Biophys. Res. Commun. *47*, 846 – 851 (1972).

[12] Product Literature, EMIT®-tox, Serum Tricyclic Antidepressants Assay, *Syva Co.,* Palo Alto, U.S.A., 1982.

2.13   Paracetamol (Acetaminophen)

N-(4-Hydroxyphenyl)acetamide

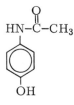

Christopher P. Price, R. Stewart Campbell, Peter M. Hammond and Tony Atkinson

General

Paracetamol is a non-steroidal compound with strong analgesic and antipyretic properties; it has little demonstrable anti-inflammatory action. It is thought to act by modulation of prostaglandin synthesis and release [1].

Paracetamol is metabolized in the liver producing a total of eight metabolites; the glucuronide and sulphate conjugates together constitute 95% of the metabolites. Microsomal oxidation produces an unstable intermediate that is normally complexed with glutathione. If an excess of drug is ingested the hepatic glutathione level becomes depleted and the unstable intermediate binds to hepatocyte proteins leading to cellular necrosis [2]. This can lead to liver failure.

The level of paracetamol in the plasma has been shown to be the best indicator of the likelihood of hepatocellular damage [3]. The liver can be protected by giving either cysteamine, methionine or N-acetylcysteine [4].

Thus, measurement of plasma paracetamol levels enables confirmation of drug ingestion and indicates the need for treatment with a protective agent.

Application of method: clinical chemistry, toxicology.

Substance properties relevant in analysis: molecular weight 151.16; maxium of UV-absorbance (in ethanol) at 250 nm; molar absorbance coefficient $\varepsilon = 13.8 \times 10^2$ $1 \times mol^{-1} \times mm^{-1}$; pK_a 9.5; substance is soluble in water to 0.5 g/l; sensitive to light.

Methods of determination: the drug can be measured using simple colorimetric techniques. One method involves nitration to give a coloured product [5]. An alternative approach is acid hydrolysis to 4-aminophenol with subsequent dye complex formation [6]. These methods are susceptible to a variety of interferences from compounds in plasma [7].

Ultraviolet spectrophotometric methods have been employed with a solvent extraction step [8]. These methods are susceptible to interference, particularly from salicylate [9].

Chromatographic methods, both high-pressure [10] and gas-liquid [11] have been reported and will achieve the desired specificity.

A homogeneous enzyme-immunoassay has been described and with appropriate equipment a result can be produced in less than ten minutes [12].

The use of an aryl acylamidase* to degrade paracetamol with quantitation of the 4-aminophenol produced has enabled the development of a rapid, specific assay yielding a clearly visible end-point [13].

International reference method and standards: no reference method or standards are known so far.

---

* Aryl acylamide amidohydrolase, EC 3.5.1.13.

2.13.1   Determination with Aryl Acylamidase

Assay

Method Design

Principle

(a)

Paracetamol   4-aminophenol

(b)

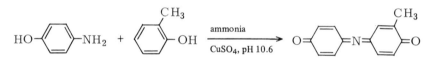

Paracetamol is degraded to 4-aminophenol and acetate in the presence of an aryl acyl-amidase. The 4-aminophenol reacts with 2-cresol in the presence of ammoniacal copper sulphate to yield a blue indophenol dye.

The formation of the indophenol dye is measured by the increase in absorbance at 615 nm and is proportional to the amount of paracetamol present.

Optimized conditions for measurement: the pH optimum of the enzyme is broad with a maximum of 8.6. Employing a 50 µl sample with 1.0 ml enzyme, a minimum of 0.4 U enzyme is required to degrade the paracetamol completely in less than two minutes at room temperature. The formation of the indophenol dye from 4-amino-phenol and 2-cresol is accelerated by the presence of ammonium and copper ions with the reaction taking place at a pH of 10.6. The indicator reaction reaches completion in less than two minutes at room temperature.

Equipment

Spectrophotometer or spectral-line photometer capable of measurement at wave-lengths between 600 and 650 nm. Manual displacement pipettes capable of delivering 50 µl and 1000 µl with a relative standard deviation of less than 1.0%.

Reagents and Solutions

Purity of reagents: the aryl acylamidase from *Pseudomonas fluorescens* should be of the best available quality; e.g. 300 U/mg at 30°C, 4-nitroacetanilide as substrate, pH 8.6. Chemicals are of analytical grade.

Preparation of solutions (for about 20 determinations): all solutions in re-purified water (cf. Vol. II, chapter 2.1.3.2). To prevent microbial contamination sterilize all containers.

1. Tris buffer (0.1 mol/l, pH 8.6):

   dissolve 242.2 mg Tris in 19 ml water, adjust pH to 8.6 with HCl, 1 mol/l. Make up to 20 ml with water.

2. Aryl acylamidase/buffer (enzyme, 1.0 kU/l; Tris, 0.1 mol/l; pH 8.6):

   reconstitute two bottles of enzyme/buffer preparation (*Porton Products,* P.O. Box 85, Wrexham, Clwyd, U.K.) with 10 ml water each.

3. Cresol/copper solution (2-cresol, 91 mmol/l; $CuSO_4$, 2 mmol/l):

   dissolve 0.2 g 2-cresol (*Sigma Chemical Co.,* 99%) and 6.5 mg anhydrous copper sulphate in 20 ml water.

4. Alkaline ammoniacal copper sulphate ($CuSO_4$, 3.93 mmol/l; Na_2CO_3, 0.6 mol/l; NH_4Cl, 91.4 mmol/l; Tris, 27 mmol/l):

   dissolve 12.76 mg anhydrous copper sulphate, 1.27 g anhydrous sodium carbonate, 97.6 mg ammonium chloride and 65.3 mg Tris base in 20 ml water. Check pH is $\geqslant 10.6$.

5. Paracetamol standard solution (paracetamol, 2 mmol/l; acetate, 50 mmol/l; pH 5.0):

   dissolve 30.2 mg paracetamol in 100 ml sodium acetate, 50 mmol/l, pH 5.0. Check the concentration of the standard: take 1 ml standard into a 100 ml volumetric flask, add 10 ml NaOH, 0.1 mol/l, make up to 100 ml with water. Measure absorbance (ΔA) at 257 nm against a suitable blank. Calculate the paracetamol concentration as follows:

$$c = \frac{\Delta A \times 10^6}{715 \times 151.16} = \Delta A \times 9.252 \quad \text{mmol/l}.$$

Stability of solutions: store all solutions, stoppered, in a refrigerator at 0°C to 4°C. The solutions (3) and (4) and the standard (5) should be stored in dark bottles; they are stable for one year. The standard (5) should be stored in a small bottle to minimize contact with air. Solution (2) is stable for at least three months after reconstitution.

Procedure

Collection and treatment of specimen

Blood: take blood from the vein without stasis. The blood may be placed in a plain, heparin, fluoride/oxalate or EDTA tube. Serum or plasma can be used for analysis.

In patients suspected of taking an overdose where management with a therapeutic agent such as *N*-acetylcysteine is being considered, the specimens should be taken between 4 and 12 h after ingestion of the drug.

If the specimen is from mildly lipaemic, haemolyzed or icteric blood a blank determination should be undertaken. In the case of serum from severely haemolyzed or lipaemic blood a fresh specimen should be sought.

Urine: use urine directly in the assay with alteration in the sample volume to accomodate the expected range of concentrations.

Stability of the substance in the sample: paracetamol is stable in serum or plasma for at least one week when stored at 0°C to 4°C. The analyte is stable for longer periods of time at $-20°C$. Any deterioration can be recognized by the fact that the sample acquires a deep brown coloration.

Assay conditions: wavelength 615 nm; light path 10 mm; incubation volume 1.05 ml; final volume 3.05 ml; room temperature; read absorbance against a reagent blank with water instead of sample.

If a sample is turbid, haemolytic or grossly icteric run a sample blank by replacing the enzyme/buffer solution (2) with buffer (1) and completing the assay in the normal way.

Calculation: correct all readings for reagent blank, yielding ΔA_{sample} and $\Delta A_{standard}$. Under the given conditions the reaction comes to completion. The substance concentration, c, and the mass concentraton, ρ, of the sample are:

$$c = \frac{\Delta A_{sample}}{\Delta A_{standard}} \times 2.0 \quad \text{mmol/l}$$

$$\rho = \frac{\Delta A_{sample}}{\Delta A_{standard}} \times 2 \times 151.16 \quad \text{mg/l}.$$

Measurement

Pipette successively into the cuvette:			concentration in assay mixture	
sample or standard (5)		0.05 ml	paracetamol	
				up to 95 μmol/l
			Tris	95 mmol/l
enzyme/buffer solution	(2)	1.00 ml	enzyme	950 U/l
mix thoroughly, let stand at room temperature for 3 min,				
2-cresol	(3)	1.00 ml	2-cresol	30 mmol/l
ammoniacal copper sulphate	(4)	1.00 ml	$CuSO_4$	1.3 mmol/l
			Na_2CO_3	0.2 mol/l
			NH_4Cl	30 mmol/l
			Tris	8.9 mmol/l
mix thoroughly, let stand at room temperature for three min and read absorbance.				

If ΔA of the sample is greater than that of the standard, dilute the sample with water and re-assay.

Interpretation: the maximum concentration of paracetamol in serum after ingestion of 1 g drug ranges from 66 to 132 μmol/l (10 to 20 mg/l) [1]; the peak is attained within one half to two hours after ingestion. The normal half-life is 2 h [3].

The treatment of paracetamol poisoning includes the removal of drugs remaining in the stomach by gastric lavage, together with management of clinical symptoms. The effect of toxic metabolites of the drug can be reduced by giving sulphydryl-containing compounds to replenish hepatic glutathione stores. Early sulphydryl compounds were found to give side-effects in some instances, and a strategy based on giving a protective agent in proportion to the serum paracetamol level was recommended [3, 4]. The protective agent now in common use is N-acetylcysteine; it has been found that treatment is only of value if given within 10 h of ingestion of the drug. Reference is made to a nomogram (as illustrated in Fig. 1) derived from clinical experience in the use of methionine in the treatment of 96 patients admitted with high serum paracetamol levels.

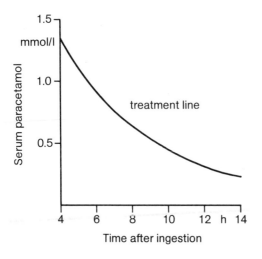

Fig. 1. Nomogram illustrating the relationship between serum paracetamol concentration and time elapsed since ingestion of drug and the susceptibility to hepatocellular damage. Patients with results above the treatment line are most susceptible to liver damage and should be given *N*-acetylcysteine.

Validation of Method

Precision, accuracy, detection limit and sensitivity: in a manual assay, relative standard deviations of 3.4% and 3.1% were observed at concentrations of 0.412 and 1.412 mmol/l, respectively (day to day) [14]. The assay when adapted for use on a centrifugal analyzer gave relative standard deviations of 1.4 and 1.3% (within batch) at 0.3 and 1.9 mmol/l and 1.7 and 1.5% (between batches) at 0.6 and 2.1 mmol/l [15].

The recovery of paracetamol added to drug-free pooled human serum ranges from 98 to 103% at concentrations of 0.5 to 2.0 mmol/l [14]. The accuracy of the method has been assessed by comparison with HPLC methods [14, 15]. The regression parameters from a comparison of results by the enzymatic procedure with HPLC on 80 patients' samples gave a slope of 0.95, intercept of 0.06, r = 0.986, n = 80.

The detection limit, defined as the mean plus three times the standard deviation of the absorbance obtained with a drug-free serum, has been found to be 50 mmol/l.

Sources of error: the protective agent *N*-acetylcysteine is known to influence the formation of the indophenol dye. At a concentration of *N*-acetylcysteine of 1.0 mmol/l in the sample (the maximum that could be expected) there is a 3% reduction in the final absorbance at a paracetamol concentration of 2.6 mmol/l.

There is no reaction with sera from patients with renal disease or liver disease (conditions in which increased levels of phenolic compounds may be found in serum) [14].

Specificity: the method is specific for paracetamol and does not measure any of the major metabolites. In addition, the method was tested with 50 commonly prescribed drugs and there was no interference [14].

2.13.2   Determination with Enzyme-multiplied Immunoassay Technique

Assay

Method Design

The method employs the technique of homogeneous enzyme-multiplied immuno-assay, first described by *Rubenstein et al.* [16].

Principle: as described in detail on pp. 7, 8, 271.

The activity of the marker enzyme glucose-6-phosphate dehydrogenase*, G6P-DH, is reduced if the enzyme-paracetamol conjugate (E-L) is coupled to the antibody (Ab). Paracetamol (L) in a sample competes with G6P-DH-labelled paracetamol (E-L) for a limited concentration of antibody (Ab). The activity of G6P-DH measured as increase in absorbance per unit time, $\Delta A/\Delta t$, is related to the paracetamol concentration (cf. Vol. I, chapter 2.7, p. 244).

Selection of assay conditions and adaptation to the individual characteristics of the reagents: as described in general on pp. 8, 9, 271. Enzyme-drug conjugate is obtained by coupling paracetamol to G6P-DH from *Leuconostoc mesenteroides*.

The desired assay range is 0.07 to 1.3 mmol/l (10 to 200 mg/l). The serum sample volume is 50 µl.

Equipment

The same equipment is used as described on p. 272.

Reagents and Solutions

Purity of reagents: as described on pp. 27, 272.

Preparation of solutions* (for 100 determinations): all solutions in re-purified water (cf. Vol. II, chapter 2.1.3.2).

---

* Reagents, calibrators and buffer are commercially available from *Syva Co.,* Palo Alto, U.S.A. The solutions contain preservatives.

1. Tris buffer (Tris, 55 mmol/l; NaN$_3$, 0.5 g/l; Triton X-100, 0.1 ml/l; pH 8.0):

   dissolve 1.996 g Tris base in 140 to 160 ml water, add 0.03 ml Triton X-100 and 0.15 g sodium azide, adjust to pH 8.0 with HCl, 1.0 mol/l; make up with water to 300 ml.

   Alternatively, dilute buffer concentrate provided with each kit to 200 ml with water.

2. Antibody/substrate solution (γ-globulin*; G-6-P, 66 mmol/l; NAD, 40 mmol/l; Tris, 55 mmol/l; NaN$_3$, 0.5 g/l; pH 5.0):

   reconstitute available lyophilized preparation with 6.0 ml water.

3. Conjugate solution (G6P-DH**; Tris, 55 mmol/l; NaN$_3$, 0.5 g/l; pH 8.0):

   this reagent must be standardized to match the antibody/substrate solution (cf. p. 272). Reconstitute available lyophilized preparation with 6.0 ml water.

4. Paracetamol standard solution (10, 25, 50, 100, 200 mg/l):

   dissolve 200 mg paracetamol in 100 ml water, with gentle warming if necessary. Dilute 1.0 ml with paracetamol-free human serum containing sodium azide (5 g/l) to 10 ml (200 mg/l). Dilute this solution 1+1, 1+3, 1+7 and 1+19 with paracetamol-free serum, yielding the following concentrations of paracetamol: 100, 50, 25, 10 mg/l (0.662, 0.331, 0.165, 0.07 mmol/l). Use paracetamol-free human serum as the zero calibrator.

   Alternatively, reconstitute each of the available lyophilized calibration preparations with 1 ml water.

Stability of solutions: after reconstitution store the antibody/substrate solution (2), conjugate solution (3) and standards (4) at room temperature (20°C to 25°C) for at least one hour before use. The reagents are then stable for 12 weeks if stored at 2°C to 8°C. The buffer solution (1) can be used for 12 weeks when stored at room temperature.

Procedure

Collection and treatment of specimen: serum or plasma can be used in this assay. Acceptable anticoagulants are heparin and EDTA. No evidence of interference was obtained when sera from slightly haemolyzed, lipaemic or icteric blood were assayed. Where sera from severely lipaemic or haemolyzed blood are obtained, a fresh sample should be sought. For further information on specimen collection cf. chapter 2.13.1, p. 366.

---

* Standard preparation from immunized sheep, concentration is titre-dependent.
** The activity concentration of the enzyme is dependent on final choice of conjugate dilution.

Stability of the substance in the sample: cf. chapter 2.13.1, p. 366.

Assay conditions: measurements of samples and zero calibrator in duplicate. Single determinations are made of the remaining standards.

A pre-dilution of samples and standard solutions is required: mix one part (0.05 ml) sample or standard solution (4) with 5 parts (0.25 ml) Tris buffer (1).

Incubation time 30 s (in a reaction cuvette); room temperature; 0.6 ml; wavelength Hg 334 or 339 nm; light path 10 mm; final volume 0.9 ml; measurement against water; 30 s at 30 °C.

Establish a calibration curve under exactly the same conditions (within the series). The zero calibrator assay serves as a reagent blank; correct all measurements for this value.

Measurement

Pipette* into a disposable beaker:			concentration in assay mixture	
Pre-diluted sample or standard solution (4)		0.05 ml	paracetamol	up to 1.9 mg/l
Tris buffer	(1)	0.25 ml	Tris	52 mmol/l
			NaN_3	470 mg/l
antibody/substrate solution	(2)	0.05 ml	γ-globulin	variable
			G-6-P	3.7 mmol/l
			NAD	2.2 mmol/l
Tris buffer	(1)	0.25 ml		
incubate for 30 s at room temperature,				
conjugate solution	(3)	0.05 ml	G6P-DH**	variable
Tris buffer	(1)	0.25 ml		
mix well; immediately aspirate into the clean spectrophotometer flow-cell; after a 15 s delay read absorbance over a period of 30 s.				

* Adequate mixing should be achieved by the addition of each reagent and no further agitation is necessary.
** Coupled to paracetamol.

If ΔA per 30 s of the sample is greater than that of the highest calibrator, dilute the specimen further with Tris buffer (1) and re-assay.

Calibration curve: plot the corrected values of the increase in absorbance per 30 s, $\Delta A/\Delta t$, of the calibrators *versus* the corresponding paracetamol concentrations (mg/l). By use of special graph paper* matched with the reagents, a linear calibration curve is obtained (Fig. 2). The construction of this curve is based on the log-logit function. So far, the log-logit model (cf. Vol. I, p. 243) has proved to be useful to fit calibration curves of this EMIT® assay [17].

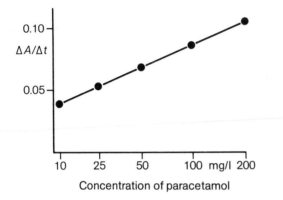

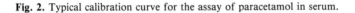

Fig. 2. Typical calibration curve for the assay of paracetamol in serum.

Calculation: read the paracetamol mass concentration of the sample from the calibration curve. To convert from mass concentration to substance concentration (mmol/l) multiply by 0.00662.

Interpretation: cf. chapter 2.13.1, p. 367.

Validation of Method

Precision, accuracy, detection limit and sensitivity: in a manual assay, relative standard deviations of 4.1 and 5.0% were observed at concentrations of 0.51 and 1.24 mmol/l, respectively (day to day). Using a totally automated procedure, relative standard deviations of 3.2, 2.0 and 3.3% were obtained at concentrations of 0.38, 0.73 and 1.04 mmol/l respectively (between days) [18].

The recovery of paracetamol added to drug-free serum ranges from 90 to 120% at concentrations between 0.3 and 1.0 mmol/l [18]. The accuracy of the method has been assessed by comparison of results with an enzymatic assay on 133 serum samples. The linear regression parameters indicated a slope of 1.00, intercept -0.05, r = 0.997, n = 133.

The detection limit, defined as the mean absorbance change plus three times the standard deviations of the zero calibrator, has been determined as 0.066 mmol/l.

* Supplied by *Syva*, for using this graph paper multiply each value of $\Delta A/\Delta t$ by 2.667.

The sensitivity of the assay varies with concentration because of the relationship between the change in absorbance per unit time and concentration and is at a maximum on the steepest part of the calibration curve. This is over the range 0.165 to 0.662 mmol/l (25 to 100 mg/l).

Sources of error: sera from severely haemolyzed, icteric or lipaemic blood may give poor reproducibility and inaccurate results.

Specificity: the method is specific for paracetamol and does not measure any of the major metabolites. The manufacturer's literature indicates a degree of cross-reactivity with the cysteine conjugate [19]; this was not observed in the automated assay [18]. In addition, the method has been tested with 30 commonly prescribed drugs and no interference was observed [18]. No interference was observed in sera from patients with chronic renal or liver diseases [18].

References

[1] *R. J. Flower, S. Moncada, J. R. Vane,* Analgesic-Antipyretics and Anti-Inflammatory Agents; Drugs Employed in the Treatment of Gout, in: *A. Goodman Gilman, L. S. Goodman, A. Gilman* (eds.), The Pharmacological Basis of Therapeutics, 6th edit., MacMillan Publishers, New York 1980, pp. 701 – 705.

[2] *J. R. Mitchell, D. J. Jollow, W. Z. Potter, D. C. Davis, J. R. Gillette, B. B. Brodie,* Acetaminophen-Induced Hepatic Necrosis, I. Role of Drug Metabolism, J. Pharmacol. Exp. Ther. *187*, 185 – 194 (1973).

[3] *L. F. Prescott, N. F. Wright, P. Roscoe, S. S. Brown,* Plasma Paracetamol Half Life and Hepatic Necrosis in Patients with Paracetamol Overdose, Lancet *I*, 519 – 522 (1971).

[4] *L. F. Prescott, J. Park, A. Ballantyne, P. Adriaenssens, A. T. Proudfoot,* Treatment of Paracetamol (Acetaminophen) Poisoning with *N*-Acetyl Cysteine, Lancet *II*, 432 – 434 (1977).

[5] *J. P. Glynn, S. E. Kendal,* Paracetamol Measurement, Lancet *I*, 1147 – 1148 (1975).

[6] *G. S. Wilkinson,* Rapid Determination of Plasma Paracetamol, Ann. Clin. Biochem. *13*, 435 – 437 (1976).

[7] *M. J. Stewart, P. I. Adriaenssens, D. R. Jarvie, L. F. Prescott,* Inappropriate Methods for the Emergency Determination of Paracetamol, Ann. Clin. Biochem. *16*, 89 – 95 (1979).

[8] *B. Dordoni, R. A. Willson, R. P. H. Thompson, R. Williams,* Reduction of Absorption of Paracetamol by Activated Charcoal and Cholestyramine: A Possible Therapeutic Measure, Br. Med. J. *3*, 86 – 87 (1973).

[9] *R. J. Spooner, P. C. Reavey, L. McIntosh,* Rapid Estimation of Paracetamol in Plasma, J. Clin. Pathol. *29*, 663 (1976).

[10] *R. A. Horvitz, P. I. Jatlow,* Determination of Acetaminophen Concentrations in Serum by High Pressure Liquid Chromatography, Clin. Chem. *23*, 1596 – 1598 (1977).

[11] *J. Grove,* Gas-Liquid Chromatography of *N*-Acetyl-p-aminophenol (Paracetamol) in Plasma and Urine, J. Chromatogr. *59*, 289 – 295 (1971).

[12] *D. S. Kabakoff, H. M. Greenwood,* Recent Advances in Homogeneous Enzyme Immunoassay, in: *K. G. M. M. Alberti, C. P. Price* (eds.), Recent Advances in Clinical Biochemistry 2, Churchill Livingstone, Edinburgh 1981, pp. 1 – 30.

[13] *P. M. Hammond, M. D. Scawen, T. Atkinson, R. S. Campbell, C. P. Price,* Development of an Enzyme-based Assay for Acetaminophen, Anal. Biochem. *143*, 152 – 157 (1984).

[14] *C. P. Price, P. M. Hammond, M. D. Scawen,* Evaluation of an Enzymic Procedure for the Measurement of Acetaminophen, Clin. Chem. *29*, 358 – 361 (1983).

[15] *R. S. Campbell, P. M. Hammond, M. D. Scawen, C. P. Price,* The Measurement of Serum Paracetamol Using a Discrete Analyser, J. Autom. Chem. *5*, 146 – 149 (1983).
[16] *K. E. Rubenstein, R. S. Schneider, E. F. Ullman,* "Homogeneous" Enzyme Immunoassay. A New Immunochemical Technique, Biochem. Biophys. Res. Commun. *47*, 846 – 851 (1972).
[17] *R. Cook, D. Wellington,* Data Handling for EMIT® Assays, *Syva Co.,* Palo Alto, U.S.A., 1980.
[18] *R. S. Campbell, C. P. Price,* Evaluation of a Homogeneous Enzyme Immunoassay for Acetaminophen on the Du Pont *aca®*, J. Clin. Chem. Clin. Biochem. *24* (1986), in press.
[19] EMIT®-tox Acetaminophen Assay (Product Literature), *Syva Co.,* Palo Alto, U.S.A., 1982.

2.14   Salicylic Acid

2-Hydroxybenzoic acid

Christopher P. Price, Peter M. Hammond, S. A. Paul Chubb, R. Stewart Campbell and Tony Atkinson

General

A family of compounds related to salicylic acid constitute the most commonly used analgesics; they also possess antipyretic and anti-inflammatory properties. The pharmacological effects are thought to be mediated through inhibition of prostaglandin synthesis [1].

Acetylsalicylic acid, the most commonly used of these compounds, is very rapidly hydrolyzed to salicylic acid in the blood; both are pharmacologically active [2]. A significant proportion of the salicylic acid is excreted unchanged; elimination of the drug is also achieved by conjugation with glycine to give salicyluric acid, conjugation with glucuronic acid to give salicyl acyl and phenolic glucuronides, with a limited amount of hydroxylation to give gentisic and gentisuric acids [3, 4].

Acetylsalicylic acid is commonly prescribed for the relief of certain types of low-intensity pain, e.g. headache, arthritis, myalgia and neuralgia. It may also be used to reduce inflammation, particularly in the joints as in rheumatoid arthritis. The anti-pyretic properties are only employed in those patients in whom fever may represent a dangerous complication. Thus, patients may be taking acetylsalicylic acid on a short- or a long-term basis [1].

Intoxication with salicylate can lead to disturbance of the central nervous system, gastro-intestinal disturbances and a significant alteration to the patient's acid-base status. In severe intoxication encephalopathy, haemorrhagic shock and acute renal failure may develop [5].

Salicylate intoxication represents an acute medical emergency; measurement of the serum salicylate concentration will give a good indication of the clinical severity and be helpful in deciding on the management of the patient.

Treatment of salicylate overdose is aimed at removing the drug from the body and treating the metabolic consequences of salicylate toxicity. Thus gastric lavage is used to remove any residual drug from the stomach. In patients with blood salicylate levels in excess of 3.6 mmol/l active elimination of the drug by the use of a forced alkaline diuresis may be considered. Treatment of the consequences of toxicity include correction of dehydration, hypokalaemia and the metabolic acidosis that is often present [5].

Application of method: clinical chemistry, toxicology.

Substance properties relevant in analysis: salicylic acid, molecular weight 138.12, is soluble in cold water to 2 g/l, in boiling water to 67 g/l. pK_a 3.5. Sodium salicylate, molecular weight 160.11, is soluble in cold water to at least 160 g/l; sensitive to light.

Methods of determination: the colorimetric methods depend on the reactions of the phenolic hydroxyl group. Several variations are based on the reaction with iron(III) salts [6, 7], whilst others employ the *Folin-Ciocalteu* reagent [8].

The dependence of these reactions on the phenolic hydroxyl group means that inter-ferences have been reported due to a variety of endogenous and exogenous com-pounds. The most notable interference with the iron(III) salt methods is due to keto-nes [9].

Chromatographic methods, particularly high-performance liquid [10] and gas-liq-uid [11] techniques have also been described and are capable of monitoring salicylic acid and its metabolites.

Enzymatic methods have been described which utilize the enzyme salicylate mono-oxygenase* [12 – 15]. Quantitation is achieved by measuring the NAD(P)H used or the product formed. In the case of methods that monitor the utilization of NAD(P)H, a blank reaction is present due to inherent NAD(P)H oxidase activity; this problem is not present when the product of the reaction, catechol, is measured.

---

* Salicylate monooxygenase, EC 1.14.13.1.

International reference method and standards: no reference method or standards are known so far.

Assay

Method Design

Principle

(a)

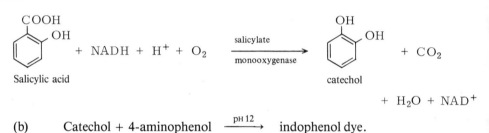

(b)   Catechol + 4-aminophenol $\xrightarrow{\text{pH 12}}$ indophenol dye.

Salicylic acid is degraded to catechol in the presence of salicylate monooxygenase. The catechol reacts with 4-aminophenol in alkaline conditions to give a purple indophenol dye.

The formation of the indophenol dye is measured by the increase in absorbance at 565 nm and is proportional to the amount of salicylic acid present.

Optimized conditions for measurement: the pH optimum of the enzyme is broad with maximum activity at pH 8.2. A minimum of 1.0 U enzyme is required to degrade the salicylic acid in a 50 µl sample of a solution with 5 mmol/l, in a total reaction volume of 1.05 ml in less than two minutes at room temperature. The formation of the indophenol dye occurs most rapidly at pH > 12.0. The indicator reaction reaches completion in less than four minutes at room temperature.

Equipment

Spectrophotometer or spectral-line photometer capable of measurement at 565 nm. Manual displacement pipettes capable of delivering 50 µl and 100 µl with a relative standard deviation of less than 1%.

Reagents and Solutions

Purity of reagents: salicylate monooxygenase should be of the best quality available, e.g. 10 U/mg at 30°C, salicylic acid as substrate, at pH 7.0; with the minimum of contamination with "NADH oxidase" activity. Chemicals are of analytical grade.

Preparations of solutions (for about 10 determinations): all solutions in re-purified water (cf. Vol. II, chapter 2.1.3.2). To prevent microbial contamination, sterilize all containers.

1. Tris buffer/BSA (Tris, 0.1 mol/l; BSA, 1.0 g/l; NaN_3, 0.5 g/l; pH 8.6):

   dissolve 182 mg Tris and 7.5 mg sodium azide in 10 ml water; adjust pH to 8.6 with HCl, 1.0 mol/l. Dissolve 15 mg bovine serum albumin in this solution and dilute to 15.0 ml.

2. Reduced nicotinamide-adenine dinucleotide solution (β-NADH, 5.0 mmol/l):

   dissolve 5.0 mg (6.40 µmol) β-NADH, disodium salt in 1.28 ml Tris/BSA solution (1).

3. Salicylate monooxygenase/reduced nicotinamide-adenine dinucleotide solution (salicylate monooxygenase, 1 kU/l; β-NADH, 0.50 mmol/l; Tris, 0.1 mol/l; BSA, 1.0 g/l; NaN_3, 0.5 g/l; pH 8.6):

   reconstitute 10 U salicylate monooxygenase (*Porton Products,* P.O.Box 85, Wrexham, Clwyd, U.K.) with 9.0 ml Tris/BSA solution (1). Immediately before use, add 1.0 ml β-NADH solution (2).

4. 4-Aminophenol/hydrochloric acid solution (4-aminophenol, 0.91 mmol/l; HCl, 0.1 mol/l):

   dissolve 15 mg 4-aminophenol in 15 ml HCl, 0.1 mol/l. Dilute 1.5 ml with 13.5 ml HCl, 0.1 mol/l.

5. Sodium hydroxide solution (NaOH, 0.25 mol/l):

   dissolve 150 mg NaOH in 15 ml water.

6. Salicylate standard solution (4.0 mmol/l):

   dissolve 64.0 mg sodium salicylate in water and dilute to 100 ml.

Stability of solutions: store all solutions, stoppered, at between 0°C and 4°C. Solutions (1), (4), (5) and (6) are stable for at least 6 months. The β-NADH solution (2) should be prepared fresh daily. Salicylate monooxygenase should be reconstituted just before use; alternatively, it can be reconstituted and stored at −20°C. The enzyme solution retains at least 90% activity for 5 weeks at 4°C and 9 weeks at −20°C.

Procedure

Collection and treatment of specimen

Blood: take blood from a vein without stasis; place it in a plain or heparinized tube. Tubes containing EDTA or fluoride/oxalate should not be used. Serum or plasma may be used in the assay.

Urine: use urine directly in the assay with alteration in the sample volume to accomodate the expected range of concentrations.

Stability of the substance in the sample: salicylic acid is stable in serum or plasma for at least five days when stored at $0°C$ to $4°C$. The analyte is stable for at least one month when stored at $-20°C$.

Assay conditions: wavelength 565 nm; light path 10 mm; incubation volume 1.05 ml; final volume 3.05 ml; room temperature; measurements against a reagent blank with water instead of sample.

If a sample is turbid, from haemolyzed or grossly icteric blood, run a sample blank by replacing the buffer/enzyme/coenzyme solution (3) with buffer and completing the assay in the normal way.

Measurement

Pipette successively into the cuvette:			concentration in assay mixture	
sample or standard (6) enzyme/buffer/coenzyme		0.05 ml	salicylate	up to 238 µmol/l
solution	(3)	1.00 ml	salicylate mono- oxygenase β-NADH Tris BSA	952 U/l 48 mmol/l 95 mmol/l 0.95 g/l
mix thoroughly, allow to stand at room temperature for four minutes,				
4-aminophenol solution (4) sodium hydroxide solution (5)		1.0 ml 1.0 ml	4-aminophenol 0.30 mmol/l NaOH 83 mmol/l	
mix thoroughly, stand at room temperature for five minutes and read absorbance.				

If a sample has a salicylic acid concentration in excess of 5.0 mmol/l dilute the sample 1:4 with 0.9% saline and repeat the assay.

Calculation: correct all readings for reagent blank, yielding ΔA_{sample} and $\Delta A_{standard}$. The substance concentration, c, and the mass concentration, ρ, of the sample is:

$$c = \frac{\Delta A_{sample}}{\Delta A_{standard}} \times 4.0 \qquad \text{mmol/l}$$

$$\rho = \frac{\Delta A_{sample}}{\Delta A_{standard}} \times 4.0 \times 138.12 \qquad \text{mg/l}.$$

Interpretation: salicylate and acetylsalicylate are readily absorbed and the maximum salicylate concentration in the blood usually occurs two hours after ingestion [1]. A therapeutic dose of 500 mg acetylsalicylic acid in an adult will yield a serum concentration of 0.36 mmol/l. The first signs of toxicity may be seen at serum concentrations between 1.5 and 2.0 mmol/l [5]; these levels can be achieved with a dose of 100 mg sodium salicylate/kg body weight [1].

Validation of Method

Precision, accuracy, detection limit and sensitivity: as a manual assay, between-days relative standard deviations of 3.2 and 2.1% were obtained at concentrations of 0.93 and 2.86 mmol/l, respectively [15]. In a manual enzymatic assay in which the depletion of NADH was monitored, between-days relative standard deviations of 4.4, 2.0 and 1.4% at concentrations of 0.91, 2.30 and 3.53 mmol/l were reported [13]. In an automated assay using tabletted reagents and monitoring the depletion of NADPH, between-days relative standard deviations of 5.5, 2.3 and 1.6% have been reported at concentrations of 0.61, 2.06 and 3.92 mmol/l, respectively [12].

The recovery of salicylic acid added to drug-free serum is between 95.1 and 96.7% of concentrations of 1.0 and 3.0 mmol/l [15]; values between 96.5 and 97.5% have also been reported for an enzymatic assay [12]. The accuracy of the method has been demonstrated by comparison with an HPLC method. The results obtained by the method described, with samples from patients who admitted having taken excessive amounts of salicylate-containing drugs, agreed well with results obtained by HPLC (enzymatic vs. HPLC: y = 0.95x + 0.08 mmol/l, r = 998, n = 120) [15].

The detection limit, defined as the mean plus three times the standard deviation of the absorbance obtained with a drug-free serum, has been shown to be 0.1 mmol/l. The method is linear to a sample salicylic acid concentration of 5.0 mmol/l and the sensitivity will essentially depend on the performance of the photometer over this range. The sensitivity would be expected to be better than 0.2 mmol/l across this range. The absorbance of the calibrator with 4.0 mmol/l is expected to be within the range 0.79 to 0.83.

Sources of error: at the present time there are no known sources of error. The potential interference from haemolysis, icterus and lipaemia has been tested to levels of 0.25 g haemoglobin, 300 µmol bilirubin and 10.0 mmol triglyceride per litre, and no effect has been observed [15].

Specificity: the method is specific for salicylic acid and does not measure any of the major metabolites. In addition, the method has been tested with 30 commonly prescribed drugs and no interference was observed [15]. Investigations of 62 common drugs with an alternative enzymatic assay have indicated no interference [13]. However, it has been noted that compounds structurally related to salicylate, e.g. benzoate, 4-hydroxybenzoate and 4-aminosalicylate, will interfere in the latter (NAD(P)H monitored) methods [12, 13]; the enzyme-mediated colorimetric method does not show this interference [15].

There is no reaction with sera from patients with renal disease or liver disease (situations in which increased levels of phenolic compounds in serum may be found) [15].

References

[1] R. J. Flower, S. Moncada, J. R. Vane, Analgesic-Antipyretics and Anti-Inflammatory Agents: Drugs Employed in the Treatment of Gout, in: A. Goodman Gilman, L. S. Goodman, A. Gilman (eds.), The Pharmacological Basis of Therapeutics, 6th edit., MacMillan Publishing, New York 1980, pp. 682 – 698.

[2] G. Levy, Pharmacokinetics of Aspirin in Man, J. Invest. Dermatol. 67, 667 – 668 (1976).

[3] G. Levy, T. Tsuchiya, L. P. Amsel, Limited Capacity for Salicyl-Phenolic Glucuronide Formation and Its Effect on the Kinetics of Salicylate Elimination in Man, Clin. Pharmacol. Ther. 13, 258 – 268 (1972).

[4] J. T. Wilson, R. L. Howell, M. W. Holladay, G. M. Brilis, J. Chrastil, J. T. Watson, D. F. Taber, Gentisuric Acid: Metabolic Formation in Animals and Identification as a Metabolite of Aspirin in Man, Clin. Pharmacol. Ther. 23, 635 – 643 (1978).

[5] T. J. Meredith, J. A. Vale, Salicylate Poisoning, in: J. A. Vale, T. J. Meredith (eds.), Poisoning: Diagnosis and Treatment, Update Books, London 1981, pp. 97 – 103.

[6] A. L. Tarnoky, V. A. L. Brews, A Simple Estimation of Salicylate in Serum, J. Clin. Pathol. 3, 289 – 291 (1950).

[7] P. Trinder, Rapid Determination of Salicylate in Biological Fluids, Biochem. J. 57, 301 – 303 (1954).

[8] M. J. H. Smith, J. M. Talbot, Estimation of Plasma Salicylate Levels, Br. J. Exp. Pathol. 31, 65 – 69 (1950).

[9] E. S. Kang, T. A. Todd, M. T. Capaci, K. Schwenzer, J. T. Tabbour, Measurement of True Salicylate Concentrations in Serum from Patients with Reye's Syndrome, Clin. Chem. 29, 1012 – 1014 (1983).

[10] B. E. Cham, D. Johns, F. Bochner, D. M. Imhoff, M. Rowland, Simultaneous Liquid Chromatographic Quantitation of Salicylic Acid, Salicyluric Acid and Gentisic Acid in Plasma, Clin. Chem. 25, 1420 – 1425 (1979).

[11] *L. J. Walter, D. F. Biggs, R. T. Coutts,* Simultaneous GLC Estimation of Salicylic Acid and Aspirin in Plasma, J. Pharm. Sci. *63,* 1754 – 1758 (1974).
[12] *R. W. Longenecker, J. E. Trafton, R. B. Edwards, A. Tableted,* Enzymic Reagent for Salicylate, for Use in a Discrete Multi-wavelength Analytical System (Paramax®), Clin. Chem. *30,* 1369 – 1371 (1984).
[13] *K. You, J. A. Bittikofer,* Quantification of Salicylate in Serum by Use of Salicylate Hydroxylase, Clin. Chem. *30,* 1549 – 1551 (1984).
[14] *A. Atkinson, R. S. Campbell, P. M. Hammond, C. P. Price, M. D. Scawen,* Method for the Estimation of Salicylates or Reduced Pyridine Nucleotides, British Patent Application 832696, 1983.
[15] *S. A. P. Chubb, P. M. Hammond, R. Ramsay, R. S. Campbell, M. D. Scawen, A. Atkinson, C. P. Price,* An Enzyme Mediated Colorimetric Assay for Salicylate, Clin. Chim. Acta (accepted for publication).

2.15 Nicotine

3-(1-Methyl-2-pyrrolidinyl)pyridine

Albert Castro and Nobuo Monji

General

Nicotine is an alkaloid present in the leaves of *Nicotiana tabacum* and *N. rustica* to the extent of 2 to 8%. During the past two decades interest in the metabolism of nicotine has been stimulated by an increased consideration of the possible role of nicotine as a reinforcer [1] in the habitual use of tobacco, and by many studies on the short half-life of nicotine in human blood plasma [2 – 4]. In addition, some of the metabolites of nicotine appear in various instances physiologically to oppose or enhance the effects of nicotine [5 – 7]. These and other considerations have been contributory factors in the development of effective methods for determining nicotine in biological fluids.

Application of method: in pharmacology and toxicology.

Substance properties relevant in analysis: nicotine is a small, volatile molecule ($M_r =$ 162) which has no property that can be used in a direct enzymatic assay. It is soluble in water and turns brown on exposure to air or light; its pK_1 (at 15 °C) is 6.16 and pK_2 (at 15 °C) is 10.96. The immunochemical approach is feasible since the nicotine molecule is large enough to produce antibody if nicotine, or a nicotine analogue, is chemically conjugated to a carrier antigen such as bovine serum albumin or thyroglobulin.

Methods of determination: gas chromatography and combined gas chromatography/ mass spectrometry have been used in the past. Limitations and advantages of these and other methods have been previously reviewed [8, 9]. Because of the specificity and sensitivity of such assays, several investigators have investigated various immunological techniques for the rapid determination of nicotine. Radioimmunoassay (RIA) for nicotine has been developed [10], but in recent years commercial preparations of [3]H- or ^{14}C-labelled nicotine of high specific activity have become increasingly scarce. Although ^{125}I-labelled nicotine of high specific activity has been prepared [10], its shelf-life is short and there is considerable lot-to-lot variation in the material. The enzyme-immunoassay (EIA) for nicotine described here avoids various disadvantages of RIA such as high cost of tracer, short shelf-life, potential radiation hazard, and waste disposal problems. EIA can be as sensitive as RIA and equally fast and efficient.

Serum concentrations of nicotine after smoking vary depending upon whether the subject is smoker or non-smoker, or inhaler or non-inhaler. Nicotine levels in human smokers ($n = 9$) after one cigarette range from 50 to 100 µg/l at 3 min after smoking and 30 to 60 µg/l at 15 min after smoking as determined by RIA [10].

International reference method and standards: neither standardization at the international level nor the existence of reference standard materials is known so far.

Assay

Method Design

This assay is based on the competitive type of enzyme-immunoassay to detect small molecules (cf. Vol. I, p. 237).

Principle

(a)

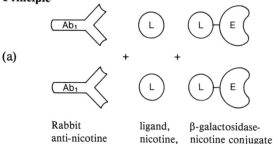

Rabbit
anti-nicotine
antibody

ligand,
nicotine,
analyte

β-galactosidase-
nicotine conjugate

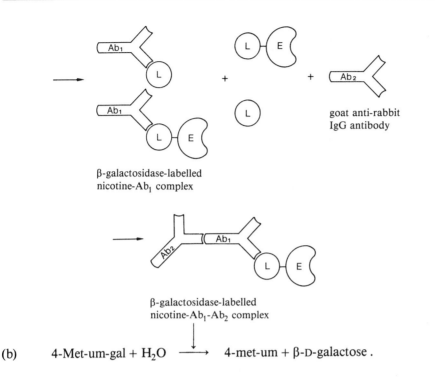

β-galactosidase-labelled
nicotine-Ab$_1$ complex

goat anti-rabbit
IgG antibody

β-galactosidase-labelled
nicotine-Ab$_1$-Ab$_2$ complex

(b)   4-Met-um-gal + H$_2$O $\longrightarrow$ 4-met-um + β-D-galactose .

The analyte analogue is conjugated chemically to an enzyme, in this case β-galactosidase*. After competitive binding of the hapten (analyte) and the enzyme-hapten conjugate to the first antibody, the first antibody is precipitated by the second antibody (antibody to the first antibody). The catalytic activity of the enzyme conjugate bound to the first antibody is measured as increase in fluorescence intensity per unit time, $\Delta F/\Delta t$.

Selection of assay conditions and adaptation to the individual characteristics of the reagents: immunochemical reactions are carried out at 25 °C. The total assay time is 150 min. The optimal quantities and concentrations of sample and reagents are dependent on the characteristics of the antibodies present in the antiserum.

Immunoreactivity of the enzyme-hapten conjugate and assay specificity vary depending on the antibody used for the assay. Assay specificity and sensitivity depend on the antibody and the hapten analogue conjugated to the enzyme. The first parameter to be examined is the range of the assay. For nicotine, the typical range is from 10 to 200 µg/l and the size of sample may vary from 20 to 100 µl per assay. The serum sample is usually 50 µl.

General procedures to examine a suitable antibody for a particular enzyme-hapten conjugate are as follows:

---

* β-D-Galactoside galactohydrolase, EC 3.2.1.23.

- carry out an antibody titration curve without the presence of analyte, by adding a fixed amount of enzyme-hapten conjugate and varying amounts of antibody;
- then carry out the antibody titration curve with addition of a fixed, excess amount of the analyte;
- examine the displacement of the enzyme-hapten conjugate in the titration curves constructed in the first two steps. The antibody dilution which shows the maximum displacement is the dilution needed for the assay;
- examine the cross-reactivity of the antibody using the analogues as well as the substances which may interfere with the assay. Select the optimal concentration of the antibody after the antibody titration curve is constructed; take the dilution which gives 50% binding of the enzyme-hapten conjugate. For enzyme-hapten conjugate, take the amount of enzyme-hapten conjugate which gives a ΔF of 1.0 per h, when the conjugate is bound to the antibody of optimal concentration.

Equipment

Spectrofluorimeter capable of measuring the fluorescence of 4-methylumbelliferone (360 nm for excitation and 450 nm for emission). A centrifuge capable of $4000\,g$ at $4\,°C$. Pipettes capable of accurately dispensing 10 to 2000 µl fluid.

Reagents and Solutions

Purity of reagents: β-galactosidase from *Escherichia coli* $(600-900\ U^*/mg)$ is available commercially in a highly purified form. (*S*)-Nicotine purchased from *Aldrich Chemical Co.*, U.S.A., is re-distilled prior to use. (*R,S*)-6-aminonicotine is prepared according to the procedure of *Tschitschibakin & Kirrasow* [11]. 4-Methylumbelliferyl β-D-galactoside should be stored dry at $4\,°C$. All other chemicals are analytical grade.

Preparation of solutions: all solutions in re-purified water (cf. Vol. II, chapter 2.1.3.2).

1. Phosphate-buffered saline, PBS (phosphate, 11 mmol/l; NaCl, 137 mmol/l; pH 7.2):

   dissolve 10.8 g $Na_2HPO_4 \cdot 12\,H_2O$, 1.6 g NaH_2PO_4 and 32.0 g NaCl in 4 l water.

2. PBS/BSA (phosphate, 11 mmol/l; NaCl, 137 mmol/l; BSA, 1 g/l; pH 7.2):

   dissolve 1 g bovine serum albumin in PBS (1), adjust pH to 7.2 with NaOH, 5 mol/l, and make up to 1 litre with water.

---

* With 2-nitrophenyl β-D-galactoside as substrate, pH 7.0, at $37\,°C$ (*Sigma*, grade VIII).

3. Assay buffer (phosphate, 11 mmol/l; NaCl, 137 mmol/l; $MgSO_4$, 2 mmol/l; β-mercaptoethanol, 2 mmol/l; $MnCl_2$, 2 mmol/l; pH 7.2):

   dissolve 0.24 g $MgSO_4$, 0.156 g β-mercaptoethanol and 0.252 g $MnCl_2$ in PBS (1), adjust pH to 7.2 with NaOH, 5 mol/l, and make up to 1 litre with water.

4. Stopping solution (glycine buffer, 100 mmol/l; pH 10.3):

   dissolve 75 g glycine in 1 litre water and adjust pH to 10.3 with NaOH, 10 mol/l.

5. β-Galactosidase-hapten conjugate (β-galactosidase, 470 U/l; phosphate, 11 mmol/l; NaCl, 137 mmol/l; BSA, 1 g/l; pH 7.2):

   dissolve 0.2 ml enzyme-hapten conjugate stock solution (cf. Appendix, p. 389), 235 kU galactosidase per litre, in 99.2 ml PBS/BSA (2).

6. Anti-nicotine antiserum (γ-globulin, 25 mg/l; phosphate, 11 mmol/l; NaCl, 137 mmol/l; BSA, 1 g/l; pH 7.2):

   dilute 0.25 ml nicotine antiserum (γ-globulin, 10 g/l) with PBS/BSA (2) to 100 ml.

7. Normal rabbit serum (γ-globulin, 500 mg/l; phosphate, 11 mmol/l; NaCl, 137 mmol/l; BSA, 1 g/l; pH 7.2):

   dilute 5.0 ml normal rabbit serum (γ-globulin, 10 g/l) with 95.0 ml PBS/BSA (2).

8. Goat anti-rabbit IgG* (125 000 units/l):

   dissolve the lyophilized antiserum (contained in 10 ml vial) in 10 ml water.

9. Substrate solution (4-met-um-gal, 300 μmol/l):

   dissolve 1.0 mg 4-methylumbelliferyl β-D-galactoside in 0.2 ml *N,N*-dimethyl-formamide and add 9.8 ml water.

10. Nicotine standard solution (0 to ca. 1.23 μmol/l; phosphate, 11 mmol/l; NaCl, 137 mmol/l; BSA, 1 g/l; pH 7.2):

    dissolve 100 mg nicotine in 1 l water (0.62 mmol/l); dilute serially with PBS/BSA (2) to make solutions of 0 to 1.23 μmol/l (0 to ~ 200 μg/l).

Stability of solutions: anti-nicotine antiserum (6) is stable at 4 °C for at least a few months if the serum is preserved with NaN_3, final concentration 1 g/l. The undiluted enzyme conjugate (5) is stable for a few months at − 18 °C. Freshly distilled nicotine should be used for the standard solution (10). The diluted antiserum (6), the diluted enzyme conjugate (5), the assay buffer (3) and the incubation buffer (2) are prepared

---

* Goat anti-rabbit IgG can be obtained from *Calbiochem Behring*, San Diego, CA 92112. Each unit of antiserum precipitates about 40 μg rabbit γ-globulin.

every week and kept at 4°C. The substrate solution (9) should be used within 3 h. Solutions (1) and (4) are stable for a few months at 4°C; solution (8) is stable for a few weeks at 4°C.

Procedure

Collection and treatment of specimen: either plasma or serum can be used in the assay. No evidence of interference was obtained when serum from slightly haemo lyzed blood is used for the assay. No data are available on sera from highly lipaemic and haemolyzed blood.

Stability of the analyte in the sample: store serum and plasma samples at $-20°C$ and assay within a few days.

Assay conditions: all measurements in duplicate.

Immunoreaction: incubation for 60 min at 37°C; for 2 h at 4°C; volume 0.3 ml and 0.5 ml, respectively.

Enzymatic indicator reaction: excitation wavelength 360 nm; emission wavelength 450 nm; light path 10 mm; measurements against substrate solution, 37.7 µmol/l; 60 min at 37°C.

Calibrate the fluorimeter with a freshly prepared 4-methylumbelliferone solution, $10-1000$ nmol/l ($M_r = 176.2$) in glycine buffer (4).

Establish a calibration curve with standard solutions (10) instead of sample under exactly the same contitions (within the series).

Calibration curve: measuring unit is the change in fluorescence intensity per 60 min, $\Delta F/\Delta t$. If $\Delta F_0/\Delta t$ is obtained at zero nicotine concentration, the ratio $\Delta F/\Delta t/\Delta F_0/\Delta t = \Delta F/\Delta F_0$ corresponds to nicotine concentration expressed in µg per litre. Plot $\Delta F/\Delta F_0$ in percent *vs.* nicotine concentration. The calibration curve is shown in Fig. 1, p. 388.

Calculation: read nicotine mass concentration, ρ (µg/l), from calibration curve. To convert from mass concentration to substance concentration, c (µmol/l), multiply by 6.1728.

Interpretation of results: in normal serum the level is usually less than 5 µg/l. Nicotine levels in smokers after one cigarette range from 50 to 100 µg/l after three minutes and from 30 to 60 µg/l after 15 minutes [10].

Measurement

Pipette into a disposable tray:			concentration in the incubation/assay mixture	
sample or standard solution (10) incubation buffer	(2)	0.05 ml 0.05 ml	nicotine phosphate NaCl BSA	up to 33 µg/l 11 mmol/l 137 mmol/l 1 g/l
enzyme conjugate anti-nicotine antiserum	(5) (6)	0.10 ml 0.10 ml	β-galactosidase γ-globulin	157 U/l 8.3 mg/l
incubate for 60 min at room temperature,				
normal rabbit serum goat anti-rabbit IgG	(7) (8)	0.10 ml 0.10 ml	γ-globulin (rabbit) IgG phosphate NaCl BSA β-galactosidase γ-globulin (anti-nicotine antiserum)	100 mg/l 25 000 units/l 8.8 mmol/l 110 mmol/l 0.8 g/l 94 U/l 5 mg/l
mix, incubate for 2 h at 4°C, centrifuge for 10 min at 2000 g; wash twice with buffer (1) by suspension and centrifugation;				
assay buffer	(3)	0.10 ml	phosphate NaCl Mg^{2+} β-mercapto-ethanol Mn^{2+}	6.7 mmol/l 91 mmol/l 1.3 mmol/l 1.3 mmol/l 1.3 mmol/l
substrate solution	(9)	0.05 ml	4-met-um-gal	100 µmol/l
incubate at 37°C for 60 min,				
stopping solution	(4)	2.5 ml	glycine	94.3 mmol/l
centrifuge at 2000 g for 10 min at 4°C; take 2 ml of the supernatant and measure fluorescence.				

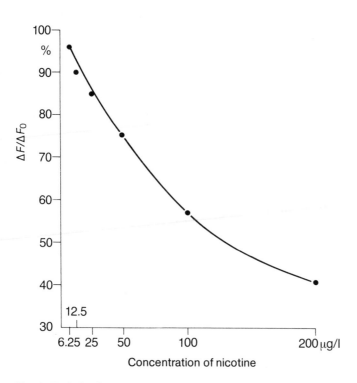

Fig. 1. Typical calibration curve for the assay of nicotine in serum.

Validation of Method

Precision, accuracy, detection limit and sensitivity: the relative standard deviation, RSD, at each point of the calibration curve, assayed in triplicate, is less that 8%. Between-days RSD is about 9%. In recovery experiments, 20, 30, 40 and 50 µg nicotine per litre pooled plasma was prepared and analyzed by EIA. The correlation coefficient of found *vs.* expected value was 0.99. The lower limit of the measuring range, defined as the concentration of nicotine measured at $\Delta F/\Delta F_0 = 90\%$, was found to be 10 µg/l.

Sources of error: since nicotine is a volatile as well as an easily oxidizable compound, it is very important to process the sample as quickly as possible. It is important to freeze serum samples as quickly as possible to ensure the stability of nicotine.

Specificity: assay specificity is dependent upon the antiserum used and the hapten analogue employed for the enzyme-conjugate preparation. In this assay, cross-reactivity (as determined by the same procedure used for nicotine RIA [12]) was 0.8% for (*S*)-nornicotine, about 0.1% for (*S*)-nicotine-*N'*-oxide, and less than 0.1% for (*S*)-cotinine, *N*-methylpyrrolidine and (*S*)-cotinine-*N*-oxide.

Appendix

Preparation of anti-nicotine antiserum: the immunogen conjugate preparation was described previously by *Castro & Prieto* [13]. Briefly, the 6-(ε-aminocapramide)-DL-nicotine derivative was conjugated to carboxyl groups of BSA as follows.

Preparation of 6-(ε-aminocapramido)-DL-nicotine: prepare the 6-(ε-aminocapramido)-DL-nicotine by the method described previously [13].

Preparation of 6-(ε-aminocapramido)-DL-nicotine-BSA conjugate: add a solution of 192 mg BSA in 3 ml water to a solution of 847 mg 6-(ε-aminocapramido)-DL-nicotine in 10 ml water, adjusted to pH 5.0 at room temperature; add 650 mg 1-ethyl-(3,8-di-methyl-aminopropyl)carbodiimide hydrochloride while stirring; maintain the pH of the reaction mixture at 5.0 throughout the first 2 h; add additional 650 mg carbodiimide and incubate for one hour at room temperature; dialyze the conjugate for 72 h against 8 l phosphate buffer, 10 mmol/l, pH 7.8; dialyze one more day against sodium chloride, 0.15 mol/l, phosphate buffer, 10 mmol/l, pH 7.8; store the dialyzed immunogen frozen at −18 °C. Using this procedure, the nicotine/BSA ratio is 11 based on tracer method calculations [13].

Immunization: rabbits are immunized intradermally (in foot pads or in the back), intramuscularly, or subcutaneously. A reasonable titre of antibody (1 : 400) can be obtained after 12 weeks.

Preparation of β-galactosidase-nicotine conjugate: the method is divided into 2 parts.

m-Maleimidobenzoyl derivative of 6-(4-aminobenzamido) nicotine (II): preparations of 3-maleimidobenzoic acid and its carbonyl chloride (I) were described previously [14]: dissolve 100 mg I in 1 ml tetrahydrofuran (THF) and add it to 2 ml THF solution containing 50 mg 6-(4-aminobenzamido) nicotine and a slurry of 10 mg Na_2CO_3; reflux the mixture for 30 min; separate the product by silica-gel chromatography using chloroform as an eluting solvent.

Conjugation of II with β-galactosidase: dissolve 0.5 mg (400 U) β-galactosidase in 1 ml phosphate-buffered saline (1); add 50 μl THF containing 20 μg II; incubate at room temperature for 60 min; dialyze for 24 h at 4 °C against the same buffer to remove excess unreacted II; apply to a Sephadex G-25 column (10 mm × 400 mm) to remove residual, unreacted II; collect the fractions containing the major enzyme activity; add NaN_3 and BSA to 0.1% and, after dividing the solution to 100 μl aliquots, store at −18 °C (once a frozen aliquot has been thawed, do not re-freeze).

References

[1] *M. E. Jarwick,* Research on Smoking Behavior, in: *M. E. Harvich, J. W. Cullen, E. R. Britz, T.M. Vogt, L. J. West* (eds.), U.S. Government Printing Office, Washington, D.C. 1977, pp. 122 – 146.

[2] *P. F. Isaac, M. J. Rand,* Cigarette Smoking and Plasma Levels of Nicotine, Nature (Lond.) *263,* 308 – 310 (1972).

[3] *J. J. Langone, H. B., Gjika, H. van Vunakis,* Nicotine and Its Metabolites; Radioimmunoassays for Nicotine and Cotinine, Biochemistry *20,* 5025 – 5030 (1973).

[4] *A. K. Armitage,* Nicotine Levels in Serum after Cigar Smoking, in: Symposium on Nicotine and Carbon Monoxide, University of Kentucky, Lexington, KY 1975, pp. 52 – 69.

[5] *U. S. von Euler, F. Haglid, F. Hedquist, I. Motelica,* Neurotransmitter Releasing Effects of Two Quarternary Nicotine, Acta Physiol. Scand. *78,* 123 – 131 (1970).

[6] *H. McKennis, Jr.,* Metabolism of Nicotine, in: *U. S. von Euler* (ed.), Tobacco Alkaloids and Related Compounds, Pergamon Press, Oxford 1965, pp. 53 – 74.

[7] *K. K. Wilson, R. S. Chang, E. R. Bowman, H. McKennis, Jr.,* Nicotine-like Actions of *cis-meta* Nicotine and *trans-meta* Nicotine, J. Pharmacol. Exp. Ther. *196,* 685 – 696 (1976).

[8] *A. Pilotti,* Nicotine and Its Metabolites; Analysis by Gas Chromatography, in: Symposium on Nicotine and Carbon Monoxide, University of Kentucky, Lexington, KY 1975, pp. 40 – 49.

[9] *A. Pilotti, C. R. Enzell, H. McKennis, Jr., E. R. Bowman, E. Dufon, B. Holmostedt,* Analysis of Nicotine and Its Metabolites by Combined Gas Chromatography/Mass Spectrometry, Beitr. Tabakforsch. *6,* 339 – 349 (1976).

[10] *A. Castro, N. Monji, H. Malkers, W. Eisenhart, H. McKennis, Jr., E. R. Bowman,* Automated Radioimmunoassay of Nicotine, Clin. Chim. Acta *95,* 473 – 481 (1979).

[11] *H. E. Tschitschibakin, A. W. Kirrasow,* Aminierung des Nicotins mit Natrium- und Kalium-Amid, Ber. Dtsch. Chem. Ges. *57,* 1163 – 1168 (1924).

[12] *A. Castro, N. Monji, H. Ali, J. M. Yi, E. R. Bowman, H. McKennis, Jr.,* Nicotine Antibodies; Comparison of Ligand Specificities of Antibodies Produced against Two Nicotine Conjugates, Eur. J. Biochem. *104,* 331 – 340 (1980).

[13] *A. Castro, I. Prieto,* Nicotine Antibody Production; Comparison of Two Nicotine Conjugates in Different Animal Species, Biochem. Biophys. Res. Commun. *63,* 583 – 589 (1975).

[14] *N. Monji, H. Malkus, A. Castro,* Maleimide Derivative of Hapten for Coupling to Enzyme; a New Method in Enzyme Immunoassay, Biochem. Biophys. Res. Commun. *85,* 671 – 675 (1978).

2.16   Sulphamethazine Residues in Animal Blood

4-Amino-*N*-(4,6-dimethyl-2-pyrimidinyl)benzenesulphonamide

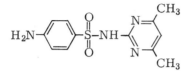

James Fleeker

General

The discovery in 1935 [1] that Prontosil had antibacterial activity led to the development of a class of synthetic drugs called the sulphonamides. These compounds are metabolic antagonists in the biosynthesis of folic acid. As such, they are useful as bacteriostatic agents in man and animals. Their effectiveness is increased when used in conjunction with trimethoprim which inhibits a different step in folic acid synthesis [2].

The major metabolic pathway of the sulphonamides is species-dependent. In the rabbit and rat the N^4-acetyl derivative is preferentially formed, whereas in the dog, the N^4-glucuronides are almost exclusively produced [3]. In the pig the major metabolites of sulphamethazine (SMT) are N^4-acetyl-SMT, the N^4-glucoside and N^4-des-amino-SMT [4]. The level of SMT in swine blood is a good indicator of the drug level in tissues and organs [4].

A major use of the sulphonamides is as a feed additive at non-therapeutic levels to prevent disease and promote growth in domestic animals. Because of this prophylactic use, regulatory agencies monitor these drugs in animal tissues before the meat is sold for human consumption. During the past decade in the United States, a significant number of swine carcasses were found to have levels (≥ 0.1 µg/g) of sulphonamide residues [5] which violated regulations. SMT was the drug usually found.

There are human health concerns over the presence of sulphonamide residues in meat. The sulphonamides are known to have an effect on the function of the thyroid gland [6]. In addition, N^4-desaminosulphonamides and some other metabolites appear to be formed in the gut from the parent sulphonamides *via* diazonium salts [7 – 9]. The diazonium compounds may be formed from nitrite present in the alimentary tract. Diazonium salts have been found to be mutagens [10, 11]. Because nitrite is present in the human digestive tract [12], the formation of these sulphonamide-diazonium derivatives in man is possible. These concerns have increased the need for rapid, selective and sensitive methods to detect sulphonamides in tissues and blood.

Application of method: in laboratories monitoring meat destined for human consumption. The immunoassay is used to screen blood or plasma for residues of SMT and metabolites of SMT having structural changes only at the N^4-position.

Substance properties relevant in analysis: SMT (molecular weight 279) is sparingly soluble in water. The solubility is enhanced in dilute acid or base. The pK values are 2.65 for the N^1-nitrogen and 7.4 for N^4 [13]. N^4-glycosyl-SMT is quickly hydrolyzed to SMT in dilute acid while SMT is rather stable in cold dilute acid. There is significant binding of SMT to serum proteins [14].

Methods of determination: SMT and other sulphonamides are screened in tissues using thin-layer chromatography [15]. These drugs may be quantitated in tissues and body fluids with the *Bratton-Marshall* method and variations of this technique [16, 17]. Both methods require the presence of an aromatic primary amine for colour development and detection. They cannot detect metabolites of sulphonamides in which the N^4 amino group is conjugated or lacking. Gas chromatography and gas chromatography-mass spectroscopy have been used to quantitate specific sulphonamides [18 – 20]. High-performance liquid chromatography has been employed to detect SMT [21, 22].

International reference method and standards: there is no known international standard method. Standard material of SMT can be obtained from the manufacturer, but no standard reference material is known.

Assay

Method Design

The method is based on competitive binding of SMT and SMT-labelled enzyme to antibody. The antibody is adsorbed to a plastic surface before the assay. The principle of the technique was first described by *van Weeman & Schuurs* [23].

Principle

(a)

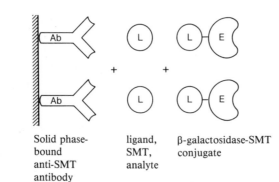

| Solid phase-
bound
anti-SMT
antibody | ligand,
SMT,
analyte | β-galactosidase-SMT
conjugate |

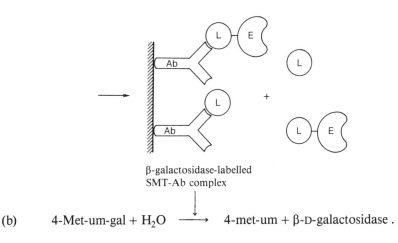

β-galactosidase-labelled
SMT-Ab complex

(b)   4-Met-um-gal + H_2O $\longrightarrow$ 4-met-um + β-D-galactosidase .

SMT is covalently bonded to β-galactosidase *via* the N^4 position, using N^4-(4-male-imidobenzoyl)SMT. Diluted antibodies to SMT are coated to polystyrene tubes. Blood extract containing SMT in buffer is introduced into coated tubes along with a specified amount of labelled enzyme. Both SMT and the labelled enzyme compete for binding to the antibodies. The greater the amount of SMT present, the smaller the amount of labelled enzyme bound to antibody. After equilibrium is reached, unbound material is removed and the amount of bound enzyme is determined with a fluorigenic substrate by measurement of the reaction rate, i.e. the increase in fluorescence intensity per unit time, $\Delta F/\Delta t$.

Selection of assay conditions and adaptation to the individual characteristics of the reagents: to screen a set of samples requires two days. The number of samples screened is limited by the extraction step. An operator with suitable equipment and experience is able to process 25 samples plus control samples in two days. Coating of the tubes with antibody can be done overnight prior to sample extraction. Sample extraction is done in the morning and part of the afternoon. Incubation of the sample extract with antibody and labelled enzyme is begun late in the afternoon and the analysis is completed on the following morning. Calculations and preparation of tubes for the next set of samples take up the remainder of the second day. Anti-SMT antibodies can be raised in rabbits using the diazonium salt of SMT conjugated to serum albumin or thyroglobulin.

Fractionation of the sera to obtain the immunoglobulin fraction is desirable for reproducibility and improved coating properties.

It is necessary to determine the coating titre for the immunoglobulin fraction in order to establish a suitable dilution for the coating step in the assay, as follows:

Prepare a 1:1000 dilution of the immunoglobulin fraction in coating buffer (3). Prepare 1:2 (v/v) serial dilutions of this solution in coating buffer. Add to 12 mm × 75 mm polystyrene tubes 1.0 ml portions of 95% ethanol. Continue as described in Appendix, p. 399, for coating the tubes, with each dilution. Add to each tube 9 mU β-galactosidase-labelled SMT (cf. Appendix, p. 400) in 0.5 ml assay buffer (1) and continue as described under Measurement. Choose a dilution of the immunoglobulin for use in the assay which binds approximately 60% of the maximum amount of enzyme conjugate [24].

Equipment

Refrigerated centrifuge with a swinging-bucket rotor taking 50 ml centrifuge tubes and capable of reaching 2500 g. A fluorimeter with excitation at 360 nm and emission at 448 nm. Quartz cuvettes for the fluorimeter. Constant-temperature bath. C_8-extraction columns (equivalent to *J. T. Baker Co.,* No. 7087-6). Polystyrene tubes, 12 mm × 75 mm. A repeating pipette capable of delivering 100 µl and even-numbered multiples of this volume up to 1.0 ml with imprecision and inaccuracy of ± 1.0%.

Reagents and Solutions

Purity of reagents: β-galactosidase from *E. coli* should be of the best available purity. Inorganic chemicals are reagent grade. SMT (99%) is recrystallized twice from ethanol/water.

Preparation of solutions (for 100 determinations): all solutions in re-purified water (cf. Vol. II, chapter 2.1.3.2).

1. Assay buffer (NaCl, 0.1 mol/l; phosphate, 0.1 mol/l; $MgSO_4$, 0.5 mmol/l; NaN_3, 3 mmol/l; DTE, 65 µmol/l; gelatin, 1 g/l; pH 7.4):

   dissolve 11.7 g NaCl, 4.6 g $NaH_2PO_4 \cdot H_2O$, 29.0 g K_2HPO_4, 0.12 g $MgSO_4$, 20 mg dithioerythritol and 400 mg NaN_3 in 900 ml water. Warm 2.0 g gelatin in 900 ml water until dissolved. Mix the two solutions and dilute to 2.0 l. Check pH.

2. Phosphate-buffered saline, PBS (NaCl, 0.15 mol/l; phosphate, 10 mmol/l; NaN_3, 3 mmol/l; pH 7.4):

   dissolve 0.88 g NaCl, 1.450 g KH_2PO_4, 230 mg $Na_2HPO_4 \cdot H_2O$ and 20 mg NaN_3 in water and dilute to 100 ml. Check pH.

3. Coating buffer (carbonate, 25 mmol/l; NaN_3, 3 mmol/l; pH 10.0):

   dissolve 0.2 g Na_2CO_3, 0.26 g $NaHCO_3$ and 40 mg NaN_3 in 200 ml water. Check pH.

4. Washing buffer (NaCl, 0.1 mol/l; phosphate, 0.1 mol/l; $MgSO_4$, 0.5 mmol/l; NaN_3, 3 mmol/l; DTE, 65 µmol/l; gelatin, 1 g/l; Tween 20, 0.5 ml/l; pH 7.4):

   dissolve 0.1 ml Tween 20 in 200 ml assay buffer (1).

5. Substrate solution 1 (4-met-um-gal, 0.3 mmol/l; NaCl, 0.1 mol/l; phosphate, 0.1 mol/l; $MgSO_4$, 0.5 mmol/l; NaN_3, 3 mmol/l; DTE, 65 μmol/l; gelatin, 1 g/l; pH 7.4):

dissolve 10 mg 4-methylumbelliferyl β-galactoside in 100 ml assay buffer (1). Filter if necessary.

6. Galactosidase-SMT conjugate solution (conjugate*; galactosidase, 90 U/l; NaCl, 0.1 mol/l; phosphate, 0.1 mol/l; $MgSO_4$, 0.5 mmol/l; NaN_3, 3 mmol/l; DTE, 65 μmol/l; gelatin, 1 g/l; pH 7.4):

use preparation according to Appendix, p. 400; dilute the stock solution of the enzyme conjugate with assay buffer (1) appropriately to 10 ml.

7. $HClO_4$ (0.34 mol/l):

dilute 102 ml perchloric acid (sp. gr. 1.67; 70%, w/w) to 3.5 l with water.

8. K_2HPO_4 (1.2 mol/l):

dissolve 314 g K_2HPO_4 in water and dilute to 1.5 l.

9. Na_2CO_3 (1.0 mol/l):

dissolve 21.2 g Na_2CO_3 in water and dilute to 200 ml.

10. Substrate solution 2 (2-NPgal, 66 mmol/l):

dissolve 200 mg 2-nitrophenyl β-D-galactoside in 10.0 ml water by gently warming.

11. Methanol (60%, v/v):

dilute 180 ml methanol with 120 ml water.

12. SMT standard solution (SMT, 36 μmol/l; 10 mg/l):

dissolve 10 mg SMT in 500 ml ethanol and dilute to 1.0 l with water.

Stability of solutions: solutions (1) – (5) stored in stoppered glass containers at 4 °C are stable for two weeks, solution (6) for one week. Prepare solution (10) when needed. Store solution (12) in the refrigerator at 4 °C and prepare every 4 weeks.

---

* Concentration is variable.

Procedure

Collection and treatment of specimen: collect 50 ml blood in a vessel with 0.2 ml PBS
(2) containing 50 units (U.S.P.) heparin. The blood can be stored at $-20\,°C$ until
assayed. If plasma is to be assayed, centrifuge the blood for 30 min at 2000 g, $10\,°C$.
Remove the plasma and store at $-20\,°C$.

The assay can be used with blood or plasma from animals other than swine. Citrate
and EDTA are acceptable as anticoagulants.

Place 5.0 ml thawed blood or plasma in 50 ml glass centrifuge tubes with screw
caps. Add 50 µl SMT standard solution (12) only to each of two drug-free blood samples (controls). Swirl to mix and allow to stand for 10 min before proceeding. Add
35 ml $HClO_4$ (7) to all tubes and swirl to mix. Allow to stand at room temperature for
10 min. Swirl again and centrifuge for 10 min at 1500 g, $10\,°C$. Decant the supernatant
solution into centrifuge tubes containing 15 ml KH_2PO_4 solution (8). Mix and centrifuge as before. pH of the supernatant solution should be between 6 and 7. Wash a 6 ml
C_8 extraction column first with 10 ml ethanol and then with 5 ml water. Pass the
neutralized extract through the column. Wash the column with 2.0 ml water. Elute
with 2.5 ml 60% methanol (11), forcing all eluting solvent through the column with
gentle air pressure. Collect the eluate in a 10 ml graduated centrifuge tube. Dilute the
eluate to 3.0 ml with water, then dilute to 6.5 ml with assay buffer (1). Use this extract
directly in the assay.

Stability of the substances in the sample: the substances detected by the assay are
stable for up to six months when stored at $-20\,°C$.

Assay conditions: measure controls and samples in duplicate under identical conditions.

Immunoreaction: incubation time 18 h; room temperature; 0.5 ml.

Enzymatic indicator reaction: incubation time 1 h; $37\,°C$; reaction volume 0.5 ml;
excitation wavelength 360 nm, emission wavelength 448 nm; slit widths 4 mm; quartz
cuvettes; measurements against a mixture of 0.5 ml substrate solution (5) and 1.0 ml
carbonate solution (9).

Check the linearity of the fluorimeter using 4-methylumbelliferone in sodium carbonate solution (9).

Calculation: average the values of $\Delta F/\Delta t$ of standards containing 0.1 mg SMT per
litre, and $\Delta F/\Delta t$ of samples. Samples containing ≥ 0.1 mg drug per litre will give a response equal to or less than the control blood fortified with SMT.

Measurement

Pipette successively into coated tubes:			concentration in incubation mixture	
extracted sample or standard solution (12) enzyme conjugate	(6)	0.4 ml 0.1 ml	SMT conjugate galactosidase phosphate NaCl Mg^{2+} NaN_3 DTE gelatin methanol	up to 45 µg/l 0.05 to 0.15 mg/l 18 U/l 0.63 mol/l 0.06 mol/l 0.32 mmol/l 1.9 mmol/l 41 µmol/l 0.63 g/l 180 ml/l
incubate for 18 h at room temperature, aspirate the solution and wash twice with (1);				
substrate solution 1	(5)	0.5 ml	4-met-um-gal phosphate NaCl Mg^{2+} NaN_3 DTE gelatin	0.3 mmol/l 0.1 mol/l 0.1 mol/l 0.5 mmol/l 3 mmol/l 65 µmol/l 1 g/l
incubate for 1 h at 37 °C;				
Na_2CO_3	(9)	1.0 ml	Na_2CO_3	0.67 mol/l
reaction is stopped; measure $\Delta F/\Delta t$.				

Validation of Method

Precision, accuracy, detection limit and sensitivity: the day-to-day and within-run relative standard deviations (RSD) are less than 9%.

The method has been checked with swine blood fortified individually with SMT, N^4-acetyl-SMT, N^4-glucosyl-SMT and N^4-desamino-SMT. The fortification level was the molar equivalent of 0.1 mg SMT per litre. SMT and the three metabolites each gave positive responses with the immunoassay (12 samples each) when the cut-off was four times the standard deviation (SD) value of the negative control. When the cut-off was three times the SD of the negative control, desamino-SMT was not reliably detected. Normally SMT is the analyte of concern. The analyst must establish with the antisera the level of SMT in blood which yields a response which is three times the SD value determined from at least 20 negative control samples.

Reliability of the assay was tested with blood obtained from 15 pigs raised on rations free from SMT. From this blood, 110 samples were prepared and fortified with SMT at 0.1 mg/l and then analyzed along with the same number of unfortified samples (negative controls). No fortified samples gave negative responses. Three negative controls gave false-positive responses in that the responses fell outside the detection limit of three times the SD of the mean response.

Recoveries of SMT, N^4-acetyl-SMT, N^4-glucosyl-SMT and N^4-desamino-SMT in the extraction procedure were $73 \pm 6\%$; $61 \pm 4\%$; $72 \pm 3\%$ and $62 \pm 2\%$, respectively. This was checked using samples fortified with ^{14}C-labelled SMT and metabolites at levels equivalent to 0.1 mg SMT per litre.

Sources of error: serious variation in response is observed if, during the extraction and clean up, all samples are not treated in the same manner. Omission of the alcohol soaking step in the coating procedure results in increased variability between duplicate tubes.

Specificity: the immunogen described in the Appendix, p. 399, gave antisera which cross-reacted poorly with all sulphonamides tested except sulphamerazine. The latter was about one-third as sensitive to the assay as SMT (Table 1).

Table 1. Cross-reactivity of sulphamethazine and other chemicals to antisera used in enzyme-immunoassay.

Compound	Cross-reactivity*, %
Sulphamethazine	100
Acetylsulphamethazine	138
Desaminosulphamethazine	78
Glucosylsulphamethazine	52
Sulphamerazine	30
Sulphadimethoxine	0.4
Sulphisomidine	0.3
Sulphadiazene	0.1
Sulphapyridine	0.5
Sulphisoxazole	0.3
Sulphathiazole	0.2
Sulphamethoxypyridazine	< 0.1
Sulphaminouracil	< 0.1
Sulphanilamide	< 0.1
Sulphaguanidine	< 0.1
2-Amino-4,6-dimethylpyrimidine	< 0.1

* Cross-reactivity is defined as the nmoles of SMT causing 50% displacement of labelled enzyme per nmole of tested compound causing the same displacement. This value, times 100, is percent cross-reactivity.

It is important to note that the metabolites of SMT having structural changes only at the N^4-position cross-react with the antisera in a similar but not identical manner to SMT (Table 1). For this reason the assay is useful for screening of SMT and these derivatives in blood, but not for quantitation purposes.

Appendix

Preparation of immunogen: recrystallize SMT twice from ethanol/water and dry the crystals. Place 57 mg (0.2 mmol) SMT in a 20 mm × 150 mm tube and add 4 ml H_2SO_4, 0.25 mol/l, and a small stirring bar. Stir and warm to dissolve the SMT. Cool the solution in ice-water. Add dropwise, with stirring, over a 3-minute period, 19 mg (0.27 mmol) $NaNO_2$ in 1 ml water. Continue stirring for 5 min. Add this solution dropwise to a cooled and stirred solution of 100 mg bovine thyroglobulin or serum albumin in 4 ml sodium carbonate (9). The solution will turn a deep red. Stir the solution for 2 h and allow it to come to room temperature. Dialyze the preparation against three changes of ammonium hydrogen carbonate, 50 mmol/l, or carry out gel chromatography using Sephadex G-75 equilibrated with the same solvent. Collect the material eluting at the void volume. Freeze-dry the resulting product. Dissolve this material in PBS (2) and mix with an adjuvant for immunizing rabbits. It is best to prepare the immunogen with both serum albumin and thyroglobulin. Alternate the use of these immunogens when immunizing.

Preparation of immunoglobulins: with stirring, slowly add to the antiserum from rabbits an equal volume of neutralized 70% saturated ammonium sulphate solution. Allow to stand overnight. Centrifuge for 25 min at 2500 g, 15 °C, discard the supernatant. Wash the pellet twice with 35% saturated ammonium sulphate solution. Take up the pellet in water and dilute to the original volume. Dialyze against several changes of PBS (2). Dilute to twice the original volume with PBS (2) and mix this solution with an equal volume of glycerol. Store at − 20 °C. The use of glycerol avoids freezing and thawing which reduces serum titre.

Coating of tubes: add 1.0 ml 95% ethanol to each tube and allow to stand for 30 min. Decant the ethanol and dry the tubes. To each tube add 0.6 ml immunoglobulin appropriately diluted (in general 1 : 10000) with coating buffer (3). Incubate for 18 h (overnight) at room temperature or 4 h at 37 °C. Aspirate the solution and rinse once with 1.0 ml washing buffer (4), then rinse twice with 1.0 ml portions of assay buffer (1). Use these tubes within 2 h.

Preparation of N^4-(4-maleimidobenzoyl)-SMT: reflux 2.6 g (12 mmol) 4-maleimidobenzoic acid in 10 ml thionyl chloride and 10 ml benzene for 40 min. Cool the solution and add 25 ml toluene. Remove the solvents under reduced pressure. Suspend the acid chloride in 25 ml toluene and take to dryness again. Take up the acid chloride in 25 ml tetrahydrofuran. Add this solution to a cooled and stirred solution of 2.8 g (10 mmol)

SMT and 1.1 g (10.9 mmol) triethylamine in 25 ml tetrahydrofuran. Stir for 30 min at room temperature and then reflux for 30 min. Cool the solution and concentrate to about 25 ml on the rotating evaporator. Warm the concentrate and add 15 ml water. Allow to stand for several hours to allow crystals to form. Collect the precipitate and recrystallize from dimethylformamide/water. About 2 g of the refined product is obtained, which melts at 228 °C to 229 °C.

Preparation of β-galactosidase-N^4-(4-maleimidobenzoyl)-SMT conjugate: dissolve 2 mg β-galactosidase from *E. coli* in 0.3 ml PBS (2) supplemented with $MgCl_2$, 0.1 mmol/l. Commercial preparations of the enzyme often contain some thiol reagents. To remove these, pass the enzyme solution through a Sephadex G-100 column (5 mm × 80 mm) equilibrated with the solvent used to dissolve the enzyme. Collect the material absorbing at 280 nm and eluting at the void volume. Dilute the eluate to 3 ml with PBS (2) supplemented with magnesium ion, 0.1 mmol/l. Add to this solution, 100 µl dimethylsulphoxide containing 1 mg N^4-(4-maleimidobenzoyl)-SMT per ml. Mix and allow to stand for 1 h at room temperature. Dialyze for 24 h against several changes of assay buffer (1). Dilute the preparation with an equal volume of glycerol and store at − 20 °C. The solution is stable for several years when stored in this manner.

It is necessary to determine the activity of the enzyme before and after derivatization and periodically during storage. Dilute the enzyme in assay buffer (1). Place 2.8 ml assay buffer and 0.1 ml diluted enzyme in a 10 mm cuvette. Warm the solution to 37 °C and add 0.1 ml substrate solution 2 (10), mix by inversion and monitor the absorbance at 410 nm and 37 °C for 3 min. One unit, U, of enzyme will catalyze the hydrolysis of 1.0 µmol 2-NPgal per min at 37 °C.

According to eqn. (k_1), Appendix 3, p. 483; the catalytic activity concentration of the analyzed sample is

$$b = \frac{\Delta A \times V \times 1000}{\varepsilon \times d \times \Delta t \times v} \quad U/l,$$

where

V   volume of reaction mixture, l
v   volume of sample assayed, l
ε   mmolar absorption coefficient, $0.35 \ l \times mmol^{-1} \times mm^{-1}$
d   light path, mm.

In general, the catalytic concentration is approximately 500 kU/l.

References

[1] *G. Domagk,* Chemotherapie bakterieller Infektionen, Dtsch. med. Wochenschr. *61*, 250 − 253 (1935).
[2] *J. J. Burchall,* Trimethoprim and Pyrimethamine, Antibiotics 3, 304 − 320 (1975).
[3] *E. Reimerdes, J. H. Thumin,* Das Verhalten der Sulfanilamine im Organismus, Arzneim. Forsch. *29*, 1171 − 1179 (1970).

[4] *G. D. Paulson, J. M. Giddings, C. H. Lamoureux, E. R. Mansager, C. B. Struble,* The Isolation and Identification of ^{14}C-Sulfamethazine Metabolites in the Tissues and Excreta of Swine, Drug Metab. Dispos. *9,* 142 – 146 (1981).

[5] *C. D. Van Houweling,* The Magnitude of the Sulfa Residue Problem, J. Am. Vet. Med. Assoc. *178,* 464 – 466 (1981).

[6] *W. S. Hoffman,* The Biochemistry of Clinical Medicine Year Book, 4th edit., Medical Publishers 1970, pp. 614 – 653.

[7] *J. L. Woolley, C. W. Sigel,* The Role of Dietary Nitrate and Nitrite in the Reductive Deamination of Sulfadiazine by the Rat, Guinea Pig, and Calf, Life Sci. *30,* 2229 – 2234 (1982).

[8] *G. D. Paulson,* The Effect of Dietary Nitrite and Nitrate on the Metabolism of Sulfamethazine in the Rat, Xenobiotica *16,* 53 – 61 (1986).

[9] *H. Endo, H. Noda, N. Kinoshita, N. Inui, Y. Nishi,* Formation of a Transplacental Mutagen, 1,3-di(4-Sulfamoylphenyl)triazene, from Sodium Nitrite and Sulfanilamide in Human Gastric Juice and in the Stomachs of Hamsters, J. Natl. Cancer Inst. *65,* 547 – 551 (1980).

[10] *B. Toth, R. Patil,* Carcinogenesis of 4-(Hydroxymethyl)benzenediazonium Ion (Tetrafluoroborate) of *Agaricus bisporus,* Cancer Res. *41,* 2444 – 2449 (1981).

[11] *G. M. Lower, S. P. Lanphear, B. M. Johnson, G. T. Bryan,* Aryl and Heterocyclic Diazo Compounds as Potential Environmental Electrophiles, J. Toxicol. Environ. Health *2,* 1095 – 1107 (1977).

[12] *I. A. Wolff, A. E. Wasserman,* Nitrates, Nitrites and Nitrosamines, Science *177,* 15 – 19 (1972).

[13] *C. Papastephanou, M. Frantz,* Sulfamethazine, Anal. Profiles Drug Subst. *7,* 401 – 421 (1978).

[14] *K. Frislid, M. Berg, V. Hansteen, P. K. M. Lunde,* Comparison of the Acetylation of Procainamide and Sulfamethazine in Man, Eur. J. Clin. Pharmacol. *9,* 433 – 438 (1976).

[15] *W. F. Phillips, J. E. Trafton,* A Screening Method for Sulfonamides Extracted from Animal Tissues, J. Assoc. Off. Anal. Chem. *58,* 44 – 47 (1975).

[16] *A. C. Bratton, E. K. Marshall, D. Babbitt, A. R. Hendrickson,* A New Coupling Component for Sulfanilamide Determination, J. Biol. Chem. *128,* 537 – 550 (1939).

[17] *F. Tishler, J. L. Sutter, J. N. Bathish, H. E. Hagman,* Improved Method for Determination of Sulfonamide in Milk and Tissues, J. Agric. Food Chem. *16,* 49 – 53 (1968).

[18] *D. P. Goodspeed, R. M. Simpson, R. B. Ashworth, J. W. Shafer, H. R. Cook,* Sensitive and Specific Gas-chromatographic-Spectrophotometric Screening Procedure for Trace Levels of Five Sulfonamides in Liver, Kidney and Muscle Tissue, J. Assoc. Off. Anal. Chem. *61,* 1050 – 1053 (1978).

[19] *A. J. Manuel, W. A. Stellar,* Gas-liquid Chromatographic Determination of Sulfamethazine in Swine and Cattle Tissues, J. Assoc. Off. Anal. Chem. *64,* 794 – 799 (1981).

[20] *A. J. Malanoski, C. J. Barnes, T. Fazio,* Comparison of Three Methods for Determination of Sulfamethazine in Pork Tissue, J. Assoc. Off. Anal. Chem. *64,* 1386 – 1391 (1981).

[21] *A. B. Vilim, L. Larocque, A. I. MacIntosh,* A HPLC Screening Procedure for Sulfamethazine Residues in Pork Tissues, J. Liq. Chromatogr. *3,* 1725 – 1736 (1980).

[22] *B. L. Cox, L. F. Krzeminski,* High-Pressure Liquid Chromatographic Determination of Sulfamethazine in Pork Tissue, J. Assoc. Off. Anal. Chem. *65,* 1311 – 1315 (1982).

[23] *B. van Weemen, A. H. W. M. Schuurs,* Immunoassay Using Antigen-enzyme Conjugates, FEBS Lett. *15,* 232 – 236 (1971).

[24] *G. H. Parsons,* Antibody-coated Plastic Tubes in Radioimmunoassay, in: *J. J. Langone, H. van Vunakis* (eds.), Methods in Enzymology, Vol. *73,* Immunological Techniques, Part B, Academic Press, New York 1981, pp. 224 – 239.

3 Pesticides

3.1   Introduction

Hans Ulrich Bergmeyer

The wide acceptance of enzyme-immunoassays in pharmacology and toxicology that is evident in the earlier parts of this volume is far from being the case in environmental analytical chemistry. All colleagues whose opinion was sought expressed keen interest in EIA and predicted important applications of the technique in the future; however, scarcely any workers in the field have attempted to compare EIA with the established "instrumental analysis" (cf. also Vol. I, chapter 1.5).

One of the reasons for this may originate in the many scientific disciplines that are involved in environmental chemistry: the water analysts, the environmental health physicians, the toxicologists and many more. Analysts working in this field have mainly been trained in classical inorganic and organic analysis, and are also well trained in modern physicochemical methods such as HPLC and GC, often combined with mass spectrometry. Therefore, they have usually had little opportunity to become familiar with biochemical or immunological procedures. Furthermore, it may not be easy for an analytical chemist to be convinced of the precision, accuracy, sensitivity and low detection limits which these basically "biological" methods can offer, while incurring low costs per analysis.

However, in the 1970s, a few scientific groups and institutions had already published first results of applying enzyme-immunoassays to research problems. For instance, it was reported in 1977 that ATPase is inhibited by chlorinated hydrocarbon pesticides and that ATPase can be reactivated by the appropriate antibody [1]. In 1978 and 1979 this antibody was isolated and used as a reagent for *in vitro* and *in vivo* studies on the effects of insecticides [2,3]. *The United States Environmental Protection Agency, Environmental Research Laboratory,* Gulfbreeze, FL 32561, published a paper entitled "Determination of the Site(s) of Action of Selected Pesticides by an Enzymatic-immunobiological Approach" [2]. In recent years some EIA have been developed for routine analysis; e.g., for the determination of Paraquat and Paraoxon as is shown in chapters 3.4 and 3.5, and, e.g., for benzoylphenylurea insecticides, diflubenzuron and BAY SIR 8514 [5,6]. *Hammock* has made a large study designed to examine the potential of immunochemical methods for solving specific problems in environmental chemistry. In 1978, ELISA techniques had already been developed at Pennsylvania State University for the analysis of pesticide residues [7] and in 1981 for environmental toxins [8]. However, one cannot imagine that these activities have reached the limits of beneficial application of these fascinating techniques.

Thus, the well-established inhibition methods, mostly using cholinesterase, are widely applied where highly sophisticated instrumentation is not available or where an investment greater than that needed for enzymatic analysis is not justified. However, it must be remembered that such enzyme-inhibition methods are not specific; they can only discriminate between negative and positive specimens, containing any compound which inhibits the enzyme used. Specificity may be improved or even obtained if these methods are combined with a separation step, e.g., with thin-layer chromatography (cf. chapter 3.2.1, p. 409) or with HPLC [9].

The existence of pesticides in almost all environmental components is a reality. Soil is a reservoir of these chemicals, from which other components of the environment can become contaminated [7, 10, 11]. Human food is the product of plants and animals, and can thus be the target of pesticide contamination through the food chain. Milk and dairy products are particularly susceptible to contamination with pesticides [12].

These health hazards have created an insatiable demand for increasing sensitivity in methods of analyses, since pesticides and their metabolites in general are toxins at very low residual levels.

This challenge has already been recognized in 1981 [8]. It is hoped that in the forthcoming years the existing EIA procedures for pesticides will at last be used, and that new methods can be offered for the most important toxins.

References

[1] *R. B. Koch, D. Desaiah, B. Glick, D. S. V. Subba Rao, R. Stinson,* Antibody Reactivation of Kepone Inhibited Brain ATPase Activities, Gen. Pharmacol. *8,* 231 – 234 (1977).

[2] *R. B. Koch, W. E. McHenry, W. E. Choate, M. Barker, B. R. Layton, R. Stinson, D. V. S. Subba Rao, B. Glick, T. N. Patil, D. Desaiah,* Determination of the Site(s) of Action of Selected Pesticides by an Enzymatic-immunobiological Approach, U.S. Environ. Prot. Agency EPA-600/3-78-093 (1978).

[3] *R. B. Koch, T. N. Patil, B. Glick, R. S. Stinson, E. A. Lewis,* Properties of an Antibody to Kelevan Isolated by Affinity Chromatography: Antibody Reactivation of ATPase Activities Inhibited by Pesticides, Pestic. Biochem. Physiol. *12,* 130 – 140 (1979).

[4] *S. I. Wie, B. D. Hammock,* Comparison of Coating and Immunizing Antigen Structure on the Sensitivity and Specificity of Immunoassays for Benzoylphenylurea Insecticides, J. Agric. Food Chem. *32,* 1294 – 1301 (1984).

[5] *B. D. Hammock, R. O. Mumma,* Potential of Immunochemical Technology for Pesticide Analysis, in: *J. Harvey Jr., G. Zweig* (eds.), Recent Advances in Pesticide Analytical Methodology, American Chemical Society Symposium Series, ACS Publications, Washington, DC 1980, pp. 321 – 352.

[6] *S. I. Wie, B. D. Hammock,* Development of Enzyme-linked Immunosorbent Assays for Residue Analysis of Diflubenzuron and BAY SIR 8514, Agric. Food Chem. *30,* 949 – 957 (1982).

[7] *A. Y. Al-Rubae,* The Enzyme-linked Immunosorbent Assay, a New Method for the Analysis of Pesticide Residues, Doctor's Thesis, The Pennsylvania State University, Department of Entomology, 1978.

[8] *R. P. Vallejo, Jr.,* The Development of Immunoassays for Pesticides and Environmental Toxicants, Doctor's Thesis, The Pennsylvania State University, Department of Chemistry, 1981.

[9] *K. A. Ramsteiner, W. D. Hörmann,* Coupling High-pressure Liquid Chromatography with a Cholinesterase Inhibition AutoAnalyzer for the Determination of Organophosphate and Carbamate Insecticide Residues, J. Chromatogr. *104,* 438 – 442 (1975).

[10] *C. B. Estep, G. N. Menon, H. E. Williams, A. C. Cole,* Chlorinated Hydrocarbon Insecticide Residues in Tennessee Honey and Beeswax, Bull. Environ. Contam. Toxicol. *17,* 168 – 174 (1977).

[11] *C. R. Harris, W. W. Sans,* Absorption of Organochlorine Insecticide Residues from Agricultural Soils by Crops Used for Animal Feed, Pest. Monit. J. *3,* 182 – 185 (1969).

[12] *W. H. Brown, J. M. Witt, F. M. Whiting, J. W. Stull,* Secretion of DDT in Fresh Milk by Cows, Bull. Environ. Contam. Toxicol. *1,* 21 – 28 (1966).

3.2   Organophosphorus and Carbamate Residues in Water

3.2.1   Radiometric Method

László Horváth

General

Because of their low persistence and high effectiveness, organophosphorus and carbamate insecticides* are used extensively throughout the world. Their toxic action is generally associated with their ability to inhibit cholinesterase**, ChE, in the central and peripheral nervous system [1]. The inhibition of cholinesterase can be effectively demonstrated *in vitro*. The activity of cholinesterase, as modified by the anti-cholinesterase compound present, can be assayed by a variety of methods [2].

Application of method: in pesticide residue analysis.

Substance properties relevant in analysis: insecticides containing a thiophosphoryl group are generally very weak inhibitors of ChE *in vitro*, and must be converted to oxon-analogues in order to enhance their anti-cholinesteratic activity. The thio-group can be oxidized with diluted bromine water and the remaining bromine does not interfere with the enzyme. Organophosphorus insecticides can inhibit cholinesterase virtually completely, but inhibition by carbamate does not go to completion. If enough time is allowed the carbamate is destroyed and the enzyme recovers totally [3].

Methods of determination: the most widely-used methods of determining pesticides are gas chromatography, HPLC and spectroscopy [4]. All of these methods require tedious extraction and clean-up procedures before instrumental analysis. Direct enzymatic assay of organophosphorus and carbamate insecticides is possible in aqueous samples. This is advantageous in monitoring pesticides in surface waters, when large numbers of samples must be handled.

Organophosphorus and carbamate insecticides can be determined enzymatically by a variety of methods which differ in the way in which the activity of the free enzyme is determined [5].

---

* E.g. parathion, paraoxon, malathion, DDVP, carbaryl; for chemical structures cf. p. 413.
** Acetylcholine acetylhydrolase, EC 3.1.1.7.

Assay

Method Design

Principle

$$\text{[}^3\text{H]Acetylcholine} + H_2O \xrightarrow[\quad]{\overset{\displaystyle\text{Inhibitor}}{\underset{\displaystyle\text{ChE}}{\downarrow}}} \text{[}^3\text{H]acetate} + \text{choline} .$$

The analyte inhibits cholinesterase. The activity of the enzyme, as modified by the pesticide, can be assayed by measuring the released acetic acid, as described by *Johnson & Russel* [6]. All steps of the assay procedure are carried out in a scintillation vial. The aqueous sample is incubated with the enzyme solution for 60 minutes. After this time [^{3}H]acetylcholine chloride is added and incubated for 10 minutes. The enzyme reaction is stopped by adding pH 2.5 buffer solution. The acetic acid is in non-dissociated form in the low pH buffer and can be extracted into the toluene-based liquid scintillation cocktail. The aqueous phase, which contains the remaining [^{3}H]acetylcholine, settles to the bottom of the vial. Because of their low energy, the β particles emitted by the unhydrolyzed [^{3}H]acetylcholine do not reach the scintillation solution: only the [^{3}H]acetic acid present in the scintillation solution is detected by the liquid-scintillation spectrometer.

Optimized conditions for measurement: the substrate concentration should be high enough to remain significantly unchanged during the course of the reaction [7]. The *Michaelis* constant of cholinesterase varies considerably according to the origin of the enzyme, but is usually in the region of 10^{-3} to 10^{-4} mol/l; the substrate concentration should, therefore, be not less than 2×10^{-3} mol/l. Much higher concentrations are not recommended because such a large excess may exert an inhibiting action.

Equipment

Liquid-scintillation counter, disposable plastic liquid-scintillation vials, constant-temperature bath, liquid dispensers, stopwatch.

Reagents and Solutions

Purity of reagents: [^{3}H]acetic acid impurity in the [^{3}H]acetylcholine must be less than 10% as measured by the scintillation method. All solutions must be prepared under sterile conditions.

Preparation of solutions: prepare all solutions with fresh re-purified water (cf. Vol. II, chapter 2.1.3.2).

1. Tris buffer (Tris, 10 mmol/l; serum albumin, 1 g/l; pH 7.4):

   dissolve 121 mg Tris in water, adjust to pH 7.4 with HCl, 1 mol/l, add 0.1 g serum albumin and dilute to 100 ml with water.

2. Cholinesterase (ChE, ca. 13 kU/l; Tris, 10 mmol/l; serum albumin, 1 g/l; pH 7.4):

   dissolve 1000 U cholinesterase from *Electrophorus electricus* (salt-free, freeze-dried powder, purified by chromatography and gel filtration, 1000 U/mg protein at 25 °C and pH 7, acetylcholine as substrate, from *Koch-Light Ltd.*) in 20 ml Tris buffer (1) and make up to 25 ml (stock solution, ca. 40 kU/l). Keep in refrigerator at 0 – 5 °C. Dilute 3 ml stock solution to 10 ml with Tris buffer (1).

3. [^{3}H]acetylcholine chloride (10 mmol/l, 37 GBq/mmol):

   dissolve 37 MBq [^{3}H]acetylcholine chloride* and 0.181 g acetylcholine chloride in water and make up to 100 ml.

4. Stopping solution ($CH_2ClCOOH$, 1 mol/l; NaOH, 0.5 ml/l; NaCl, 2 mol/l; pH 2.5):

   dissolve 47.25 g monochloroacetic acid, 10 g NaOH and 58.44 g NaCl in water, make up to 500 ml.

5. Scintillation solution:

   dissolve 0.3 g POPOP, 4 g PPO and 100 ml *i*-amyl alcohol in scintillation grade toluene and make up to 1000 ml.

6. Standard solution (1 mg/l):

   dissolve 1 mg authentic pesticide in water and make up to 1000 ml. Make a series of dilutions from it.

Stability of solutions: store Tris buffer (1), cholinesterase (2) and [^{3}H]acetylcholine (3) solutions, stoppered, in a refrigerator at 0 – 5 °C. Solutions (2) and (3) are stable for 3 months. Solution (1) is usable indefinitely. Solutions (1), (2) and (3) should be protected from growth of micro-organisms. Stopping solution (4) and scintillation solution (5) are stable indefinitely at room temperature.

---

*   The [^{3}H]acetylcholine chloride can be purchase with 18 – 74 GBq/mmol specific radioactivity, usually in 37 MBq quantity.

Procedure

Collection and treatment of specimen: for monitoring of organophosphorus and carbamate insecticide residues in water of rivers or lakes, screen those specimens which contain detectable amount of residue. For this purpose filter through a folded filter-paper (e.g. *Macherey-Nagel*) about 5 ml water. Use 100 µl for assay.

If various anti-cholinesterase compounds are present in the sample and specific determinations are required, combine the method with thin-layer chromatography. Extract 100 ml aqueous specimen with 3×20 ml dichloromethane in a 200 ml separating funnel. Shake the extract with 10 g anhydrous sodium sulphate and filter through a folded filter-paper. Concentrate the extract to about 3 ml and transfer to a 5 ml graduated flask. Make up solution to 5 ml. Apply 100 µl sample to a thin-layer chromatogram plate (e.g. *Merck* DC Alufolien Kieselgel 60/Kieselgur F254). Develop the chromatogram in an appropriate solvent system (e.g. benzene : ethylacetate, 4 : 1; petroleum ether : acetone, 3 : 1, and n-hexane : acetone, 4 : 1, systems are advisable). After developing the solvent front to 15 cm, dry the plate and divide it into 1 cm² parts and scrape off the silica into scintillation vials. Add 100 µl water to each vial and carry out the assay procedure.

Thio- or dithiophosphates can be oxidized to their "oxon" analogues which are usually more powerful inhibitiors of cholinesterase than the parent compounds: add 0.1 ml saturated bromine water to 10 ml filtered aqueous specimen. Use 100 µl for assay.

Details for measurement in other samples: anti-cholinesterase pesticides in soil or food can be analyzed after extraction with an appropriate organic solvent (e.g. petroleum ether, n-heptane, dichloromethane). Make a thin-layer chromatogram and assay by the enzymatic method.

Assay conditions: use constant-temperature bath furnished with a rack for scintillation vials. Label the pipettes for cholinesterase, [³H]acetylcholine and stopping solution. Number 30 disposable plastic vials and pre-warm in the 35 °C water-bath.

Pipette sample and solutions (2) – (4) at intervals of exactly 15 s into all vials and take care to pipette to the bottom of the vials.

Incubation volume 200 µl; 35 °C.

For measurement of non-inhibited ChE, use water instead of sample.

For blank (background level of radioactivity) use water instead of sample and add stopping solution (4) before all other solutions.

Establish a calibration curve for each compound.

Calibration curve: calculate the remaining enzyme activity, z, as described under Calculation. Plot z (%) (linear scale) against concentration of pesticide, c_i (logarithmic scale).

Measurement

Pipette successively into the vials:			concentration in assay mixture	
sample or standard (6) ChE solution (2)		100 µl 50 µl	pesticide up to 0.5 mg/l ChE ca. 300 U/l Tris 2.5 mmol/l serum albumin 0.25 g/l	
mix, incubate for 60 min at 35 °C,				
[^{3}H]acetylcholine solution (3)		50 µl	[^{3}H]acetylcholine 2.5 mmol/l	
mix, incubate for 10 min at 35 °C;				
stopping solution (4)		200 µl	$CH_2ClCOOH$ 0.5 mol/l NaOH 0.25 mol/l NaCl 1 mol/l	
mix well, allow to reach room temperature,				
scintillation solution (5)		10 ml		
mix well, let stand for 10 min, count radioactivity.				

Calculation: from the measured counting rates, the percentage remaining enzyme activity can be calculated:

$$z = \frac{C_i - C_b}{C_0 - C_b} \times 100 \quad \% \, ,$$

where

z enzyme activity
C_b background counting rate, cpm,
C_0 counting rate due to the reaction of non-inhibited enzyme, cpm,
C_i counting rate due to the reaction of inhibited enzyme, cpm.

Quantitative determination is only possible when the sample contains an identified anti-cholinesteratic compound. If the sample contains various anti-cholinesteratic insecticides the result may be given as "equivalent value" (e.g., "malathion equivalent" means that sum of pesticides in the sample inhibits the acetylcholinesterase to the same extent as the given concentration of malathion).

If the kind of pesticide in the sample is known, the concentrations are obtained by means of the calibration curve.

Validation of Method

Precision, accuracy, detection limit and sensitivity: precision of measurements depends on the precision with which the solutions are prepared and pipetted. Well trained personnel is necessary. To determine the concentration of an inhibitor, three values are needed: C_i, C_0 and C_b. These values were measured in 21 repetitions and the relative standard deviation of enzyme activity was found to be $\pm 0.678\%$. $\Delta 2\,P_{min} = 98.62\%$ enzyme activity can be measured with a 99.45% confidence limit.

The detection limits for some insecticides are:

parathion	8.88×10^{-7} mol/l
paraoxon	4.29×10^{-10} mol/l
malathion	2.19×10^{-7} mol/l
DDVP	1.28×10^{-7} mol/l .

The accuracy of the method is greatest when standards are assayed within series together with samples.

The sensitivity of the method depends on the insecticide to be determined and on the kind of enzyme used in the assay.

Sources of error: the reagents and solutions must be kept clean and sterile: on contamination with micro-organisms the enzyme solution loses its activity, the [³H]acetylcholine solution is partly hydrolyzed, and the background rises. If the background is more than 10% of the C_0 value, the vessel containing the [³H]acetylcholine should be re-sterilized and fresh solution (3) prepared.

Specificity: the enzymatic determination of anti-cholinesteratic insecticides is highly sensitive, but not selective. To overcome this shortcoming, combine the method with thin-layer chromatography.

References

[1] *R. D. O'Brien,* Insecticides: Action and Metabolism, Academic Press, New York 1967, pp. 39 – 54, 86 – 95.
[2] *D. C. Villeneuve,* A Review of Enzymatic Techniques Used for Pesticide Residue Analysis, in: *R. F. Gould* (ed.), Advances in Chemistry, Series *104*, American Chemical Society, Washington D.C. 1971, pp. 27 – 38.
[3] *F. P. W. Winteringham, K. S. Fowler,* Substrate and Dilution Effects on the Inhibition of Acetylcholinesterase by Carbamates, Biochem. J. *101*, 127 – 139 (1966).

[4] *J. S. Sherma,* Manual of Analytical Quality Control for Pesticides and Related Compounds in Human and Environmental Samples, EPA-600/2-81-059.
[5] *G. G. Guilbauldt,* Enzymatic Methods of Analysis, Pergamon Press, Oxford 1970, pp. 44 – 45.
[6] *C. D. Johnson, R. L. Russel,* A Rapid, Simple Radiometric Assay for Cholinesterase, Suitable for Multiple Determinations, Anal. Biochem. *64,* 229 – 241 (1975).
[7] *M. Eto,* Organophosphorus Pesticides: Organic and Biological Chemistry, CRC Press, Cleveland 1974, pp. 134 – 136.
[8] *L. Horváth, T. Forster,* Radiometric Enzymic Method for Determining Organophosphorus and Carbamate Residues in Water, in Agrichemical Residue-Biota Interactions in Soil and Aquatic Ecosystems, IAEA STI/PUB/548 Vienna 1980.

3.2.2   Colorimetric Method

Gyöngyi Kovács Huber

General

Organophosphates and carbamate insecticides represent a large number of pesticides. Their mode of action has been explained by their ability to inhibit acetylcholinesterase [1]. The determination of these molecules in residues can be performed by various chemical methods [2] and by enzymatic methods, the latter having recently achieved popularity due to their sensitivity and specifity. There is a correlation between the quantity of pesticide residue and the degree of cholinesterase inhibition which can form the basis of residue analysis. A large number of analyses can be carried out routinely using automatic instrumentation.

Application of method: in environmental researches, and for screening (residue analysis in water).

Methods of determination: the analysis of organophosphorus and carbamate insecticide residues by enzyme inhibition can be performed with different substrates using various analytical techniques [3]. Thin-layer chromatography is the most simple and rapid method of estimating the residue level in samples of different origins [4 – 7]. The analysis can also be performed spectrophotometrically by measuring the quantity of unreacted acetylcholine [8], and by following the turnover of different substrates by cholinesterase inhibited by the insecticides [9 – 11].

The inhibition of cholinesterase can be determined by gas chromatography [12] and by electrochemical methods using specific enzyme electrodes [13 – 15]. Some automatic systems for the determination of pesticide residue have also been described [16 – 18].

Assay

Method Design

Principle

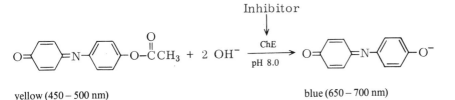

yellow (450 – 500 nm)   blue (650 – 700 nm)

$$+ \; CH_3COO^- \; + \; H_2O$$

Indophenyl acetate, (*N*-4'-acetoxyphenyl)-4-quinone imine, is hydrolyzed by cholinesterase*, ChE, in buffer, pH 8.0, producing an intensely blue-coloured compound. Indophenyl acetate was used by *Kramer & Gamson* [19] to determine the acetylcholinesterase activity. In the presence of cholinesterase-inhibiting pesticides the rate of hydrolysis of indophenyl acetate is reduced. The change in absorbance at 650 nm per unit time, $\Delta A / \Delta t$, compared to that of the non-inhibited reaction, is a measure of the amount of pesticide residues.

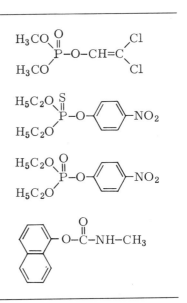

DDVP
O,O-dimethyl-*O*-(2,2-dichlorovinyl) phosphate

Parathion
O,O-diethyl-*O*-(4-nitrophenyl) phosphorothioate

Paraoxon
O,O-diethyl-*O*-(4-nitrophenyl) phosphate

Carbaryl
1-naphthyl-*N*-methyl carbamate

---

* Acylcholine acylhydrolase, EC 3.1.1.8.

Optimized conditions for measurement: the rate of hydrolysis depends on pH, temperature, and substrate and inhibitor concentrations. The conversion is maximum at pH 8, thus the analysis is carried out in alkaline phosphate buffer (pH 8) at 40 °C. The hydrolysis is measured over 30 min; 2 h pre-incubation is necessary to assure optimal reaction between enzyme and inhibiting pesticides.

The optimum enzyme concentration is 380 U/l in the assay mixture. Increasing this concentration decreases the sensitivity of the method.

The validity of *Lambert-Beer's* law depends on the pesticides. In the case of DDVP, carbaryl, parathion and paraoxon the linear range is 0.01 to 0.2 mg/l.

Equipment

Pye Unicam AC 60 Chemical Processing Unit coupled to an SP 8000 spectrophotometer (200 – 800 nm). The analyzer contains 120 sample places. The temperature, reaction time and volume of solutions can be controlled automatically.

Reagents and Solutions

Purity of reagents: all chemicals are analytical grade. Cholinesterase (pseudocholinesterase) is a lyophilized preparation from horse serum (*Reanal*, Budapest), 1 U/mg at 37 °C, butyrylcholine as substrate.

Indophenyl acetate is from *Merck*, Darmstadt. The insecticide standard materials are from *PolyScience Corporation*, P.O. Box 791, Evanston, IL, U.S.A.

Preparation of solutions (for about 40 determinations): use only re-purified water (cf. Vol. II, chapter 2.1.3.2).

1. Phosphate buffer (50 mmol/l, pH 8.0):

   dissolve 400 mg NaOH pellets in 100 ml water and 1.4 g KH_2PO_4 in 100 ml water; mix 46.8 ml NaOH solution with 50 ml KH_2PO_4 solution (and 2 ml freshly saturated bromine water, if necessary, cf. Assay conditions) and make up to 100 ml with water.

2. Cholinesterase solution (ChE, 875 U/l; phosphate, 50 mmol/l; pH 8.0):

   dissolve 87.5 mg cholinesterase (and 60 mg sodium thiosulphate, if necessary, cf. Assay conditions) in 100 ml phosphate buffer (1).

3. Indophenyl acetate solution (2.5 mmol/l):

   dissolve 60 mg indophenyl acetate in 100 ml absolute ethanol.

4. Insecticide standard solutions (10, 50, 100, 150 and 200 µg/l):

dissolve 0.1 to 2.0 mg DDVP, carbaryl, parathion and paraoxon in 100 ml water. Dilute with water to obtain the desired solutions.

Stability of solutions: store all solutions stoppered in a refrigerator at 0°C. Prepare solutions freshly every week, the enzyme solution (2) every day.

Procedure

Collection and treatment of specimen: collect 50 ml water in a plastic bottle and store in a refrigerator at 0°C until analysis.

Stability of the analytes in the sample: 5 to 7 days after collection the pesticide content of water has decreased by 15 to 20% at room temperature and by 1 to 2% at 0°C.

Assay conditions: wavelength 635 nm; light path 10 mm; final volume 4.6 ml; 40°C; pre-incubation at 40°C.

Run a "zero" with water instead of sample after every 10th sample, yielding $(\Delta A/\Delta t)_0$.

Establish a calibration curve with standard solutions (4) instead of sample.
    In the case of phosphate compounds containing P → S groups the oxidation is carried out with bromine (in solution (1)), and the excess of bromine is removed with thiosulphate (in solution (2)).

Measurement

Pipette (inject) successively into tubes:			concentration in assay mixture	
sample or standards (4)		0.3 ml	pesticide	up to 13 µg/l
phosphate buffer	(1)	2.0 ml	phosphate	43 mmol/l
cholinesterase	(2)	2.0 ml	ChE	380 U/l
pre-incubate for 2 h at 40°C,				
indophenyl acetate	(3)	0.3 ml	indophenyl acetate	0.16 mmol/l
incubate for exactly 30 min at 40°C, measure absorbance; use the increases in absorbance per 30 min, $\Delta A/\Delta t$, for calibration curve.				

Calibration curve: plot the difference of $(\Delta A/\Delta t)_0 - (\Delta A/\Delta t)_{standard}$ *versus* pesticide concentrations (μg/l) on linear graph paper (Fig. 1).

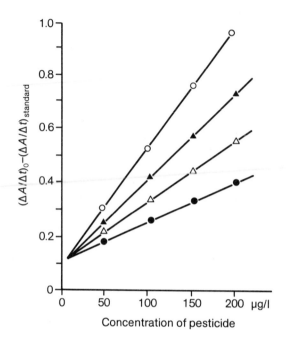

Fig. 1. Typical calibration curves. O——O parathion, ▲——▲ DDVP, dichlorvos, △——△ carbaryl, ●——● paraoxon.

Calculation: read for $(\Delta A/\Delta t)_0 - (\Delta A/\Delta t)_{sample}$ the corresponding insecticide concentration of sample from the calibration curve.

Validation of Method

Precision, accuracy, detection limit and sensitivity: the relative standard deviation at a residue level of 0.1 mg/l for DDVP is 3.7%. The detection limit is 0.01 mg/l water. The sensitivity of this method is $\Delta A/\Delta t = 0.002/30$ min.

Sources of error: using the automatic process unit, the possibility of errors is minimal, if temperature, reaction time and volumes of solutions are correctly adjusted at the instrument, if careful reagent preparation and storage are ensured, and if the linear range is used for calibration.

Specificity: indophenyl acetate is a specific substrate of cholinesterase at pH 8. This reaction can also be used to determine enzyme activity.

References

[1] *I. R. Corbett,* The Biochemical Mode of Action of Pesticides, Academic Press, New York 1974, p. 540.

[2] *K. A. McCully,* Report on Organophosphorus Pesticides, J. Assoc. Off. Anal. Chem. *64,* 401 – 405 (1981).

[3] *M. H. Sadar, S. S. Kuan, G. G. Guilbault,* Trace Analysis of Pesticides Using Cholinesterase from Human Serum, Rat Liver, Electric Eel, Bean Leaf Beetle and White Fringe Beetle, Anal. Chem. *42,* 1770 – 1774 (1970).

[4] *H. Achermann,* Dünnschichtchromatographisch-enzymatischer Nachweis Phosphororganischer Insektizide, J. Chromatogr. *44,* 414 – 418 (1969).

[5] *G. E. Mendoza, J. B. Schields,* Determination of Some Carbamates by Enzyme Inhibition Techniques Using Thin Layer Chromatography and Colorimetry, J. Agric. Food Chem. *21,* 178 – 181 (1973).

[6] *G. F. Ernst, F. Schuring,* A Modified Enzymatic Detection for Thin-layer Chromatograms of Pesticides, J. Chromatogr. *49,* 325 – 328 (1970).

[7] *Á. Ambrus, É. Hargitai, G. Károly, A. Fülöp, J. Lantos,* General Method for Determination of Pesticide Residues in Samples of Plant Origin, Soil and Water, J. Assoc. Off. Anal. Chem. *64,* 743 – 750 (1981).

[8] *D. Palut, H. Jonczyk, T. Syrowatka,* Enzymatic Assays of Thionophosphate Insecticides and of Their Activated Products, Rocz, Panstw. Zakl. Hig. *18,* 527 – 535 (1967).

[9] *T. E. Archer,* Direct Colorimetric Analysis of Cholinesterase Inhibiting Insecticides with Indophenyl Acetate, J. Agric. Food Chem. *7,* 179 – 181 (1959).

[10] *G. G. Guilbault, D. N. Kramer,* Fluorometric System Employing Immobilized Cholinesterase for Assaying Anticholinesterase Compounds, Anal. Chem. *37,* 1675 – 1680 (1965).

[11] *N. V. Kumar,* Colorimetric Determination of Methyl Parathion and Oxygen Analog, J. Assoc. Off. Anal. Chem. *63,* 536 – 538 (1980).

[12] *J. Kovac, V. Markova, P. Králik,* Gas-Chromatographic Determination of Cholinesterase Inhibition, J. Chromatogr. *100,* 171 – 174 (1974).

[13] *G. Baum, F. Ward,* Ion-selective Electrode Procedure for Organophosphate Pesticide Analysis, Pestic. Chem. Proc. Int. Congr. Pestic. Chem. 2nd edit., Vol. *4,* 1971, pp. 215 – 225.

[14] *K. L. Crochet, J. G. Montalvo,* Enzyme Electrode System for Assay of Serum Cholinesterase, Anal. Chim. Acta *66,* 259 – 269 (1973).

[15] *M. Granmer, A. Peoples,* Determination of Trace Quantities of Anticholinesterase Pesticides, Anal. Biochem. *55,* 255 – 265 (1973).

[16] *F. A. Gunther, D. E. Ott,* Automated Pesticide Residue Analysis and Screening, Res. Rev. *14,* 12 – 16 (1966).

[17] *D. C. Leegwater, H. W. Van Gend,* Automated Differential Screening Method for Organophosphorus Pesticides, J. Sci. Food Agric. *19,* 513 – 518 (1968).

[18] *B. H. Chin,* Automated Method for Determining *in vitro* Cholinesterase Inhibition by Experimental Insecticide Candidates, J. Agric. Food Chem. *28,* 1342 – 1344 (1980).

[19] *D. N. Kramer, R. M. Gamson,* Colorimetric Determination of Acetylcholinesterase Activity, Anal. Chem. *30,* 251 – 254 (1958).

3.3   Pesticides in Plants

Anna Wędzisz

General

A common feature of the organophosphorus compounds and carbamate insecticides used in protection of cultivated plants is their ability to inhibit hydrolases, including cholinesterase*, ChE, not only in organisms but also *in vitro*. This property can be used for analytical purposes. The enzyme from horse blood serum used in the method is able to hydrolyze acetylcholine and other acylcholines.

Application of method: in pesticide residue analysis, especially to observe the dynamics of their decay on protected cultivated plants.

Substance properties relevant in analysis: derivatives of phosphoric acid are strong inhibitors of cholinesterase whereas thiono-organophosphorus compounds must first be transformed into an oxon-form. Fresh peracetic acid [1] is used as the desulphurizing agent in the method. Inhibition constants for organophosphorus compounds are of the order of 10^{-7} mol/l, and for carbamates 10^{-5} mol/l.

Methods of determination: the methods of determination of small quantities of pesticides in vegetable material can be divided into three main groups: 1. colorimetric reactions based on individual chemical properties of pesticides, 2. physicochemical reactions – mainly gas chromatography, 3. enzymatic reactions, based on the anti-cholinesterase properties of the compounds. Individual colorimetric reactions for pesticides are rarely used because of their low sensitivity which limits their application to the determination of larger amounts of the compounds. Gas chromatography is frequently used in pesticide residue analysis but it requires precise purification of the extract, as well as good laboratory equipment; a monograph by *Zweig & Sherma* [2] has been devoted to this method. For rapid control of pesticide residues in vegetable material, thin-layer chromatography has been used combined with enzymatic detection based on inhibition of esterase action on various substrates. However, these are semi-quantitative methods [3, 4].

The quantitative enzymatic method presented here makes use of the anticholinesterase properties of pesticides. It can be applied without purification of an extract or, when purification is necessary, the method may be combined with thin-layer chromatography.

---

* Acylcholine acylhydrolase, EC 3.1.1.8.

Assay

Method Design

Principle

$$\text{(a)} \qquad \text{Acetylcholine} + H_2O \xrightarrow[\text{}]{\text{ChE}} \text{acetate} + \text{choline}$$

with Inhibitor acting (↓) on the ChE step.

$$\text{(b)} \qquad \text{Acetylcholine} + \text{hydroxylamine} \longrightarrow \text{acetylhydroxamic acid}$$

$$\text{(c)} \qquad \text{Acetylhydroxamic acid} + Fe^{3+} \longrightarrow \text{coloured complex}.$$

The analyte inhibits cholinesterase, i.e., the velocity of reaction (a) is reduced. If stopped by addition of alkali after a defined incubation time, when acetylcholine has been converted to only a limited extent, the remaining amount of acetylcholine is a measure of the amount of inhibitor.

The unreacted acetylcholine and hydroxylamine form acetylhydroxamic acid; this reacts with ferric ions resulting in a coloured complex measured at 510 nm.

Optimized conditions for measurement: optimization of the method concerns the choice of the source of the enzyme and the substrate used for it, the ratio of substrate concentration to enzyme concentration, the reaction time, pH of the reaction, and the parameters of final determination. For cholinesterase from horse serum, acetylcholine has been chosen as the most suitable non-specific substrate.

Optimal reaction conditions found experimentally are as follows: enzyme/substrate ratio 1 U : 30 µmol acetylcholine; reaction time 40 min (progress-curve is rectilinear); pH 7.2. Absorbance maximum of the complex of acetylhydroxamic acid and Fe^{3+} is at 510 nm.

From the various inhibitor solvents miscible with water, acetone has been chosen for preparation of inhibitor solutions. A concentration of acetone of 0.5% does not disturb the enzyme reaction.

Equipment

Spectrophotometer, water-bath with heating jacket, test-tube centrifuge, mixer, mechanized shaker, vacuum evaporator.

Reagents and Solutions

Purity of reagents: all standard substances, the substrate and the reagents are of *pro analysi* purity. The purity of the enzyme preparation is not checked; its quality is evaluated on the basis of its activity. Independent of the producer's certificate, the activity control is included in the procedure itself, by means of the investigation of the course of enzyme reaction.

Preparation of solutions: prepare all solutions with fresh re-purified water (cf. Vol. II, chapter 2.1.3.2).

1. Phosphate buffer (0.66 mol/l, pH 7.2):

   dissolve 16.72 g $Na_2HPO_4 \cdot 12\ H_2O$ and 2.72 g KH_2PO_4 in 1000 ml water.

2. Cholinesterase (ChE, 640 U/l; phosphate, 0.66 mol/l):

   dissolve a portion of the enzyme containing ca. 100 U ChE in 25 ml phosphate buffer (1). Immediately before use dilute 4 ml (16 U) enzyme stock solution to 25 ml with phosphate buffer (1).

3. Acetylcholine chloride (ACh, 19 mmol/l; acetate, 1 mmol/l):

   dissolve 1.086 g acetylcholine chloride in sodium acetate, 1 mmol/l, to 100 ml (60 mmol/l); dilute 8 ml stock solution to 25 ml with water.

4. Hydroxylamine hydrochloride (2.0 mol/l):

   dissolve 139 g $NH_2OH \cdot HCl$ in 1000 ml water.

5. Sodium hydroxide (3.5 mol/l):

   dissolve 140 g NaOH in 1000 ml water.

6. Alkaline hydroxylamine (NH_2OH, 1 mol/l):

   mix before use equal volumes of solutions (4) and (5).

7. Hydrochloric acid (4 mol/l):

   dilute 50 ml conc. HCl with 100 ml water.

8. $FeCl_3$ solution (0.37 mol/l):

   dissolve 100 g $FeCl_3 \cdot 6\ H_2O$ in 1000 ml HCl, 0.1 mol/l.

9. Acetone solution (5 ml/l):

   dilute 5 ml acetone with water to 1 litre.

10. Peracetic acid:

mix before use 50 ml glacial acetic acid with 10 ml 30% (w/v) hydrogen peroxide.

11. Saturated sodium sulphate solution:

dissolve 44 g $Na_2SO_4 \cdot 7\ H_2O$ in 100 ml water at 20°C.

12. Inhibitor standard solution of phosphoric acid derivatives:

dissolve inhibitor (cf. Table 1) in 100 ml acetone (solution a); dilute 1 ml solution a to 100 ml with acetone (solution b); dilute 0.5 ml solution b with water to 100 ml.

13. Inhibitor standard solution of thiophosphoric acid derivatives:

dissolve inhibitor (cf. Table 1) in 100 ml benzene (solution a); dilute 1.0 ml solution a with 4.0 ml benzene and perform desulphurization (cf. p. 422). Evaporate benzene, dissolve the dry residue in 10 ml acetone (solution b); dilute 0.5 ml solution b with water to 100 ml.

14. Inhibitor standard solution of carbamate derivatives:

dissolve inhibitor (cf. Table 1) in 100 ml acetone (solution a); dilute 0.5 ml solution a with water to 100 ml.

Table 1. Amounts to weigh in and final concentrations of standard solutions.

Solu- tion	Compound	Chemical name	Amount	Final concentration µmol/l	mg/l
(12)	Tetrachlorvinphos	2-Chloro-1-(2,4,5-trichlorophenyl)- ethenyl dimethyl phosphate	100 mg	0.1350	0.05
	Chlorphenvinphos	2-Chloro-1-(2,4-dichlorophenyl)- ethenyl diethyl phosphate	100 mg	0.1375	0.05
	Bromphenvinphos	2-Bromo-1-(2,4-dichlorophenyl)- ethenyl diethyl phosphate	100 mg	0.1235	0.05
	Naled	1,2-Dibromo-2,2-dichloroethyl dimethyl phosphate	100 mg	0.1310	0.05
	Dichlorphos	2,2-Dichloroethenyl dimethyl phosphate	200 mg	0.4520	0.10
(13)	Fenitrothion	*O,O*-Dimethyl *O*-(3-methyl-4-nitro- phenyl) phosphorothioate	100 mg	1.80	0.50
	Malathion	Diethyl [(dimethoxyphosphino- thioyl)thio]butanedioate	300 mg	4.65	1.50
(14)	Carbaryl	1-Naphthyl methylcarbamate	50 mg	12.50	2.50
	Propoxur	2-(1-Methylethoxy)phenyl methyl- carbamate	200 mg	47.80	10.00

Stability of solutions: cholinesterase solution (2) and acetylcholine stock solution (3) should be stored at 0°C to 5°C. No change in enzyme activity was found during one week, or in acetylcholine stock solution during a month. The acetone stock solutions of inhibitors, when stored in the ice-box, are stable for at least 1 month; standard solutions are prepared *ex tempore*. Other solutions used in the analysis are stable.

Procedure

Collection and treatment of specimen

Extraction of the inhibitor from vegetable material: place 50 or 100 g (depending on the amount of insecticides expected) disintegrated specimen of vegetable in a vessel with a ground-in stopper, add anhydrous sodium sulphate*, mix, add 100 or 200 ml benzene**. Shake on a shaker for 1 h. Filter through a fluted filter paper with 1 to 2 g anhydrous sodium sulphate on it. Evaporate benzene from 50 or 100 ml extract (corresponding to 25 or 50 g vegetable) in a rotating vacuum evaporator at 40°C. Dissolve the dry residue in 5 ml benzene.

If the inhibitor is a derivative of thiophosphoric acid, perform desulphurization as described below.

Desulphurization of thionorganophosphorus compounds: to a flask containing 5 ml benzene extract add 3 ml freshly prepared peracetic acid (10). Heat in a 75°C water-bath for 20 min, transfer the mixture into a 50 ml separatory funnel, wash 3 times with 1 ml saturated sodium sulphate solution (11) and 3 times with 3 ml water. Transfer the benzene solution quantitatively to a round-bottom flask, pour through a filter with anhydrous sodium sulphate and wash with a small amount of benzene. Evaporate benzene in a rotating vacuum evaporator at 40°C. Dissolve the dry residue in 5 ml benzene.

For assay evaporate 0.1 to 0.5 ml of the benzene solution to dryness, add 1 ml acetone solution (9), put into water-bath of 35°C for 5 minutes, add 1 ml phosphate buffer (1).

Stability of the analytes in the sample: the residue after benzene evaporation can be stored until the next day at 0°C to 5°C.

Assay conditions: wavelength 510 nm; light path 10 mm; incubation volume 4.0 ml (enzyme reaction); final volume 12.0 ml (colour reaction); measurement against a reference solution, containing all reagents except the inhibitor***.

---

* Dried for 1 h at 130°C.

** Benzene may be replaced by toluene.

*** The reference sample is coloured. Should it not be possible to zero the instrument, measure all absorbances against a blank not containing the enzyme, and subtract the value of reference solution from all readings.

Establish a calibration curve with standard solutions (12) to (14) instead of sample. 5 calibration points should be measured.

For all blanks, standards and samples composition of assay mixture for colour reaction is identical.

Since substrate and enzyme solutions may be unstable, check as follows.

- *Substrate assay:* replace enzyme solution (2) by buffer (1). Measured absorbance must be ca. 1.3; if necessary adjust by added amounts accordingly.

- *Enzyme assay:* replace sample by acetone solution (a) (E + S assay). If the reading does not conform with expected value (ΔA = 0.600 to 0.650), prepare fresh enzyme solution.

Measurement

Pipette successively into test tubes:			concentration in assay mixture	
sample or standard (12), (13), (14)		0.2 – 1.0 ml	pesticide	variable*
acetone solution	(9)	0.8 – 0.0 ml	acetone	ca. 5 ml/l
phosphate buffer	(1)	1.0 ml	phosphate	ca. 0.33 mol/l
keep in 35 °C water-bath for 5 min				
enzyme solution	(2)	1.0 ml	ChE	160 U/l
substrate solution	(3)	1.0 ml	ACh	5 mmol/l
			acetate	0.25 mmol/l
mix well and put into 35 °C water-bath for exactly 40 min				
alkaline hydroxylamine solution	(6)	4.0 ml	NH_2OH	0.33 mol/l
hydrochloric acid	(7)	2.0 ml	HCl	0.67 mol/l
$FeCl_3$ solution	(8)	2.0 ml	Fe^{3+}	62 mmol/l
mix well and measure absorbance at 10 to 30 min from time when colour appeared.				

* Cf. solutions (12) to (14), table 1 and calibration curves (1) to (3).

Calibration curve: correct absorbances of the standards for those of E + S assay. Plot ΔA_{510} *versus* ng or µg pesticide in 4 ml assay mixture.

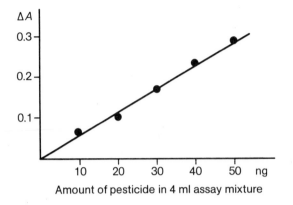

Fig. 1. Typical calibration curve for the assay of tetrachlorvinphos; number of determinations $n = 45$; standard deviation SD $= \pm 0.36$ (for $10-50$ ng in assay mixture).

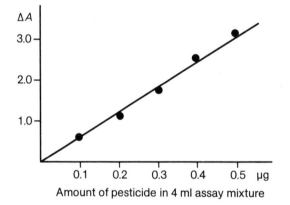

Fig. 2. Typical calibration curve for the assay of fenitrothion (after desulphurization); number of determinations $n = 45$; standard deviation SD $= \pm 0.46$ (for $0.1-0.5$ µg in assay mixture).

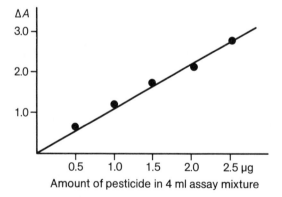

Fig. 3. Typical calibration curve for the assay of carbaryl; number of determinations $n = 45$; standard deviation SD $= \pm 0.73$ (for $0.5-2.5$ µg in assay mixture).

Calculation: take the amount of inhibitor in the sample from the calibration curve; or calculate the amount m_I by use of a calibration coefficient α, i.e. the ratio of $\Delta A_I : m_I$, calculated for the whole range of contents and averaged. It is

$$m_I = \frac{\Delta A_I}{\alpha} \text{ ng } (\mu g) \ .$$

Validation of Method

Precision, accuracy, detection limit and sensitivity: imprecision of the method for various inhibitors is ca. 5 to 10%. The method allows for the differentiation of solutions differing from one another by 10%. Detection limit depends on the inhibitory properties of the compounds investigated. Accuracy and sensitivity of the method were not determined.

Sources of error: errors may be brought about by the decrease in the activity of cholinesterase or by improper storage of acetylcholine chloride (a highly hygroscopic substance).

Specificity: the method is sensitive but not specific. It may be combined with thin-layer chromatography used as the method for inhibitor identification or extract purification.

References

[1] *P. A. Giang,* Organophosphorus and Carbamate Insecticides, in: *H. U. Bergmeyer* (ed.), Methods of Enzymatic Analysis, Vol. *4*, 2nd edit., Verlag Chemie, Weinheim, and Academic Press, New York 1974, pp. 2249 – 2259.
[2] *G. Zweig, J. Sherma,* Gas Chromatographic Analysis, in: *G. Zweig* (ed.), Analytical Methods for Pesticides and Plant Growth Regulators, Vol. *6*, Academic Press, New York 1972, pp. 191 – 233.
[3] *C. E. Mendoza,* Analysis of Pesticides by the Thinlayer Chromatographic Enzyme-inhibition Technique, Residue Rev. *43*, 105 – 142 (1972).
[4] *C. E. Mendoza,* Analysis of Pesticides by the Thin-layer Chromatographic Enzyme-inhibition Technique, Part II, Residue Rev. *50*, 43 – 72 (1974).
[5] *W. Hardegg,* Cholinesterasen, in: *Hoppe-Seyler/Thierfelder, K. Lang, E. Lehnartz* (eds.), Handbuch der Physiologisch- und Pathologisch-Chemischen Analyse, Vol. *6*, Part B, Enzyme, Springer-Verlag, Heidelberg 1966, pp. 921 – 962.

3.4   Paraoxon in Body Fluids

O,O-Diethyl-*O*-(4-nitrophenyl)phosphate

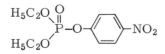

Kenneth W. Hunter, Jr., David E. Lenz and Alan A. Brimfield

General

The detection and quantitation of human exposure to insecticides relies for the most part on physicochemical techniques of analysis such as gas chromatography or mass spectrometry [1]. While these techniques for insecticide analysis are certainly sensitive and specific, they are beyond the reach of most hospital laboratories. The instrumentation required is very expensive and requires highly trained operators (cf. Vol. I, chapter 1.5). Moreover, considerable pre-analysis preparation of samples is often required which, together with running and interpretation time, make these techniques time-consuming and labour-intensive. In instances of acute human exposure, the response time for analysis is inappropriate.

As an alternative or adjunct to the use of physicochemical methods of analysis, several groups have reported on immunoassays for insecticides [2 – 6]. The value of antibodies as analytical tools has been appreciated since the birth of serology at the end of the nineteenth century. Immunological techniques, now applied throughout the biological sciences, identify and quantitate substances ranging from low-molecular weight chemical species to whole micro-organisms.

Exposure to organophosphorus insecticides such as parathion can lead to illness or death. The toxic effects of parathion arise *via* its metabolite, paraoxon, by phosphorylation of the active site of acetylcholinesterase and inactivation of the enzyme. The resultant build-up of unmetabolized acetylcholine causes disruption of normal synaptic transmission at parasympathetic (muscarinic) and neuromuscular (nicotinic) sites leading to characteristic symptoms.

Therapy for organophosphate poisoning involves sequential massive doses of the muscarinic blocker atropine, artificial respiration to compensate for muscular paralysis and administration of a pralidoxime preparation to reactivate phosphorylated acetylcholinesterase. Atropine and pralidoxime are also highly toxic compounds.

Emergency treatment of acute poisoning is based on symptoms and circumstances. The physician makes use of the obvious neurotoxin symptomatology plus information from containers found in the vicinity of the incident and gleaned from co-workers or the patient himself. In the absence of the circumstantial component, as in attempted suicide, homicide or an unaccompanied comatose patient, the physician must rely on symptomatology alone. Unfortunately, organophosphates produce symptoms similar to those caused by other compounds, such as parasympathomimetic drugs and carbamate insecticides. Organophosphate-antidote therapy in cases involving these compounds could add atropine and/or oxime poisoning to a patient's list of problems.

The use of compound specific immunoassays in cases of neurotoxicosis could guide therapy as well as providing potentially useful forensic data. Additionally, the identification of the actual organophosphate involved can rule out the possibility of delayed neurotoxicity which appears 14 to 21 days after exposure to certain organophosphorus compounds.

Application of method: in diagnosis of acute exposure to parathion.

Substance properties relevant in analysis: molecular weight is 275.2. Because the toxic principle in preparation poisoning is the metabolite paraoxon, the compounds against which an antibody must be prepared is paraoxon. Antibodies raised against paraoxon cross-react only weakly with parathion [7]. Caution! Paraoxon is a potent acetylcholinesterase inhibitor. Special precautions must be taken to prevent inhalation and skin contamination when using this and other organophosphorus compounds.

Methods of determination: exposure to organophosphorus compounds can be detected by a variety of cumbersome, time-consuming physicochemical techniques [8], but parathion poisoning is usually diagnosed by measuring the levels of serum or erythrocyte cholinesterase [9]. This indirect technique is not specific for parathion, and the normal variation in cholinesterase levels between individuals and in the same individual at different times make cholinesterase measurements unreliable. We have developed an enzyme-immunoassay to measure levels of paraoxon in body fluids directly and specifically [5]. The sensitivity afforded by enzymatic amplification of the detection signal, coupled with the specificity of antibodies, renders this technique precise and quantitative. The competitive-inhibition enzyme-immunoassay (CIEIA) for paraoxon can specifically detect levels as low as 10^{-9} mol/l in serum.

International reference method and standards: no generally agreed reference method or standards are available yet for enzyme-immunoassays. The present method should be calibrated against an established gas-chromatographic method.

Assay

Method Design

Principle

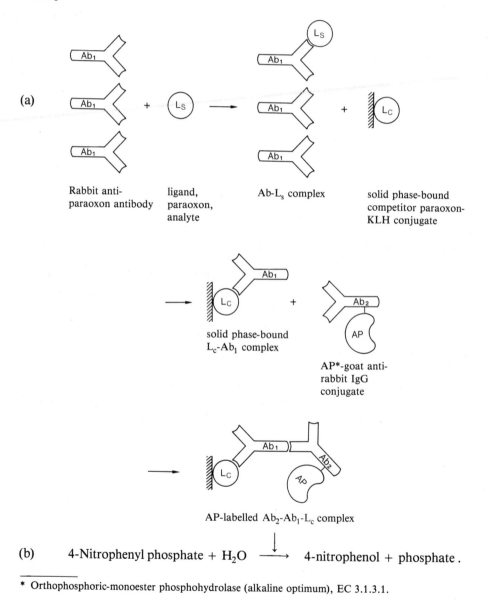

(a)

Rabbit anti-   ligand,   Ab-L$_s$ complex   solid phase-bound
paraoxon antibody   paraoxon,   competitor paraoxon-
                    analyte   KLH conjugate

solid phase-bound
L$_c$-Ab$_1$ complex

AP*-goat anti-
rabbit IgG
conjugate

AP-labelled Ab$_2$-Ab$_1$-L$_c$ complex

(b)   4-Nitrophenyl phosphate + H$_2$O $\longrightarrow$ 4-nitrophenol + phosphate .

* Orthophosphoric-monoester phosphohydrolase (alkaline optimum), EC 3.1.3.1.

When a paraoxon-specific antibody is held in constant limiting concentration in solution, added paraoxon from the sample will reduce the concentration of antibody available for subsequent solid-phase binding in a reproducible manner that is dependent on paraoxon concentration in the sample.

We use a two-step system in which, in the first step, free antibody is allowed to react with paraoxon from the sample. In the second step the antibody remaining unreacted in the assay mixture is allowed to bind to immobilized paraoxon-protein complex; free antibody and free antibody-paraoxon complex are washed away and the immobilized antibody is measured with a second antibody conjugated to the enzyme alkaline phosphatase. The amount of bound enzyme is inversely related to the paraoxon concentration in the sample. The enzyme activity is measured as increase in absorbance at 405 nm per unit time, $\Delta A/\Delta t$.

Selection of assay conditions and adaptation to the individual characteristics of the reagents: this assay is designed to be both sensitive and rapid. Therefore, the conditions chosen represent a compromise that yields a sufficiently low detection limit (10^{-9} mol/l) in an appropriate assay time (2.5 h).

It is important to determine the optimum concentration of paraoxon-protein conjugate for coating the solid phase [10, 11]. The goal is to obtain a unimolecular layer of hydrophobically adsorbed conjugate. A detailed description of the coating procedure is given in the Appendix, p. 437.

Immunization of rabbits with paraoxon chemically conjugated to the protein keyhole limpet haemocyanin, KLH, yields high-titre antisera that contain antibodies reactive with paraoxon (cf. p. 437). It is essential that the solid phase-bound paraoxon-protein conjugate used to analyze the antiserum be composed of non-cross-reacting carrier protein so that the only antibodies detected are those against paraoxon. We prefer to use affinity-purified antibodies [5] from a high-titre source of anti-paraoxon antibodies of the highest possible affinity in order to achieve sufficient sensitivity.

The incubation time of anti-paraoxon antibody with paraoxon from the sample is another variable of importance. It appears not to be necessary for the reactions to attain equilibrium for the derivation of useful and reproducible data [12, 13]. We find that half-hour incubations are sufficient in the CIEIA.

It is critical to determine a non-saturating concentration of anti-paraoxon for the CIEIA. If anti-paraoxon is in excess, reaction with paraoxon from the sample may not reduce antibody binding to solid phase-bound paraoxon-protein conjugate sufficiently, thereby decreasing sensitivity of the assay. The first step in setting up a CIEIA is to titrate the anti-paraoxon antibodies. Note that at high antibody concentrations the titration curve begins to flatten out due to saturation of available binding sites on the solid-phase antigen (Fig. 1). It is important to choose a concentration near the mid-point of the linear portion of the calibration curve for performing the CIEIA.

For the paraoxon immunoassay, we employ an indirect or second antibody procedure in which the enzyme alkaline phosphatase is not directly conjugated to the anti-paraoxon antibody but rather to an antibody directed to this anti-paraoxon antibody (cf. Appendix, p. 436). Besides the advantage of having a second antibody that can be used in assays using different first antibodies, the binding of a second antibody provides a significant amplification of the signal (more than one second antibody binds to each first antibody). We routinely employ a commercially available affinity-purified alkaline phosphatase-labelled second antibody at a concentration between 1 and 10 ng/l (depending on the potency of the batch).

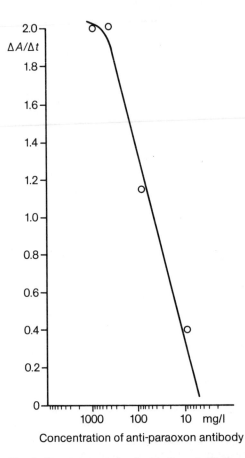

Fig. 1. Titration curve for the binding of affinity-purified rabbit anti-paraoxon to solid phase-bound paraoxon-KLH. Binding sites were saturated at antibody concentrations above 450 mg/l. (Reprinted with permission from Life Sciences [5], Copyright 1982, Pergamon Press, Ltd.)

In the CIEIA system, 50 µl solution of various concentrations of free paraoxon is incubated with 50 µl anti-paraoxon antibody, then a 50 µl aliquot of the mixture is added to a paraoxon-protein coated microtitre plate. The amount of anti-paraoxon that binds to paraoxon-protein is inversely related to the concentration of the paraoxon inhibitor.

Equipment

Linbro EIA microtitre plates, 96-well, flat-bottomed, 0.35 ml capacity; Titertek®
(*Flow*) Multiskan micro-ELISA reader; Titertek 12-channel pipettes, 50 to 200 μl and
5 to 50 μl capacities and tips; *Gilson* Pipetman adjustable pipetters, 0 to 200 μl and
0 to 1000 μl capacities, and tips; reagent troughs for use with multichannel pipettes;
acetate sealers for sealing plates; *Dynatech* microtitre plates, disposable, 96-well,
round-bottom, 0.25 ml capacity for use in the antibody-free hapten reaction; *Nunc*
ELISA plate-washing system; glassware for preparing solutions.

Reagents and Solutions

Purity of reagents: all chemicals should be of analytical grade. Alkaline phosphatase
(e.g. from *Boehringer Mannheim* or *Sigma*) should be of highest quality, e.g. specific
activity 2500 U/mg at 37 °C, 4-nitrophenyl phosphate as the substrate.

Preparation of solutions: all solutions in re-purified water (cf. Vol. II, chapter
2.1.3.2).

1. Phosphate-buffered saline, PBS (phosphate, 10 mmol/l; NaCl, 0.14 mol/l;
   pH 7.4):

   dissolve 8.2 g NaCl, 1.14 g Na_2HPO_4 and 0.2 g KH_2PO_4 in 900 ml water; adjust
   pH to 7.4 with NaOH, 1 mol/l; make up to 1 litre with water. Store at 4 °C or add
   0.5 g NaN_3 to retard microbial growth.

2. Phosphate-buffered saline, PBS (phosphate, 10 mmol/l; NaCl, 0.14 mol/l;
   pH 6.8):

   dissolve 8.2 g NaCl, 1.14 g Na_2HPO_4 and 0.2 g KH_2PO_4 in 900 ml water; adjust
   pH to 6.8 with concentrated HCl; make up to 1 litre with water; store at 4 °C.

3. Phosphate-buffered saline/Tween, PBS/T (phosphate, 10 mmol/l; NaCl,
   0.14 mol/l; Tween 20, 0.5 ml/l; pH 7.4):

   add 0.5 ml Tween 20 (polyoxyethylene sorbitan monolaurate) to 1 litre PBS (1) and
   store at 4 °C.

4. Coating (Tris) buffer (0.1 mol/l, pH 8.0):

   dissolve 15.76 g Tris hydrochloride in 900 ml water; adjust pH to 8.0 with NaOH,
   1 mol/l; make up to 1 litre with water.

5. Substrate solution (DEA, 0.95 mol/l; 4-NPP, 13.5 mmol/l; pH 9.8):

   add 97 ml diethanolamine to 900 ml water; adjust pH to 9.8 with HCl, 1 mol/l; dissolve 5 g 4-nitrophenyl phosphate, disodium salt, hexahydrate, make up to 1 litre with water and store at 4°C in a brown bottle.

6. Rabbit anti-paraoxon antibody (IgG, 2 mg/l; phosphate, 10 mmol/l; NaCl, 0.14 mol/l; Tween 20, 0.5 ml/l; pH 7.4):

   use preparation according to Appendix, p. 437; dilute with buffer (3).

7. Alkaline phosphatase-goat anti-rabbit IgG conjugate (IgG, 1 to 10 ng/l; AP*; phosphate, 10 mmol/l; NaCl, 0.14 mol/l; Tween 20, 0.5 ml/l; pH 7.4):

   dilute commercial preparation from *Sigma* (IgG, 10 mg/l)with buffer (3).

8. Paraoxon standard solutions (10^{-4} to 10^{-10} mol/l):

   prepare a log dilution series from 10^{-4} to 10^{-10} mol/l with either normal human serum or PBS (2) just before use.

9. Paraoxon-keyhole limpet haemocyanin conjugate (paraoxon-KLH, 90 mg/l; Tris, 0.1 mol/l; pH 8.0):

   use paraoxon-KLH according to Appendix, p. 436; dilute stock solution to 90 mg/l with buffer (4).

Stability of solutions: buffers (1), (2), (3) and (4) are stable and useful for months if concentration by evaporation or microbial contamination are avoided. Solution (6) must be prepared freshly each time, and solution (5) should be protected from light.

Procedure

Collection and treatment of specimen: serum is the sample of choice for paraoxon analysis, although plasma can be used. Anticoagulants such as heparin and sodium citrate have no effect. Analysis of urine is more likely to yield 4-nitrophenol and diethylphosphate, breakdown products of paraoxon that cannot be detected with anti-paraoxon antibody. It is not necessary to extract the paraoxon from the serum for clinical or forensic work, but environmental samples would require the use of organic extracts.

Culture media represent aqueous systems and are compatible with the CIEIA so long as the pH lies in the range of 6.0 to 8.0 upon final sample dilution with antibody in PBS/T (3).

---

* Activity concentration of AP is not known.

However, when the sample to be analyzed is an organic extract we recommend the use of various water-miscible organic solvents to dissolve hydrophobic paraoxon analogues prior to CIEIA analysis. Dioxane, acetonitrile, methanol, ethanol, isopropanol, acetone, ethyl acetate, dimethyl sulphoxide and 1-ethoxyethanol have been tested. The results indicate that the CIEIA system is completely functional in the presence of at least 5% of all the solvents except dioxane and ethyl acetate. Compatibility with dioxane was limited to 2.5% and the use of ethyl acetate is not recommended. Ethyl cellosolve was useful at concentrations up to 25%, the highest concentration tested.

Stability of the analyte in the sample: paraoxon undergoes hydrolysis in serum; if the sample cannot be directly analyzed, it should be frozen.

The limiting factor in the use of the CIEIA with buffered aqueous media is actually the base-lability of the paraoxon rather than the sensitivity of the antibodies. At a pH above 7.0 paraoxon begins to be converted to its basic hydrolysis product, and this transformation greatly reduces sensitivity of CIEIA. We suggest the use of a slightly acid buffer (2) to minimize hydrolysis.

Assay conditions: all measurements in triplicate. Measure the sample undiluted and 1 : 10, 1 : 100 and 1 : 1000 diluted with PBS/T (3) to be certain that its inhibition value falls on the linear part of the calibration curves.

Immunoreactions: perform the 1 h step of reaction between antibody and paraoxon in the sample and the subsequent reaction of the enzyme-antibody conjugate with the solid phase-bound antibody (30 min) at room temperature.

Enzymatic indicator reaction: incubation for 30 min (or until a specified absorbance value is obtained) at room temperature; wavelength 405 nm.

Establish a calibration curve with standard solutions (8) instead of sample under exactly the same conditions.

Calibration curve: calculate the mean $\Delta A/\Delta t$ at each standard paraoxon concentration. Plot these mean values *versus* concentration on semi-log graph paper (Fig. 2) or calculate the calibration curve according to the weighted logit-log regression programme described by *Davis et al.* [14]. Fig. 2 represents a typical calibration curve.

Measurement

Pipette into disposable uncoated plastic plate:		concentration in incubation/ assay mixture	
sample, control or standard (8)	0.05 ml	paraoxon	up to 50 µmol/l
rabbit anti-paraoxon antibody (6)	0.05 ml	IgG	1 mg/l
		phosphate	10 mmol/l
		NaCl	0.14 mol/l
		Tween 20	0.25 ml/l
seal plates; incubate for 1 h at room temperature;			
transfer 0.05 ml antibody-paraoxon mixture to paraoxon-KLH coated plate; incubate 30 min at room temperature; wash* with PBS/T (3);			
AP-goat anti-rabbit IgG conjugate (7)	0.05 ml	IgG	1 to 10 ng/l
		AP	unknown
		phosphate	10 mmol/l
		NaCl	0.14 mol/l
		Tween 20	0.5 ml/l
incubate for 30 min at room temperature; wash as above;			
substrate solution (5)	0.1 ml	4-NPP	13.5 mmol/l
		DEA	0.95 mmol/l
incubate for 30 min at room temperature, read absorbance.			

* Five cycles of filling and emptying each well with PBS/T using a *Nunc* immunowash or similar device.

Paraoxon was added to a serum sample and analyzed in the CIEIA. When these results are compared with those using paraoxon in PBS/T as the inhibitor, it is evident that the system is about 10 times less sensitive when serum is employed. This is probably due to the binding of paraoxon to proteins in the serum itself (e.g., cholinesterases) which effectively lowers the concentration in the sample.

Calculation: take values for paraoxon in samples by interpolation from the calibration curve.

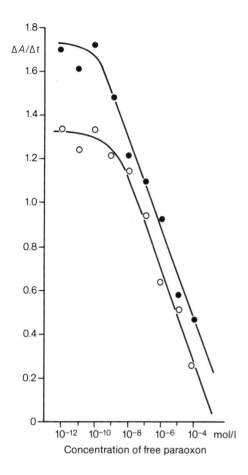

Fig. 2. Typical calibration curve for the assay of paraoxon in serum. Inhibition of binding of affinity-purified rabbit anti-paraoxon to solid phase-bound paraoxon-KLH by various concentrations of paraoxon diluted in PBS/T (●) or human serum (○). (Reprinted with permission from Life Sciences [5]. Copyright 1982, Pergamon Press, Ltd.)

Validation of Method

Precision, accuracy, detection limit and sensitivity: we routinely subject our data to analysis by the quality-control system described by *Rodbard* [15]. This programme calculates the within-assay and between-assays variance, which for the paraoxon assay is less than 7% of the mean. The detection limit for the paraoxon assay is determined by the method of *Miles* [16]; for paraoxon in serum, the limit was 1 nmol/l.

Because paraoxon undergoes some hydrolysis in body fluids, and because paraoxon molecules bind to proteins in body fluids, the accuracy of the method is biased toward low estimates of concentration. The sensitivity of the method reveals concentration changes resulting from 1 : 2 dilutions of known standards.

Sources of error: because paraoxon binds to serum protein such as the cholinesterases, the values obtained by CIEIA probably represent low estimates.

Specificity: the specificity of an immunoassay depends on the cross-reactivity of the antibodies. However, antibodies can provide a level of specificity rivalling that attainable with sophisticated physicochemical techniques. For instance, the cross-reactivity with parathion, determined by comparing the molar concentration of paraoxon and parathion that inhibits 50% of anti-paraoxon antibody binding to the solid phase, is less than 1% though these compounds differ by only a single atom [7]. A similar situation has been reported by *Brimfield et al.* [17] for another organophosphorus compound.

Appendix

Preparation of KLH-paraoxon conjugates: this is a three-step procedure involving the reduction of the aromatic nitro group of paraoxon to an amine, the formation of a diazonium derivative of the reduced paraoxon, and coupling of this derivative to a carrier protein through a diazo linkage [18]. Caution! Paraoxon is a potent acetylcholinesterase inhibitor. Special precautions must be taken to prevent inhalation and skin contamination when using this and other organophosphorus compounds.

Briefly, dissolve 1 g paraoxon (*Sigma Chemical Co.,* St. Louis, MO) in 10 to 20 ml diethylether, add to a flask equipped with a stirring bar and containing 2 g zinc powder (*Aldrich Chemical Co.,* Milwaukee, WI). Begin stirring, fit a reflux condenser, and add dropwise glacial acetic acid/concentrated hydrochloric acid (9:1, v/v) to a total of 10 ml. The reaction proceeds vigorously with acid addition, so this step must be carried out carefully. After acid addition is complete, reflux the reaction mixture at room temperature for 1 h.

At 1 h, filter the contents of the reaction mixture into a separatory funnel and wash two times with 10 ml chloroform. Add the chloroform washes to the ethereal solution and wash the combined organic phases with water until the aqueous wash becomes neutral to pH paper. Dry the organic phase over granular sodium sulphate and remove the solvent on a rotating evaporator. Confirm reduction by thin-layer chromatography on a silica gel plate (*E. Merck,* Darmstadt) containing fluorescent indicator using hexane/acetone (4:1, v/v) as a solvent system. The product is a colourless liquid.

Add 0.1 ml aminoparaoxon to 25 ml hydrochloric acid, 40 mmol/l, with stirring at room temperature, followed by dropwise addition of sodium nitrite solution, 0.1 mol/l. Continue to add sodium nitrite until a persistent excess of nitrous acid is present as indicated by monitoring with starch iodide test paper (*American Scientific Products,* McGaw Park, IL). Decompose the excess nitrous acid by addition of small quantities of urea until no further gas evolution is observed. No further purification of the aminoparaoxon diazonium compound is carried out because of the inherent instability of diazonium ions.

Dissolve 400 mg protein to be derivatized in 25 ml borate buffer, 0.1 mol/l, pH 9, and bring the temperature of this solution to 0 °C by stirring in an ice-bath. Add the solution of aminoparaoxon diazonium derivative to the protein incrementally until it has all been added and stir the mixture for 2 h. The protein-paraoxon conjugate solution is orange to deep reddish brown.

When the reaction is complete, transfer the reaction mixture to dialysis tubing and dialyze for 24 h against two times 4 l phosphate-buffered saline (1). At the end of this period filter the dialyzed protein to remove any precipitated material and aliquot into sterile serum vials through a syringe filtration apparatus for sealing and store in solution at 4 °C or frozen. At this stage the material is ready for use after determination of protein concentration [19] (cf. Vol. II, pp. 88 – 92) and calculation of the epitope density by the method of *Fenton & Singer* [20].

Rabbit immunization and antibody purification: inoculate New Zealand white rabbits intramuscularly with 0.5 ml paraoxon-KLH (0.7 mg protein/ml) over several months. Bleed the animals, collect the serum, and precipitate the immunoglobulin by saturation with $(NH_4)_2SO_4$ to 33%. Dialyze exhaustively against PBS (1), prepare the IgG fraction by DE-52 anion-exchange chromatography. Isolate IgG anti-paraoxon antibodies by affinity chromatography on Sepharose 4B coupled with paraoxon-BSA. Dialyze the eluted antibodies exhaustively against PBS (1) and store at 4 °C until use. For long-term storage, the antibodies can be frozen (– 20 °C), but the protein concentration should be greater than 1 g/l.

Coating of microtitre plates: it is essential to perform preliminary experiments to determine the optimum concentration of paraoxon-protein conjugate (9) for coating the microtitre plate wells. We have found 100 ng paraoxon-KHL per litre to be optimal for this assay.

The volume used for coating is not critical, but must be equivalent to the volumes of the reagents added subsequently. Dilute the conjugate with buffer (4), and add 100 µl to each well. Although it appears that as little as 2 h at temperatures from 4 °C to 37 °C yields optimum binding, we routinely use an overnight incubation at 4 °C. When coating is terminated remove residual unbound conjugate by repeated washing with buffer (3). Note that Tween 20 of buffer (3) is a surfactant that reduces non-specific binding of the various antibody reagents to the plastic surface.

References

[1] *M. Eto,* Organophosphorus Pesticides: Organic and Biological Chemistry. CRC Press, Boca Raton, FL, 1979.
[2] *G. J. Hass, E. J. Guardia,* Production of Antibodies against Insecticide-Protein Conjugates, Proc. Soc. Exp. Biol. Med. *129,* 546 – 551 (1968).
[3] *E. R. Centeno, W. J. Johnson, A. H. Sehon,* Antibodies to Two Common Pesticides, DDT and Malathion, Int. Arch. Allergy *37,* 1 – 13 (1970).
[4] *J. J. Langone, H. van Vunakis,* Radioimmunoassay for Dieldrin and Aldrin, Res. Commun. Chem. Pathol. Pharmacol. *10,* 163 – 171 (1975).
[5] *K. W. Hunter, D. E. Lenz,* Detection and Quantification of the Organophosphate Insecticide Paraoxon by Competitive Inhibition Enzyme Immunoassay, Life Sci. *30,* 355 – 361 (1982).
[6] *Z. Niewola, C. Hayward, B. A. Symington, R. T. Robson,* Quantitative Estimation of Paraquat by an Enzyme-Linked Immunosorbent Assay Using a Monoclonal Antibody, Clin. Chim. Acta *148,* 149 – 156 (1985).
[7] *D. E. Lenz, T. Clow, T. Bonsack,* Specificity and Toxicological Potential of Antiparaoxon Antibodies, American College of Toxicology, Annual Meeting 30 November – 2 December 1983, Washington, DC, Abstr. 143 – 149.

[8] *P. A. Giang,* Organophosphorus and Carbamate Insecticides, in: *H. U. Bergmeyer* (ed.), Methods of Enzymatic Analysis, Vol. *IV*, Verlag Chemie, Weinheim and Academic Press, New York 1981, pp. 2249 – 2259.

[9] *E. P. W. Winteringham, R. W. Disney,* Radiometric Assay of Acetylcholinesterase, Nature *195*, 1303 (1962).

[10] *O. P. Lehtonen, M. K. Vilganen,* Antigen Attachment in ELISA, J. Immunol. Methods *34*, 61 – 70 (1980).

[11] *L. A. Cantarero, J. E. Butler, J. W. Osborne,* The Absorptive Characteristics of Proteins for Polystyrene and Their Significance in Solid Phase Immunoassays, Anal. Biochem. *105*, 375 – 382 (1980).

[12] *A. A. Brimfield, D. E. Lenz, C. Graham, K. W. Hunter,* Mouse Monoclonal Antibodies against Paraoxon: Potential Reagents for Immunoassay with Constant Immunochemical Characteristics, J. Agric. Food Chem. *33*, 1237 – 1242 (1985).

[13] *A. Zettner, P. E. Duly,* Principles of Competitive Binding Assays (Saturation Analyses). II. Sequential Saturation, Clin. Chem. *20*, 5 – 14 (1974).

[14] *S. E. Davis, P. J. Munson, M. L. Jaffe, D. Rodbard,* Radioimmunoassay Data Processing with a Small Programmable Calculator, J. Immunoassay *1*, 15 – 25 (1980).

[15] *D. Rodbard,* Statistical Quality Control and Routine Data Processing for Radioimmunoassay and Immunoradiometric Assay, Clin. Chem. *20*, 1255 – 1270 (1974).

[16] *L. E. M. Miles,* Immunoradiometric Assay and Two Site IRMA Systems (Assay of Soluble Antigens Using Labelled Antibodies; in: *G. E. Abraham* (ed.), Handbook of Radioimmunoassay, Part 1, Marcel Dekker Inc., New York 1977, pp. 131 – 177.

[17] *A. A. Brimfield, K. W. Hunter, D. E. Lenz, H. P. Benschop, C. Van Dijk, L. P. A. de Jong,* Structural and Stereochemical Specificity of Mouse Monoclonal Antibodies to the Organophosphorus Cholinesterase Inhibitor Soman, Mol. Pharmacol. *28*, 32 – 39 (1985).

[18] *A. Grossberg, D. Pressman, in: H. N. Eisen* (ed.), Methods in Medical Research, Vol. *10*, Yearbook Medical Publishers, Inc., Chicago 1964, pp. 103 – 105.

[19] *O. H. Lowry, N. J. Rosebrough, A. L. Farr, P. Randall,* Protein Measurement with the *Folin* Reagent, J. Biol. Chem. *193*, 265 – 275 (1951).

[20] *J. W. Fenton, S. J. Singer,* Affinity Labelling of Antibodies to the p-Azophenyltrimethylammonium Hapten, and a Comparison of Affinity-labelled Antibodies of Two Different Specificities, Biochemistry *10*, 1429 – 1437 (1971).

3.5   Atrazine in Water

6-Chloro-*N*-ethyl-*N*′-(1-methylethyl)-1,3,5-triazine-2,4-diamine

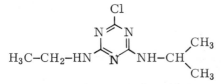

Sigmund J. Huber and Bertold Hock

General

Atrazine is widely used as a herbicide for the selective control of annual broadleaf weeds and some grass weeds, primarily in maize cultures. s-Triazine herbicides such as atrazine interfere with the photosynthetic electron transfer by binding and modifying the electron carrier Q_B on the reducing site of photosystem II. Detoxification of s-triazine herbicides by hydroxylation (caused by benzoxazinone derivatives) or by conjugation to glutathione is restricted to corn plants and is therefore of immense economic importance [1].

Contamination of fresh water close to cornfields can be due to soil erosion or to wind-drift during application. The persistence of s-triazines in fresh water is documented by several aquatic outdoor studies [2].

Application of method: in environmental studies. The method described here for the analysis of atrazine in fresh water and other aqueous solutions should contribute to a rapid and effective environmental monitoring of the herbicide.

The specific binding of antibodies to their antigens in aqueous solution allows a direct assay of specimens from lakes or ponds after filtering and correcting pH.

Substance properties relevant in analysis: the relatively small organic atrazine molecule with a molecular weight of 214.5 cannot evoke an immunogenic response. Therefore, it must be covalently linked to an immunogenic carrier, e.g. a protein. Antibodies are produced to a conjugate of ametryn, a structural analogue of atrazine, and the keyhole limpet pigment, haemocyanin. Ametryn contains a $-SCH_3$ group at position 2 (instead of Cl as in atrazine) which can be easily reacted with proteins after sulphoxidation of the methylthio group. This procedure and the antibody preparation are described in the Appendix, p. 448.

Methods of determination: several analytical methods for quantitative determination of atrazine and other s-triazines are available, e.g. GC, GLC, and HPLC.

The commonly used GC analysis involves a twofold extraction with chloroform, drying with sodium sulphate, evaporation to dryness and solution of the residue in hexane/ethanol [3]. The minimal amount which can be detected by this method is 0.2 µg/l. Similar clean-up procedures are necessary for GLC analysis. A detection limit of 0.5 µg/l can be reached [4]. The HPLC analysis of s-triazines lies in the same detection range [5]. It is evident that these procedures require extensive and time-consuming clean-up steps.

The enzyme-immunoassay involves a minimum of handling, and it requires only standard laboratory equipment. It requires no pre-treatment, for example extraction or clean-up procedures: water samples can be analyzed directly. The EIA provides a rapid and efficient method which allows large numbers of samples to be analyzed simultaneously. In addition, portable EIA systems have recently become commercially available: therefore, water samples can be assayed and screened close to their sources.

The detection limit reached in the EIA variant with an affinity-purified antiserum and polystyrene spheres as antibody carriers lies far below that of other techniques, and is in the range of ng/l.

International reference method and standards: no reference method has been proposed yet. Reference material like atrazine and related s-triazines are available from *Ciba-Geigy Ltd.,* Basel, in a purity of > 99%.

Assay

Method Design

Two different types of solid-phase enzyme-immunoassays are described for the s-triazines, atrazine and ametryn. Polystyrene microtitre plates and polystyrene spheres are used as different antibody-adsorbing systems. The ratio of sample volume to moistened surface is fixed in the case of microtitre wells. It can be varied, however, with polystyrene spheres. In the latter case, a higher sample volume can be analyzed. Consequently, lower sample concentrations are detected with polystyrene spheres as antibody-adsorbent surfaces because of the higher dilutions [6].

Principle

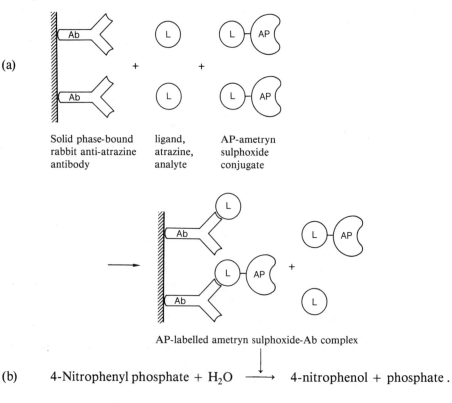

(a)

Solid phase-bound   ligand,   AP-ametryn
rabbit anti-atrazine atrazine,   sulphoxide
antibody   analyte   conjugate

AP-labelled ametryn sulphoxide-Ab complex

(b)   4-Nitrophenyl phosphate + H_2O $\longrightarrow$ 4-nitrophenol + phosphate .

Solid phase-bound anti-atrazine antibodies bind atrazine (if present) from the sample. The enzyme tracer, an alkaline phosphatase*-hapten conjugate, occupies the free antibody-binding sites. All unbound molecules are removed. The amount of bound enzyme is determined with 4-nitrophenyl phosphate as the substrate. The reaction rate is determined by measuring the increase in absorbance at 405 nm per unit time, $\Delta A / \Delta t$. The readings are inversely related to the atrazine concentration in the sample.

Selection of assay conditions and adaptation to the individual characteristics of the reagents: the coating concentrations for the two different polystyrene surfaces (microtitre plate, sphere) as well as the incubation times and temperatures were examined in preliminary studies [6, 7]. If the described incubation conditions are exactly followed, the method is robust and reproducible. However, an incubation temperature of 4 °C was found to be optimal for the present test systems.

Equipment

For EIA in microtitre plates: polystyrene microtitre plates (10 × 10 wells, 0.5 ml capacity, ELISA-*Pollähne,* D-3015 Wennigsen); vertical light path photometer capable of measurement at 405 nm (e.g. Manureader S, Meku, *Pollähne,* D-3015 Wennigsen, which is compatible with the microtitre plates described).

For EIA with polystyrene spheres: polystyrene spheres (w/spec. finish, diameter 6.4 mm, *Precision Plastic Ball Co.,* Chicago, IL); glass scintillation vials (25 ml) and glass tubes (diameter 8 mm); optional: microcomputer for the construction of calibration curves and the calculation of atrazine concentrations from the absorbance data; the RIA-programme "Inhibition" [8] can be used for calculation of EIA data.

Reagents and Solutions

Purity of reagents: all chemicals must be of analytical reagent grade.

Preparation of solutions: prepare all solutions in re-purified water (cf. Vol. II, chapter 2.1.3.2.).

1. Coating carbonate buffer (carbonate, 50 mmol/l; NaN$_3$, 0.1 g/l; pH 9.6):

   dissolve 1.59 g Na$_2$CO$_3$, 2.94 g NaHCO$_3$, and 0.1 g NaN$_3$ in water and make up to 1 litre.

2. Coating solution (Ab**; carbonate, 50 mmol/l; NaN$_3$, 0.1 g/l; pH 9.6):
   dissolve the affinity-purified and lyophilized antibodies (prepared according to Appendix, p. 448) in coating buffer (1).

---

* AP, orthophosphoric-monoester phosphohydrolase (alkaline optimum), EC 3.1.3.1.
** For coating microtitre plates: 150 µg/ml; for coating polystyrene spheres: 450 µg/ml.

3. Washing solution, Tris-buffered saline, TBS (Tris, 25 mmol/l; NaCl, 5 mmol/l; $MgCl_2$, 0.5 mmol/l; Tween 20, 0.5 ml/l; pH 7.8):

dissolve 3.0 g tris(hydroxymethyl)aminomethane, 102 mg $MgCl_2 \cdot 6\,H_2O$, 292 mg NaCl, and 0.5 ml polyoxyethylene sorbitan monolaurate (Tween 20) in water, adjust pH to 7.8 with HCl, 8 mol/l; and make up to 1 litre with water.

4. Incubation buffer, TBS (Tris, 50 mmol/l; NaCl, 10 mmol/l; $MgCl_2$, 1 mmol/l; pH 7.8):

dissolve 6.1 g tris(hydroxymethyl)aminomethane, 203 mg $MgCl_2 \cdot 6\,H_2O$, 584 mg NaCl in water and make up to 1 litre.

5. Alkaline phosphatase-hapten conjugate (ca. 2 mg/ml; AP, ca. 4000 U/ml; Tris, 50 mmol/l; NaCl, 10 mmol/l; $MgCl_2$, 1 mmol/l; pH 7.8):

dilute the stock solution (cf. Appendix, p. 450), 1 : 1000 with TBS (4).

6. Substrate solution (4-NPP, 4 mmol/l; carbonate, 50 mmol/l; NaN_3, 0.1 g/l; pH 9.6):

dissolve 149 mg 4-nitrophenyl phosphate, disodium salt, hexahydrate, in 100 ml carbonate buffer (1).

7. Stopping solution (KOH, 5 mol/l):

dissolve 280.5 g KOH pellets in water and make up to 1 litre.

8. Atrazine standard solutions (0.01 to 33.0 µg/l; 0.1 to 550 µg/l):

  a) Stock I (220 mg/l):
     dissolve 2.2 mg pure crystalline atrazine in absolute ethanol and make up with ethanol to 10 ml.
  b) Stock II (1 mg/l): dilute 100 µl stock I with 22 ml TBS (4). For establishing a calibration curve dilute with TBS buffer (4).

Stability of reagents: store at 4 °C in the dark; the buffer solutions (1), (3) and (4) are stable for 4 weeks, the enzyme tracer conjugate (stock solution) for 6 months. The atrazine stock solutions should be prepared weekly; store at 4 °C in the dark. The substrate solution (6) is made freshly on the day of use. The coating solution is stable for only 3 – 4 days and must be prepared freshly for each coating process. Re-use of the coating solution is not recommended.

Procedure

Collection and treatment of specimen: filter water specimens through a double layer of paper filters (e.g. *Whatman* GS/C) before analysis and adjust to pH 7.8 with KOH or HCl, 0.1 mol/l, as required. These samples can be directly analyzed unless a dilution with incubation buffer (4) is required.

Stability of the analyte in the sample: water samples which cannot be analyzed within the following two days should be kept frozen in order to avoid microbial destruction of atrazine.

Assay conditions: all measurements in triplicate.

Immunoreaction: first incubation at 4 °C in 200 µl for 3 h (microtitre plate), or in 20 ml for 5 h (spheres); second incubation at 4 °C in 220 µl (21 ml) for 30 min.

Enzymatic indicator reaction: incubation at 37 °C for 60 min; wavelength 405 nm; light path in assay with spheres 10 mm. Measurement against unreacted substrate solution.

Run a reagent blank omitting sample or standard.

Establish a calibration curve with at least 10 dilutions of standard solution (8b) instead of sample (covering the measuring range of 0.01 to 33.0 µg/l in the microtitre assay, and 0.1 to 550 µg/l in the assay using coated spheres); additionally run atrazine-free sample and one sample with an excess (e.g. 500 µg/l) of atrazine.

Measurement

With coated polystyrene spheres

Immunoreaction

Pipette into (place in) glass scintillation vials:		concentration* in incubation mixture	
sample or standards (8) coated spheres	20 ml 3	atrazine Tris NaCl MgCl$_2$	up to 550 ng/l 50 mmol/l 10 mmol/l 2 mmol/l
incubate for 5 h at 4 °C; invert the vials carefully every 30 min;			
AP-hapten conjugate (5)	1 ml	conjugate AP	variable variable
incubate for 30 min at 4 °C; discard contents of the vials, wash 3 times with 20 ml washing solution (3), then 3 times with 50 ml water; transfer the balls individually to another clean glass tube.			

Enzymatic indicator reaction

Pipette into (place in) a 8 mm diameter glass tube:		concentration in assay mixture	
coated ball after primary reaction substrate solution (6)	1 0.5 ml	4-NPP carbonate NaN$_3$	4 mmol/l 50 mmol/l 0.1 g/l
incubate for 60 min at 37°C, remove the sphere, read absorbance.			

* Only valid for the standards or when samples are diluted with buffer (4).

In microtitre plate

Pipette into the wells of a coated microtitre plate:		concentration in incubation*/assay mixture	
sample or standard (8)	200 µl	atrazine Tris NaCl MgCl$_2$	up to 33 µg/l 50 mmol/l 10 mmol/l 1 mmol/l
incubate for 3 h at 4°C,			
AP-hapten conjugate (5)	20 µl	conjugate AP	variable variable
incubate for 30 min at 4°C; discard contents of the wells, wash 3 times with 0.5 ml washing solution (3), then 3 times with 3 ml water;			
substrate solution (6)	200 µl	4-NPP carbonate NaN$_3$	4 mmol/l 50 mmol/l 0.1 g/l
incubate for 60 min at 37°C,			
KOH (7)	50 µl	KOH	1 mol/l
reaction is stopped; read absorbance.			

* Only valid for the standards or when samples are diluted with buffer (4).

Calibration curve: from all readings subtract reading of tube with excess atrazine, yielding $\Delta A/\Delta t$. Calculate the ratio of $\Delta A/\Delta t$ for standards (samples) and $\Delta A_0/\Delta t$ for zero atrazine control, $(\Delta A/\Delta t)/(\Delta A_0/\Delta t) = \Delta A/\Delta A_0$.

Linear calibration curves are obtained by logit transformation of $100 \times \Delta A/\Delta A_0$ [9]. The logit transformation is given by the formula

$$\text{logit}\,(100 \times \Delta A/\Delta A_0) = \ln \frac{100 \times \Delta A/\Delta A_0}{100 - 100\,\Delta A/\Delta A_0}$$

The sigmoidal calibration curve of the EIA can be linearized by plotting logit $100 \times \Delta A/\Delta A_0$ *vs.* log concentration.

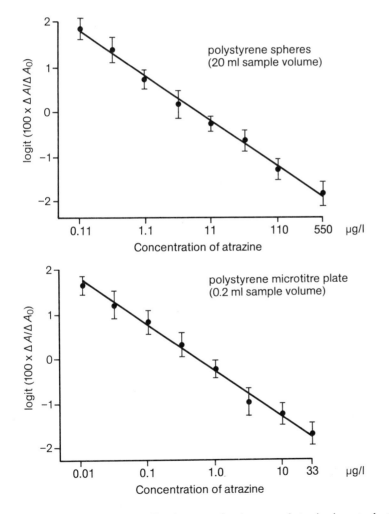

Fig. 1. Typical linearized calibration curve for the assay of atrazine in water by the two EIA systems.

Calculation: sample concentrations are directly calculated from the calibration curve, e.g. by the RIA-programme "Inhibition" for personal computers [8].

Validation of Method

Precision, accuracy, detection limit and sensitivity: no data about imprecision and inaccuracy are available. The measuring range of this EIA is defined as the range over which the calibration curve can be linearized. The detection limit is defined as the lowest concentration which can be detected within the measuring range.

The microtitre plate EIA yields a measuring range of 10 to 30000 fmol atrazine/assay, corresponding to 2.1 to 6300 pg/assay. With a sample volume of 0.2 ml, this test system allows atrazine concentrations from 0.011 to 33 µg/l in aqueous solutions to be determined.

The test system with affinity-purified antiserum and polystyrene spheres proved to be more sensitive. Its measuring range, from 10 to 50000 fmol atrazine/assay, corresponding to 2.1 to 10500 pg/assay, is even wider than that of the procedure with microtitre plates. In addition, the large sample volume permits a detection range from 0.11 ng/l to 550 ng/l.

In Tab. 1, the advantages and disadvantages of the two different EIA systems are compared.

Table 1. Advantages and disadvantages of the two different EIA systems

Solid-phase EIA in the polystyrene microtitre plates as antibody carrier	Solid-phase EIA with polystyrene spheres as antibody carriers
Constant ratio between sample volume and moistened polystyrene surface in the wells	Variable ratio between sample volume and polystyrene surface
Increase of sensitivity by an increase of sample volume not possible	Increase of sensitivity by an increase of sample volume possible within a determined range
Suitable for samples, which are provided in small volumes (200 µl/assay)	Suitable for samples which are provided in larger volume ranges (20 ml/assay), e.g. water samples
2 h incubations of sample are sufficient	5 h incubations of samples are necessary
Test system can be easily mechanized and automated	Test system difficult to mechanize and automate

Sources of error: due to the low detection limit of the assay, it is essential to use absolutely clean glassware. In order to remove traces of atrazine, e.g. from previous assays, it is recommended to wash the glassware in the following order: acetone, Deconex, acetone, re-purified water.

If linearized calibration curves (logit $100 \times \Delta A / \Delta A_0$ vs. concentration) are used, small deviations of the logit values yield large errors in the results. Therefore, non-linearized calibration curves (logit $100 \times \Delta A / \Delta A_0$ vs. concentration) are recommended for critical cases. Computer programmes are available for best fits.

Specificity: the cross-reactivities of the affinity-purified antibodies were estimated with respect to other triazines. For this purpose, 550 ng of each s-triazine per litre was assayed in the microtitre-plate EIA, and the apparent concentrations were evaluated from the linear calibration curve of the atrazine EIA and compared to atrazine. The corresponding per cent values of the reference triazines are listed in Tab. 2.

Table 2. Cross-reactivities of different s-triazines in the atrazine EIA in microtitre plates (affinity-purified antibodies according to [7]).

s-Triazine	Apparent concentration in the atrazine EIA*		Relative amount as compared to atrazine
	ng/l	nmol/l	%
Atrazine	550	2.50	100
Propazine	319	1.45	57
Simazine	297	1.35	55
Terbuthylazine	66	0.30	12
Ametryn	583	2.65	106
Prometryn	308	1.40	56
Simetryn	418	1.74	75
Terbutryn	77	0.35	14

* 0.2 ml were used per assay.

Except for ametryn, which was used for the synthesis of the immunogenic conjugate, simetryn is the most effective cross-reacting triazine, followed by a group with mean cross-reactivities consisting of simazine, propazine, and prometryn. The cross-reactivities of terbutryn and terbuthylazine are however very low. Therefore, only water samples free from propazine, prometryn, simazine and simetryn can be analyzed properly. Contamination with terbuthylazine, or terbutryn is negligible. However, several s-triazines are unlikely to be applied simultaneously.

Appendix

Preparation of antigen: couple keyhole limpet haemocyanin to ametryn sulphoxide. The reaction mechanism is based upon a nucleophilic substitution which is described in principle by *Hamboeck et al.* [10]. The sulphoxide can be reacted with free NH_2- and SH groups of the carrier protein.

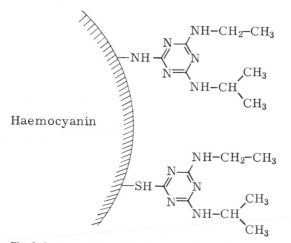

Fig. 2. Immunogenic ametryn-haemocyanin conjugate

- Dissolve separately 226 mg ametryn and 207 mg 3-chloroperbenzoic acid in 1 ml chloroform, combine the solutions and stir for 2 h at room temperature in the dark.
- Separate the ametryn-sulphoxide derivative from unreacted compounds by thin-layer chromatography (plates KG 60, F_{254}; *Merck*) in chloroform/acetone (7:3);

  $R_{f\ (unreacted\ ametryn)}$: 0.5
  $R_{f\ (ametryn\ sulphoxide)}$: 0.3
  $R_{f\ (3\text{-}chloroperbenzoic\ acid)}$: 0.7

  the compounds are detected by their fluorescence under UV-light (254 nm).
- Elute the ametryn sulphoxide band from the gel with chloroform and evaporate to dryness with a rotating evaporator.
- To 50 mg keyhole limpet haemocyanin, dissolved in carbonate buffer (1), add 10 mg crystalline ametryn sulphoxide, dissolved in 200 µl abs. ethanol.
- Incubate the reaction mixture for 5 days at 37 °C under stirring in the dark.
- Dialyze the reaction mixture three times against 5 l NaCl solution, 9 g/l, pH 7.5. The reaction product can be lyophilized and stored in a crystalline state at 4 °C for 6 months.

Preparation of anti-atrazine antibodies: dissolve 1 mg ametryn-haemocyanin conjugate in 0.5 ml sterilized saline (NaCl, 9 g/l) and emulsify with 0.5 ml *Freund's* complete adjuvant (*Difco,* Detroit, U.S.A.). Immunize 12 weeks old New Zealand rabbits by injecting 1 ml of this emulsion subcutaneously into the neck area and intradermally into the caudal area, distributed among five punctations. Repeat the procedure three times at weekly intervals. An intravenous booster injection into the ear vein (1 mg conjugate dissolved in 0.5 ml saline) follows one week later.

Bleed the rabbits two weeks after the booster injection. For this purpose, tranquillize the animal with Combelen (*Bayer,* Leverkusen, West Germany), 0.05 ml/kg injected subcutaneously. After 40 min, insert a disposable indwelling canula of Teflon

("Microcath", *Braun*, Melsungen, West Germany) into the ear artery. The blood pressure in the artery allows a collection of 60 to 80 ml blood within a period of 15 min. Thereafter, inject about 50 ml infusion solution Inosteril (*Fresenius,* Bad Homburg, West Germany) through the same canula within further 15 min. Then, another 10 to 20 ml blood can be collected without health hazard for the rabbit. Six weeks after this procedure, another booster injection can be made, and blood can be collected two weeks later in the same manner. The booster injection can be repeated for a third time. However, the danger of an anaphylactic shock increases with each subsequent booster injection, and the antibody titre decreases with time. Table 3 shows the immunization scheme.

Table 3. Preparation of antiserum

Day	Immunogenic conjugate	Application	Blood collection
1	1 mg dissolved in 0.5 ml saline (NaCl, 9 g/l) and emulsified with 0.5 ml *Freund's* complete adjuvant	intradermal	
8	as above	intradermal	
16	as above	intradermal	
24	1 mg dissolved in 0.5 ml saline	intravenous	
38	–	–	blood collection
85	1 mg dissolved in 0.5 ml saline	intravenous	
99	–	–	blood collection

Leave ca. 80 to 90 ml collected blood for ca. 1 h at room temperature for coagulation; store overnight at 4 °C. Separate the coagulated part from the serum on a funnel, and centrifuge the serum for 10 min at 4300 g to remove coagulated particles.

Precipitate the immunoglobulin fraction from the serum by addition of 3.5 ml Rivanol* solution, 4 g/l, followed by saturated ammonium sulphate solution according to [11] and lyophilize. For this purpose, add 35 ml 0.4% Rivanol solution per 10 ml antiserum and stir gently on a magnetic stirrer for 30 min at room temperature. Decant the supernatant and decolorize it by filtering through a double-layered paper filter, coated with activated charcoal. Repeat this procedure until the yellow Rivanol residue is removed. To 10 ml of the clear solution add 6 ml saturated ammonium sulphate solution dropwise under gentle stirring. Stir at room temperature for 30 min after adding the entire ammonium sulphate solution. Centrifuge the precipitated IgG at 12000 g for further 30 min. Decant the supernatant and dissolve the pellet in 0.9% saline (pH 7.5) according to the original volume of the serum. Repeat the ammonium sulphate precipitation and the centrifugation as described above. Then dissolve the pellet in NaCl solution (original serum volume) and dialyze three times against 5 l saline. Lyophilize the dialyzed IgG fraction.

Lyophilization of the IgG solution is the optimal procedure for storage (4 °C in a dry state) and for further handling of the antibodies. The lyophilized IgG fraction can be exactly weighed out and re-suspended in the appropriate buffer.

---

* 2-Ethoxy-6,9-diaminoacridine lactate

Affinity-purified anti-atrazine antibodies are obtained by affinity chromatography with the hapten as ligand. The specific antibodies are eluted according to [10]:

- 3 g aminohexyl Sepharose-4B gel (*Pharmacia*) is swollen in 30 ml carbonate buffer (1) overnight.
- Dissolve 55 mg ametryn sulphoxide in 2 ml abs. ethanol and react it with the Sepharose gel for 5 days at 37 °C under slight shaking in a water-bath.
- Place the gel into a short column (20 mm diameter; 80 mm gel-bed height).
- Wash the gel with an excess of TBS (4) containing 100 ml abs. ethanol per litre.
- Wash the gel with elution buffer (glycine, 0.1 mol/l, pH adjusted to 2 with HCl 80 mmol/l), followed by a large excess of TBS (4).
- Pass 10 ml re-suspended IgG precipitate (5 mg/ml) slowly over the bed.
- Wash the gel with a large excess of TBS (4) until no protein can be detected in the washing buffer (at 280 nm).
- Elute the specific antibodies with 50 ml glycine/HCl (pH 2.0).
- Quickly adjust pH of each 2 ml fraction with 100 μl NaOH, 5 mol/l, to pH 8.5.
- Determine the protein concentration of each fraction ($A_{280\,nm}$) and dialyze the protein fractions (ca. 15 ml) against 2 l carbonate buffer (1).
- Lyophilize the dialyzed fractions. Re-suspend the antibodies in carbonate buffer (1) for coating the solid phases.

Preparation of alkaline phosphatase-hapten conjugate: the marker enzyme alkaline phosphatase (grade I, specific activity 2 500 U/mg at 37 °C, 4-nitrophenyl phosphate as the substrate; *Boehringer Mannheim*) is coupled to the hapten (ametryn sulphoxide).

- Dissolve 5 mg ametryn sulphoxide (preparation as described above) in 40 μl absolute ethanol.
- Dilute 60 μl (0.6 mg) alkaline phosphatase with 250 μl carbonate buffer (1).
- Combine the two solutions dropwise and incubate the reaction mixture for 5 days at 37 °C.
- Dialyze the reaction product against TBS (4) containing 100 ml ethanol per litre and then three times against TBS (4).
- Store the dialyzed enzyme tracer at 4 °C for up to 6 months.

Coating of the solid phase

Microtitre plates: fill the plates with 0.3 ml coating solution (2) and incubate for 24 h at 4 °C. After the coating process, empty the plates. Rinse each well with 500 μl washing solution (3) and thereafter with a large excess of water (ca. 10 ml/well). Store the air-dried plates at 4 °C for not more than 3 days.

Polystyrene spheres: incubate the spheres for 48 h with the corresponding coating solution (2), 0.5 ml per sphere, at 4 °C in *Erlenmeyer* flasks (100 ml) without shaking. The spheres must be totally covered by the solution. The vessel should only be slightly moved during the coating process. Thereafter, rinse the spheres with 1 ml washing solution (3) per sphere, followed by 10 ml water per sphere.

References

[1] *H. O. Esser, G. Dupuis, E. Ebert, G. J. Marco, C. Vogel*, s-Triazines, in: *P. C. Kearney, D. D. Kaufman* (eds.), Herbicides, Chemistry, Degradation, and Mode of Action, Marcel Dekker, New York, Basel 1975, pp. 129 – 189.

[2] *L. Peichl, J. P. Lay, F. Korte,* Wirkung von Dichlobenil und Atrazin auf die Populationsdichte von Zooplanktern in einem aquatischen Freilandsystem, Z. Wasser-Abwasser-Forsch. *17*, 134 – 145 (1984).

[3] *K. Ramsteiner, W. D. Hörman, D. O. Eberle,* Multiresidue Method for the Determination of Triazine Herbicides in Field-Grown Agricultural Crops, Water and Soils, J. Assoc. Off. Anal. Chem. *57*, 192 – 201 (1974).

[4] *D. C. G. Muir,* Determination of Terbutryn and Its Degradation Products in Water, Sediments, Aquatic Plants, and Fish, J. Agric. Food Chem. *28*, 714 – 719 (1980).

[5] *D. Paschal, R. Bicknell, K. Siebenmann,* Determination of Atrazine in Runoff Water by High Performance Liquid Chromatography, J. Environm. Sci. Health, Part B, *13*, 105 – 115 (1978).

[6] *S. J. Huber, B. Hock,* Solid-Phase-Enzymimmunoassay zum Nachweis von Pflanzenschutzmitteln in Gewässern, GIT, Fachz. f. d. Lab. *29*, 969 – 977 (1985).

[7] *S. J. Huber,* Improved Solid-Phase Enzyme-Immunoassay Systems in the ppt Range for Atrazine in Fresh Water, Chemosphere *14*, 1795 – 1803 (1985).

[8] *R. B. Barlow,* Biodata Handling with Microcomputers, Elsevier Science Publishers, Amsterdam 1983, pp 135 – 145.

[9] *E. W. Weiler, P. S. Jourdan, W. Conrad,* Levels of Indole-3-acetic Acid in Intact and Decapitated Coleoptiles as Determined by a Specific and Highly Sensitive Solid-Phase Enzyme-Immunoassay, Planta *153*, 561 – 571 (1981).

[10] *H. Hamboeck, R. W. Fischer, E. E. Di Iorio, K. H. Winterhalter,* The Binding of s-Triazine Metabolites to Rodent Hemoglobins Appears Irrelevant to Other Species, Mol. Pharmacol. *20*, 579 – 584 (1981).

[11] *B. A. L. Hurn, S. M. Chantler,* Production of Reagent Antibodies, in: *H. Van Vunakis, J. J. Langone* (eds.), Methods in Enzymology, Vol. *70*, Part A, Academic Press, New York 1980, pp. 104 – 141.

3.6 Paraquat in Soil

1,1'-Dimethyl-4,4'-bipyridinium

$$H_3C-\overset{+}{N}\langle\!\!\!\!=\!\!\!\!\rangle\!-\!\!\langle\!\!\!\!=\!\!\!\!\rangle\overset{+}{N}-CH_3$$

Jill P. Benner and Zbigniew Niewola

General

Paraquat is a non-specific contact herbicide which rapidly kills green plant material. Photosynthetic activity in plants reduces paraquat to the corresponding stable free radical. This radical subsequently reacts with oxygen, regenerating paraquat and giving rise to a peroxide radical or hydrogen peroxide, which is believed to be the actual phytotoxic agent.

One of the principal properties of paraquat, which makes it so useful in agriculture, is its ability to kill unwanted vegetation quickly without leaving biologically active residues. When paraquat reaches the soil it is very rapidly and strongly adsorbed to clay minerals and in this form exerts no effect on plant growth, crop yield or soil organisms.

Application of method: in environmental studies. The method described here is intended for the analysis of paraquat residues in soil. Similar assays for the analysis of paraquat in serum and urine are described in the literature [1, 2].

Routine environmental studies are carried out on soil samples from locations around the world to correlate levels and patterns of usage of paraquat with observed soil-bound residues. Specialized trials are also conducted which are intended to give an improved understanding of the physical and chemical properties of soil bound paraquat. Both types of study may generate hundreds of samples and the enzyme-linked immunosorbent assay, ELISA, provides a rapid and efficient analytical method which can handle large numbers of samples simultaneously.

Substance properties relevant in analysis: paraquat is a relatively small and simple organic molecule (molecular weight of the paraquat cation is 186). It cannot be used directly in an enzymatic assay and the ELISA relies entirely upon the specific reaction of an antibody with paraquat. The compound itself is incapable of inducing an immunogenic response because of its low molecular weight and must therefore be covalently linked to a protein moiety. Antibodies are produced to a conjugate of paraquat and bovine serum albumin, PQ-BSA*, the synthesis of which is described in the Appendix, p. 463.

Methods of determination: several different analytical methods for paraquat are currently available and could be applied to the analysis of soil, but all require extensive and time-consuming clean-up steps. One of the most widely used methods relies on cation-exchange chromatography and spectrophotometry [3]. The light absorbing species is the free radical obtained by reaction of paraquat with alkaline dithionite. Two of the techniques most commonly applied to pesticide analysis, gas-liquid chromatography, GC, and high-performance liquid chromatography, HPLC, may be used in the determination of paraquat. The GC method requires reduction of paraquat prior to analysis [4].

A number of HPLC methods have been reported, e.g. [5], but in some cases peak shape may be rather poor and careful clean-up procedures are necessary. A sensitive radioimmunoassay is available for the analysis of paraquat in serum and urine [6], but

---

*   Abbreviations
    PQ-BSA   Paraquat-bovine serum albumin conjugate
    PQ-KLH   Paraquat-keyhole limpet haemocyanin conjugate
    POD-RAM   Rabbit anti-mouse immunoglobulin G labelled with horseradish peroxidase.

the ELISA method, relying on enzymatic labelling rather than radioisotopes, is more convenient for general use by a diverse range of personnel in laboratories all over the world.

International reference method and standards: there is no specific international reference method for the analysis of paraquat. The compound may be readily synthesized in bulk quantity and substantial quantities of analytical grade reagent, for use as a reference standard, may be prepared by further purification of the commercial grade material.

Assay

Method Design

Several different types of enzyme-immunoassay for the analysis of small molecules, such as drugs, have been described in the literature [7 – 9]. Many of these assays are dependent upon the specialized synthesis of enzyme-labelled reagents. The method described here for the analysis of paraquat avoids this requirement by employing an indirect ELISA modified to detect paraquat by an inhibition technique.

Principle

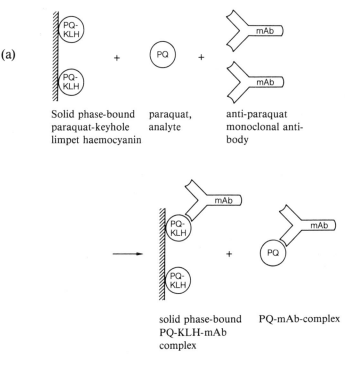

(a)

Solid phase-bound   paraquat,   anti-paraquat
paraquat-keyhole   analyte   monoclonal anti-
limpet haemocyanin   body

solid phase-bound   PQ-mAb-complex
PQ-KLH-mAb
complex

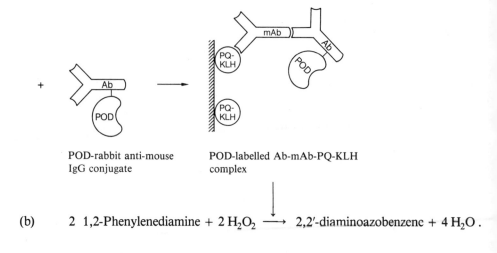

| POD-rabbit anti-mouse IgG conjugate | POD-labelled Ab-mAb-PQ-KLH complex |

(b)   2 1,2-Phenylenediamine + 2 H$_2$O$_2$ $\longrightarrow$ 2,2′-diaminoazobenzene + 4 H$_2$O .

Paraquat-keyhole limpet haemocyanin, PQ-KLH, is passively adsorbed by polystyrene plates which are then incubated with the samples containing paraquat and a known amount of the monoclonal antibody. The residual monoclonal antibody which has not reacted with free paraquat in the sample combines with PQ-KLH on the plate. Estimation of the fixed antibody is achieved by incubating the washed plates with rabbit anti-mouse immunoglobulin G labelled with horseradish peroxidase*, POD-RAM. Further reaction with a chromogenic substrate enables the enzyme activity of the solid phase to be determined from measurements of the increase in absorbance per unit time, $\Delta A/\Delta t$. The paraquat concentration is inversely related to $\Delta A/\Delta t$.

Selection of assay conditions and adaptation to the individual characteristics of the reagents: a series of preliminary studies are required to determine the optimum reagent concentrations, incubation times and temperatures for the various steps (described in references [1, 2]). The assay is intended to be rapid but also convenient for the operator; for example, all incubation stages are carried out at room temperature ($\sim 22\,^{\circ}$C) as elevated or reduced temperatures appear to offer few advantages. Details of the recommended conditions are described here, but the method is relatively robust and amenable to some degree of flexibility.

The neutral soil extracts contain high concentrations of sodium sulphate and the effects of this on the calibration curve have been evaluated. An increase in sodium sulphate concentration produces a parallel shift of the curve to the right, (Fig. 1). On the basis of this observation, 50% saturated sodium sulphate is included in the incubation buffer to optimize the accuracy of the method. For the analysis of water samples or body fluids, sodium sulphate is omitted from the incubation buffer thus leading to improved detectability.

---

* Donor: hydrogen-peroxide oxidoreductase, EC 1.11.1.7.

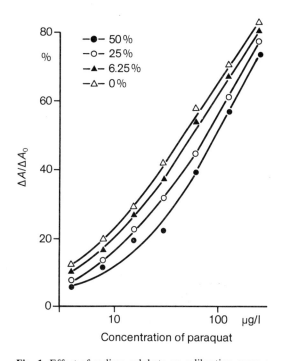

Concentration of paraquat

Fig. 1. Effect of sodium sulphate on calibration curve.
Plot of $\Delta A/\Delta A_0$ *versus* paraquat concentration for solutions of incubation buffer containing 0%, 6.25%, 25% and 50% sodium sulphate.

Equipment

Polystyrene microtitre plates (12 × 8 wells, capacity 0.4 ml) (*Nunc* Immuno Plate I). Adjustable 12-channel digital pipette (0.05 – 0.2 ml). 12-Channel plate washer; pH meter. Vertical light path spectrophotometer which can simultaneously measure the absorbances (at 450 nm) of the contents of 8 wells (1 column) of a microtitre plate (e.g. Titertek Multiskan MC, *Flow Laboratories*). Microcomputer interfaced to Multiskan MC and appropriate software (computer programmes).

Reagents and Solutions

Purity of reagents: all chemicals are of analytical grade. The POD-RAM should be of the best available purity (supplied by *Miles-Yeda Limited*) and is stored at − 20°C. 1,2-Phenylenediamine dihydrochloride should be stored in the dark to prevent deterioration.

Preparation of solutions: all solutions are prepared in re-purified water (cf. Vol. II, chapter 2.1.3.2).

1. Coating buffer (carbonate, 50 mmol/l; NaN$_3$, 0.2 g/l; pH 9.6):

   dissolve 1.59 g Na$_2$CO$_3$, 2.93 g NaHCO$_3$ and 0.2 g NaN$_3$ in 1 l water.

2. PQ-KLH coating solution (PQ-KLH, 1 mg/l; carbonate, 50 mmol/l; NaN$_3$, 0.2 g/l; pH 9.6):

   use the solution of PQ-KLH prepared according to Appendix, p. 462; dilute with coating buffer (1).

3. Phosphate-buffered saline/Tween 20, PBS/T (phosphate, 9.6 mmol/l; NaCl, 137 mmol/l; Tween 20, 0.5 ml/l; pH 7.4):

   dissolve 0.2 g KH$_2$PO$_4$, 2.9 g Na$_2$HPO$_4$ · 12 H$_2$O, 8 g NaCl and 0.5 ml polyoxyethylene sorbitan monolaurate (Tween 20) in 1 l water.

4. Incubation buffer (phosphate, 9.6 mmol/l; NaCl, 137 mmol/l; Tween 20, 0.5 ml/l; gelatin, 1 g/l; Na$_2$SO$_4$, 2 mol/l; pH 7.4):

   dissolve 0.1 g KH$_2$PO$_4$, 1.45 g Na$_2$HPO$_4$ · 12 H$_2$O, 4 g NaCl and 0.25 ml Tween 20 in 250 ml water; heat 50 ml at ca. 60°C, dissolve 0.1 g gelatin, add 50 ml saturated sodium sulphate solution.

5. Antibody solution (mAb, 0.3 mg/l; phosphate, 9.6 mmol/l; NaCl, 137 mmol/l; Tween 20, 0.5 ml/l; pH 7.4):

   use the antibody prepared according to Appendix, p. 463. Dilute the stock solution 1 : 1000 with buffer (3).

6. POD-RAM solution (conjugate, ca. 2.5 – 3 mg/l; IgG, ca. 0.5 – 0.67 mg/l; POD, 0.21 kU/l; phosphate, 9.6 mmol/l; NaCl, 137 mmol/l; Tween 20, 0.5 ml/l; BSA, 5 g/l; pH 7.4):

   dilute the rabbit anti-mouse POD-IgG reagent 1 : 4000 with buffer (3) containing 5 g/l bovine serum albumin.

7. Citrate/phosphate buffer (citrate, 24 mmol/l; phosphate, 47 mmol/l; pH 5.0):

   dissolve 5.0 g citric acid monohydrate and 17 g Na$_2$HPO$_4$ · 12 H$_2$O in 1 l water.

8. Hydrogen peroxide solution (H$_2$O$_2$, 400 mmol/l; citrate, 24 mmol/l; phosphate, 47 mmol/l; pH 5.0):

   dissolve 1 g urea hydrogen peroxide (1 tablet from *BDH*, Poole, U.K.) in 25 ml buffer (7).

9. Substrate solution (1,2-phenylenediamine, 4.4 mmol/l; H$_2$O$_2$, 2 mmol/l; citrate, 24 mmol/l; phosphate, 47 mmol/l; pH 5.0):

   dissolve 80 mg 1,2-phenylenediamine dihydrochloride and 0.5 ml solution (8) in 100 ml buffer (7).

10. Citric acid (0.5 mol/l):

   dissolve 105 g citric acid, monohydrate, in 1 l water.

11. Paraquat standard solutions

   a) Stock solution (10 g/l):
      dissolve 3.453 g pure paraquat dichloride in water and make up to 250 ml in water.

   b) Solution in PBS/gel (100 mg/l):
      dissolve 0.2 g gelatin in 200 ml hot (~ 60 °C) PBS (3). The preparation of PBS/gel is carried out as for buffer (4) but omitting Tween 20. Dilute 1.0 ml stock solution (11a) to 100 ml in PBS/gel.

   c) Solution in PBS/gel (1 mg/l):
      dilute 1.0 ml solution (11b) to 100 ml with PBS/gel.

Stability of reagents and solutions: coating buffer (1), PBS/T (3) and citrate/phosphate buffer (7) are stable for 2 weeks at room temperature. Citric acid solution (10) is stable for approximately 6 months. Incubation buffer (4) and the solutions of PQ-KLH (2), antibody (5) and POD-RAM (6) should be prepared on the day of use. The incubation buffer (4) should only be used if the solution is perfectly clear. Urea hydrogen peroxide solution (8) is stable for 6 months when stored in a refrigerator at 4 °C. The substrate solution (9) must be used within 5 min of preparation as it undergoes very rapid deterioration. The paraquat solution (11a) has very long term stability at room temperature (several years); it should be stored in the dark. Store solutions (11b) and (11c) in a refrigerator at 4 °C and discard when cloudiness develops due to microbial contamination (typically after several weeks).

Procedure

Collection and treatment of specimen: remove a number of soil cores (typically ca. 20) from the area of interest. Bulk together all samples from the same depth range, then air-dry the soil and pass it through a 2 mm sieve to remove stones and bulky organic matter. Thoroughly mix the prepared sample and take out a representative 10 g aliquot. Boil the sample under reflux with 40 ml sulphuric acid, 6.0 mol/l, for 5 h. Filter off the solid material and remove a 5 ml aliquot of the filtrate solution. Stir the aliquot magnetically and cautiously add 3.5 ml sodium hydroxide solution, 12.5 mol/l. Continue to add sodium hydroxide until pH of the extract shows a significant rise, pH > 3. Buffer the solution by addition of 0.5 ml EDTA solution, 50 g/l, then adjust pH to 6.0 to 7.0 with dilute solutions of sodium hydroxide. Dilute the solution to a known volume with water. Samples containing up to 50 mg paraquat per kg should be diluted to 10 ml and those containing in excess of 50 mg/kg should be diluted to 50 ml.

Stability of the analyte in the sample: analysis of the extract should be carried out within 3 days of the neutralization step. If longer periods of storage are required (up to several weeks), the extracts should be left as acidic solutions and stored at 4°C. Paraquat is extremely stable in acid but is degraded in alkali. It is therefore very important that during neutralization of the extract pH is not allowed to rise above 7.0 for any significant period of time.

Assay conditions: the general layout in the microtitre plate is shown in Fig. 1. Measure 4 samples and 1 standard solution (11c) in duplicate. Several standards of varying concentrations are needed if a calibration curve must be established. For the analysis of water or body fluids omit Na_2SO_4 from incubation buffer (cf. p. 454).

Immunoreaction: incubations for 30 min and 1 h at 20°C.

Enzymatic indicator reaction: incubation for exactly 10 min at 20°C; wavelength 450 nm.

Run reagent blanks without sample and antibody (5), and zero standards without sample.

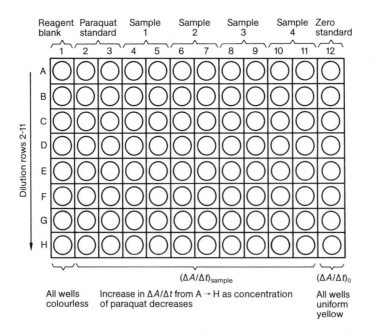

Fig. 2. Layout of microtitre plate for the paraquat ELISA.

Measurement

Prepare 1:2 serial dilutions of samples and standard down plate from row A to row H.

Pipette successively into the wells of a coated plate:			concentration in incubation/ assay mixture	
incubation buffer	(4)	0.1 ml	phosphate NaCl Na$_2$SO$_4$ Tween gelatin	9.6 mmol/l 0.14 mol/l 0.7 mol/l 0.33 ml/l 0.67 g/l
sample or standard (11c)		0.1 ml	paraquat	up to 0.33 mg/l
antibody solution	(5)	0.1 ml	antibody	ca. 0.1 mg/l
incubate for 30 min at 20°C; discard incubation mixture, wash* 3 times with > 0.4 ml PBS/T (3), drain plates to remove solution as completely as possible;				
POD-RAM solution	(6)	0.2 ml	IgG POD phosphate NaCl BSA Tween	0.5 – 0.67 mg/l 0.21 kU/l 9.6 mmol/l 0.14 mol/l 5 g/l 0.5 ml/l
incubate for 1 h at 20°C; discard the solution, wash and drain as above;				
substrate solution	(9)	0.2 ml	1,2-phenylene- diamine H$_2$O$_2$ citrate phosphate	 4.4 mmol/l 2 mmol/l 24 mmol/l 47 mmol/l
incubate for 15 min at 20°C;				
citric acid	(10)	0.5 ml	citrate	0.36 mol/l
reaction is stopped, read absorbance.				

* Preferably using a 12-channel washing device.

If $\Delta A/\Delta t$ of the sample is greater than that of the highest standard dilute the sample with water and re-assay.

Calculation: measuring unit is the POD indicator reaction rate, $\Delta A/\Delta t$. The more paraquat the assay mixture contains, the larger is the term

$$\frac{(\Delta A/\Delta t)_0 - (\Delta A/\Delta t)_{sample}}{(\Delta A/\Delta t)_0} = \Delta A/\Delta A_0$$

where

$(\Delta A/\Delta t)_0$ reaction rate of zero standard, corrected for blank

$(\Delta A/\Delta t)_{sample}$ reaction rate of sample, corrected for blank.

Plot log $\Delta A/\Delta A_0$ *versus* logarithm of the paraquat concentration (mg/l) for calibration curve (cf. Fig. 1, p. 455).

Since this method was designed to be rapid and convenient with the facility for handling large numbers of samples, computer programmes were developed which store the measured data and then calculate the concentration of paraquat in the samples. The output of data from the calculation programme includes a listing of the $\Delta A/\Delta t$ of each well, the concentration of paraquat expressed in mg/l and the concentration of paraquat after correction for the sequential dilution steps. The need for tedious manual calculations is thus eliminated.

Validation of Method

Precision, accuracy, detection limit and sensitivity: the suitability of the ELISA for soil analysis was initially tested by assaying a number of control soil samples fortified with paraquat in the range of 10 – 300 mg/kg after extraction and neutralization. The results confirmed that natural soil components did not interfere with the assay and justified further refinement of the method for soil analysis.

Subsequent studies using the optimum assay conditions, in conjunction with extraction, measured the accuracy and precision of the ELISA. Control soil samples which had been fortified pre-extraction with known concentrations of paraquat, typically 1.0 and 10 mg paraquat per kg of soil, were extracted, neutralized and assayed on consecutive days. A relative standard deviation of 14.5% was obtained at the 1.0 mg/kg level ($n = 13$, mean $= 0.97$, SD $\pm$ 0.14), at the 10 mg/kg level it was 15.7% ($n = 51$, mean $= 9.6$, SD $\pm$ 1.5). At least one untreated control sample and several accurately fortified recovery samples were analyzed alongside each batch of test soils. Where necessary, the figures generated for the test samples were then corrected for the mean recovery value.

The limit of detection of the assay, defined as the lowest concentration which can be conclusively recorded as a positive paraquat residue, is at present 0.02 mg/l which corresponds to 0.2 mg paraquat per kg soil for a typical extraction procedure. This level is satisfactory for virtually all analyses of paraquat in soil and lower limits of detection are rarely required.

Sources of error: a wide variety of different soil types including coarse sands, clays and peats have been analyzed by the ELISA method. Soil composition has no overall effect upon the assay but elevated soil concentrations and natural soil components may lead to anomalous values for the top wells on the plate, i.e. row A. For this reason it is usually advisable not to use the data from row A.

Specificity: the cross-reactivity of the monoclonal antibody with a series of compounds related to paraquat was examined. Cross-reactivity was measured in terms of the concentration of the compound required to produce 50% inhibition of binding of the antibody to the solid phase PQ-KLH conjugate in the ELISA. The results are summarized in Table 1.

Cross-reactivity with diethyl paraquat was very high (>100%) but was much less with monoquat. Diquat cross-reactivity was only 0.007% and this is particularly advantageous. Diquat is used extensively in agriculture, either on its own or as a mixture with paraquat, and the ELISA is therefore of considerable value for the analysis of paraquat in paraquat/diquat mixtures. Traditional absorption spectroscopy is not amenable to the measurement of paraquat in paraquat/diquat mixtures due to significant overlap of the spectra generated by the two compounds. Cross-reactivity with all the other compounds tested was low and in many cases negligible.

Table 1. Cross-reactivity of the monoclonal antibody.

Compound		% Cross-reactivity*
H_3C-N⟨⟩⟨⟩$N-CH_3$	Paraquat	100
H_5C_2-N⟨⟩⟨⟩$N-C_2H_5$	Diethylparaquat	113
H_3C-N⟨⟩⟨⟩N	Monoquat	5.7
⟨⟩$N+$⟨⟩$N-CH_3$ CH_3		0.016
⟨⟩⟨⟩	Diquat	0.007
H_3C-N⟨⟩⟨⟩$N-CH_3$		0.002
⟨⟩⟨⟩		0.002
H_3C-N⟨⟩⟨⟩$N-CH_3$		<0.001

Table 1 (Continued)

Compound	% Cross-reactivity*
(4,4'-bipyridine structure)	< 0.001
(N,N'-dimethyl-2,2'-bipyridinium structure, H₃C CH₃)	< 0.001
(N-methyl-2,2'-bipyridinium structure, CH₃)	< 0.001
(dione bipyridyl structure, O...O)	< 0.001
(2,2'-bipyridine structure)	< 0.001
(pyridine-COOH structure)	< 0.001
H_3C-N^+ *(pyridinium)*—COOH	< 0.001
(pyridinium NH O structure)	< 0.001

* % Cross-reactivity is calculated from the molar concentration of compound required to produce 50% inhibition in the binding of antibody to the solid phase PQ-KLH conjugate relative to paraquat.

Appendix

Preparation of antigens: keyhole limpet haemocyanin, KLH, and bovine serum albumin are coupled to the adduct (II) derived from 6-bromohexanoic acid and monoquat (I) via a carbodiimide reaction as reported previously by *Niewola, Walsh & Davies* [1].

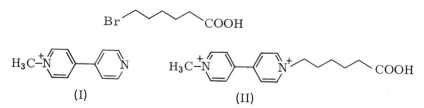

$$Br\diagup\diagdown\diagup\diagdown COOH$$

(I)   (II)

- Suspend 2.58 g monoquat in 4 ml isopropanol and stir with 5.36 g 6-bromohexanoic acid.
- Heat the mixture under reflux for 4 h.

- Cool the solution and filter off the resulting solid.
- Re-crystallize the solid product from isopropanol; yield 3.6 g.
- Dissolve 115 mg compound (II) in 1 ml dimethyl sulphoxide.
- Stir with 85 mg dicyclohexylcarbodiimide and 45 mg N-hydroxysuccinimide for 10 min.
- Centrifuge, then mix the supernatant with an equal volume of water and centrifuge for a second time.
- Add the clear supernatant to 2.5 ml KLH solution, 22 mg/ml, previously made basic by the addition of 0.02 ml sodium hydroxide solution, 5 mol/l.
- Maintain pH at 8 – 9 by further addition of small quantities of sodium hydroxide solution, 5 mol/l.
- Stir the solution for 4 h.
- Separate the protein conjugate from free paraquat derivative by chromatography on a Sephadex G-25 column.

The presence of paraquat in the high-molecular weight fractions is judged by the colorimetric dithionite assay for paraquat:

- add 0.3 ml conjugate to a dithionite solution, 1 g/l, in sodium hydroxide solution, 1 mol/l;
- prepare a series of standard solutions of compound (II) ranging from 0.005 to 0.25 g/l;
- estimate the amount of paraquat bound to the protein by comparison with these standards.

The fractions containing the paraquat derivative are combined and the protein concentration established by a standard *Lowry* assay [10] (cf. also Vol. II, chapter 1.3.4).

The preparation of PQ-BSA conjugate follows the same basic procedure as the PQ-KLH method already described.

Coating of the microtitre plates: pipette into each well 0.2 ml PQ-KLH coating solution (2), incubate for 2 h at 20 °C; discard solution (2), wash plate (preferably using a 12-channel washing device) three times with >0.4 ml PBS/T (3), 3 min per wash; drain plates to remove solution as completely as possible.

Preparation of monoclonal antibody to paraquat-BSA: the procedure described by *Niewola et al.* [2] is used. Spleen cells from Balb/c mice which have been immunized with PQ-BSA are fused with the NS.1 myeloma cell line using a protocol based on the method of *Köhler & Milstein* [11]. Antibody secreting clones are detected using an immunoradiometric assay and paraquat-specific hybrids are cloned three times by limiting dilution. Large quantities of monoclonal antibody are produced by injecting 5×10^6 hybrid cells intraperitoneally into mice which have been primed with pristane (2,6,10,14-tetramethylpentadecane) 10 days earlier. After 10 – 14 days the ascites are collected, separated from cells and stored at – 20 °C until required.

References

[1] *Z. Niewola, S. T. Walsh, G. E. Davies,* Enzyme Linked Immunosorbent Assay (ELISA) for Paraquat, Int. J. Immunopharmacol. *5*, 211 – 218 (1983).

[2] *Z. Niewola, C. Hayward, B. A. Symington, R. T. Robson,* Quantitative Estimation of Paraquat by an Enzyme Linked Immunosorbent Assay Using a Monoclonal Antibody, Clin. Chim. Acta *148*, 149 – 156 (1985).

[3] *A. Calderbank, S. H. Yuen,* An Ion-exchange Method for Determining Paraquat Residues in Food Crops, The Analyst *90*, 99 – 106 (1965).

[4] *S. Kawase, S. Kanno, S. Ukai,* Determination of the Herbicides Paraquat and Diquat in Blood and Urine by Gas Chromatography, J. Chromatogr. *283*, 231 – 240 (1984).

[5] *R. Gill, S. C. Qua, A. C. Moffatt,* High-Performance Liquid Chromatography of Paraquat and Diquat in Urine with Rapid Sample Preparation Involving Ion-Pair Extraction on Disposable Cartridges of Octadecyl Silica, J. Chromatogr. *255*, 483 – 490 (1983).

[6] *T. Levitt,* Radioimmunoassay for Paraquat, Lancet *1*, 358 (1977).

[7] *M. J. O'Sullivan, J. W. Bridges, V. Marks,* Enzyme Immunoassay: A Review, Ann. Clin. Biochem. *16*, 221 – 239 (1979).

[8] *E. Engvall,* Enzyme Immunoassay ELISA and EMIT, in: *H. Van Vunakis, J. J. Langone* (eds.), Methods in Enzymology, Vol. *70*, Academic Press, New York, London 1980, pp. 419 – 439.

[9] *A. J. O'Beirne, H. R. Cooper,* Heterogeneous Enzyme Immunoassay, J. Histochem. Cytochem. *27*, 1148 – 1162 (1979).

[10] *O. H. Lowry, N. J. Rosebrough, A. L. Farr, R. J. Randall,* Protein Measurement with the *Folin* Phenol Reagent, J. Biol. Chem. *193*, 265 – 275 (1951).

[11] *G. Köhler, C. Milstein,* Continuous Cultures of Fused Cells Secreting Antibody of Predefined Specificity, Nature *256*, 495 – 497 (1975).

Appendix

1 Symbols, Quantities, Units and Constants

Units and symbols in ⟨ ⟩ should not be used.

		Prefixes for unit symbols		
m	Metre, m			
g	Gram, g	Prefix	Symbol	Factor
l	Litre, 10^{-3} m³			
		exa	E	10^{18}
h	Hour	peta	P	10^{15}
min	Minute	tera	T	10^{12}
s	Second	giga	G	10^9
		mega	M	10^6
t	Time, s, (h, min)	kilo	k	10^3
Δt	Interval between			
	measurements, s (h, min)	⟨ hecto	h	10^2 ⟩
		⟨ deca	da	10^1 ⟩
Pa	Pascal, N/m²	⟨ deci	d	10^{-1}⟩
		⟨ centi	c	10^{-2}⟩
t	Temperature, °C	milli	m	10^{-3}
T	Temperature, K	micro	μ	10^{-6}
K	Kelvin	nano	n	10^{-9}
		pico	p	10^{-12}
		femto	f	10^{-15}
		atto	a	10^{-18}

V	Volume (usually volume of assay mixture), ml, l
v	Volume (usually volume of sample in assay mixture), ml, l
φ	Volume fraction of sample in assay mixture
m_s	Mass of substance, g
n_c	Amount of substance, mol
c	Substance concentration, mol/l
ρ	Mass concentration, g/l
n_c/m_s	Substance content, mol/kg
w	Mass fraction
⟨ ppm	Parts per million⟩
⟨ ppb	Parts per billion⟩
%	Percentage
% (v/v)	Percentage, volume related to volume
% (v/w)	Percentage, volume related to weight
% (w/v)	Percentage, weight related to volume
% (w/w)	Percentage, weight related to weight

M_r Molecular weight, relative molecular mass

v Rate of reaction, $mol \times l^{-1} \times s^{-1}$ ($\mu mol \times ml^{-1} \times min^{-1}$)
V Maximum rate of reaction, $mol \times l^{-1} \times s^{-1}$
$\dot{\xi}$ Rate of conversion, $mol \times s^{-1}$ ($\mu mol \times min^{-1}$)
v_i Stoichiometric coefficient

kat   Katal, $mol \times s^{-1}$
U*   International unit (for enzymes), $\mu mol \times min^{-1}$
EU   Endotoxin unit
Inh.U   Inhibitor unit, $\mu mol \times min^{-1}$
I.U.**   International unit (for others than enzymes)
KU   Biological kallikrein unit
KIU   Biological kallikrein inhibitor unit

z Catalytic activity, U, kat
b Catalytic activity concentration, U/l, kat/l
z_c/m_s Catalytic activity content (specific catalytic activity), U/g, kat/kg

ε Linear molar absorption coefficient, $l \times mol^{-1} \times mm^{-1}$
d Light path, mm
A Absorbance
F Fluorescence intensity
I Light intensity
T   Transmission

Bq   Becquerel, s^{-1}
$\langle$ Ci   Curie, $s^{-1} \rangle$
cpm   Counts per minute, min^{-1}
dpm   Disintegrations per min, min^{-1}
C Counting rate of the radioactive product, cpm
η Counting efficiency
X Specific radioactivity, $Bq \times mol^{-1}$ $\langle Ci \times mol^{-1} \rangle$
Y Number of counts measured
Z Decay rate, disintegrations per minute, dpm

k Reaction constant
K Equilibrium constant
K' Apparent equilibrium constant
K_m Michaelis constant, mol/l
K_I Inhibitor constant, mol/l
pH   Negative logarithm of the hydrogen ion concentration
pK   Negative logarithm of the dissociation constant, $-\log K$

---

* There is no abbreviation used for arbitrary unit, it is expressed as "unit" in this series.
** In general defined by WHO or other authorities for antigens, cf. Vol. IX, chapter 1.1.

$[\alpha]_D^{20}$	Specific rotation (sodium D-line at 20°C)
sp.gr.	Specific gravity at 20°C relative to water at 4°C
g	Acceleration due to gravity; 9.81 m × s^{-2}
rpm	Revolutions per minute, min^{-1}

J	Joule, m^2 × kg × s^{-2}
R	Gas constant, 8.312 J × mol^{-1} × K^{-1}

h	*Planck's* constant
v	Frequence of emitted light

$\bar{x}$	Mean value
s, SD	Standard deviation
RSD	Relative standard deviation
⟨ CV	Coefficient of variation⟩

2   Abbreviations for Chemical and Biochemical Compounds

It is unavoidable with the numerous abbreviations in use that one abbreviation occasionally is used for different compounds. In such cases the correct meaning can be obtained from the text. Only the unequivocal abbreviations are used in the book without further explanation.

A	Adenosine
4-AAP	4-Aminoantipyrine
Ab	Antibody
mAb	monoclonal antibody
pAb	polyclonal antibody
ABTS® *	2,2'-Azinobis-[3-ethyl-2,3-dihydrobenzothiazole-6-sulphonate]
Acetoacetyl-CoA	Acetoacetyl coenzyme A
Acetyl-CoA	Acetyl coenzyme A
ACE	Angiotensin converting enzyme
AChE	Acetylcholinesterase
AcP	Acid phosphatase
ACS	Acetyl-CoA synthetase
ACTH	Adrenocorticotropic hormone
ADA	Adenosine deaminase
ADH	Alcohol dehydrogenase
ADP	Adenosine 5'-diphosphate
ADPglucose	Adenosine 5'-diphosphoglucose
AFP	α_1-Foetoprotein
Ag	Antigen
AHB	5-Amino-2-hydroxybenzoic acid
ALAT (ALT, AlaAT)	Alanine aminotransferase
Alcohol-OD	Alcohol oxidase
ALD	Aldolase
Al-D	δ-Amino laevulinate dehydratase
AK	Acetate kinase
AK	Adenylate kinase
Ammediol	2-Amino-2-methylpropane-1,3-diol
AMP	Adenosine 5'-monophosphate
A-2-MP	Adenosine 2'-monophosphate
A-3-MP	Adenosine 3'-monophosphate
A-3(2)-MP	Adenosine 3'(2')-monophosphate
A-3:5-MP (cAMP)	Adenosine 3':5'-monophosphate, cyclic
AOD	Amino-acid oxidase
AP	Alkaline phosphatase

---

* Also in use: 2,2'-azino-di-(3-ethylbenzthiazoline-6-sulphonate).

α_2-AP	α_2-Antiplasmin, α_2-plasmin inhibitor
Ap$_5$A	Diadenosine 5'-pentaphosphate
APAD	Acetylpyridine-adenine dinucleotide
APADH	Acetylpyridine-adenine dinucleotide, reduced
APTT	Activated partial thromboplastin time
ASAT (AST; AspAT)	Aspartate aminotransferase
ASC	ascorbic acid
Ascorbate-OD	Ascorbate oxidase
AT	Antithrombin
ATP	Adenosine 5'-triphosphate
ATPase	Adenosine 5'-triphosphatase
BAEE	Benzoyl-L-arginine ethyl ester
BAPNA	N-Benzoyl-arginine-4-nitroanilide
Benzoyl-CoA	Benzoyl-coenzyme A
BES	Bis(2-hydroxyethylamino)ethanesulphonic acid
Bicine	N,N-Bis(2-hydroxyethyl)glycine
BME	β-Mercaptoethanol
BMTD	6-Benzamido-4-methoxy-3-toluidinediazonium chloride
BOC	t-Butyloxycarbonyl
BSA	Bovine serum albumin
Bz	Benzoyl
C	Cytidine
Cbz	Carbobenzoxy (benzyloxycarbonyl)
CCPN	N-3-(Carboxylpropionyl)-L-phenyl-alanino-4-nitroanilide
CDP	Cytidine 5'-diphosphate
CDPglucose	Cytidine 5'-diphosphoglucose
CDPG-P	CDPglucose pyrophosphorylase
CE	Cholesterol esterase
CEA	Carcinoembryonic antigen
Cellosolve	Ethyleneglycol monomethyl ether
CF	Complement fixation
hCG	Human chorionic gonadotrophin
CHA	Cyclohexylammonium
ChE	Cholinesterase
ChOD	Cholesterol oxidase
CK	Creatine kinase
CL	Citrate lyase
pCMB(4-CMB)	4-Chloromercuribenzoate
CMP	Cytidine 5'-monophosphate
C-2-MP	Cytidine 2'-monophosphate
C-2:3-MP	Cytidine 2':3'-monophosphate, cyclic
C-3-MP	Cytidine 3'-monophosphate
C-3(2)-MP	Cytidine 3'(2')-monophosphate

CNBr	Cyanogen bromide
CoA	Coenzyme A
COMT	Catechol-*O*-methyltransferase
COX	Cytochrome c oxidase
CPS	Capsular polysaccharide
CS	Citrate synthase
CT	Calcitonin
CTP	Cytidine 5'-triphosphate
Cyt-c	Cytochrome c
d	deoxy (prefix)
dansyl(Dns)	5-Dimethylaminonaphthalene-1-sulphonyl
DAP	Dihydroxyacetone phosphate
DEA	Diethanolamine
DEAE	Diethylaminoethyl
DFP	Diisopropyl fluorophosphate
DHA	Dehydroascorbic acid
DHF	Dihydrofolate
DHFR	Dihydrofolate reductase
DIG-ELISA	Diffusion-in-gel ELISA
DMAB	Dimethylaminobenzaldehyde
DM-POPOP	2,2'-(1,4-Phenylene)bis-(4-methyl-5-phenyloxazole)
DMSO	Dimethylsulphoxide
DNase	Deoxyribonuclease
DNP	Dinitrophenylhydrazine
DTE	Dithioerythritol
DTNB	5,5'-Dithiobis-(2-nitrobenzoic acid)
DTT	Dithiothreitol
EDTA	Ethylenediamine tetraacetate
EGTA	Ethyleneglycol-bis-(β-aminoethyl ether) *N,N'*-tetraacetic acid
EHT	*N*-Ethyl-*N*-2-hydroxymethyl-3-toluidine
EIA	Enzyme-immunoassay
ELA	Enzyme-labelled antigen
ELFA	Enzyme-linked fluorescence assay
ELISA	Enzyme-linked immunosorbent assay
ELISPOT	Enzyme-linked immunospot
ETF	Electron-transferring flavoprotein
FA	Furanacryloyl
Fab	Antigen-binding fragment of antibody
FAD	Flavin-adenine dinucleotide
FAS	Fatty acid synthetase
FBS	Foetal bovine serum
Fc	Cristallizable fragment of antibody
FCS	Foetal calf serum

FDH	Formate dehydrogenase
FDH	Formaldehyde dehydrogenase
FDH	Fucose dehydrogenase
FDNB	1-Fluoro-2,4-dinitrobenzene
FH_2	Dihydrofolate
FH_4	Tetrahydrofolate
FiGlu	N-Formimino-L-glutamate
FITC	Fluorescein iso-thiocyanate
FMN	Flavin mononucleotide
FNR	Formate-nitrate reductase
F-1-P	Fructose 1-phosphate
$F-1,6-P_2$	Fructose 1,6-bisphosphate
$F-2,6-P_2$	Fructose 2,6-bisphosphate
F-6-P	Fructose 6-phosphate
F-6-PK	Fructose-6-phosphate kinase
FUDPgal	5-Fluoro-UDPgalactose
FUDglc	5-Fluoro-UDPglucose
FS	Formyltetrahydrofolate synthetase
FSH	Follicle-stimulating hormone
G	Guanosine
GABAse	Mixture of GAB-AT and SS-DH
GAB-AT	γ-Aminobutyrate aminotransferase
Gal-DH	Galactose dehydrogenase
Gal-OD	Galactose oxydase
GAP	Glyceraldehyde phosphate
GAPDH	D-Glyceraldehyde-3-phosphate dehydrogenase
GDH	L-Glycerol-3-phosphate dehydrogenase/glycerol-1-phosphate dehydrogenase/α-glycerophosphate dehydrogenase
GDP	Guanosine 5′-diphosphate
hGH	Human growth hormone
Glc-DH	Glucose dehydrogenase
Glc-1-P	Glucose 1-phosphate
GlcUAI	D-Glucuronate isomerase
Gl-I	Glyoxylase I
GlDH	L-Glutamate dehydrogenase
Gly-DH	Glycerate dehydrogenase
Gly-R	Glyoxylate reductase
GMP	Guanosine 5′-monophosphate
G-2-MP	Guanosine 2′-monophosphate
G-3-MP	Guanosine 3′-monophosphate
G-3(2)-MP	Guanosine 3′(2′)-monophosphate
G-3:5-MP (cGMP)	Guanosine 3:5-monophosphate, cyclic
GMPK	GMP kinase
GOD	Glucose oxidase

GOT	Glutamate-oxaloacetate transaminase/Aspartate aminotransferase
G-1-P	Glucose 1-phosphate
G-1,6-P$_2$	Glucose 1,6-bisphosphate
G-6-P	Glucose 6-phosphate
G6P-DH	Glucose-6-phosphate dehydrogenase
GPT	Glutamate-pyruvate transaminase/Alanine aminotransferase
GR	Glutathione reductase
GSH	Glutathione, reduced
GSSG	Glutathione, oxidized
γ-GT	γ-Glutamyltransferase
GTP	Guanosine 5'-triphosphate
h	human
Hb	Haemoglobin
HBsAg	Hepatitis B surface antigen
HBV	Hepatitis B virus
HDL	High-density lipoprotein
Hepes	N-2-Hydroxyethyl piperazine-N'-2-ethanesulphonic acid
HK	Hexokinase
KHL	Keyhole limped
HMG-CoA	3-Hydroxy-3-methylglutaryl coenzyme A
H-Y	Sex-associated antigen
I	Inosine
ICDH	Isocitrate dehydrogenase
IDP	Inosine 5'-diphosphate
IFAT	Indirect fluorescent antibody test
IFN	Interferon
Ig	Immunoglobulin
s-Ig	Secretory immunoglobulin
IMP	Inosine 5'-monophosphate
INT	2-(4-Iodophenyl)-3-(4-nitrophenyl)-5-phenyltetrazolium-chloride
ITP	Inosine 5'-triphosphate
KLH	Key-hole limpet haemocyanin
hLA	Human leukocyte antigen
LAL	Limilus amoebocyte lysate
LDH (L-LDH)	L-Lactate dehydrogenase
D-LDH	D-Lactate dehydrogenase
LDL	Low-density lipoprotein
LEAP	Linked enzyme assay procedure
LF	Lactoferrin
LH	Luteotropic hormone
LPS	Lipopolysaccharide

α_1-m	α_1-Microglobulin
α_2-M	α_2-Macroglobulin
β_2-m	β_2-Microglobulin
MCA	7-Amino-4-methylcoumarin
MDH	L-Malate dehydrogenase
ME	Mercaptoethanol
MES	2-(N-Morpholino)ethanesulphonic acid
4-Met-um	4-Methylumbelliferon/7-hydroxy-4-methylcoumarin (4-methylumbelliferyl)
Met-um-gal	4-Methylumbelliferyl β-D-galactoside
MK	Myokinase/adenylate kinase
Mops	3-(N-Morpholino)propanesulphonic acid
MPDH	Mannitol-1-phosphate dehydrogenase
4-NA	4-Nitroaniline
NAC	N-Acetylcysteine
NAD	Nicotinamide-adenine dinucleotide
NADH	Nicotinamide-adenine dinucleotide, reduced
NADP	Nicotinamide-adenine dinucleotide phosphate
NADPH	Nicotinamide-adenine dinucleotide phosphate, reduced
NANA	N-Acetylneuraminic acid
NBT	Nitro-BT-tetrazolium salt; 2,2'-bis(4-nitrophenyl)-5,5'-diphenyl-3,3'-(-dimethoxy-4,4'-diphenylene)-ditetrazolium chloride
NBTH	N-Methyl-2-benzothiazolone hydrazone
NDPK	Nucleoside diphosphate kinase
NMN	Nicotinamide mononucleotide
NMPK	Nucleoside monophosphate kinase
NP	Nucleoside phosphorylase
4-NP	4-Nitrophenol (4-Nitrophenyl)
4-NP-G	4-Nitrophenyl glucuronide
4-NP-G_1	4-Nitrophenyl α-D-glucopyranoside
4-NP-G_2	4-Nitrophenyl [α-D-glucopyranosyl-$(1 \to 4)$-α-D-glucopyranoside]/4-Nitrophenyl α-D-maltoside
4-NP-G_3	4-Nitrophenyl {di[α-D-glucopyranosyl-$(1 \to 4)$]-α-D-glucopyranoside}/4-Nitrophenyl α-D-maltotrioside
4-NP-G_4	4-Nitrophenyl {tri[α-D-glucopyranosyl-$(1 \to 4)$]-α-D-glucopyranoside}/4-Nitrophenyl α-D-maltotetraoside
4-NP-G_5	4-Nitrophenyl {tetra[α-D-glucopyranosyl-$(1 \to 4)$]-α-D-glucopyranoside}/4-Nitrophenyl α-D-maltopentaoside
4-NP-G_6	4-Nitrophenyl {penta[α-D-glucopyranosyl-$(1 \to 4)$]-α-D-glucopyranoside}/4-Nitrophenyl α-D-maltohexaoside
4-NP-G_7	4-Nitrophenyl {hexa[α-D-glucopyranosyl-$(1 \to 4)$]-α-D-glucopyranoside}/4-Nitrophenyl α-D-maltoheptaoside
2-NPgal	2-Nitrophenyl galactoside

4-NPgal	4-Nitrophenyl galactoside
4-NPglc	4-Nitrophenyl glucoside
4-NPman	4-Nitrophenyl mannoside
4-NPP	4-Nitrophenyl phosphate
NTA	Nitriloacetate
OMP	Orotidine 5′-monophosphate
OxOD	Oxalate oxidase
OVA	Ovalbumin
PAG	Polyacrylamide gel
PAGE	Polyacrylamide gel electrophoresis
PALP/PLP	Pyridoxal 5′-phosphate
PAMP	Pyridoxamine 5′-phosphate
PAPP-A	Pregnancy-associated plasma protein A
PAT	Phosphate acetyltransferase (PTA)
PBS	Phosphate-buffered saline
PCA	Perchloric acid
PDC	Pyruvate decarboxylase
PDE	Phosphodiesterase
PEG	Polyethylene glycol
PEP	Phosphoenolpyruvate
PG	Prostaglandin
6-PGDH	6-Phosphogluconate dehydrogenase
PGK	3-Phosphoglycerate kinase
PGluM	Phosphoglucomutase
PGM	Phosphoglycerate mutase
P_i	Inorganic phosphate
Pipes	Piperazine-N,N′-bis(2-ethanesulphonic acid)
PK	Pyruvate kinase
PL-A	Phospholipase A
PL-C	Phospholipase C
PL-D	Phospholipase D
PMS	Phenazine methosulphate
PMSF	Phenylmethanesulphonyl fluoride
POD	Peroxidase
POPOP	2,2′-(1,4-Phenylene)bis[5-phenyloxazole]
POX	Pyruvate oxidase
PPase	Pyrophosphatase, inorganic
PP_i	Inorganic pyrophosphate
PP_i-PFK	Pyrophosphate: fructose-1,6-phosphotransferase
PPO	2,5-Diphenyloxazole
PRA	Plasma renin activity
PRI	Phosphoriboisomerase
PRK	Phosphoribulosekinase

PRPP	5-Phospho-α-D-ribose 1-diphosphate
PT	Prothrombin time
PVP	Polyvinylpyrrolidone
PyDC	Pyruvate decarboxylase
QPRTase	Quinolate phosphoribosyltransferase
RBP	Retinol-binding protein
RDH	Ribitol dehydrogenase
RF	Rheumatoid factor
RNA	Ribonucleic acid
rRNA	Ribosomal ribonucleic acid
tRNA	Transfer ribonucleic acid
RNase	Ribonuclease
R-5-P	Ribose 5-phosphate
Ru-5-P	Ribulose 5-phosphate
Ru-1,5-P_2	Ribulose 1,5-bisphosphate
RuP_2CO	Ribulose-1,5-bisphosphate carboxylase
RVV	*Russel's* viper venom
RZ	Reinheitszahl (purity number) for peroxidase
SAH	*S*-Adenosylhomocysteine
SAM	*S*-Adenosylmethionine
SDH	Sorbitol dehydrogenase
SDS	Sodium dodecyl sulphate
SOD	Superoxide dismutase
SPDP	3-(2-Pyridyldithio)propionic acid *N*-hydroxysuccinimide ester
SP1	Pregnancy-specific β_1-glycoprotein
SS-DH	Succinate-semialdehyde dehydrogenase
ST	Sialyltransferase
SBTI	Soybean trypsin inhibitor
SUPHEPA	*N*-Succinyl-L-phenylalanine-4-nitroanilide
T	Ribosylthymidine
T_3	L-Triiodothyronine
T_4	L-Tetraiodothyronine, L-thyroxine
dT	Thymidine
TA	Transaldolase
TagUAR	D-Tagaturonate reductase
TAT (TyrAT)	Tyrosine aminotransferase
TBG	Thyroxone-binding globulin
TBP	Thyroxine-binding proteins
TBPA	Thyroid-binding pre-albumin
TDC	Tyrosine decarboxylase
TEMED	*N,N,N,N*-Tetramethylethylenediamine

dTDP	Thymidine 5′-diphosphate
dTDPglucose	Thymidine 5′-diphosphoglucose
TCA	Trichloroacetic acid
TEA	Triethanolamine
Tg	Thyroglobulin
THF	Tetrahydrofolic acid
TIM	Triosephosphate isomerase
TK	Transketolase
TPP	Thiamine pyrophosphate
Tricine	N,N,N-Tris(hydroxymethyl)methylglycine
Tris	Tris(hydroxymethyl)aminomethane
TSH	Thyroid-stimulating hormone
dTTP	Thymidine 5′-triphosphate
TX	Thromboxane
U	Uridine
UDP	Uridine 5′-diphosphate
UDPG (UDPglucose)	Uridine 5′-diphosphoglucose
UDPAG	Uridine 5′-diphospho-N-acetylglucosamine
UDPGA	Uridine 5′-diphosphoglucuronate
UDPgal	Uridine 5′-diphosphogalactose
UDPG-DH	UDPglucose dehydrogenase
UDPG-P(P)	UDPglucose pyrophosphorylase
UMP	Uridine 5′-monophosphate
U-2-MP	Uridine 2′-monophosphate
U-2:3-MP	Uridine 2′:3′-monophosphate, cyclic
U-3-MP	Uridine 3′-monophosphate
U-3(2)-MP	Uridine 3′(2′)-monophosphate
UT	Uridyl transferase
UTP	Uridine 5′-triphosphate
X	Xanthosine
XOD	Xanthine oxidase
Xu-5-P	Xylulose 5-phosphate

3 Formulae

Experimental data must be converted to concentrations or contents; only then one can speak of a result (cf. Vol. II, chapter 3.1). According to international recommendations [1, 2] the following symbols and units are equivalent:

Metabolites		*Enzymes*	
Substance concentration c	mol/l mmol/l	Catalytic activity concentration b	kat/l; kU/l mkat/l; U/l
Mass concentration ρ	g/l mg/l		μkat/l nkat/l
Substance content n_c/m_s	mol/kg mmol/g	Catalytic activity content z_c/m_s	kat/kg (U/g, U/mg)
Mass fraction w	1 0.001		mkat/g μkat/g

The use of the basic kind of quantities of the SI (cf. Vol. I, pp. 10, 22) involves some changes in the symbols, quantities and units formerly used. For example, the base unit of length is the metre (m) and thousands or thousandths of it. The path length of a photometer cuvette is in general 10 mm (not 1 cm). Accordingly, the unit of the absorption coefficient also changes ($1 \times \text{mol}^{-1} \times \text{mm}^{-1}$, instead of $1 \times \text{mol}^{-1} \times \text{cm}^{-1}$).

For example, for NADH at 339 nm $\varepsilon = 6.3 \times 10^2\, 1 \times \text{mol}^{-1} \times \text{mm}^{-1}$ instead of $6.3 \times 10^3\, 1 \times \text{mol}^{-1} \times \text{cm}^{-1}$. For practical purposes it is proposed to use the path length $d = 10$ mm and the concentration c in mmol/l (μmol/ml). Then for NADH at 339 nm $\varepsilon \times d = 6.3\, 1 \times \text{mmol}^{-1}$ applies.

The following symbols, kind of quantities and units are customary in this context. Throughout this book we use v, V for reaction rates and v, V for volumes (deviating from international recommendations). We also write T for transmission of light and T for thermodynamic temperature.

Symbols in radiometry are still not completely coherent with those in other laboratory sciences, also for various commonly used terms apparently no official symbols exist (cf. [3]).

General

c	substance concentration, mol/l
ρ	mass concentration, g/l
n_c/m_s	substance content, mol/kg
w	mass fraction, 1

MW   mass of one millimole or mole of substrate, mg/mmol, g/mol
V   assay volume, l
v   volume of sample used in assay, l
φ volume fraction of sample in assay (incubation) mixture, v/V
t   time, s (min, h)
z   catalytic activity, kat, U
b   catalytic activity concentration, kat/l, U/l
z_c/m_s catalytic activity content (specific catalytic activity), kat/kg, U/g
v_i stoichiometric coefficient

Photometry, fluorimetry, luminometry

A   absorbance
ε linear millimolar absorption coefficient, $l \times mmol^{-1} \times mm^{-1}$
I   light intensity
F   fluorescence intensity
d   light path, mm

For experimental data obtained by fluorimetric methods, F and ΔF are used instead of A and ΔA. In luminometry I and ΔI are used.

Radiometry

Y   number of counts measured
Z   decay rate, disintegrations per minute, dpm
C_b background counting rate, cpm
C_p sample (probe) counting rate, cpm
C_n net counting rate, cpm
η counting efficiency, cpm/dpm
X   specific radioactivity, Bq/mol (Ci/mol)
v_r volume taken for scintillation counting, l
cpm   counts per minute
dpm   disintegrations per minute
2.22×10^{12} factor for conversion of decay rate Ci to dpm ($1\ Ci = 2.22 \times 10^{12}$ dpm)
3.7×10^{10} factor for conversion of Ci to Bq ($1\ Ci = 3.7 \times 10^{10}$ Bq)
2.7×10^{-11} factor for conversion of Bq to Ci ($1\ Bq = 2.7 \times 10^{-11}$ Ci)

Metabolites

Photometry, fluorimetry, luminometry

From *Lambert-Beer*'s law (cf. Vol. I, p. 283) follows

(a) $\qquad c = \dfrac{\log I_0/I}{\varepsilon \times d} = \dfrac{A}{\varepsilon \times d}$ mmol/l.

For chemical reactions this gives

$$c_1 - c_2 = \frac{A_1 - A_2}{\varepsilon \times d} \; ; \qquad \Delta c = \frac{\Delta A}{\varepsilon \times d} \quad \text{mmol/l}$$

and for complete conversion $(c_2 = 0)$

(a$_1$) $c = \dfrac{\Delta A}{\varepsilon \times d}$ mmol/l   (in the cuvette) .

For the determination of the concentration of the analyte in the sample the ratio of assay volume : sample volume (V : v) or the volume fraction of sample in the assay (v/V) = φ), respectively, must be considered:

(b) $c = \dfrac{\Delta A \times V}{\varepsilon \times d \times v} = \dfrac{\Delta A}{\varepsilon \times d \times \varphi}$ mmol/l   (in the sample)

(b$_1$) $\rho = \dfrac{\Delta A \times V \times MW}{\varepsilon \times d \times v} = \dfrac{\Delta A \times MW}{\varepsilon \times d \times \varphi}$ mg/l   (in the sample) .

The substance content of the sample n_c/m_s (i.e. of the analyte in the material under investigation) is

(c) $n_c/m_s = \dfrac{\Delta A \times V}{\varepsilon \times d \times v \times \rho_{\text{sample}}} = \dfrac{\Delta A}{\varepsilon \times d \times \varphi \times \rho_{\text{sample}}}$ mmol/g .

The mass fraction of the analyte in the sample (g/g) is

(c$_1$) $w = \dfrac{\Delta A \times V \times MW}{\varepsilon \times d \times v \times \rho_{\text{sample}}} = \dfrac{\Delta A \times MW}{\varepsilon \times d \times \varphi \times \rho_{\text{sample}}} .$

If two or more moles of light-absorbing reaction products (e.g. NADH) are formed or consumed per mole substrate that reacts, the corresponding stoichiometric coefficient v_i (cf. Vol. I, p. 11) also appears in the denominator of eqns. (b) and (c). If several moles of one substrate go to form the unit of substance on which the measurement is based, the corresponding factor appears in the numerator.

The result for the sample in relation to a standard is

(d) $c = \dfrac{c_{\text{sample (measured)}}}{c_{\text{standard (measured)}}} \times c_{\text{standard (weighed out)}}$ mmol/l

or

(d$_1$) $c = \dfrac{A_{\text{sample (standard)}}}{A_{\text{standard (measured)}}} \times c_{\text{standard (weighed out)}}$ mmol/l .

From the concentration of the substance in the sample solution (e.g. tissue extract), the content of the substance in the material under investigation is calculated by relating to its mass concentration in the standard solution:

(e) $\qquad n_c/m_s = \dfrac{c_{\text{sample (measured)}}}{\rho_{\text{standard (weighed out)}}} \dfrac{\text{mmol/l}}{\text{mg/l}} = \text{mmol/mg}$,

(f) $\qquad w = \dfrac{\rho_{\text{sample (measured)}}}{\rho_{\text{standard (weighed out)}}} \dfrac{\text{mg/l}}{\text{mg/l}} = 1$.

Example

Determination of fructose 1,6-bisphosphate (molecular weight: 340) in rat liver. Enzymatic analysis with aldolase/triosephosphate isomerase/glycerophosphate dehydrogenase. Two moles NADH are oxidized per mole fructose-1,6-P_2 ($\nu_i = 2$). To prepare an "extract", 1 g fresh liver was homogenized in 7.25 ml $HClO_4$. With a value of 75% (w/w) for the fluid content of the liver, the volume of the extract is $7.25 + 0.75 = 8.00$ ml. To neutralize and remove perchlorate, 0.2 ml K_2CO_3 solution was added to 6 ml extract. The volume of the perchlorate-free extract is thus 6.2 ml. The dilution factor for the extract is $6.2/6.0 = 1.033$, and that for the tissue is $8 \times 6.2/6.0 = 8.267$. The experimental data must be multiplied by these values to express the results per 1 ml acid extract or per 1 g tissue.

The measured change in absorbance at 339 nm ($\varepsilon = 0.63 \, 1 \times \text{mmol}^{-1} \times \text{mm}^{-1}$) was $\Delta A = 0.120$; the volume of the assay solution was 3×10^{-3} l, the volume of sample was 1.5×10^{-3} l, and the light path was 10 mm. The concentration in the perchlorate-free sample used for the assay was:

according to eqn. (b)

$$c = \frac{0.120 \times 3 \times 10^{-3}}{0.63 \times 10 \times 1.5 \times 10^{-3} \times 2} = 0.0190 \, \text{mmol/l}$$

or according to eqn. (b$_1$)

$$\rho = \frac{0.120 \times 3 \times 10^{-3} \times 340}{0.63 \times 10 \times 1.5 \times 10^{-3} \times 2} = 6.48 \, \text{mg/l}.$$

Multiplication by the dilution factor gives the concentration in the acid extract:

$$c = 0.0190 \times 1.033 = 0.0196 \, \text{mmol/l}$$

$$\rho = 6.46 \times 1.033 = 6.69 \, \text{mg/l}.$$

The substance content of fructose-1,6-P_2 in the tissue is

$$n_c/m_s = 0.0190 \times 8.267 = 0.157 \text{ mmol/kg}$$

or its mass fraction (g/g)

$$w = 6.46 \times 10^{-6} \times 8.267 = 5.4 \times 10^{-5}.$$

Radiometry

For calculation of the analyte concentration and content from the net counting rate C_n of sample and blank, the following relationships are valid.

Substance concentration

(g) $$c = \frac{\Delta C_n}{\eta \times 2.22 \times 10^{12} \times X \times \varphi \times v_r} \quad \text{mol/l (in the sample)}$$

(units: C_n in cpm; X in Ci/mol or Bq/mol; $\varphi = v/V$ in l/l; v_r in l).

Mass concentration

(g_1) $$\rho = \frac{\Delta C_n \times MW}{\eta \times 2.22 \times 10^{12} \times X \times \varphi \times v_r} \quad \text{g/l (in the sample)}$$

(units as for eqn. (g); MW in g/mol).

Substance content

(h) $$n_c/m_s = \frac{\Delta C_n}{\eta \times 2.22 \times 10^{12} \times X \times \varphi \times v_r \times \rho_{sample}} \quad \text{mol/g (in the sample)}$$

(units as for eqn. (g); ρ_{sample} in g/l).

Mass fraction

(h_1) $$w = \frac{\Delta C_n \times MW}{\eta \times 2.22 \times 10^{12} \times X \times \varphi \times v_r \times \rho_{sample}}$$

(units as for eqn. (g_1); ρ_{sample} in g/l).

Enzymes

Photometry, fluorimetry, luminometry

For measurement of the catalytic activity z of enzymes the rate of the catalyzed substrate conversion per time unit is used, μmol/min or mol/s.

Concerning the stoichiometric coefficient v_i, cf. p. 480.

According to eqn. (a)

catalytic activity (conversion in mol* per time unit) is

(i) $\qquad z = \dfrac{\Delta c \times V}{\Delta t} = \dfrac{\Delta A \times V}{1000 \times \varepsilon \times d \times \Delta t}$ mol/s (kat)

(units: ε in $l \times$ mmol$^{-1} \times$ mm^{-1}; V of the assay volume in l; d in mm; t in s).

(i$_1$) $\qquad z = \dfrac{\Delta c \times V}{\Delta t} = \dfrac{\Delta A \times V \times 1000}{\varepsilon \times d \times \Delta t}$ μmol/min (U)

(units as for eqn. (i); t in min).

The catalytic activity concentration in the sample is

(k) $\qquad b = \dfrac{\Delta A \times V}{1000 \times \varepsilon \times d \times \Delta t \times v} = \dfrac{\Delta A}{1000 \times \varepsilon \times d \times \Delta t \times \varphi}$ mol$\times$ s$^{-1} \times$ l^{-1} (kat/l))

(units as for eqn. (i); v in l; $\varphi = $ v/V in l/l).

(k$_1$) $\qquad b = \dfrac{\Delta A \times V \times 1000}{\varepsilon \times d \times \Delta t \times v} = \dfrac{\Delta A \times 1000}{\varepsilon \times d \times \Delta t \times \varphi}$ μmol $\times$ min$^{-1} \times$ l^{-1} (U/l)

(units as for eqn. (i$_1$); v in l; $\varphi = $ v/V in l/l).

The catalytic activity related to the mass of protein, catalytic activity content z_c/m_s (specific activity) is

(l) $\qquad z_c/m_s = \dfrac{\Delta A \times V}{1000 \times \varepsilon \times d \times \Delta t \times v \times \rho_{protein}} = \dfrac{\Delta A}{1000 \times \varepsilon \times d \times \Delta t \times \varphi \times \rho_{sample}}$ kat/g

(units as for eqn. (k); $\rho_{protein}$ in g/l).

(l$_1$) $\qquad z_c/m_s = \dfrac{\Delta A \times V \times 1000}{\varepsilon \times d \times \Delta t \times v \times \rho_{protein}} = \dfrac{\Delta A \times 1000}{\varepsilon \times d \times \Delta t \times \varphi \times \rho_{protein}}$ U/g

(units as for eqn. (k$_1$); $\rho_{protein}$ in g/l).

---

* $c \times$ V (mol/l) $\times$ l = mol.

Example

Determination of the (specific) catalytic activity of an enzyme. The measurements were made at 339 nm; $\Delta A/\Delta t = 0.063/60$ s in a 3 ml assay mixture ($V = 3 \times 10^{-3}$ l). The volume of the sample was 2×10^{-4} l. The sample diluted over thousandfold for measurement contained 10 g enzyme protein per litre.

In the assay mixture the catalytic activity is according to eqn. (i) or (i_1), respectively

$$z = \frac{0.063 \times 3 \times 10^{-3}}{10^3 \times 0.63 \times 10 \times 60} = 5 \times 10^{-10} \text{ kat} = 0.5 \text{ nkat}$$

or

$$z = \frac{0.063 \times 3 \times 10^{-3} \times 10^3}{0.63 \times 10 \times 1} = 0.03 \text{ U}.$$

The catalytic activity concentration in the sample solution according to eqn. (k) or (k_1), respectively, is

$$b = \frac{0.063 \times 3 \times 10^{-3}}{10^3 \times 0.63 \times 10 \times 60 \times 2 \times 10^{-4}} = 2500 \text{ nkat/l}$$

or

$$b = \frac{0.063 \times 3 \times 10^{-3} \times 10^3}{0.63 \times 10 \times 1 \times 2 \times 10^{-4}} = 150 \text{ U/l}.$$

Related to the mass of protein (dilution factor 1000), the catalytic activity content is

$$z_c/m_s = \frac{2500 \times 1000}{10} \frac{\text{nkat/l}}{\text{g/l}} = 250 \,\mu\text{kat/g}$$

or

$$z_c/m_s = \frac{150 \times 1000}{10} \frac{\text{U/l}}{\text{g/l}} = 15 \text{ U/g}.$$

The result for the sample in relation to an enzyme standard solution is

(m) $\qquad b = \dfrac{b_{\text{sample (measured)}}}{b_{\text{standard (measured)}}} \times b_{\text{standard (indicated)}} \quad \text{nkat/l (U/l)}$

or simply

(m_1) $\qquad b = \dfrac{A_{\text{sample (measured)}}}{A_{\text{standard (measured)}}} \times b_{\text{standard (indicated)}} \quad \text{nkat/l (U/l)}.$

Radiometry

According to eqns. (k) and (k₁), respectively, and using eqns. (g) and (h)

catalytic activity is

(n) $$z = \frac{\Delta C_n \times V}{\eta \times 2.22 \times 10^{12} \times 10^{-9} \times X \times v_r \times \Delta t} \quad \text{nmol/s (nkat)}$$

(units: C_n in cpm; X in Ci/mol or Bq/mol; V and v_r in l; t in s).

(n₁) $$z = \frac{\Delta C_n \times V}{\eta \times 2.22 \times 10^{12} \times 10^{-6} \times X \times v_r \times \Delta t} \quad \text{µmol/min (U)}$$

(units as for eqn. (n); t in min).

The catalytic activity concentration in the sample is

(o) $$b = \frac{\Delta C_n}{\eta \times 2.22 \times 10^{12} \times 10^{-9} \times X \times \varphi \times v_r \times \Delta t} \quad \text{nkat/l}$$

(units as in eqns. (n), (n₁); $\varphi = v/V$ in l/l).

(o₁) $$b = \frac{\Delta C_n}{\eta \times 2.22 \times 10^{12} \times 10^{-6} \times X \times \varphi \times v_r \times \Delta t} \quad \text{U/l}$$

(units as in eqn. (o); t in min).

The catalytic activity related to the mass of protein, catalytic activity content z_c/m_s (specific activity), is

(p) $$z_c/m_s = \frac{\Delta C_n}{\eta \times 2.22 \times 10^{12} \times 10^{-9} \times X \times \varphi \times v_r \times \Delta t \times \rho_{\text{protein}}} \quad \text{nkat/g}$$

(units as for eqn. (o); ρ_{protein} in g/l).

(p₁) $$z_c/m_s = \frac{\Delta C_n}{\eta \times 2.22 \times 10^{12} \times 10^{-6} \times X \times \varphi \times v_r \times \Delta t \times \rho_{\text{protein}}} \quad \text{U/g}$$

(units as for eqn. (o₁); ρ_{protein} in g/l).

Statistics

For parameters and sample statistics, for mathematical evaluation of experiments (mean value, standard deviation, imprecision, inaccuracy) and for statistical evaluation of results, especially for regression analysis, cf. Vol. II, chapter 3.2.

Refering the Experimental Results to Biological Material

In the analysis of organ extracts, blood, serum, etc., the dilution resulting from deproteinization of specimens must also be taken into account along with the fluid content of the specimen. The following values give reasonable accuracy: blood 80% (w/w), tissue (liver, kidney, muscle, heart) 75% (w/w). Tissue specimens are weighed out; a sp. gr. of 1.06 is used for the conversion of blood volumes into mass.

Apart from the volume (in the case of serum, plasma, blood, urine, etc.), other reference quantities that may be used for biological material are the fresh weight, dry weight, total nitrogen, protein content, protein nitrogen; cell count, e.g. erythrocyte count; haemoglobin content, cytochrome c content, and dry weight of the cell-free sample solution.

Examples

Determination of the dilution factor for blood.

The specific gravity of blood is taken as 1.06, and its fluid content is taken as 80% (w/w). 2 ml blood are deproteinized with 3 ml perchloric acid and centrifuged, and 2.5 ml supernatant is neutralized with 1 ml K_2CO_3 solution. The blood is therefore diluted by the following factor:

$$F = \frac{2 \times 1.06 \times 0.8 + 3}{2} \times \frac{2.5 + 1}{2.5} = \frac{1.696 + 3}{2} \times \frac{3.5}{2.5} = 3.29 \,.$$

The experimental result obtained with the neutralized blood extract must be multiplied by this factor to obtain the content of the metabolite in blood.

Calculation of results for multi-stage assays.

In the determination of a disaccharide in biological fluids, the various reaction steps have different pH optima. The assay begins with an incubation: 0.5 ml sample with 0.5 ml buffer and 0.02 ml hydrolase solution ($v_1 = 0.5$ ml, $V_1 = 1.02$ ml, $\varphi_1 = 0.5/1.02 = 0.49$). After inactivation of the enzyme, the second step is the determination of glucose in $v_2 = 0.2$ ml incubation solution (total assay volume is $V_2 = 3.42$ ml, $\varphi_2 = v_2/V_2 = 0.058$). Since 1 disaccharide $\triangleq$ 2 glucose, the stoichiometric coefficient is $v_i = 2$ (cf. p. 480).

Therefore, e.g. eqn. (a_1), p. 480, must be extended as follows:

$$c = \frac{\Delta A}{\varepsilon \times d \times \varphi_1 \times \varphi_2 \times v_i} = \frac{\Delta A \times 17.6}{\varepsilon \times d} \qquad \text{mmol/l} \,.$$

To calculate the metabolite content of the cells of a tissue, the metabolite content of the blood in this tissue must be taken into account.

The mass fraction of blood in the tissue is determined according to *Bücher et al.* [4] from absorbance measurements at 578, 560, and 540 nm. Assuming that the proportion of oxyhaemoglobin (HbO_2) in the circulating blood and the tissue is approximately the same, it follows that the fraction of blood w in the tissue is

$$w = \frac{\Delta A_{HbO_2} \times F_1 \times d_1}{\Delta A'_{HbO_2} \times F_2 \times d_2} \quad (g/g)$$

where

ΔA_{HbO_2} is absorbance difference for tissue extract
$\Delta A'_{HbO_2}$ is absorbance difference for blood dilution
F_1 and F_2 are dilution factors
d_1 and d_2 are light paths of the cuvettes

ΔA_{HbO_2} and $\Delta A'_{HbO_2}$ are calculated [4] from the absorbance measurements at 578, 560, and 540 nm, according to the formula:

$$\Delta A_{HbO_2} \quad \text{or} \quad \Delta A'_{HbO_2} = (A_{578} - A_{560}) + [(A_{540} - A_{578}) \times 0.47].$$

If the metabolite concentration in the tissue specimen is to be referred to the true volume of cellular fluid in order to give the physiological concentration, the result (in mmol per litre tissue extract) is multiplied by the following factor (*G. Michal,* unpublished):

$$F = \frac{V_{\text{after neutralization}}}{V_{\text{before neutralization}}} \times \frac{\dfrac{m_{\text{tissue}}}{\text{sp. gr.}} \times \varphi + V_d}{\dfrac{m_{\text{tissue}}}{\text{sp. gr.}} \times \varphi}$$

φ is the volume fraction of liquid in the tissue volume; m_{tissue} is tissue wet weight; specific gravity relates to the sample. The latter quantity may be taken as unity in most cases. V_d is volume of reagent solution (e.g. $HClO_4$) used for deproteinization.

A similar calculation for erythrocytes has been published by *Bürgi* [5] (cf. Vol. II, chapter 1.1.1.3, p. 13).

References

[1] International Union of Pure and Applied Chemistry (IUPAC) and International Federation of Clinical Chemistry (IFCC): Approved Recommendation (1978), Quantities and Units in Clinical Chemistry, Clin. Chim. Acta *96*, 157F – 183F (1979).

[2] Expert Panel on Nomenclature and Principles of Quality Control in Clinical Chemistry; Committee on Standards (IFCC): Approved Recommendation (1978) on Quality Control in Clinical Chemistry, Part 1: General Principles and Terminology, J. Clin. Chem. Clin. Biochem. *18*, 69 – 77 (1980).

[3] International Commission on Radiation Units and Measurements: ICRU Report 33 "Radiation Quantities and Units", 1980, 7910 Woodmont Ave., Washington D.C. 20014, USA.

[4] *H. J. Hohorst, F. H. Kreutz, Th. Bücher,* Über Metabolitgehalte und Metabolit-Konzentrationen in der Leber der Ratte, Biochem. Z. *332*, 18 – 46 (1950).

[5] *W. Bürgi,* The Volume Displacement Effect in Quantitative Analysis of Red Blood Cell Constituents, J. Clin. Chem. Clin. Biochem. *7*, 458 – 460 (1969).

4   Absorption Coefficients of NAD(P)H

The absorption curves and absorption coefficients depend on the temperature, pH, and the ionic strength of the solution. The best-investigated cases [1 – 4] are NADH and NADPH. At $\lambda = 334$ nm the temperature dependence of ε is approximately zero. The value of ε falls with rising temperature at wavelengths $\lambda > 334$ nm (including the maximum of the absorption curve) and increases at $\lambda < 334$ nm; the absorption maximum is shifted accordingly.

Table 1. Molar decadic absorption coefficients ($1 \times mol^{-1} \times mm^{-1}$) for β-NADH and β-NADPH (measured in triethanolamine/HCl buffer, 0.1 mol/l; pH 7.6) [1].

	°C	Hg 334 nm	339 nm	340 nm	339.85 nm[+]	Hg 365 nm
β-NADH	25	6.176×10^2	no measurement	$6.317 \times 10^{2++}$	6.292×10^2	3.441×10^2
		6.182×10^2 *	cf. Table 2	–	6.298×10^2 *	3.444×10^2 *
						3.427×10^2 *
β-NADH	30	6.187×10^2 *	–	–	–	3.532×10^2 *
β-NADPH	25	6.178×10^2 *	–	–	–	3.515×10^2 *
β-NADPH	30	6.186×10^2 *	–	–	–	

$^+$ Checking the photometer revealed 339.85 nm to be correct, instead of 340 nm.
$^{++}$ Acc. to [2] in Tris buffer, 0.1 mol/l; pH 7.8.
* Values were not corrected for beam convergence or intrinsic absorbance of the oxidized coenzyme.

The above mentioned investigations [1 – 4] on NAD(P)H show:

- ε is different for NADH and NADPH,
- ε is temperature-dependent,
- ε depends on the pH and the ionic strength of the solution to be measured,
- ε cannot be determined sufficiently accurately at 37 °C because of the instability of the coenzyme,
- the absorption maximum of NAD(P)H is not located at exactly 340 nm; 339 nm can be taken as a first approximation,
- the absorption maximum is temperature-dependent,
- the differences in the value of ε due to the factors mentioned above are smallest at Hg 334 nm.

All of the influences mentioned above lead to deviations of the value of ε, not exceeding 0.5% at Hg 334 nm. Consequently, it is best, since practically independent of the conditions of measurement, to perform measurements at this wavelength. Moreover, the values of ε are identical here for both coenzymes.

However, the values of ε at 25 °C and 30 °C at 340 nm (or 339 nm) and Hg 365 nm are also sufficiently close for practical purposes (<1% error at 340 (339) nm; about 2% at Hg 365 nm) and are independent of the other conditions of measurement. For measurement at Hg 365 nm, however, one must distinguish between the values of ε for NADH and NADPH.

In a routine laboratory it would not be practical to use the exact values of the absorption coefficients given above. It would then be necessary, under certain circumstances, to determine the exact value of ε for each experiment. The figures recommended for practical purposes in the routine laboratory vary, according to temperature and other measurement conditions, by less than 1 to 2%, which is within the limits of error attainable for routine enzymatic analyses.

Molar decadic absorption coefficients $(l \times mol^{-1} \times mm^{-1})$ for NADH and NADPH at temperatures of 25 °C and 30 °C [3] are for practical use:

	Hg 334 nm (334.15 nm)	340 nm (339 nm)	Hg 365 nm (365.3 nm)
NADH	6.18×10^2	6.3×10^2	3.4×10^2
NADPH	6.18×10^2	6.3×10^2	3.5×10^2

References

[1] *J. Ziegenhorn, M. Senn, T. Bücher,* Molar Absorptivities of β-NADH and β-NADPH, Clin. Chem. *22,* 151 – 160 (1976).
[2] *R. B. McComb, L. W. Bond, R. W. Burnett, R. C. Keech, G. N. Bowers, jr.,* Determination of the Molar Absorptivity of NADH, Clin. Chem. *22,* 141 – 150 (1976).
[3] *H. U. Bergmeyer,* Neue Werte für die molaren Extinktions-Koeffizienten von NADH und NADPH zum Gebrauch im Routine-Laboratorium, J. Clin. Chem. Clin. Biochem. *13,* 507 – 508 (1975).
[4] *Th. Bücher, G. Lüsch, H. Krell* in: *G. Anido, E. J. van Kampen, S. B. Rosalki, M. Rubin* (eds.), Temperature Dependence of Difference-Absorption Coefficients of NADH Minus NAD$^+$ and NADPH Minus NADP$^+$ in the Near Ultraviolet, Quality Control in Clinical Chemistry, Walter de Gruyter, Berlin, New York 1975, pp. 301 – 310.

Index